OXFORD ENGINEERING SCIENCE SERIES

General Editors

THE OXFORD ENGINEERING SCIENCE SERIES

1 D. R. RHODES: *Synthesis of planar antenna sources*
2 L. C. WOODS: *Thermodynamics of fluid systems*
3 R. N. FRANKLIN: *Plasma phenomena in gas discharges*
4 G. A. BIRD: *Molecular gas dynamics*
5 H.-G. UNGER: *Planar optical waveguides and fibres*
6 J. HAYMAN: *Equilibrium of shell structures*
7 K. H. HUNT: *Kinematic geometry of mechanism*
8 D. S. JONES: *Methods in electromagnetic wave propagation* (Two volumes)
9 W. D. MARSH: *Economics of electric utility power generation*
10 P. HAGEDORN: *Non-linear oscillators* (Second edition)
11 R. HILL: *Mathematical theory of plasticity*
12 D. J. DAWE: *Matrix and finite element displacement analysis of structures*
13 N. W. MURRAY: *Introduction to the theory of thin-walled structures*
14 R. I. TANNER: *Engineering theology*
15 M. F. KANNINEN: *Advanced fracture mechanics*
16 R. H. T. BATES and M. J. McDONNELL: *Image restoration and reconstruction*
18 K. HUSEYIN: *Multiple parameter stability theory and its applications*
19 R. N. BRACEWELL: *The Hartley transform*
20 J. WESSON: *Tokamaks*
21 P. B. WHALLEY: *Boiling, condensation and gas-liquid flow*
22 C. SAMSON, M. Le BORGNE, and B. ESPIAU: *Robot control: the task function approach*
23 H. J. RAMM: *Fluid dynamics for the study of transonic flow*
24 R. R. A. SYMS: *Practical volume holography*
25 W. D. McCOMB: *The physics of fluid turbulence*
26 Z. P. BAZANT and L. CEDOLIN: *Stability of structures: elastic, inelastic, fracture and damage theories*
27 J. D. THORNTON: *Science and practice of liquid–liquid extraction* (Two volumes)
28 J. VAN BLADEL: *Singular electromagnetic fields and sources*
29 M. O. TOKHI and R. R. LEITCH: *Active noise control*
30 I. V. LINDELL: *Methods for electromagnetic field analysis*
31 J. A. C. KENTFIELD: *Nonsteady, one-dimensional, internal, compressible flows*
32 W. F. HOSFORD: *Mechanics of crystals and polycrystals*
33 G. S. H. LOCK: *The tubular thermosyphon: variations on a theme*
34 A. LIÑÁN and F. A. WILLIAMS: *Fundamental aspects of combustion*
35 N. FACHE, D. DE ZUTTER, and F. OLYSLAGER: *Electromagnetic and circuit modelling of multiconductor transmission lines*
36 A. N. BERIS and B. J. EDWARDS: *Thermodynamics of flowing systems: with internal microstructure*
37 K. KANATANI: *Geometric computation for machine vision*
38 J. G. COLLIER and J. R. THOME: *Convective boiling and condensation* (Third edition)
39 I. I. GLASS and J. P. SISLIAN: *Nonstationary flows and shock waves*
40 D. S. JONES: *Methods in electromagnetic wave propagation* (Second edition)
42 G. A. BIRD: *Molecular gas dynamics and the direct simulation of gas flows*
43 G. S. M. LOCK: *Latent heat transfer: an introduction to fundamentals*
44 C. E. BRENNEN: *Cavitation and bubble dynamics*
45 T. C. T. TING: *Anisotropic elasticity: theory and applications*
46 K. ISHIHARA: *Soil dynamics in earthquake engineering*
47 A. CHANDRA and S. MUKHERJEE: *Boundary element methods in manufacturing*

Boundary Element Methods in Manufacturing

Abhijit Chandra

Subrata Mukherjee

New York Oxford
OXFORD UNIVERSITY PRESS
1997

Oxford University Press

Oxford New York
Athens Auckland Bangkok Bogota Bombay Buenos Aires
Calcutta Cape Town Dar es Salaam Delhi Florence Hong Kong
Istanbul Karachi Kuala Lumpur Madras Madrid Melbourne
Mexico City Nairobi Paris Singapore Taipei Tokyo Toronto

and associated companies in
Berlin Ibadan

Published by Oxford University Press, Inc.
198 Madison Avenue, New York, New York 10016

Library of Congress Cataloging-in-Publication Data
Chandra, Abhijit.
Boundary element methods in manufacturing / Abhijit Chandra and Subrata Mukherjee.
p. cm.
Includes bibliographical references.
ISBN 0-19-507921-3
1. Manufacturing processes—Mathematical models. 2. Boundary element methods. I. Mukherjee, Subrata. II. Title.
TS183.C44 1996
670.42′01′5118—dc20 95-49816

9 8 7 6 5 4 3 2 1

Printed in the United States of America
on acid-free paper

To our parents,
who lighted in us the flame for the quest of knowledge
and to our families,
Dolly, Koushik, and Shoma, and Yu, Ananda, and Alok
who gave us the support and the courage to keep that flame burning

Preface

Manufacturing processes have existed, in some form, since the dawn of civilization. Modeling and analysis of the mechanics of manufacturing processes, however, are of fairly recent vintage. It is only in the last few decades that the knowledge of the fundamental mechanics involved in these processes has evolved to a stage where it is now possible to propose reliable mathematical models for the evolution of the state (in terms of microstructure, stresses, displacements, temperatures, and other variables) of a workpiece during its process of manufacture. Concurrent dramatic improvements in the power of computers have made it possible to carry out numerical simulations, based on these detailed models, to chart the evolution of the state of a real-life manufacturing process.

Our capabilities for designing the manufacturing processes, however, usually lag significantly behind our understanding of the basic mechanics involved in these processes. In this arena of design integration, we have not yet utilized the insights gained from analyses to their fullest extent.

Accordingly, analysis of manufacturing processes and its integration into the design cycle of these processes are the dual themes of this book. The boundary element method (BEM) provides a unifying fabric of rigorous and powerful numerical techniques for this purpose.

There exist numerical methods of various different kinds. Of these, the finite difference method (FDM) and the finite element method (FEM) have been very popular for several years. Another very powerful general-purpose method is the boundary element method (BEM). The BEM is based on an integral equation formulation of the governing equations of a problem. These equations are discretized and the resulting algebraic equations are solved numerically. Numerical results obtained by this method are usually very accurate and secondary variables can be obtained as accurately as the primary ones. This has significant ramifications for nonlinear problems where many time steps need to be taken and the secondary variable (e.g., stress) at a certain time step drives the following step. Moreover, since the primary unknowns are usually defined only on the boundary, rather than over the entire domain of the body, the numerical schemes resulting from the boundary element method are generally very efficient.

Following the introductory chapter, the second and third chapters are devoted to providing an in-depth review of the fundamental mechanics associated with various manufacturing processes. Chapter 2 reviews the fundamentals of fully nonlinear problems involving both material and geometric nonlinearities. Such problems arise frequently in manufacturing. Almost every forming and machining process gives rise to large strains and displacements. The fundamentals of thermal problems that are frequently observed in manufacturing are discussed in chapter 3. Such problems arise frequently in machining and solidification processes. The issues of design sensitivities and procedures for their determination are discussed in chapter 4. Design sensitivity coefficients, as discussed in chapter 4, are the rates of change of variables of interest (e.g., stress, plastic strain, or temperature) for a manufacturing process with respect to various geometric, material, or process parameters. The design sensitivity studies provide an avenue for utilizing the insights gained from analysis, in the context of design synthesis. Chapters 5 through 9 are devoted to detailed discussions of a broad range of manufacturing processes. Chapters 5 and 6 cover, respectively, planar and axisymmetric forming processes. Solidification processes are discussed in chapter 7. Material removal processes are discussed in chapters 8 and 9. Chapter 8 focuses on thermal aspects of these processes, while chapter 9 introduces hybrid micro–macro BEM formulations for investigations of ceramic grinding processes.

This book is directed toward researchers, practicing engineers, scientists, and postgraduates who are interested in manufacturing and computational mechanics. The reader is expected to be familiar with the general areas of solid mechanics and heat transfer as well as with the fundamental techniques of applied mathematics and numerical methods.

We wish to thank a number of people and organizations who have contributed in various ways to making this book possible. Our graduate students have contributed significantly to the research work that led to this book. In particular, we want to mention Kai Xiong Hu, Xin Wei, Harindra Rajiyah, Liang-Jenq Leu, and Qing Zhang, who have contributed immensely to the work reported here. We can never thank them enough. We are also indebted to many of our colleagues, Nicholas Zabaras at Cornell, Cholik Chan at the University of Arizona, Young Huang at the Michigan Technological University, and Sunil Saigal at Carnegie Mellon University, with whom we have collaborated on various aspects of the research work reported here. This book would not have been possible without the technical and financial research support of the National Science Foundation, General Motors Research and Development Center, ALCOA Technical Center, and Hughes Aircraft Company. In addition to their financial support, we gratefully acknowledge and appreciate the interest and advice of many individuals in these organizations: Owen Richmond, Michael Wenner, Don Rudy, and others. These have been very important and very special to us. We learnt much of manufacturing from them. We are thankful to Glaucio Paulino, Kangping Wang and Wei Wang for their careful reading and suggestions regarding some of the chapters, and to Rudra Pratap, Sohel Anwar, and Wenjing Ye for their help with the writing and illustrations of this book; and last but certainly not least, our sincere thanks are due to Xiao-Lin Ling for her cheerful and accurate preparation of the manuscript.

Houghton, Michigan A.C.
Ithaca, New York S.M.

August 1996

Contents

Boundary Element Methods in Manufacturing

1

Introduction

The word *manufacturing* comes from the Latin *manu factus*, which means *made by hand*. In the modern sense, manufacturing involves making products from raw materials through various processes. These manufacturing processes transform the raw material into the desired product. In a mechanical sense, these transformations may be classified into three broad categories: (1) deformation processes, (2) material removal processes, and (3) phase-change processes.

1.1 Deformation Processes

In deformation or forming processes, the material is loaded beyond its elastic regime and the desired shape is obtained due to the irreversibility of the plastic strains. Mass conservation (and volume conservation in most cases) between the raw material and the finished part is a unique feature of the forming process. Forming may be carried out under room temperature or at elevated temperatures. Historically, forming has been used for a wide variety of metallic parts. Recently, however, forming techniques, particularly at elevated temperatures, have also been used for polymers and various ceramic composites. Two major advantages of forming processes are the strength enhancement due to process-induced strain hardening and higher efficiency in material utilization. About 90 percent of manufactured products contain a significant portion of formed parts. Accordingly, optimum design of forming processes plays a crucial role in improving the effectiveness and efficiency of product realization procedures.

Forming processes (extrusion, rolling, forging, sheet forming, etc.) typically involve large strains and rotations along with material nonlinearities. Moreover, the boundary conditions at tool–workpiece interfaces for many forming problems can be quite involved and the complete set of boundary conditions in some instances (e.g., rolling) may not be known a priori. Such features make process modeling of forming processes a challenging task. Over the years, slab (e.g., von Kármán 1925, Orowan 1943, Hill 1959, 1978), slip-line (e.g., Alexander 1972), and upper-bound techniques (e.g., Avitzur 1980) have been used widely in modeling of forming processes. These techniques are very useful in determining overall quantities like power requirements, but cannot deliver detailed histories of deformation and stress fields at desired neighborhoods of material points. In the 1980s, various FEM approaches were also developed for investigations of forming problems. Among them, the rigid-plastic (e.g., Onate and Zienkiewicz 1983, Dawson 1984, 1987, Kobayashi et al. 1989) formulation, which views the forming process as a viscous flow, is very cost-effective. An elastoplastic FEM formulation (e.g., Lee et al. 1977, Chandra and Mukherjee 1984a, 1984b, Chandra 1986) is employed when details of residual stresses and springbacks are also needed from the analyses. Today, there also exist several general purpose FEM codes (e.g., ABAQUS, MARC, NIKE2D, NIKE3D) that are capable of carrying out detailed elastoplastic analyses of forming processes. Recently, Chandra and Tvergaard (1993) have also pursued a unit-cell approach using FEM to investigate the effects of microdefects during an extrusion process.

The FEM has been applied very successfully in analyzing various forming problems. However, the sensitivities of finite elements to degradations in aspect ratios necessitate frequent re-meshing at a considerable computational burden. In recent years, mixed Eulerian–Lagrangian formulations (e.g., Benson 1992) have been proposed to alleviate this drawback. The secondary variables (e.g., stresses in displacement formulations) obtained through numerical differentiation in a finite element technique are also inherently less accurate than the primary variables. Yet, in most forming problems, secondary variables like stresses at the current time step essentially drive the problem through the future time steps. This may cause accumulation of errors over the time steps. The rigid-plastic formulations can effectively avoid this drawback and are very cost-effective. However, it is extremely difficult to accurately recover critical quantities like residual stresses from such analyses.

It is interesting to note that a forming process is essentially effected through the interactions of a relatively rigid forming tool (e.g., die or rolls) with the workpiece. Accordingly, the interactions at and near the tool–workpiece interfaces are of crucial importance and the accuracy of any model of a forming process depends, to a large extent, on its capability of representing such interactions. This becomes even more crucial in the designing of forming processes, since the only avenue for altering the characteristics of a forming process (for a given workpiece material) is through alterations in the associated interactions at tool–workpiece interfaces.

1.2 Material Removal Processes

Material removal through mechanical means essentially involves inducing a stress field in the workpiece that will eventually cause fracture and generate new surfaces. Accordingly, a material removal process produces scrap in the form of chips and its efficiency in material utilization is usually inferior to that of forming processes. However, the surface quality achievable through processes of this class is currently unparalleled by any other technique. As a result, most precision parts requiring a high degree of surface quality are prepared through material removal processes and in many cases the finishing operations represent a sizable portion of the total cost of the product. Thus, optimum design of material removal processes is of significant concern for ensuring product quality as well as for containing the manufacturing costs.

For ductile materials, such as metals, the chip formation occurs through large plastic deformations in a zone of concentrated shear. It has been observed from various quick stop machining tests (Shaw 1984), that there is generally no crack extending in front of the tool point and the deformed and undeformed regions are separated by a very thin zone (von Turkovich 1972). Early experiments by Kececioglu (1958) established that the thickness of this zone is about 100 times less than its other dimensions. Thus, this zone may be collapsed to a plane, which, in machining literature, is normally referred to as the *shear plane*. The deformation rate in the shear plane during a machining operation is also very high, typically of the order of 10^3/s–10^4/s (Bhattacharyya 1984, Trent 1984). Plastic deformation induces strain hardening and the plastic work is dissipated in the form of heat. In metal cutting, the plastic deformation occurs at such a fast rate that the heat does not have time to dissipate through diffusion into the workpiece, the tool, and the chip, or through convection and radiation into the surrounding. As a result, the material in the cutting zone is heated very rapidly. As the homologous temperature (actual temperature normalized by the melting point of the material) of the material rises, it undergoes thermal softening. The softening, in turn, accommodates further plastic deformation prior to fracture, and further plastic deformation produces further heating. Thus, the material in the cutting zone can reach very high levels of strain before actually reaching its ultimate strength during a metal cutting process (Shaw 1984). The temperature in this zone can also rise by 500–700°C in a typical metal cutting process.

For brittle materials, such as ceramics and ceramic composites, the chip formation takes place through brittle fracture. Accordingly, the quality and integrity of the finished surface is primarily governed by interactions and resulting propagations of microcracks (Malkin and Ritter 1989, Hu and Chandra 1993). Even for such materials, very high temperatures have been observed by Huerta and Malkin (1976a, 1976b) directly under the abrasive grit, which may be attributable to high strain rate, to large hydrostatic stress states due to large negative rake angles in abrasive grits, as well as to frictional effects.

Accordingly, thermomechanical effects are of crucial importance in machining processes. The thermomechanical effects are further intensified by material transport along a heavily loaded region of relative motion between the chip and the tool. Furthermore, the wear mechanisms, for various tool materials under high-speed machining conditions, are predominantly governed by diffusion. Thus, the thermal field influences the chip formation mechanisms as well as the mechanisms controlling tool wear.

Over the years, many models (e.g., Bhattacharyya 1984, Shaw 1984, Trent 1984, and references therein) have been developed for investigating the mechanical and thermal aspects of metal cutting processes. These investigations have significantly enhanced our understanding of the fundamental mechanics of machining processes. Most of the existing models of machining today are analytical or semiempirical in nature, whose parameters are adjusted through experimental observations. The "partition factor" (representing the portion of the total heat entering a particular domain, workpiece, chip, or tool) that is typically used in thermal analyses of machining processes is an example of such experimentally adjusted parameters. The existence of such parameters makes design of machining processes for new materials very difficult and expensive, since the process designer is forced to go through expensive experimental trial-and-error iterations. Historically, the existence of these parameters has been necessitated by the simplifications needed in formulations of boundary value problems to make the solution tractable through analytical or semianalytical approaches. In such cases, the necessary parameters are usually determined numerically through detailed numerical simulations (e.g., in forming processes). However, unlike many other areas of mechanics, where (particularly since the 1960s) the power of numerical methods such as FDM and FEM has been utilized extensively to solve complete boundary value problems, there exists only a handful of numerical investigations of machining (e.g., Stevenson et al. 1983, Dawson and Malkin 1984) where attempts have been made to solve the complete boundary value problem, so as to make the results directly usable in process design. Among many difficulties associated with detailed numerical models of machining processes, perhaps the existence of extreme gradients near the cutting zone has been the most formidable. Only recently have researchers started to develop numerical techniques (Marusich and Ortiz 1994) capable of handling such problems adequately.

It is also interesting to note that accurate representations of interactions at the tool–chip–workpiece interfaces must be a key element of any model of a machining process. This becomes even more crucial in optimal designs of machining processes, since alterations in these interface interactions are the only available avenues for altering process characteristics.

1.3 Phase Change Processes

In many manufacturing processes, the transformation from raw material to the desired product is effected through a change of phase in the raw material. Casting or solidification processes, welding processes, as well as many nonconventional manufacturing processes such as electrochemical machining (ECM), electrodischarge machining (EDM), and laser processing techniques, rely on the phase change characteristics of constituent materials.

Over the years, many excellent models of phase change processes have been developed (e.g., O'Neill 1983, Roose and Storrer 1984, Huppert and Worster 1985, Sadegh et al. 1985, Worster 1986, Thomson and Szekely 1988, Zabaras and Mukherjee 1987, Zabaras et al. 1988, Zabaras 1990, to name a few). These models have provided crucial insights into several salient features of phase change processes that are characterized by moving and continuously changing interfaces across which the phase change takes place. In a solidification process, this interface refers to the interface between the solid and the liquid phases. In many cases, this interface at a macro-scale may not be sharp (such as mushy zones in solidification processes). Such macro-scale diffused interfaces, however, are made-up of many sharp interfaces in the micro-scale. Accordingly, modeling of the process kinetics across such interfaces plays a crucial role in investigations of phase change processes.

For any phase change process, the quality of the final product and also the efficiency of the process depend intimately on the spatial and temporal characteristics of the motion of the interfaces between different constituent phases. In most manufacturing processes involving phase change, the material properties are usually determined by product design requirements. The designer of the manufacturing process then attempts to achieve the desired development of interfaces by appropriately controlling the boundary conditions of the process. Thus, two kinds of boundaries play crucial roles in manufacturing processes involving phase change. The first set refers to interphase boundaries which may exist in macro- or micro-scales and the second set refers to macro-scale product boundaries along which process design parameters are varied. And the interactions of evolutions at or near these boundaries profoundly influence the quality and efficiency of the particular process.

1.4 Salient Features of Manufacturing Processes and the Boundary Element Method

It is evident from numerous investigations that an understanding of the evolutions at or near boundaries is crucial to the understanding of salient features of most manufacturing processes. For many processes, these boundaries may be physical macro-scale boundaries between the tool and the workpiece, while, for others, these boundaries may also involve micro-scale interfaces (e.g., those of microfeatures or different phases).

This brings us to another very interesting, and crucial, aspect of real-life manufacturing processes. For almost every manufacturing process, the process design and control is usually carried out at a spatial scale that is comparable to overall macro-scale dimensions of the finished product. For most mechanical processes, this dimension is of the order of 10–100 mm. However, the strength and life of the finished product, under intended service conditions, depend to a large extent on the micro-scale evolutions induced during the manufacturing operations. These microfeatures (e.g., phase interfaces, microcracks, inclusions or voids) are typically 1–100 μm in dimension. Accordingly, crucial insights into a manufacturing process (e.g., development of texture in forming) can only be gained if the underlying model is capable of representing the effects of macro-scale design decisions on such micro-scale evolutions. In many instances, the microfeatures will first interact among themselves, new microdefects may nucleate and existing microdefects may coalesce. Finally, the coalesced microdefects will manifest themselves as macroscopic failure modes of the component.

Such a demand for representing various phenomena at two widely different scales within the context of a single analysis poses a tremendous burden on numerical techniques. Attempts to utilize existing FEM or FDM methodology for such multiscale problems result in proliferations in degrees of freedom. Of greater importance is the fact that these numerical techniques can capture the interactions among microfeatures only through numerical discretization. Thus, their capability of representing such interactions deteriorates rapidly as the spacing between defects becomes smaller. For predictions of coalescence, however, the model must accurately capture interactions at very close spacings (10^{-2} to 10^{-3} of crack lengths for coalescence of microcracks). In many instances, the singular fields associated with individual microdefects are significantly modified due to such interactions, and a reliable model must accurately account for such modifications.

The boundary element method (BEM—also called the boundary integral equation method) is a powerful general purpose procedure for the solution of boundary value problems in many branches of science and engineering. With its roots in classical integral equation approaches, the BEM is particularly powerful in capturing detailed descriptions of rapidly varying quantities at or near boundaries, in linear as well as nonlinear problems, a feature which is very advantageous for modeling of a wide variety of manufacturing processes. The BEM can also deliver secondary quantities, such as stresses, as accurately as the primary ones. This is particularly important for nonlinear problems where the secondary variables obtained in the current time step essentially drive the problem through future time steps. Other advantages of BEM over the widely used FEM lie in the fact that the BEM for linear problems requires only boundary (or surface) discretizations of the unknowns in the resultant algebraic systems. The FEM, for comparison, requires discretizations of unknowns over the entire domain of the problem. This advantage in numerical effort becomes more and more significant with increase in the size of the problem. The resultant matrices from the BEM are fully populated but tend to be numerically well conditioned. This arises from the fact that the singular kernels in the integral equations weight the unknown quantities near a singular point more heavily than those that are farther away. Accordingly, the coefficient matrix becomes diagonally dominant.

The BEM, however, is not without its limitations. Historically, the most significant limitation of BEM has been the requirement that an appropriate fundamental solution, for the adjoint governing equation under point load in an infinite domain, must be known a priori. The actual solution for the problem at hand is then constructed in terms of this fundamental solution. While fundamental solutions for certain differential operators in homogeneous media are well known in the applied mathematics literature, they are difficult to obtain for various other cases. Over the past decade, many such problems involving nonhomogeneous media, material nonlinearity, and geometric nonlinearity, have been solved using iterative techniques in conjunction with fundamental solutions corresponding to highest order differential operators in appropriate governing equations.

The heart of the BEM technique lies in the integral equation formulation for a given boundary value problem. The mathematical basis of this approach, of course, is classical and numerous applications of Green's functions have been reported in the literature. As would be expected, initial applications of the BEM have been confined to linear problems in various branches of mechanics. Applications of BEM to nonlinear problems are of more recent vintage, dating back about 20 years.

Currently, there is a great deal of activity in applications of BEM to a wide class of problems. These activities may be broadly classified into two groups. The first vein of activities includes various BEM applications, particularly to nonlinear problems (e.g., Mukherjee 1982), utilizing iterative techniques in conjunction with fundamental solutions for highest order terms in appropriate governing equations. Most BEM applications to nonlinear problems have followed this approach, and over the last decade several manufacturing processes (e.g., forming) have been investigated using this approach. Such an approach necessitates domain integrals in the BEM formulations, requiring domain discretizations. It should be emphasized, however, that the nature of domain discretization and construction of internal cells is quite different from domain discretization in FEM. Unlike FEM, the internal cells in the BEM, for certain classes of problems, do not contain unknowns that appear in the resultant algebraic system and no intercell continuity requirement needs to be imposed in BEM. Accordingly, the internal cells in BEM can be disjoint and need not cover the entire domain. Such properties of internal cells are very useful for problems where the nonlinearity is mostly restricted to a part of the domain.

Pursuing the other vein, researchers have always looked for appropriate fundamental solutions for their intended problems. BEM researchers following this trend have discovered many interesting fundamental solutions in applied mathematics and brought them to bear on numerous useful engineering applications, and, in many cases, broadened the areas of applications (e.g., Cruse 1988, Banerjee 1994, Chandra and Chan 1994).

Historically, the BEM research exercise along this second vein remained one of discovery rather than invention, and the association of BEM with the fundamental solution came to be viewed as one of its drawbacks. Some very recent research along this second vein (Chandra et al. 1995), however, has attempted to change that notion. Utilizing integral equation techniques in conjunction with known analytical fundamental solutions that are applicable at the micro-scale, researchers have been able to obtain numerically (as discussed in chapter 9) fundamental solutions applicable to the overall macro-scale problem involving interacting microfeatures in a finite body. Such an approach decomposes the overall problem into one lying entirely in the micro-scale while the other involves only macro-scale parameters. The fundamental solution is utilized to incorporate the effects due to micro-scale evolutions into an otherwise completely macro-scale problem! Previous attempts to capture numerically singular behaviors of fundamental solutions have mostly met with failure due to inaccuracy. The recent approach, however, separates the singular and regular portions of the fundamental solution. The singular part is handled analytically, while the regular part is obtained and integrated numerically. Such an approach utilizes the augmented fundamental solution or the Green's function to incorporate the micro-scale effects within a macro-scale computational strategy. Thus, the association of BEM with fundamental solutions, which has traditionally been viewed as one of its weaknesses, is utilized as one of its unique features to bridge the micro-scale with the macro-scale. Unlike most numerical techniques, such as FEM and FDM, that are unable to utilize known analytical solutions, the fundamental solution provides BEM with a readily available conduit for introducing analytical advance. The resulting multiscale BEM formulations also decomposes the overall problem into subproblems residing solely at particular scales. This facilitates parallelization and enhances computational efficiency.

The purpose of this book is the presentation of applications of the BEM to investigations of various manufacturing processes. It is important to note here that the physics of certain classes of manufacturing processes makes them particularly amenable to BEM. The following chapters discuss applications along both of the recent directions (discussed above) in applications of BEM to various manufacturing problems. Domain integrals and internal cells are used in chapters 5 and 6 to investigate the forming processes involving large strains and displacements. Existing time-dependent fundamental solutions are utilized in chapters 7 and 8 to investigate, respectively, solidification and thermal aspects of machining processes. Chapter 9 describes hybrid micro–macro BEM formulations and their application to investigations of ceramic grinding processes.

Issues of manufacturing process design are addressed along with analyses of particular processes. A quantitative approach is pursued for this purpose. A review of design sensitivity studies is presented in chapter 4, and the design sensitivities with respect to various material, geometric, and process parameters are utilized as a bridge between analysis and design synthesis. Starting from a nominal design of a manufacturing process, an analysis of its salient features is performed first. Design sensitivity studies are then pursued to investigate the effects of variations in various design parameters. Utilizing the results from the analyses and the design sensitivity studies, the design problem can be cast quantitatively in two alternate ways. Following the first path, the design considerations may be posed as an optimization problem with appropriate objective functions. Such an approach, based on the minimum plastic work path (Hill 1986), is discussed in chapter 5. Alternatively, the information gained from analyses and design sensitivity studies may be used to cast the design issues as a regularized inverse problem

(Tikhonov and Arsenin 1977). Following analyses of particular manufacturing processes, issues regarding their design are also discussed along the above perspectives.

Thus, in essence, this book aims to present a comprehensive and up-to-date account of applications of the BEM to analyses and design sensitivity studies of various manufacturing processes. In addition, it also highlights a few potential research directions (e.g., hybrid multiscale formulations, inverse problems) where further developments of BEM can potentially make significant contributions to the understanding and design of broad classes of manufacturing processes.

References

Alexander, J. M. (1972). "On the Theory of Rolling," *Proc. Royal Soc., London, Series A*, **326**, 535–563.

Avitzur, B. (1980). *Metal Forming: The Applications of Limit Analysis*. Marcel Dekker, New York.

Banerjee, P. K. (1994). *The Boundary Element Methods in Engineering*. McGraw-Hill, London.

Benson, D. J. (1992). "Computational Methods in Lagrangian and Eulerian Hydrocodes," *Computational Methods in Applied Mechanics and Engineering*, **99**, 235–394.

Bhattacharyya, A. (1984). *Metal Cutting Theory and Practice*. Central Book Publisher, Calcutta.

Chandra, A. (1986). "A Generalized Finite Element Analysis of Sheet Metal Forming with an Elastic-Viscoplastic Material Model," *J. Eng. Ind., ASME*, **108**, 9–15.

Chandra, A. and Chan, C. L. (1994). "Thermal Aspects of Machining: A BEM Approach," *Int. J. Solids Structures*, **31**, 1657–1693.

Chandra, A. and Mukherjee, S. (1984a). "A Finite Element Analysis of Metal Forming Problems with an Elastic-Viscoplastic Material Model," *Int. J. Num. Meth. Eng.*, **20**, 1613–1628.

Chandra, A. and Mukherjee, S. (1984b). "A Finite Element Analysis of Metal Forming Processes with Thermomechanical Coupling," *Int. J. Mech. Sci.*, **26**, 661–676.

Chandra, A. and Tvergaard, V. (1993). "Void Nucleation and Growth During Plane Strain Extrusion," *Int. J. Damage Mech.*, **2**, 330–348.

Chandra, A., Huang, Y., Wei, X., and Hu, K. X. (1995). "A Hybrid Micro-Macro BEM Formulation for Micro–Crack Clusters in Elastic Components," *Int. J. Num. Meth. Eng.*, **38**, 1215–1236.

Cruse, T. A. (1988). *Boundary Element Analysis in Computational Fracture Mechanics*. Kluwer Academic Publishers, Dordrecht, The Netherlands.

Dawson, P. R. (1984). "A Model for the Hot or Warm Forming of Metals with Special Use of Deformation Mechanism Maps," *Int. J. Mech. Sci.*, **26**, 227–244.

Dawson, P. R. (1987). "On Modeling Mechanical Property Changes During Flat Rolling of Aluminum," *Int. J. Solids Structures*, **23**, 947–968.

Dawson, P. R. and Malkin, S. (1984). "Inclined Moving Heat Source Model for Calculating Metal Cutting Temperatures," *J. Eng. Ind., ASME*, **106**, 179–186.

Hill, R. (1959). "Some Basic Principles in the Mechanics of Solids Without Natural Time," *J. Mech. Phys. Solids*, **7**, 209–225.

Hill, R. (1978). "Aspects of Invariance in Solid Mechanics," *Advances in Applied Mechanics*, **18**, ed. Chia-Shun Yih. Academic Press, New York.

Hill, R. (1986). "External Paths of Plastic Work and Deformation," *J. Mech. Phys. Solids*, **34**, 511–518.

Hu, K. X. and Chandra, A. (1993). "A Fracture Mechanics Approach to Modeling Strength Degradation in Ceramic Grinding Processes," *J. Eng. Ind., ASME*, **115**, 73–84.

Huerta, M. and Malkin, S. (1976a). "Grinding of Glass: the Mechanics of the Process," *J. Eng. Ind., ASME*, **98**, 459–467.

Huerta, M. and Malkin, S. (1976b). "Grinding of Glass: Surface Structure and Fracture Strength," *J. Eng. Ind., ASME*, **98**, 468–473.

Huppert, H. E. and Worster, M. G. (1985). "Dynamic Solidification of a Binary Melt," *Nature*, **314**, 703-707.

Kármán, T. von (1925). "On the Theory of Rolling," *Z. Angew. Math. Mech.*, **5**, 130–141.

Kececioglu, D. (1958). "Shear-Strain Rate in Metal Cutting and Its Effects on Shear-Flow Stress," *Trans. ASME*, **80**, 158–168.

Kobayashi, S., Oh, S.-I., and Altan, T. (1989). *Metal Forming and the Finite Element Method*. Oxford University Press, New York.

Lee, E. H., Mallet, R. L., and Yang, W. H. (1977). "Stress and Deformation Analysis of Metal Extrusion Processes," *Comp. Meth. Appl. Mech. Eng.*, **10**, 339–353.

Malkin, S. and Ritter, J. E. (1989). "Grinding Mechanisms and Strength Degradation for Ceramics," *J. Eng. Ind., ASME*, **111**, 167–174.

Marusich, T. D. and Ortiz, M. (1994). "Finite Element Simulation of High Speed Machining," *Mechanics in Materials Processing and Manufacturing* (ed. T. J. Moon and M. N. Ghasemi Nejhad). AMD-194, ASME, New York, pp. 137–150.

Mukherjee, S. (1982). *Boundary Element Methods in Creep and Fracture*. Elsevier Applied Science Publishers, Barking, Essex, U.K.

Onate, E. and Zienkiewicz, O. C. (1983). "A Viscous Shell Formulation for the Analysis of Thin Sheet Metal Forming," *Int. J. Mech. Sci.*, **25**, 305–335.

O'Neill, K. (1983). "Boundary Integral Equation Solution of Moving Boundary Phase Change Problems," *Int. J. Num. Meth. Eng.*, **19**, 1825–1850.

Orowan, E. (1943). "The Calculation of Roll Pressure in Hot and Cold Flat Rolling," *Proc. Inst. Mech. Eng.*, **150**, 140–167.

Roose, J. and Storrer, W. O. (1984). "Modelization of Phase Changes by Fictitious Heat Flow," *Int. J. Num. Meth. Eng.*, **20**, 217–225.

Sadegh, A., Jiji, L. M., and Weinbaum, S. (1985). "Boundary Integral Equation Technique with Application to Freezing Around a Buried Pipe," ASME, paper presented at the Winter Annual Meeting, Miami Beach, Florida, 17–21 November.

Shaw, M. C. (1984). *Metal Cutting Principles*. Clarendon Press, Oxford.

Stevenson, M. G., Wright, P. K., and Chow, J. G., (1983). "Further Developments in Applying the Finite Element Method to the Calculation of Temperature Distributions in Machining and Comparisons with Experiment," *J. Eng. Ind., ASME*, **105**, 149–154.

Thomson, M. E. and Szekely, J. (1988). "Mathematical and Physical Modelling of Double-Diffusive Convection of Aqueous Solutions Crystallizing at a Vertical Wall," *J. Fluid Mech.*, **187**, 409–433.

Tikhonov, A. N. and Arsenin, V. Y. (1977). *Solution of Ill-posed Problems*. V. H. Winston, Washington, D.C.

Trent, E. M. (1984). *Metal Cutting*. Butterworths, London.

Von Turkovich, B. F. (1972). "On a Class of Thermo-Mechanical Processes During Rapid Plastic Deformation (with Special Reference to Metal Cutting)," *Ann. CIRP*, **21**, 15–16.

Worster, M. G. (1986). "Solidification of an Alloy from a Cooled Boundary," *J. Fluid Mech.*, **167**, 481–501.

Zabaras, N. (1990). "Inverse Finite Element Techniques for the Analysis of Solidification Processes," *Int. J. Num. Meth. Eng.*, **29**, 1569–1587.

Zabaras, N. and Mukherjee, S. (1987). "An Analysis of Solidification Problems by the Boundary Element Method," *Int. J. Num. Meth. Eng.*, **24**, 1879–1900.

Zabaras, N., Mukherjee, S., and Richmond, O., (1988). "An Analysis of Inverse Heat Transfer Problems with Phase Changes Using an Integral Method," *J. Heat Transfer*, **110**, 554–561.

2

Problems Involving Large Strains and Rotations

This chapter begins with a brief review of the fundamentals of continuum mechanics for solid mechanics problems with large strains and large rotations. The review is limited to material that is relevant to the rest of the discussion in this book. This is followed by a presentation of boundary element formulations for hypoelastic materials subjected to small elastic strains but arbitrarily large inelastic strains and rotations. Finally, finite element formulations for the same class of problems are briefly reviewed. Since this book is devoted to boundary element methods, the purpose of the last section of this chapter is not to discuss details of various finite element formulations but to present a brief comparison between boundary element and finite element formulations for this class of problems.

2.1 Continuum Mechanics Fundamentals

This section first presents a discussion of kinematics and kinetics using Cartesian coordinates. Next, these concepts are revisited, using general tensors. Finally, a discussion of objective rates of general tensors is presented.

2.1.1 Kinematics in Cartesian Coordinates

One-dimensional strains. The simplest starting point is a one-dimensional problem. Let a bar of initial length L be stretched to a current length l. Then, one can define

$$\text{Stretch } \lambda = \frac{\text{current length}}{\text{original length}} = \frac{\text{gauge length}}{\text{initial gauge length}} = \frac{l}{L}$$

Various measures of strain, for one-dimensional problems, can be defined as follows:

$$\text{Engineering strain } \varepsilon^{(e)} = \frac{l - L}{L} = \lambda - 1 \tag{2.1}$$

$$\text{Logarithmic strain } \varepsilon^{(ln)} = \int_L^l \frac{dl}{l} = \ln\left(\frac{l}{L}\right) = \ln \lambda \tag{2.2}$$

$$\text{Green strain } \varepsilon^{(G)} = \frac{1}{2}\left(\frac{l^2 - L^2}{L^2}\right) = \frac{1}{2}(\lambda^2 - 1) \tag{2.3}$$

$$\text{Almansi strain } \varepsilon^{(A)} = \frac{1}{2}\left(\frac{l^2 - L^2}{l^2}\right) = \frac{1}{2}\left(1 - \frac{1}{\lambda^2}\right) \tag{2.4}$$

Three-dimensional kinematics. The general three-dimensional problem is approached here from a Lagrangian point of view in which a material point is followed through a deformation process. Let a body with a reference configuration B_0 be deformed into a current configuration B (figure 2.1). A material point p_0 with some material element dL (chosen gauge length in the reference configuration) moves to a point p with gauge length dl in the current configuration. Here, p_0 and p represent the same material point *and are viewed from a spatially fixed basis of mutually perpendicular unit vectors* $(\boldsymbol{e}_1, \boldsymbol{e}_2, \boldsymbol{e}_3)$. Now, $\lambda = dl/dL$ is associated with a particular material point and chosen direction.

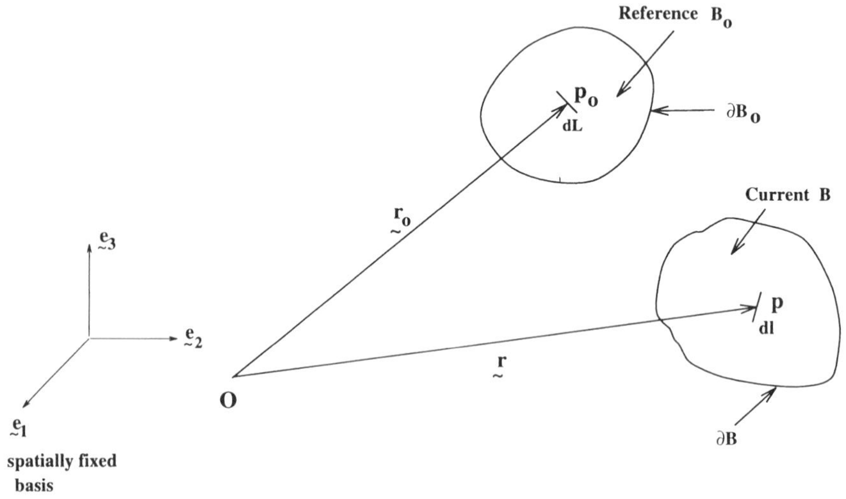

Figure 2.1. Large deformation from a Lagrangian viewpoint.

As shown in figure 2.1, the position vectors of p_0 and p are

$$\boldsymbol{r}_0 = \boldsymbol{X} = X_1\boldsymbol{e}_1 + X_2\boldsymbol{e}_2 + X_3\boldsymbol{e}_3 \tag{2.5}$$

$$\boldsymbol{r} = \boldsymbol{x} = x_1\boldsymbol{e}_1 + x_2\boldsymbol{e}_2 + x_3\boldsymbol{e}_3 \tag{2.6}$$

The deformation gradient tensor $\boldsymbol{F}$ is defined as

$$\boldsymbol{F} = \frac{\partial \boldsymbol{x}}{\partial \boldsymbol{X}} \tag{2.7}$$

The polar decomposition theorem states that any deformation gradient tensor $\boldsymbol{F}$ can be decomposed according to

$$\boldsymbol{F} = \boldsymbol{R} \cdot \boldsymbol{U} = \boldsymbol{V} \cdot \boldsymbol{R} \tag{2.8}$$

where $\boldsymbol{R}$ is an orthogonal tensor and $\boldsymbol{R}^{\mathrm{T}}$ is its transpose ($\boldsymbol{R} \cdot \boldsymbol{R}^{\mathrm{T}} = \boldsymbol{I}$, the identity tensor), and $\boldsymbol{U}$ and $\boldsymbol{V}$ are symmetric positive definite tensors. Physically, this means that, at each material point, there exist three orthogonal directions (unit vectors $\boldsymbol{M}_i$ in the reference and $\boldsymbol{m}_i$ in the current configuration) that undergo pure rigid-body rotation represented by the rotation tensor $\boldsymbol{R}$, i.e.,

$$\boldsymbol{m}_i = \boldsymbol{R} \cdot \boldsymbol{M}_i \qquad i = 1, 2, 3 \tag{2.9}$$

A general deformation at a material point can be decomposed into eigenstretches λ_i in the eigendirections $\boldsymbol{M}_i$ followed by a rotation, i.e.,

$$d\boldsymbol{x} = \boldsymbol{R} \cdot (\lambda_i \boldsymbol{M}_i \, dX_i) \tag{2.10}$$

$$= \boldsymbol{R} \cdot (\lambda_i \boldsymbol{M}_i \boldsymbol{M}_i) \cdot d\boldsymbol{X} \tag{2.11}$$

where the relationship $dX_i = \boldsymbol{M}_i \cdot d\boldsymbol{X}$ is used.

Thus, $\boldsymbol{F} = \boldsymbol{R} \cdot \boldsymbol{U}$, where the right stretch tensor

$$\boldsymbol{U} = \lambda_1 \boldsymbol{M}_1 \boldsymbol{M}_1 + \lambda_2 \boldsymbol{M}_2 \boldsymbol{M}_2 + \lambda_3 \boldsymbol{M}_3 \boldsymbol{M}_3 \tag{2.12}$$

is symmetric and has eigenvalues λ_i and eigenvectors $\boldsymbol{M}_i$.

The relationship $\boldsymbol{F} = \boldsymbol{V} \cdot \boldsymbol{R}$ can be proved in a similar way. Here, the left stretch tensor is

$$\boldsymbol{V} = \lambda_1 \boldsymbol{m}_1 \boldsymbol{m}_1 + \lambda_2 \boldsymbol{m}_2 \boldsymbol{m}_2 + \lambda_3 \boldsymbol{m}_3 \boldsymbol{m}_3 \tag{2.13}$$

It is useful to note that the right Cauchy–Green tensor is

$$C = F^{\mathrm{T}} \cdot F = U \cdot U = \lambda_1^2 M_1 M_1 + \lambda_2^2 M_2 M_2 + \lambda_3^2 M_3 M_3 \tag{2.14}$$

and this allows one to compute λ_i and M_i directly from the deformation gradient F, through the spectral decomposition of $F^{\mathrm{T}} \cdot F$. Here, F^{T} is the transpose of F.

Three-dimensional strains. Various strain measures for three-dimensional deformation are defined as

$$\text{Green strain:} \quad E^{(G)} = \tfrac{1}{2}(F^{\mathrm{T}} \cdot F - I) = \tfrac{1}{2}(\lambda_i^2 - 1) M_i M_i \tag{2.15}$$

$$\text{Almansi strain:} \quad E^{(A)} = \tfrac{1}{2}(I - F^{\mathrm{T}} \cdot F^{-1}) = \tfrac{1}{2}(I - (V \cdot V)^{-1})$$

$$= \tfrac{1}{2}\left(1 - \frac{1}{\lambda_i^2}\right) m_i m_i \tag{2.16}$$

It is interesting to note that it is natural to write the Green strain in the reference basis and the Almansi strain in the current basis. This issue is discussed further in section 2.1.3.

Finally, for completeness, the logarithmic strain in three dimensions is

$$\text{Hencky (}or\text{ logarithmic strain):} \quad E^{(ln)} = \ln U \tag{2.17}$$

Rates of kinematic variables. Measures of rates of kinematic variables are often used in large-deformation analysis. The most common are

$$\text{Velocity gradient:* } \quad h = \nabla v = \dot{F} \cdot F^{-1} \tag{2.18}$$

$$\text{Symmetric velocity gradient:} \quad d = \tfrac{1}{2}(h + h^{\mathrm{T}}) \tag{2.19}$$

$$\text{Spin:* } \quad \omega = \tfrac{1}{2}(h - h^{\mathrm{T}}) \tag{2.20}$$

In the above, v is the velocity of a material point and ∇ is the usual gradient operator in *current coordinates* defined as $\nabla(\cdot) = \partial/\partial x_i(\cdot) e_i$. The symmetric velocity gradient d is sometimes called the rate of deformation tensor or the stretching tensor. Also,

$$d = \tfrac{1}{2}(\dot{F} \cdot F^{-1} + F^{-\mathrm{T}} \cdot \dot{F}^{\mathrm{T}}) \tag{2.21}$$

in terms of the deformation gradient F and its time derivative $\dot{F}$.

*In Cartesian tensor notation $h_{ij} = v_{i,j}$ and $\omega_{ij} = \tfrac{1}{2}(v_{i,j} - v_{j,i})$.

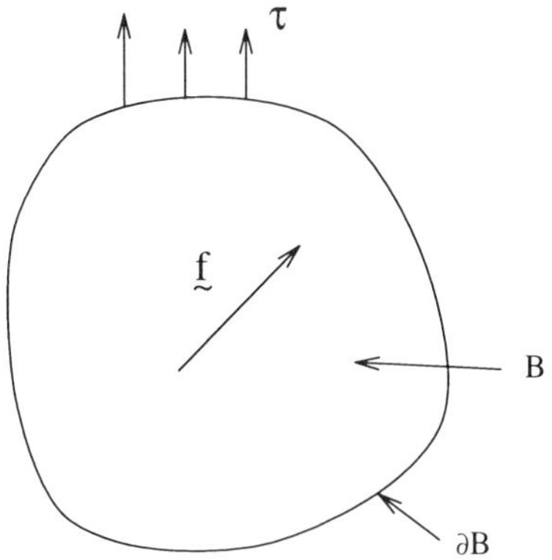

Figure 2.2. Forces acting on a body in its current configuration.

2.1.2 *Kinetics in Cartesian Coordinates*

Equilibrium. The starting point here is the equation of equilibrium in the current configuration B. Consider a current configuration of the body, subjected to surface tractions $\boldsymbol{\tau}$ per unit current surface area and body force $\boldsymbol{f}$ per unit current volume (figure 2.2). Equilibrium of forces demands that

$$\int_{\partial B} \boldsymbol{\tau}\, ds + \int_{B} \boldsymbol{f}\, dv = \mathbf{0} \tag{2.22}$$

The Cauchy stress tensor $\boldsymbol{\sigma}$ is defined as

$$\boldsymbol{n} \cdot \boldsymbol{\sigma} = \boldsymbol{\tau} \tag{2.23}$$

where $\boldsymbol{n}$ is the current normal (unit vector) at a point on the boundary ∂B.

Using the above and Gauss' theorem, one can transform equation (2.22) into

$$\int_{B} (\boldsymbol{\nabla} \cdot \boldsymbol{\sigma} + \boldsymbol{f})\, dv = \mathbf{0} \tag{2.24}$$

which gives the equilibrium equation in the current configuration as

$$\boldsymbol{\nabla} \cdot \boldsymbol{\sigma} + \boldsymbol{f} = \mathbf{0} \tag{2.25}$$

It can be shown by considering moment equilibrium that the tensor $\boldsymbol{\sigma}$ is symmetric.

Principle of virtual work. The principle of virtual work can be obtained from a weighted residual equation,

$$\int_B (\sigma_{ij,i} + f_j)\tilde{v}_j \, dv = 0 \tag{2.26}$$

where $\tilde{\boldsymbol{v}}$ is a kinematically compatible but otherwise arbitrary velocity field. The above equation can be written as

$$\int_{\partial B} \tau_j \tilde{v}_j \, ds + \int_B f_j \tilde{v}_j \, dv = \int_B \sigma_{ij} \tilde{v}_{j,i} \, dv = \int_B \sigma_{ij} \tilde{d}_{ij} \, dv \tag{2.27}$$

where Gauss' theorem and the symmetry of the Cauchy stress have been used. In direct notation,

$$\int_{\partial B} \boldsymbol{\tau} \cdot \tilde{\boldsymbol{v}} \, ds + \int_B \boldsymbol{f} \cdot \tilde{\boldsymbol{v}} \, dv = \int_B \boldsymbol{\sigma} : \tilde{\boldsymbol{d}} \, dv \tag{2.28}$$

where $\boldsymbol{a}:\boldsymbol{b} = a_{ij}b_{ij}$ for any two second-rank tensors $\boldsymbol{a}$ and $\boldsymbol{b}$.

Equation (2.28) means that the rate of external work done by the surface and body forces on the virtual velocity field is equal to the rate of internal work of the stress field $\boldsymbol{\sigma}$. Here, as mentioned before, $\boldsymbol{\sigma}$, $\boldsymbol{\tau}$, and $\boldsymbol{f}$ satisfy equilibrium and $\tilde{\boldsymbol{v}}$ and $\tilde{\boldsymbol{d}}$ are compatible. Of course, the actual velocity field $\boldsymbol{v}$ and stretching $\boldsymbol{d}$ for a given problem can also be used instead of $\tilde{\boldsymbol{v}}$ and $\tilde{\boldsymbol{d}}$ in equation (2.28).

Alternate stress measures. It is easiest to introduce alternate stress measures through the concept of work conjugacy of stress and strain rate measures. For example,

$$\int_B \boldsymbol{\sigma} : \boldsymbol{d} \, dv = \int_{B_0} \boldsymbol{\sigma} : \boldsymbol{d} \left| \frac{dv}{dV} \right| dV = \int_{B_0} (J\boldsymbol{\sigma}) : \boldsymbol{d} \, dV \tag{2.29}$$

where the internal work is written in reference coordinates and B_0 denotes the reference configuration. This leads to the definition of the Kirchhoff stress $\boldsymbol{K}$ as

$$\boldsymbol{K} = J\boldsymbol{\sigma} \tag{2.30}$$

which is work conjugate to the stretching $\boldsymbol{d}$. Here,

$$J = \left| \frac{dv}{dV} \right| = \det(\boldsymbol{F}) = \det(\boldsymbol{U}) = \lambda_1 \lambda_2 \lambda_3 \tag{2.31}$$

in terms of the eigenvalues of $\boldsymbol{U}$. Also, det denotes the determinant of a tensor.

The nominal or Lagrange stress $\boldsymbol{S}$ can be obtained by writing the internal work expression in an alternate way. Thus,

$$\begin{aligned} J\boldsymbol{\sigma}:\boldsymbol{d} &= J(\boldsymbol{\sigma}:\boldsymbol{\nabla v}) = J\boldsymbol{\sigma}:(\dot{\boldsymbol{F}}\boldsymbol{F}^{-1}) = J\boldsymbol{\sigma}_{ij}\frac{\partial v_i}{\partial X_k}\frac{\partial X_k}{\partial x_j} \\ &= J\sigma_{ij}(\boldsymbol{F}^{-1})_{kj}(\dot{\boldsymbol{F}})_{ik} = (J\boldsymbol{\sigma}\cdot\boldsymbol{F}^{-1}):\dot{\boldsymbol{F}} \end{aligned} \tag{2.32}$$

This leads to the definition of the Piola–Kirchhoff stress of the first kind, which is conjugate to $\dot{\boldsymbol{F}}$:

$$\boldsymbol{P}^{(1)} = J\boldsymbol{\sigma}\cdot\boldsymbol{F}^{-\mathrm{T}} \tag{2.33}$$

The transpose of $\boldsymbol{P}^{(1)}$ is called the nominal or Lagrange stress $\boldsymbol{S}$. Thus,

$$\boldsymbol{S} = J\boldsymbol{F}^{-1}\cdot\boldsymbol{\sigma} \tag{2.34}$$

Since $\boldsymbol{F}$ contains both stretch and rotation, and is not symmetric, neither is $\boldsymbol{S}$. The nominal stress plays an important role in the equilibrium equations written in a reference configuration. This matter will be addressed soon. Finally, one can show by similar calculations that the Piola–Kirchhoff stress of the second kind, $\boldsymbol{P}^{(2)}$, is work conjugate to the rate of the Green strain, $\dot{\boldsymbol{E}}^{(G)}$. This stress is

$$\boldsymbol{P}^{(2)} = J(\boldsymbol{F}^{-1}\cdot\boldsymbol{\sigma}\cdot\boldsymbol{F}^{-\mathrm{T}}) \tag{2.35}$$

which is symmetric. It will be shown later that this stress measure is associated with quantities in the reference configuration. It is specially useful for the study of problems with small strains and large rotations, but such problems are not of primary interest in this book.

Equilibrium in a reference configuration. The last topics in this section are discussions of the equations of equilibrium and the rate of equilibrium in a reference configuration. The virtual work equation in the reference configuration can be written as [see equations (2.28) and (2.32)]

$$\int_{B_0} \boldsymbol{P}^{(1)}:\dot{\boldsymbol{F}}\,dV = \int_{\partial B_0} \bar{\boldsymbol{\tau}}\cdot\tilde{\boldsymbol{v}}\,dS + \int_{B_0} \bar{\boldsymbol{f}}\cdot\tilde{\boldsymbol{v}}\,dV \tag{2.36}$$

where $\bar{\boldsymbol{\tau}}$ and $\bar{\boldsymbol{f}}$ are surface traction and body force per unit reference surface area and unit reference volume, respectively. Thus,

$$\bar{\boldsymbol{\tau}} = \boldsymbol{\tau}\left|\frac{ds}{dS}\right|, \qquad \bar{\boldsymbol{f}} = J\boldsymbol{f} \tag{2.37}$$

The left-hand side of equation (2.36) can be written as

$$\begin{aligned}\int_{B_0} P_{ij}^{(1)}\frac{\partial \tilde{v}_i}{\partial X_j}\,dV &= \int_{\partial B_0} N_j P_{ij}^{(1)}\,\tilde{v}_i\,dS - \int_{B_0}\frac{\partial P_{ij}^{(1)}}{\partial X_j}\tilde{v}_i\,dV \\ &= \int_{\partial B_0} N_i S_{ij}\tilde{v}_j\,dS - \int_{B_0}\frac{\partial S_{ij}}{\partial X_i}\tilde{v}_j\,dV \\ &= \int_{\partial B_0} \boldsymbol{N}\cdot\boldsymbol{S}\cdot\tilde{\boldsymbol{v}}\,dS - \int_{B_0}\bar{\nabla}\cdot\boldsymbol{S}\cdot\tilde{\boldsymbol{v}}\,dV\end{aligned} \tag{2.38}$$

where $\boldsymbol{N}$ is the unit normal vector at a point on the boundary ∂B_0 of the reference configuration B_0 and $\bar{\boldsymbol{\nabla}}$ is the gradient operator defined as $\bar{\boldsymbol{\nabla}}(\cdot) = \boldsymbol{e}_i\,\partial/\partial X_i(\cdot)$.

Since $\tilde{\boldsymbol{v}}$ is arbitrary, equations (2.36) and (2.38) yield the equilibrium equations in the reference configuration as

$$\bar{\boldsymbol{\nabla}}\cdot\boldsymbol{S} + \bar{\boldsymbol{f}} = \boldsymbol{0} \tag{2.39}$$

together with the surface relationship

$$\boldsymbol{N}\cdot\boldsymbol{S} = \bar{\boldsymbol{\tau}} \tag{2.40}$$

The nominal or Lagrange stress $\boldsymbol{S}$ has a "dual" identity. Its divergence with respect to reference coordinates gives the negative of the body force per unit reference volume, but written in the current basis. Similarly, its dot product with the reference normal gives the traction vector per unit reference surface area, but also in the current basis. This is because the "natural" bases for $\boldsymbol{S}$ involve a reference basis on the left and a current basis on the right! This fact becomes clear when some of these equations and variables are written in general tensor notation. This is done in section 2.1.3.

Rates of equilibrium equations. The equilibrium equations in rate form are as follows. In the absence of body forces (for simplicity), equation (2.39) takes the rate form

$$\bar{\boldsymbol{\nabla}}\cdot\dot{\boldsymbol{S}} = \boldsymbol{0} \tag{2.41}$$

since the reference coordinates are time independent. Now,

$$\mathbf{0} = \int_{B_0} \bar{\boldsymbol{\nabla}} \cdot \dot{\boldsymbol{S}}\, dV = \int_{\partial B_0} \boldsymbol{N} \cdot \dot{\boldsymbol{S}}\; dS \tag{2.42}$$

Nanson's formula (Nanson 1877–78) relates the unit normal in the reference and current configurations as

$$\frac{\boldsymbol{n} \cdot \boldsymbol{F}}{J}\, ds = \boldsymbol{N}\, dS \tag{2.43}$$

so that

$$\mathbf{0} = \int_{\partial B} \frac{\boldsymbol{n} \cdot \boldsymbol{F} \cdot \dot{\boldsymbol{S}}}{J}\, ds = \int_{\partial B} \tilde{\boldsymbol{\nabla}} \cdot \left[\frac{\boldsymbol{F} \cdot \dot{\boldsymbol{S}}}{J} \right] ds \tag{2.44}$$

with $\tilde{\boldsymbol{\nabla}}(\,\cdot\,) = \boldsymbol{e}_i\, \partial/\partial x_i(\,\cdot\,)$.

Thus, the rates of equilibrium equations in the current configuration have the form

$$\tilde{\boldsymbol{\nabla}} \cdot \left[\frac{\boldsymbol{F} \cdot \dot{\boldsymbol{S}}}{J} \right] = \mathbf{0} \tag{2.45}$$

or, in indicial notation,

$$\left[\frac{F_{im} \dot{S}_{mj}}{J} \right]_{,i} = \mathbf{0} \tag{2.46}$$

It is interesting to note that the above procedure for the usual problem (i.e., starting with $\bar{\boldsymbol{\nabla}} \cdot \boldsymbol{S} = \mathbf{0}$) gives the well-known equation (2.25 with $\boldsymbol{f} = \mathbf{0}$)

$$\tilde{\boldsymbol{\nabla}} \cdot \left[\frac{\boldsymbol{F} \cdot \boldsymbol{S}}{J} \right] = \boldsymbol{\nabla} \cdot \boldsymbol{\sigma} = 0$$

Here, $\tilde{\boldsymbol{\nabla}} \cdot \boldsymbol{\sigma} = \boldsymbol{\nabla} \cdot \boldsymbol{\sigma}$ since the Cauchy stress is symmetric.

Finally, using Nanson's formula, equation (2.43), one can write

$$\boldsymbol{N} \cdot \dot{\boldsymbol{S}}\, dS = \frac{\boldsymbol{n} \cdot \boldsymbol{F} \cdot \dot{\boldsymbol{S}}}{J}\, ds \tag{2.47}$$

A scaled Lagrange traction rate, which is used later, is defined in current coordinates as

$$\boldsymbol{\tau}^{(s)} = \frac{\boldsymbol{n} \cdot \boldsymbol{F} \cdot \dot{\boldsymbol{S}}}{J} \tag{2.48}$$

and

$$\boldsymbol{\tau}^{(s)}\, ds = \boldsymbol{N} \cdot \dot{\boldsymbol{S}}\, dS \tag{2.49}$$

where $\boldsymbol{N} \cdot \dot{\boldsymbol{S}}$ is usually referred to as the Lagrangian traction rate.

2.1.3 Kinematics and Kinetics in General Curvilinear Coordinates

The main purpose of this section is to present representations of various kinematic and kinetic variables in general tensor notation in order to clearly show the "natural" (or "preferred") bases associated with each of these variables. This approach is also useful for the derivations of various objective rates of tensors. In view of the fact that the rest of the book uses Cartesian tensor notation, details of general tensor manipulations are omitted in this section—only the results are given here. For a discussion of tensor analysis, the reader is referred to Sokolnikoff (1964). This section and the next section (2.1.4) are primarily based on the papers by Rubinstein and Atluri (1983) and Atluri (1984).

Three-dimensional kinematics. Referring to figure 2.3, let ξ^J be arbitrary coordinates in B_0 and η^i arbitrary coordinates in B. Also, ξ^J in B are convected or intrinsic coordinates.

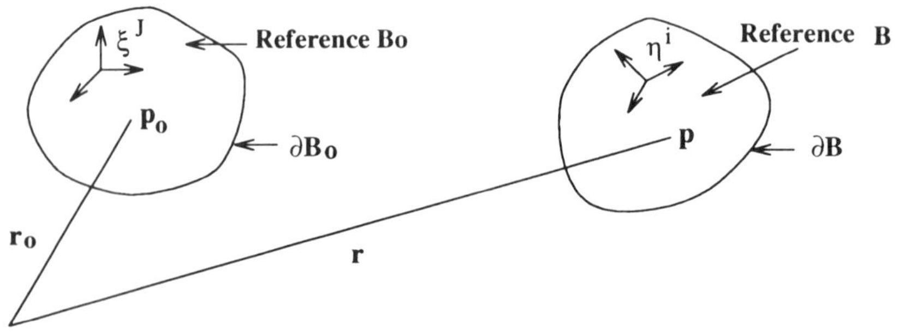

Figure 2.3. Kinematics in curvilinear coordinates.

Basis vectors can be defined as

$$G_J = \frac{\partial r_0}{\partial \xi^J}, \qquad g_i = \frac{\partial r}{\partial \eta^i}, \qquad g_J = \frac{\partial r}{\partial \xi^J} \tag{2.50}$$

where g_J are the convected basis vectors. The bases G_J and g_i are arbitrary and, therefore, each could be the Cartesian basis e_i (as used in the previous sections of this chapter). The convected basis g_J is, in general, curvilinear.

Tensor quantities introduced in previous sections of this chapter can, of course, be expanded in any basis. However, physical quantities are best expanded in certain "natural" (or "preferred") bases in order to clearly see their meaning. Thus, for example, the deformation gradient

$$F = \frac{\partial r}{\partial r_0} = \frac{\partial r}{\partial \eta_i}\frac{\partial \eta_i}{\partial \xi^J}\frac{\partial \xi^J}{\partial r_0} = \frac{\partial \eta^i}{\partial \xi^J} g_i G^J = F^i_{.J} g_i G^J \tag{2.51}$$

This interpretation has been used so far with $g_i = G^J = e_i$, the Cartesian unit vectors, and $\xi^J = X^J$, $\eta^i = x^i$. Note, however, that the convected basis

$$g_M = \frac{\partial r}{\partial r_0} \cdot \frac{\partial r_0}{\partial \xi^M} = F \cdot \mathbf{G}_M = F^i_{.J} g_i G^J \cdot G_M = F^i_{.M} g_i \tag{2.52}$$

so that F becomes

$$F = F^i_{.J} g_i G^J = g_J G^J \tag{2.53}$$

in terms of the convected basis g_J and reference basis G^J.

Thus, the "preferred" bases for F are the current basis vectors on the left and the reference basis vectors on the right. If the current basis is the convected one, the matrix representation of F becomes the identity since the deformation is represented by the convected basis.

For convenience, it is recorded that

$$F^{\mathrm{T}} = F^{\mathrm{T}I}_{.j} G_I g^j = F_j^{\ I} G_I g^j$$

$$F^{-1} = F^{-1I}_{.j} G_I g^j, \qquad F^{-\mathrm{T}} = F^{-\mathrm{T}i}_{.J} g_i G^J \tag{2.54}$$

Strain measures. The Green and Almansi strains become

$$E^{(G)} = \tfrac{1}{2}(F^{\mathrm{T}} \cdot F - I) = \tfrac{1}{2}(F^{\mathrm{T}I}_{.k} F^k_{.J} - \delta^I_{.J}) G_I G^J \tag{2.55}$$

$$E^{(A)} = \tfrac{1}{2}(I - F^{-\mathrm{T}} \cdot F^{-1}) = \tfrac{1}{2}(\delta^i_{.j} - F^{-\mathrm{T}i}_{.K} F^{-1K}_{.j}) g_i g^j \tag{2.56}$$

so that the Green strain appears in the reference and the Almansi strain in the current basis. It can be shown that the Almansi strain becomes zero in the convected basis.

Finally, the stretching tensor $\boldsymbol{d}$ has the form

$$\boldsymbol{d} = d^{i}_{.j}\boldsymbol{g}_i\boldsymbol{g}^j \tag{2.57}$$

with the current basis as the "preferred" one.

Stress measures. The different stress measures, which were introduced in the previous section, can be written as follows:

$$\text{Cauchy stress:} \qquad \boldsymbol{\sigma} = \sigma^{i}_{.j}\boldsymbol{g}_i\boldsymbol{g}^j \tag{2.58}$$

$$\text{Kirchhoff stress:} \qquad \boldsymbol{K} = K^{i}_{.j}\boldsymbol{g}_i\boldsymbol{g}^j \tag{2.59}$$

$$\text{Lagrange stress:} \qquad \boldsymbol{S} = S^{I}_{.j}\boldsymbol{G}_I\boldsymbol{g}^j \tag{2.60}$$

$$\text{Piola–Kirchhoff one:} \qquad \boldsymbol{P}^{(1)} = P^{(1)i}_{.J}\boldsymbol{g}_i\boldsymbol{G}^J \tag{2.61}$$

$$\text{Piola–Kirchhoff two:} \qquad \boldsymbol{P}^{(2)} = P^{(2)I}_{.J}\boldsymbol{G}_I\boldsymbol{G}^J \tag{2.62}$$

It is important to note that each of the conjugate pairs $\boldsymbol{K}$ and $\boldsymbol{d}$, $\boldsymbol{P}^{(1)}$ and $\dot{\boldsymbol{F}}$, and $\boldsymbol{P}^{(2)}$ and $\dot{\boldsymbol{E}}^{(G)}$ have the same preferred bases.

Equilibrium in a reference configuration. Equations (2.39) and (2.40) can now be written in general tensor notation as follows [note that now $\bar{\boldsymbol{\nabla}}(\cdot) = \boldsymbol{G}^{(M)}\,\partial/\partial\xi^M(\cdot)$]:

$$(S^{I}_{.j})_{,I}\boldsymbol{g}^j + \bar{f}_j\boldsymbol{g}^j = 0 \tag{2.63}$$

$$N_I S^{I}_{.j}\boldsymbol{g}^j = \bar{\tau}_j\boldsymbol{g}^j \tag{2.64}$$

where I denotes the covariant derivative with respect to ξ^I. This equation clearly shows that $\bar{f}_j$ and $\bar{\tau}_j$ are components of the body force and surface traction vectors per unit reference volume and reference surface area, respectively, but expanded in the current basis.

The various stress measures are displayed in table 2.1. The Lagrange rule is a consequence of equation (2.64). The Kirchhoff rule, which involves the second Piola–Kirchhoff stress, can be proved from the equation (see equation 2.40)

$$\boldsymbol{N}\cdot\boldsymbol{P}^{(2)} = \boldsymbol{N}\cdot\boldsymbol{S}\cdot\boldsymbol{F}^{-\mathrm{T}} = \bar{\boldsymbol{\tau}}\cdot\boldsymbol{F}^{-\mathrm{T}} \tag{2.65}$$

Table 2.1. Stress measures

Stress measure	Symbol	Symmetric ?	Normal	Traction in configuration	Traction/unit surface area	Equation
Cauchy	$\boldsymbol{\sigma}$	Yes	Current	Current	Current	$n_j \sigma_{ji}\, dS = d\tau_i$
Nominal or Lagrange	$\boldsymbol{S} = J\boldsymbol{F}^{-1} \cdot \boldsymbol{\sigma}$	No	Reference	Current	Reference	$N_j S_{ji}\, dS = d\tau_i^{(L)} = d\bar{\tau}_i$ Lagrange rule
PKI	$\boldsymbol{P}^{(1)} = J\boldsymbol{\sigma} \cdot \boldsymbol{F}^{-\mathrm{T}}$					
PKII	$\boldsymbol{P}^{(2)} = J\boldsymbol{F}^{-1} \cdot \boldsymbol{\sigma} \cdot \boldsymbol{F}^{-\mathrm{T}}$	Yes	Reference	Reference	Reference	$N_j P_{ji}^{(2)}\, dS = d\tau_i^{(K)} = \dfrac{\partial X_i}{\partial x_j}\, d\bar{\tau}_j$ Kirchhoff rule

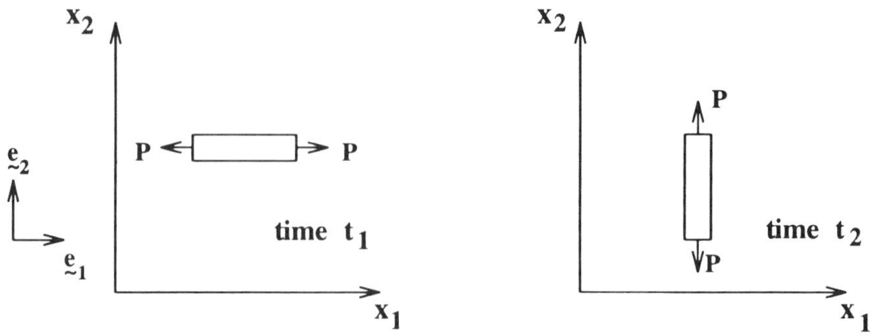

Figure 2.4. Rotation of a one-dimensional stress state by 90 degrees.

2.1.4 *Objective Rates of Tensors*

Spin and rotation. Figure 2.4 shows a bar, with applied axial force P, being rotated by 90°. This process is viewed from a global Cartesian basis $\boldsymbol{G}_i = \boldsymbol{g}_i = \boldsymbol{e}_i$. At time t_1, the only nonzero component of the Cauchy stress is $\sigma_{11} = P/A$ (where A is the cross-sectional area of the bar) and, at time t_2, the only nonzero stress component is $\sigma_{22} = P/A$. Thus, during this time interval, $\dot{\sigma}_{11} \neq 0$ and $\dot{\sigma}_{22} \neq 0$. However, this change of stress is not related to material constitutive behavior.

Note that, in this example, the Green strain $\boldsymbol{E}^{(G)}$ and the second Piola–Kirchhoff stress $\boldsymbol{P}^{(2)}$, which are material based (i.e., typically written in the basis $\boldsymbol{G}_i$), stay constant during the above rotation. The Cauchy stress components σ_{ij}, however, which are typically written in the current basis $\boldsymbol{g}_i$, undergo a change purely due to rotation (when $\boldsymbol{g}_i = \boldsymbol{G}_i \equiv \boldsymbol{e}_i$). In general, the change in $\boldsymbol{\sigma}$ can be attributed to two factors—one associated with the stress–strain law and the other with rotation. In this example, an observer fixed to the rod and rotating with it does not see a change in stress. Thus, *in the above example*, the stress rate that is related to strain rate through a constitutive law must be zero since all of it comes from rotation.

Such a rate is said to be "objective" with respect to rigid body rotation. While it is clear which rate should be chosen in the rigid body rotation example above, such is not the case in general because there are many objective rates of a tensor.

Rates of the Cauchy stress. The starting point here is writing the Cauchy stress $\boldsymbol{\sigma}$ in a convected basis as

$$\boldsymbol{\sigma} = \sigma^{ML} \boldsymbol{g}_M \boldsymbol{g}_L \tag{2.66}$$

For notational simplicity, σ^{ML} (in this section) denotes the components of $\boldsymbol{\sigma}$ in a convected basis; previously, upper case super- and subscripts have been used to represent components of tensors in the reference basis $\boldsymbol{G}_I$.

To find the Jaumann derivative, one uses a rotating basis $\boldsymbol{g}_I = \hat{\boldsymbol{g}}_I$, which moves with a material particle and spins rigidly with a material element with speed $\boldsymbol{\omega}$. This is a special case of a convected basis.

The quantity $d\hat{\boldsymbol{g}}_L/dt$ is the time derivative of $\boldsymbol{g}_L$ for a fixed material particle (i.e., following a particle, ξ^J = constant)

$$\frac{d\hat{\boldsymbol{g}}_L}{dt} = \boldsymbol{\Omega} \times \hat{\boldsymbol{g}}_L = \boldsymbol{\omega} \cdot \hat{\boldsymbol{g}}_L \tag{2.67}$$

where $\boldsymbol{\Omega}$ is a vector with components Ω_i and $\boldsymbol{\omega}$ is the antisymmetric spin tensor (equation 2.20). In Cartesian coordinates, one has the relations $\omega_{12} = -\Omega_3$, $\omega_{13} = \Omega_2$, and $\omega_{23} = -\Omega_1$.

The material time derivative of $\boldsymbol{\sigma}$ is

$$\begin{aligned}\frac{D\boldsymbol{\sigma}}{Dt} &= \frac{D}{Dt}(\sigma^{ML}\hat{\boldsymbol{g}}_M\hat{\boldsymbol{g}}_L)\\ &= \frac{D\sigma^{ML}}{Dt}\hat{\boldsymbol{g}}_M\hat{\boldsymbol{g}}_L + \sigma^{ML}\frac{d\hat{\boldsymbol{g}}_M}{dt}\hat{\boldsymbol{g}}_L + \sigma^{ML}\hat{\boldsymbol{g}}_M\frac{d\hat{\boldsymbol{g}}_L}{dt}\end{aligned} \tag{2.68}$$

With

$$\boldsymbol{\omega} = w^J_{\cdot L}\hat{\boldsymbol{g}}_J\hat{\boldsymbol{g}}^L \tag{2.69}$$

and equation (2.67), it is easy to show that

$$\frac{d\hat{\boldsymbol{g}}_M}{dt} = \omega^J_{\cdot M}\hat{\boldsymbol{g}}_J \tag{2.70}$$

so that the second and third terms on the right-hand side of equation (2.68) become $\boldsymbol{\omega} \cdot \boldsymbol{\sigma}$ and $-\boldsymbol{\sigma} \cdot \boldsymbol{\omega}$, respectively. Also, it can be shown that the four quantities

$$\frac{D\sigma^{MN}}{Dt}\hat{\boldsymbol{g}}_M\hat{\boldsymbol{g}}_N, \qquad \frac{D\sigma_{MN}}{Dt}\hat{\boldsymbol{g}}^M\hat{\boldsymbol{g}}^N,$$

$$\frac{D\sigma^M_{\cdot N}}{Dt}\hat{\boldsymbol{g}}_M\hat{\boldsymbol{g}}^N \quad \text{and} \quad \frac{D\sigma_M^{\cdot N}}{Dt}\hat{\boldsymbol{g}}^M\hat{\boldsymbol{g}}_N$$

form one and the same tensor, $\hat{D}\boldsymbol{\sigma}/Dt$. Finally, one gets the equation

$$\frac{\hat{D}\boldsymbol{\sigma}}{Dt} = \frac{D\boldsymbol{\sigma}}{Dt} - \boldsymbol{\omega}\cdot\boldsymbol{\sigma} + \boldsymbol{\sigma}\cdot\boldsymbol{\omega} \tag{2.71}$$

Here, $\hat{D}\boldsymbol{\sigma}/Dt$ (also called $\hat{\boldsymbol{\sigma}}$) is the corotational (Zaremba, Noll, Jaumann, rigid body) derivative, which is the rate of change of the stress in a basis spinning rigidly along with a material element and moving with it. As will be shown shortly, the Jaumann rate of a tensor is objective but the material rate $D\boldsymbol{\sigma}/Dt$ is not.

The usual material rate $D\boldsymbol{\sigma}/Dt$ (also called $\dot{\boldsymbol{\sigma}}$) is the rate of change of $\boldsymbol{\sigma}$ as experienced by a material particle identified by ξ^J. This can be evaluated from the equation

$$\begin{aligned}\frac{D\boldsymbol{\sigma}}{Dt} &= \left.\frac{d\boldsymbol{\sigma}}{dt}\right|_{\xi_J} = \left.\frac{d}{dt}(\sigma^{mn}\boldsymbol{g}_m\boldsymbol{g}_n)\right|_{\xi_J} \\ &= \left[\left.\frac{\partial\sigma^{mn}}{\partial t}\right|_{\eta^i} + \frac{\partial\sigma^{mn}}{\partial\eta^l}v^l\right]\boldsymbol{g}_m\boldsymbol{g}_n + \sigma^{mn}\frac{d\boldsymbol{g}_m}{dt}\boldsymbol{g}_n + \sigma^{mn}\boldsymbol{g}_m\frac{d\boldsymbol{g}_n}{dt}\end{aligned} \tag{2.72}$$

where the velocity of a material particle

$$\boldsymbol{v} = \frac{d\boldsymbol{r}}{dt} = \frac{\partial\boldsymbol{r}}{\partial\eta^i}\frac{d\eta^i}{dt} = \boldsymbol{g}_i v^i \tag{2.73}$$

Also,

$$\frac{d\boldsymbol{g}_m}{dt} = \frac{\partial\boldsymbol{g}_m}{\partial\eta^j}v^j = \left\{\begin{matrix} e \\ m\ j \end{matrix}\right\} v^j\boldsymbol{g}_e \tag{2.74}$$

since

$$\left.\frac{\partial\boldsymbol{g}_m}{\partial t}\right|_{\eta_i} = 0 \tag{2.75}$$

by virtue of the fact that the basis vectors $\boldsymbol{g}_m$ are not explicit functions of time. In equation (2.74), the quantity within brackets represents the Christoffel symbols for the spatial coordinates η^j at p (figure 2.3).

Using equation (2.74) in (2.72), one obtains

$$\frac{D\boldsymbol{\sigma}}{Dt} = \left[\left.\frac{\partial\sigma^{mn}}{\partial t}\right|_{\eta_i} + v^k\sigma^{mn}{}_{,k}\right]\boldsymbol{g}_m\boldsymbol{g}_n \tag{2.76}$$

$$= \frac{\delta\sigma^{mn}}{\delta t}\boldsymbol{g}_m\boldsymbol{g}_n \tag{2.77}$$

The comma in equation (2.76) represents the covariant derivative and $\delta(\cdot)/\delta t$ in equation (2.77) represents the intrinsic derivative, respectively, of the contravariant components of $\boldsymbol{\sigma}$ in the basis $\boldsymbol{g}_i$.

Other rates of $\boldsymbol{\sigma}$ can be derived by using $\boldsymbol{g}_M$ and $\boldsymbol{g}_N$ instead of $\hat{\boldsymbol{g}}_M$ and $\hat{\boldsymbol{g}}_N$. Now,

$$\frac{d\boldsymbol{g}_M}{dt} = \boldsymbol{h} \cdot \boldsymbol{g}_M, \qquad \frac{d\boldsymbol{g}^N}{dt} = - \boldsymbol{h}^{\mathrm{T}} \cdot \boldsymbol{g}^N \tag{2.78}$$

where $\boldsymbol{h}$ is the velocity gradient (equation 2.18)

$$\boldsymbol{h} = v_{m,n} \boldsymbol{g}^m \boldsymbol{g}^n \tag{2.79}$$

Starting from

$$= \sigma^{MN} \boldsymbol{g}_M \boldsymbol{g}_N = \sigma^{M}_{.\,N} \boldsymbol{g}_M \boldsymbol{g}^N, \quad \text{etc.}$$

one can show that

$$\begin{aligned}
\frac{D^{(1)}\boldsymbol{\sigma}}{Dt} &= \frac{D\boldsymbol{\sigma}}{Dt} - \boldsymbol{h} \cdot \boldsymbol{\sigma} - \boldsymbol{\sigma} \cdot \boldsymbol{h}^{\mathrm{T}} \\
\frac{D^{(2)}\boldsymbol{\sigma}}{Dt} &= \frac{D\boldsymbol{\sigma}}{Dt} + \boldsymbol{h}^{\mathrm{T}} \cdot \boldsymbol{\sigma} + \boldsymbol{\sigma} \cdot \boldsymbol{h} \\
\frac{D^{(3)}\boldsymbol{\sigma}}{Dt} &= \frac{D\boldsymbol{\sigma}}{Dt} - \boldsymbol{h} \cdot \boldsymbol{\sigma} + \boldsymbol{\sigma} \cdot \boldsymbol{h} \\
\frac{D^{(4)}\boldsymbol{\sigma}}{Dt} &= \frac{D\boldsymbol{\sigma}}{Dt} + \boldsymbol{h}^{\mathrm{T}} \cdot \boldsymbol{\sigma} - \boldsymbol{\sigma} \cdot \boldsymbol{h}^{\mathrm{T}}
\end{aligned} \tag{2.80}$$

where

$$\begin{aligned}
\frac{D^{(1)}\boldsymbol{\sigma}}{Dt} &= \frac{D\sigma^{MN}}{Dt} \boldsymbol{g}_M \boldsymbol{g}_N \\
\frac{D^{(2)}\boldsymbol{\sigma}}{Dt} &= \frac{D\sigma_{MN}}{Dt} \boldsymbol{g}^M \boldsymbol{g}^N \\
\frac{D^{(3)}\boldsymbol{\sigma}}{Dt} &= \frac{D\sigma^{M}_{.\,N}}{Dt} \boldsymbol{g}_M \boldsymbol{g}^N \\
\frac{D^{(4)}\boldsymbol{\sigma}}{Dt} &= \frac{D\sigma_{M}^{\cdot\,N}}{Dt} \boldsymbol{g}^M \boldsymbol{g}_N
\end{aligned} \tag{2.81}$$

Table 2.2. Various stress rates and their properties

	Derivative	Objective	Symmetric
Material	$\frac{D\boldsymbol{\sigma}}{Dt}$	No	Yes
Jaumann	$\frac{\hat{D}\boldsymbol{\sigma}}{Dt}$	Yes	Yes
Truesdell	$\frac{D^{(1)}\boldsymbol{\sigma}}{Dt}$	Yes	Yes
Rivlin	$\frac{D^{(2)}\boldsymbol{\sigma}}{Dt}$	Yes	Yes
?	$\frac{D^{(3)}\boldsymbol{\sigma}}{Dt}$	Yes	No
?	$\frac{D^{(4)}\boldsymbol{\sigma}}{Dt}$	Yes	No

The derivatives of different types of components of $\boldsymbol{\sigma}$ give different tensors and different stress rates. The first rate above is called the Truesdell or Oldroyd rate of $\boldsymbol{\sigma}$ and the second is called the Cotter–Rivlin rate of $\boldsymbol{\sigma}$.

As discussed in the next section, all these rates are objective. Other objective rates have been used by different researchers. While there is agreement on the need for use of an objective stress rate in a constitutive equation, the question of which one to use in a particular circumstance has been the subject of a lively debate among continuum mechanists for at least the past decade.

The various stress rates that have been discussed in this section are summarized in table 2.2.

Objectivity. Consider a second observer whose reference frame rotates with respect to the first with a time-dependent rigid rotation $\boldsymbol{Q}(t)$. Here, $\boldsymbol{Q}(t)$ is an orthogonal tensor. Let the quantities, as seen by the second observer, be identified by a superposed bar. Tensors that transform as

$$\bar{\boldsymbol{T}} = \boldsymbol{Q} \cdot \boldsymbol{T} \cdot \boldsymbol{Q}^{\mathrm{T}} \tag{2.82}$$

are said to be objective.

Similarly, a rate $\mathscr{D}\boldsymbol{A}/\mathscr{D}t$ (which is a generic rate) is objective if

$$\frac{\mathscr{D}\boldsymbol{A}}{\mathscr{D}t} = \boldsymbol{Q} \cdot \frac{\mathscr{D}\boldsymbol{A}}{\mathscr{D}t} \cdot \boldsymbol{Q}^{\mathrm{T}} \tag{2.83}$$

The material derivative of $\boldsymbol{\sigma}$ is not objective since

$$\begin{aligned}\frac{D}{Dt}\bar{\boldsymbol{\sigma}} &= \dot{\boldsymbol{Q}} \cdot \boldsymbol{\sigma} \cdot \boldsymbol{Q}^{\mathrm{T}} + \boldsymbol{Q} \cdot \dot{\boldsymbol{\sigma}} \cdot \boldsymbol{Q}^{\mathrm{T}} + \boldsymbol{Q} \cdot \boldsymbol{\sigma} \cdot \dot{\boldsymbol{Q}}^{\mathrm{T}} \\ &\neq \boldsymbol{Q} \cdot \dot{\boldsymbol{\sigma}} \cdot \boldsymbol{Q}^{\mathrm{T}}\end{aligned} \tag{2.84}$$

The Jaumann rate, however, is objective, since

$$\frac{\hat{D}\boldsymbol{\sigma}}{Dt} = \frac{D\boldsymbol{\sigma}}{Dt} - \boldsymbol{\omega} \cdot \boldsymbol{\sigma} + \boldsymbol{\sigma} \cdot \boldsymbol{\omega} \tag{2.85}$$

$$\begin{aligned}\frac{\hat{D}\bar{\boldsymbol{\sigma}}}{Dt} &= \frac{D\bar{\boldsymbol{\sigma}}}{Dt} - \bar{\boldsymbol{\omega}} \cdot \bar{\boldsymbol{\sigma}} + \bar{\boldsymbol{\sigma}} \cdot \bar{\boldsymbol{\omega}} \\ &= \boldsymbol{Q}[\dot{\boldsymbol{\sigma}} - \boldsymbol{\omega} \cdot \boldsymbol{\sigma} + \boldsymbol{\sigma} \cdot \boldsymbol{\omega}]\boldsymbol{Q}^{\mathrm{T}} \\ &= \boldsymbol{Q} \cdot \frac{\hat{D}\boldsymbol{\sigma}}{Dt} \cdot \boldsymbol{Q}^{\mathrm{T}}\end{aligned} \tag{2.86}$$

The relationship

$$\bar{\boldsymbol{\omega}} = \boldsymbol{Q} \cdot \boldsymbol{\omega} \cdot \boldsymbol{Q}^{\mathrm{T}} + \dot{\boldsymbol{Q}} \cdot \boldsymbol{Q}^{\mathrm{T}} \tag{2.87}$$

is used to prove equation (2.86).

Various quantities of interest transform as shown in table 2.3.

Table 2.3. Summary of transformations

$\boldsymbol{F}, \boldsymbol{R}, \boldsymbol{U}$	$\boldsymbol{Q}\cdot\boldsymbol{F},\ \boldsymbol{Q}\cdot\boldsymbol{R},\ \boldsymbol{U}$
$\boldsymbol{d}, \boldsymbol{\omega}$	$\boldsymbol{Q}\cdot\boldsymbol{d}\cdot\boldsymbol{Q}^{\mathrm{T}},\ \boldsymbol{Q}\cdot\boldsymbol{\omega}\cdot\boldsymbol{Q}^{\mathrm{T}}+\dot{\boldsymbol{Q}}\cdot\boldsymbol{Q}^{\mathrm{T}}$
$\boldsymbol{P}^{(2)}, \boldsymbol{E}^{(G)}, \dot{\boldsymbol{P}}^{(2)}, \dot{\boldsymbol{E}}^{(G)}$	$\boldsymbol{P}^{(2)}, \boldsymbol{E}^{(G)}, \dot{\boldsymbol{P}}^{(2)}, \dot{\boldsymbol{E}}^{(G)}$
$\boldsymbol{P}^{(1)}, \boldsymbol{F}$	$\boldsymbol{Q}\cdot\boldsymbol{P}^{(1)},\ \boldsymbol{Q}\cdot\boldsymbol{F}$
$\boldsymbol{S}, \boldsymbol{F}^{\mathrm{T}}$	$\boldsymbol{S}\cdot\boldsymbol{Q}^{\mathrm{T}},\ \boldsymbol{F}^{\mathrm{T}}\cdot\boldsymbol{Q}^{\mathrm{T}}$
$\dot{\boldsymbol{P}}^{(1)}, \dot{\boldsymbol{F}}$	$\dot{\boldsymbol{Q}}\cdot\boldsymbol{P}^{(1)}+\boldsymbol{Q}\cdot\dot{\boldsymbol{P}}^{(1)},\ \dot{\boldsymbol{Q}}\cdot\boldsymbol{F}+\boldsymbol{Q}\cdot\dot{\boldsymbol{F}}$
$\dot{\boldsymbol{S}}, \dot{\boldsymbol{F}}^{\mathrm{T}}$	$\boldsymbol{S}\cdot\dot{\boldsymbol{Q}}^{\mathrm{T}}+\dot{\boldsymbol{S}}\cdot\boldsymbol{Q}^{\mathrm{T}},\ \boldsymbol{F}^{\mathrm{T}}\cdot\dot{\boldsymbol{Q}}^{\mathrm{T}}+\dot{\boldsymbol{F}}^{\mathrm{T}}\cdot\boldsymbol{Q}^{\mathrm{T}}$
$\boldsymbol{\sigma}, \boldsymbol{\tau}, \boldsymbol{E}^{(A)}, \boldsymbol{V}$	$\boldsymbol{Q}\cdot\boldsymbol{\sigma}\cdot\boldsymbol{Q}^{\mathrm{T}},\ \boldsymbol{Q}\cdot\boldsymbol{\tau}\cdot\boldsymbol{Q}^{\mathrm{T}}$
	$\boldsymbol{Q}\cdot\boldsymbol{E}^{A}\cdot\boldsymbol{Q}^{\mathrm{T}},\ \boldsymbol{Q}\cdot\boldsymbol{V}\cdot\boldsymbol{Q}^{\mathrm{T}}$
$\hat{\boldsymbol{\sigma}}, \hat{\boldsymbol{K}}, \hat{\boldsymbol{E}}^{(A)}, \hat{\boldsymbol{V}}$	$\boldsymbol{Q}\cdot\hat{\boldsymbol{\sigma}}\cdot\boldsymbol{Q}^{\mathrm{T}},\ \boldsymbol{Q}\cdot\hat{\boldsymbol{K}}\cdot\boldsymbol{Q}^{\mathrm{T}}$
	$\boldsymbol{Q}\cdot\hat{\boldsymbol{E}}^{A}\cdot\boldsymbol{Q}^{\mathrm{T}},\ \boldsymbol{Q}\cdot\hat{\boldsymbol{V}}\cdot\boldsymbol{Q}^{\mathrm{T}}$

This concludes the discussion on the fundamentals of continuum mechanics. For detailed discussions of many of these topics, the reader is referred to books devoted to the subject of continuum mechanics, such as Gurtin (1981) and Ogden (1984).

2.2 Boundary Element Formulations

A boundary element method (BEM) formulation for solid mechanics problems with large strains and large rotations is presented in this section. This formulation is the basis for several chapters in this book that are concerned with metal forming processes.

In view of applications in metal forming, elastic strains in a deforming body are assumed to be small, while nonelastic strains and rotations are allowed to be large. Therefore, a hypoelastic constitutive model is assumed to be valid in this work. Also, the Jaumann rate of the Cauchy stress is chosen as the objective rate in the hypoelastic model. Other objective rates should be considered in problems where both the following situations prevail (see Chandra and Mukherjee 1986): (a) the shearing strains become very large; (b) the nonelastic deformation process is strongly anisotropic.

A BEM formulation for the above class of problems was first presented by Chandra and Mukherjee [see Chandra and Mukherjee (1983, 1984a) and Mukherjee and Chandra (1984, 1987)]. An updated Lagrangian formulation, in which the configuration of a deforming body at time t is used as the reference for a time step t to $t + \Delta t$ (where Δt is small), was used to derive the BEM formulation in the above publications. Recently (Zhang et al. 1992b), the formulation has been re-derived in the current configuration without making the simplifying assumptions inherent in the updated Lagrangian approach. Although the final results are essentially the same, the derivation in the current configuration, which is more general, is presented in this chapter. Numerical results, which are presented later in this chapter, are obtained from an updated Lagrangian approach. An explicit integration scheme is used in this work, and the velocities on the boundary are treated as primary variables.

Recent related research on applications of the BEM in finite-deformation elastoplastic problems has been carried out by, among others, Jin et al. (1989), Okada et al. (1990), and Chen and Ji (1990). Jin et al. (1989) proposed an implicit integration scheme for the constitutive equations. Okada et al. (1990) presented a full tangent stiffness field BEM in which the velocity field variables both inside and on the boundary of a body are treated as primary variables. They also carry out objective stress integration and present numerical results for problems of diffuse necking in initially perfect elastoplastic bars under tension.

The remainder of this section is organized as follows. A three-dimensional BEM formulation is presented first. This is followed by plane strain, plane stress, and axisymmetric formulations. Next, a derivative boundary element method (DBEM) formulation is discussed for planar problems. This DBEM formulation is the basis for design sensitivity analysis, which is the subject of the next chapter.

2.2.1 Constitutive Assumptions

The starting point here is figure 2.1. A spatially fixed Cartesian basis is used in this section. Unless otherwise specified, the range of indices is 1,2,3 for three-dimensional and 1,2 for two-dimensional problems.

The key constitutive assumptions, which are valid in the current configuration, are as follows. The first assumption is that the tensor $\boldsymbol{d}$ can be additively decomposed into an elastic and a nonelastic part:

$$\boldsymbol{d} = \boldsymbol{d}^{(e)} + \boldsymbol{d}^{(n)} \tag{2.88}$$

The second is that the material is homogeneous and isotropic, and the elastic field obeys the hypoelastic relationship (Fung, 1965)

$$\hat{\boldsymbol{\sigma}} = \lambda \operatorname{tr} (\boldsymbol{d}^{(e)})\boldsymbol{I} + 2G\boldsymbol{d}^{(e)} \tag{2.89}$$

where $\boldsymbol{\sigma}$ is the Cauchy stress, λ and G are the Lamé constants, tr denotes the trace of a tensor, and $\boldsymbol{I}$ is the identity tensor. A hat (^) over $\boldsymbol{\sigma}$ denotes one of its objective rates. The Jaumann rate (equation 2.71), written as

$$\hat{\boldsymbol{\sigma}} = \dot{\boldsymbol{\sigma}} + \boldsymbol{\sigma} \cdot \boldsymbol{\omega} - \boldsymbol{\omega} \cdot \boldsymbol{\sigma} \tag{2.90}$$

is adopted here.

A general form of a nonelastic constitutive law governing the behavior of $\boldsymbol{d}^{(n)}$ is

$$\boldsymbol{d}^{(n)} = \boldsymbol{f}(\boldsymbol{\sigma}, \boldsymbol{q}_1, \boldsymbol{q}_2, \ldots, \boldsymbol{q}_k) \quad \text{and} \quad \hat{\boldsymbol{q}}_p = \boldsymbol{g}_p(\boldsymbol{\sigma}, \boldsymbol{q}_1, \boldsymbol{q}_2, \ldots, \boldsymbol{q}_k) \tag{2.91}$$

where $\boldsymbol{q}_p$, $p = 1, 2, \ldots, k$ (where k is usually a small integer), are suitably defined state variables, which can be scalars or tensors. A general discussion of such unified viscoplastic constitutive models, using state variables, can be found in Mukherjee (1982). A conventional plasticity model for material behavior can also be employed if desired. In that case, $\boldsymbol{d}^{(n)}$ is also a function of the rate of stress. Discussion of specific constitutive models is deferred until later in this chapter.

2.2.2 *Three-Dimensional BEM Formulation for Velocities*

The relationships between stresses, which were developed in section 2.1, are restated below:

$$\boldsymbol{K} = J\boldsymbol{\sigma} \tag{2.92}$$

$$\boldsymbol{S} = \boldsymbol{F}^{-1} \cdot \boldsymbol{K} \tag{2.93}$$

It is easy to prove the following relationships:

$$\dot{J} = J \operatorname{tr} \boldsymbol{d} \tag{2.94}$$

$$\boldsymbol{d} + \boldsymbol{\omega} = \dot{\boldsymbol{F}} \cdot \boldsymbol{F}^{-1} \tag{2.95}$$

Using equations (2.90) and (2.92) through (2.95), some straightforward algebraic manipulations reveal the relationship

$$\hat{\boldsymbol{\sigma}} = \frac{\boldsymbol{F} \cdot \dot{\boldsymbol{S}}}{J} + \boldsymbol{g} \tag{2.96}$$

where

$$\boldsymbol{g} = \boldsymbol{\sigma} \cdot \boldsymbol{\omega} + \boldsymbol{d} \cdot \boldsymbol{\sigma} - (\operatorname{tr} \boldsymbol{d})\, \boldsymbol{\sigma}$$

Also, in the current configuration,

$$\boldsymbol{\tau}^{(c)} = \boldsymbol{n} \cdot \hat{\boldsymbol{\sigma}} \tag{2.97}$$

and

$$\boldsymbol{\tau}^{(s)} = \boldsymbol{\tau}^{(c)} - \boldsymbol{n} \cdot \boldsymbol{g} = \boldsymbol{n} \cdot \frac{\boldsymbol{F} \cdot \dot{\boldsymbol{S}}}{J} \tag{2.98}$$

where $\boldsymbol{\tau}^{(c)}$ and $\boldsymbol{\tau}^{(s)}$ are the Cauchy and scaled Lagrange traction rates, respectively, and $\boldsymbol{n}$ is the unit outward normal to a body at a point on its boundary. A useful formula for $\boldsymbol{n}$, which is easily derived from Nanson's formula (2.43), is

$$\boldsymbol{n} = \frac{\boldsymbol{N} \cdot \boldsymbol{F}^{-1}}{|\boldsymbol{N} \cdot \boldsymbol{F}^{-1}|} \tag{2.99}$$

where $|\,|$ denotes the absolute value of the vector.

The starting point of the BEM formulation is Betti's reciprocal theorem (Betti 1872). In the current configuration (Mukherjee and Chandra 1987),

$$\int_B \hat{\boldsymbol{\sigma}} : \boldsymbol{\varepsilon}^{(R)}\, dv = \int_B \boldsymbol{\sigma}^{(R)} : \boldsymbol{d}^{(e)}\, dv \tag{2.100}$$

The variables with a superscript R are reference field variables due to a point force $\boldsymbol{F}^{(R)}$ in an infinite (three-dimensional) linear elastic isotropic solid undergoing infinitesimal deformation. This force, acting in an (arbitrary) direction $\boldsymbol{a}$ (a unit vector), can be written as

$$\boldsymbol{F}^{(R)} = \Delta(p,q)\,\boldsymbol{a} \qquad \text{or} \qquad F_i^{(R)} = \Delta(p,q)\,a_i \tag{2.101}$$

where Δ is the Dirac delta function and p and q are source and field points, respectively, inside the body. The reference stress and strain fields are assumed to satisfy the usual Hooke's law with the same Lamé constants λ and G as in equation (2.89). Thus,

$$\boldsymbol{\sigma}^{(R)} = \lambda\, \mathrm{tr}\,(\boldsymbol{\varepsilon}^{(R)})\boldsymbol{I} + 2G\boldsymbol{\varepsilon}^{(R)} \tag{2.102}$$

In view of equations (2.89) and (2.102), the integrands on either side of equation (2.100) are equal.

It is convenient to write

$$u_i^{(R)} = U_{ij} a_j \tag{2.103}$$

$$\tau_i^{(R)} = T_{ij} a_j \tag{2.104}$$

where $\boldsymbol{u}^{(R)}$ and $\boldsymbol{\tau}^{(R)}$ are the reference displacement and traction fields due to the point force defined in equation (2.101). The traction field is defined on a boundary ∂B, which is identical to that of the physical body. The Kelvin kernels (Thomson 1848) are available in many references (e.g., Mukherjee 1982). They are

$$U_{ij} = \frac{1}{16\pi(1-\nu)Gr}[(3-4\nu)\delta_{ij} + r_{,i}r_{,j}) \tag{2.105}$$

$$T_{ij} = -\frac{1}{8\pi(1-\nu)r^2}\left([(1-2\nu)\delta_{ij} + 3r_{,i}r_{,j}]\frac{\partial r}{\partial n} + (1-2\nu)(r_{,i}n_j - r_{,j}n_i)\right) \tag{2.106}$$

where δ_{ij} is the Kronecker delta, $r(p,q)$ is the distance from a source point p to a field point q and n_i are the components of the unit outward normal to ∂B at a point Q on it. The convention used here is that lower-case letters p and q denote points inside the body B and capital letters P and Q denote points on the boundary ∂B. A comma denotes a derivative with respect to a field point, that is,

$$r_{,i} = \frac{\partial r}{\partial \xi_i} = \frac{\xi_i - x_i}{r} \tag{2.107}$$

where $\boldsymbol{\xi}$ and $\boldsymbol{x}$ are the coordinates of a field and a source point, respectively.

Using the definition of $\boldsymbol{\varepsilon}^R$ in terms of $\boldsymbol{u}^R$,

$$\varepsilon_{ij}^{(R)} = \tfrac{1}{2}(u_{i,j}^{(R)} + u_{j,i}^{(R)}) \tag{2.108}$$

together with equations (2.96) and (2.103), the left-hand side of equation (2.100) becomes

$$\int_B \hat{\boldsymbol{\sigma}} : \boldsymbol{\varepsilon}^{(R)}\, dv = a_k \int_B \left[\frac{\boldsymbol{F} \cdot \dot{\boldsymbol{S}}}{J}\right]_{ij} U_{jk,i}\, dv + a_k \int_B U_{jk,i} g_{ij}\, dv \tag{2.109}$$

Using the divergence theorem, together with equation (2.98), one gets

$$a_k \int_B \frac{F_{im}\dot{S}_{mj}}{J} U_{jk,i}\, dv = a_k \int_{\partial B} U_{jk} \tau_j^{(s)}\, ds - a_k \int_B \left[\frac{F_{im}\dot{S}_{mj}}{J}\right]_{,i} U_{jk}\, dv \tag{2.110}$$

The second term on the right of equation (2.110) vanishes as a consequence of the rate of equilibrium equations and Nanson's formula (see equation 2.46). Consequently,

$$\int_B \hat{\boldsymbol{\sigma}} : \boldsymbol{\varepsilon}^{(R)}\, dv = a_j \int_{\partial B} U_{ij} \tau_i^{(s)}\, ds + a_j \int_B U_{ij,m} g_{mi}\, dv \tag{2.111}$$

Using equation (2.88), the right-hand side of equation (2.100) is written as

$$\int_B \sigma_{ij}^{(R)} d_{ij}^{(e)}\, dv = \int_B \sigma_{ij}^{(R)} (d_{ij} - d_{ij}^{(n)})\, dv \tag{2.112}$$

Now, using equilibrium equations for the reference stress, both inside and on the boundary of the body, that is,

$$\sigma_{ji,j}^{(R)} = -F_i^{(R)} = -\Delta(p,q)\, a_i \tag{2.113}$$

$$n_j \sigma_{ji}^{(R)} = \tau_i^{(R)} \tag{2.114}$$

together with equations (2.19) and (2.104) and the divergence theorem, leads to

$$\int_B \sigma_{ij}^{(R)} d_{ij} dv = a_j \int_{\partial B} T_{ij} v_i \, ds + a_j \int_B \Delta(p, q) v_j \, dv \tag{2.115}$$

Also, using equations (2.102), (2.108), and (2.103), one gets

$$\int_B \sigma_{ij}^{(R)} d_{ij}^{(n)} \, dv = a_j \int_B [\lambda U_{ij,i} d_{kk}^{(n)} + 2G U_{ij,k} d_{ik}^{(n)}] \, dv \tag{2.116}$$

Finally, upon collecting the equations (2.100), (2.111), (2.112), (2.115), and (2.116), an integral equation is obtained for the velocity at an internal point p as

$$\begin{aligned} v_j(p) = & \int_{\partial B} [U_{ij}(p, Q) \tau_i^{(s)}(Q) - T_{ij}(p, Q) v_i(Q)] \, ds(Q) \\ & + \int_B \lambda U_{ij,i}(p, q) \, d_{kk}^{(n)}(q) \, dv(q) + \int_B 2G U_{ij,k}(p, q) \, d_{ik}^{(n)}(q) \, dv(q) \\ & + \int_B U_{ij,m}(p, q) g_{mi}(q) \, dv(q) \end{aligned} \tag{2.117}$$

where one uses the fact that the direction $\boldsymbol{a}$ (in equation 2.101) is arbitrary.

A boundary version of the above equation is obtained by taking the limit as an internal point p in B approaches a boundary point P on ∂B. This is

$$\begin{aligned} C_{ij}(P) v_i(P) = & \int_{\partial B} [U_{ij}(P, Q) \tau_i^{(s)}(Q) - T_{ij}(P, Q) v_i(Q)] \, ds(Q) \\ & + \int_B \lambda U_{ij,i}(P, q) d_{kk}^{(n)}(q) \, dv(q) \\ & + \int_B 2G U_{ij,k}(P, q) \, d_{ik}^{(n)}(q) \, dv(q) \\ & + \int_B U_{ij,m}(P, q) g_{mi}(q) \, dv(q) \end{aligned} \tag{2.118}$$

where $C_{ij}(P)$ is the "corner" tensor (Mukherjee 1982, Guiggiani and Gigante 1990).

Comparing equation (2.118) with corresponding BEM formulations for small-strain elasticity and small-strain elasto-viscoplasticity (Mukherjee 1982), it is seen that the first term on the right-hand side of the above equation is analogous to the formulation for small-strain elasticity, and that the first three terms together are analogous to small-strain elasto-viscoplasticity. The last term is the so-called "geometric correction" term, which arises due to finite deformations and rotations inside the body.

Usual boundary conditions involve prescribed velocities and tractions on the boundary ∂B. While kinematic boundary conditions, which are common in metal forming applications, are straightforward, traction boundary conditions need some elaboration.

Two common types of boundary loading are a dead load and a follower load. A dead load has a fixed direction in space, and a follower load has a fixed direction relative to the deforming boundary. The most common follower load is pressure that remains normal to a deforming boundary.

Dead load. First, it is useful to note that

$$\boldsymbol{\tau}\, ds = \boldsymbol{n} \cdot \boldsymbol{\sigma}\, ds = \frac{\boldsymbol{n} \cdot \boldsymbol{F} \cdot \boldsymbol{S}}{J}\, ds = \boldsymbol{N} \cdot \boldsymbol{S}\, dS \tag{2.119}$$

where equations (2.34) and Nanson's formula (2.43) have been used.

Let a dead load $\boldsymbol{F}^{(D)}$ on a surface element be

$$\boldsymbol{F}^{(D)}(t) = f(t)\boldsymbol{a}, \qquad \dot{\boldsymbol{F}}^{(D)}(t) = \dot{f}(t)\,\boldsymbol{a} \tag{2.120}$$

where $\boldsymbol{a}$ is a fixed direction in space. Now,

$$\boldsymbol{F}^{(D)}(t) = \boldsymbol{N} \cdot \boldsymbol{S}(t)\, dS \tag{2.121}$$

so that

$$\dot{\boldsymbol{F}}^{(D)}(t) = \dot{f}(t)\boldsymbol{a} = \boldsymbol{N} \cdot \dot{\boldsymbol{S}}(t)\, dS = \boldsymbol{\tau}^{(s)}\, ds \tag{2.122}$$

where equation (2.49) has been used.

Given $\dot{\boldsymbol{F}}^{(D)}(t)$, $\boldsymbol{\tau}^{(s)}$ in equations (2.117) and (2.118) can be found from equations (2.122) at each stage of the deformation process. As expected, the scaled Lagrangian traction rate $\boldsymbol{\tau}^{(s)}$ is the rate of the dead load on a surface element per unit current area.

Follower load. Let a pressure-type follower load be defined as

$$\boldsymbol{F}^{(F)} = p(t)\,\boldsymbol{n}\,ds \tag{2.123}$$

where $p(t)$ is given. The goal here is to relate $\dot{p}(t)$ to the traction rates $\boldsymbol{\tau}^{(s)}$ and $\boldsymbol{\tau}^{(c)}$.

Two formulas that are useful for this exercise are (Hill 1978)

$$\dot{n}_j = (n_j n_l - \delta_{jl})\, v_{k,l} n_k \tag{2.124}$$

and

$$d\dot{s} = (v_{k,k} - v_{j,l} n_j n_l)\, ds \tag{2.125}$$

where the quantities on the left-hand side of the above equations are the rates of the current normal and area of the current surface element, respectively.

One starts with

$$p(t) n_i = \sigma_{ij} n_j \tag{2.126}$$

and

$$\dot{p}(t) n_i + p(t)\dot{n}_i = \dot{\sigma}_{ij} n_j + \sigma_{ij}\dot{n}_j \tag{2.127}$$

Using equation (2.124), together with equations (2.90) and (2.97), in equation (2.127), one can show that

$$\boldsymbol{\tau}^{(c)} = \dot{p}(t)\boldsymbol{n} - \boldsymbol{d}\cdot\boldsymbol{\sigma}\cdot\boldsymbol{n} + \boldsymbol{\sigma}\cdot\boldsymbol{d}\cdot\boldsymbol{n} \tag{2.128}$$

For rigid-body-rotation problems, $\boldsymbol{d} = 0$ and one obtains $\boldsymbol{\tau}^{(c)} = p\boldsymbol{n}$, which is consistent with the definition of the Jaumann rate. This assumption is often employed for geometrically nonlinear analyses, especially in nonlinear structural mechanics problems including thin beams, plates, and shells. For metal forming problems, however, the full equation (2.128) should be used since the stretch tensor can be large.

Also, using the definition of $\boldsymbol{\tau}^{(s)}$ (see equation 2.98),

$$\boldsymbol{\tau}^{(s)} = \boldsymbol{\tau}^{(c)} - \boldsymbol{n}\cdot\boldsymbol{g} \tag{2.129}$$

one can replace $\boldsymbol{\tau}^{(c)}$ in equation (2.128) by $\boldsymbol{\tau}^{(s)}$ to get (Hill 1978)

$$\boldsymbol{\tau}^{(s)} = \dot{p}(t)\boldsymbol{n} - p(t)\boldsymbol{n} \cdot \boldsymbol{\nabla v} + p(t)\boldsymbol{n} \operatorname{tr} \boldsymbol{d} \tag{2.130}$$

where $(\boldsymbol{n} \cdot \boldsymbol{\nabla v})_i = n_k v_{k,i}$.

Equation (2.128) is easier to interpret physically, while equation (2.130) is easier to implement in the BEM equations (2.117) and (2.118) since they already contain $\boldsymbol{\tau}^{(s)}$.

It should be noted that equations (2.128) and (2.130) appear, among other places, in Chen and Ji (1990). Also, while equation (2.125) is not used in this derivation, it is useful if one wishes to calculate the rate of the follower force $\boldsymbol{F}^{(F)}$.

The presence of velocity gradients in equation (2.130) [and therefore in the BEM equations (2.117) and (2.118)] is sometimes referred to as "load correction" and is a consequence of the change in the area of a surface element and the rotation of a unit normal, at a point on it, during the large-deformation process. It should be pointed out that the unknown velocity gradient appears in the surface integral, as well as in the last domain integral (through $\boldsymbol{g}$) in equations (2.117) and (2.118). Thus, iterations become necessary within each time step in order to solve the BEM equation (2.118). This matter is discussed further later in this chapter.

2.2.3 *Stress Rates and Velocity Gradients on the Boundary*

The most convenient way to find stress rates and velocity gradients, at a point P on the boundary ∂B where it is locally smooth, is to use the following equations (Mukherjee and Chandra 1991):

$$\overset{\wedge}{\sigma}_{ij} = \lambda(v_{k,k} - d_{kk}^{(n)})\delta_{ij} + G(v_{i,j} + v_{j,i}) - 2Gd_{ij}^{(n)} \tag{2.131}$$

$$\tau_i^{(c)} = n_j \overset{\wedge}{\sigma}_{ji} \tag{2.132}$$

$$\frac{\partial v_i}{\partial s_1} = v_{i,j} t_j^{(1)} \tag{2.133}$$

$$\frac{\partial v_i}{\partial s_2} = v_{i,j} t_j^{(2)} \tag{2.134}$$

Equation (2.131) is a restatement of the hypoelastic law (2.89) in terms of velocity gradients, and $\partial v_i/\partial s_1$ and $\partial v_i/\partial s_2$ are tangential derivatives of the velocity field at P along two surface directions given by the unit vectors $\boldsymbol{t}^{(1)}$ and $\boldsymbol{t}^{(2)}$. Once the velocity field is known, these tangential derivatives can be determined by suitably differentiating the shape functions for $\boldsymbol{v}$ on ∂B. Now, the system of equations, given $d_{ij}^{(n)}$, has 15 linear scalar equations for the 15 scalar unknowns $v_{i,j}$ and $\hat{\sigma}_{ij}$. The material rate of the Cauchy stress can then be obtained from its Jaumann rate from equation (2.90). The stretch $\boldsymbol{d}$ and spin $\boldsymbol{\omega}$ can be easily obtained from $h_{ij} = v_{i,j}$.

In practice, it is easier to solve equations (2.131)–(2.134) by writing them in local orthogonal coordinates with one direction normal to ∂B at P.

2.2.4 *Internal Stress Rates and Velocity Gradients*

Analytical differentiation of equation (2.117) at an internal source point p is the best way to obtain velocity gradients at internal points. To this end, a differentiated version of equation (2.117) is [here, $(,\bar{l}) \equiv \partial/\partial x_l(p)$]

$$\begin{aligned} v_{j,\bar{l}}(p) = & \int_{\partial B} [U_{ij,\bar{l}}(p,Q)\, \tau_i^{(s)}(Q) - T_{ij,\bar{l}}(p,Q)\, v_i(Q)]\, ds(Q) \\ & + \frac{\partial}{\partial x_{\bar{l}}} \int_B \lambda U_{ij,i}(p,q)\, d_{kk}^{(n)}(q)\, dv(q) \\ & + \frac{\partial}{\partial x_{\bar{l}}} \int_B 2G U_{ij,k}(p,q)\, d_{ik}^{(n)}(q)\, dv(q) \\ & + \frac{\partial}{\partial x_{\bar{l}}} \int_B U_{ij,m}(p,q) g_{mi}(q)\, dv(q) \end{aligned} \tag{2.135}$$

While the boundary integrals in the above equation are not singular, the domain integrands are $O(1/r^2)$ singular in three-dimensional problems. Differentiation of these domain integrals must be handled with care.

Various strategies have been tried for the evaluation of these integrals. In early work on planar problems (e.g., Chandra and Mukherjee 1983), the volume integrals were evaluated analytically on polygonal internal cells (assuming piecewise constant $d_{ij}^{(n)}$ and $v_{i,j}$ on these cells) and differentiated at an arbitrary internal source point p. Later, for axisymmetric problems (Rajiyah and Mukherjee 1987, 1989), convected derivatives were calculated by dividing the domain B into $B - B_\eta(p)$ and $B_\eta(p)$, respectively, where $B_\eta(p)$ is a torus obtained by revolving a circle of small radius η centered at p. The resulting Cauchy singular integral [$O(1/r^2)$ in two dimensions] was evaluated by subtracting and adding the term with the strongest singularity from the kernel $U_{ij,kl}$. Some details of this procedure are given later in this chapter.

Currently, it is felt that the best approach is to carry out a convected derivative followed by regularization, as suggested by Huang and Du (1988). This procedure is illustrated for the integral

$$I_j = \int_B U_{ij,k}(p,q)\, d_{ik}^{(n)}(q)\, dv(q) \tag{2.136}$$

Differentiating the above after first decomposing it into two integrals over $B_\eta(p)$ and $B - B_\eta(p)$ [for three-dimensional problems, $B_\eta(p)$ is a small sphere, of radius η, centered at p], one gets [see Bui (1978) and Mukherjee (1982)]

$$\begin{aligned} I_{j,\bar{l}} = & \int_{B-B_\eta(p)} U_{ij,k\bar{l}}(p,q)\, d_{ik}^{(n)}(q)\, dv(q) \\ & + \int_{\partial B_\eta(p)} U_{ij,k}(p,Q)\, d_{ik}^{(n)}(Q) n_l(Q)\, ds(Q) \end{aligned} \tag{2.137}$$

where $\boldsymbol{n}$ in the last integral is exterior to the region $B - B_\eta$, that is, the inward normal to the small sphere $B_\eta(p)$.

Ordinarily, the last term on the right-hand side of equation (2.137) would lead to a free term. Here, however, regularization is carried out to yield

$$\begin{aligned} I_{j,\bar{l}} = & \int_{B-B_\eta(p)} U_{ij,k\bar{l}}(p,q)[d_{ik}^{(n)}(q) - d_{ik}^{(n)}(p)]\, dv(q) \\ & + d_{ik}^{(n)}(p) \int_{B-B_\eta(p)} (U_{ij,k\bar{l}}(p,q)\, dv(q) \\ & + \int_{\partial B_\eta(p)} U_{ij,k}(p,Q)\, d_{ik}^{(n)}(Q) n_l(Q)\, ds(Q) \end{aligned} \tag{2.138}$$

Applying Gauss' theorem to the second integral on the right-hand side of (2.138), one gets (using $U_{ij,k\bar{l}} = -U_{ij,kl}$)

$$\begin{aligned} \int_{B-B_\eta(p)} U_{ij,k\bar{l}}(p,q)\, dv(q) = & -\int_{\partial B} U_{ij,k}(p,Q) n_l(Q)\, ds(Q) \\ & - \int_{\partial B_\eta(p)} U_{ij,k}(p,Q) n_l(Q)\, ds(Q) \end{aligned} \tag{2.139}$$

In the limit as $\eta \to 0$ [using a Taylor series expansion for $d_{ik}^{(n)}(Q)$ about $d_{ik}^{(n)}(p)$], the last integral on the right-hand side of equation (2.139), multiplied by $d_{ij}^{(n)}(p)$, cancels the last integral on the right-hand side of equation (2.138). In other words, the integrals on the surface of the small sphere, centered at p, cancel out. One is finally left with

$$I_{j,\bar{l}} = \int_B U_{ij,k\bar{l}}(p,q)[d_{ik}^{(n)}(q) - d_{ik}^{(n)}(p)]\,dv(q)$$
$$- d_{ik}^{(n)}(p) \int_{\partial B} U_{ij,k}(p,Q) n_l(Q)\,ds(Q) \tag{2.140}$$

Applying this idea to all the domain integrals in equations (2.135) results in the equation

$$v_{j,\bar{l}}(p) = \int_{\partial B} [U_{ij,\bar{l}}(p,Q)\tau_i^{(s)}(Q) - T_{ij,\bar{l}}(p,Q)v_i(Q)]\,ds(Q)$$
$$- \lambda d_{kk}^{(n)}(p) \int_{\partial B} U_{ij,i}(p,Q) n_l(Q)\,ds(Q)$$
$$- 2G\, d_{ik}^{(n)}(p) \int_{\partial B} U_{ij,k}(p,Q) n_l(Q)\,ds(Q)$$
$$- g_{mi}(p) \int_{\partial B} U_{ij,m}(p,Q) n_l(Q)\,ds(Q)$$
$$+ \int_B \lambda U_{ij,i\bar{l}}(p,q)[d_{kk}^{(n)}(q) - d_{kk}^{(n)}(p)]\,dv(q)$$
$$+ \int_B 2G U_{ij,k\bar{l}}(p,q)[d_{ik}^{(n)}(q) - d_{ik}^{(n)}(p)]\,dv(q)$$
$$+ \int_B U_{ij,m\bar{l}}(p,q)[g_{mi}(q) - g_{mi}(p)]\,dv(q) \tag{2.141}$$

The domain integrals in the above equations are now weakly $O(1/r)$ singular since quantities like $d_{ik}^{(n)}(q) - d_{ik}^{(n)}(p)$ are $O(r)$.

Once $\boldsymbol{h}$ is known at an internal point p, $\boldsymbol{d} = (\boldsymbol{h} + \boldsymbol{h}^{\mathrm{T}})/2$, $\boldsymbol{\omega} = (\boldsymbol{h} - \boldsymbol{h}^{\mathrm{T}})/2$, and the hypoelastic law

$$\hat{\boldsymbol{\sigma}} = \lambda\,\mathrm{tr}\,(\boldsymbol{d} - \boldsymbol{d}^{(n)})\boldsymbol{I} + 2G(\boldsymbol{d} - \boldsymbol{d}^{(n)}) \tag{2.142}$$

can be used to calculate the Jaumann rate of the Cauchy stress at p. Finally, the material rate of $\boldsymbol{\sigma}$ can be obtained from equation (2.90).

2.2.5 Plane Strain

Plane strain equations can be obtained from the three-dimensional ones by setting $v_3 = 0$ and $\partial/\partial x_3 = 0$, and using the corresponding reference solutions with $u_3^{(R)} = 0$. This means that the left-hand side of equation (2.100) is now summed over $i, j = 1, 2$. The right-hand side of (2.100) still retains the term $\sigma_{33}^{(R)} d_{33}^{(e)}$. (Note that, for plane strain, $d_{33} = 0$ but its elastic and nonelastic components are not separately zero.) The resulting equation, corresponding to the three-dimensional equation (2.117) is [with the range of indices 1, 2; see Mukherjee and Chandra (1987)]

$$
\begin{aligned}
v_j(p) = &\int_{\partial B} [U_{ij}(p, Q)\tau_i^{(s)}(Q) - T_{ij}(p, Q)v_i(Q)]\, ds(Q) \\
&+ \int_B \lambda U_{ij,i}(p, q)\eta^{(n)}(q)\, da(q) + \int_B 2G U_{ij,k}(p, q)\, d_{ik}^{(n)}(q)\, da(q) \\
&+ \int_B U_{ij,m}(p, q) g_{mi}(q)\, da(q)
\end{aligned} \tag{2.143}
$$

where $\eta^{(n)} = d_{11}^{(n)} + d_{22}^{(n)} + d_{33}^{(n)}$.

The plane strain kernels must be used in the above equation. These are [see, for example, Mukherjee (1982)]

$$U_{ij} = -\frac{1}{8\pi(1-\nu)G}[(3-4\nu)\ln r\delta_{ij} - r_{,i}r_{,j}] \tag{2.144}$$

$$T_{ij} = -\frac{1}{4\pi(1-\nu)r}\left([(1-2\nu)\delta_{ij} + 2r_{,i}r_{,j}]\frac{\partial r}{\partial n} + (1-2\nu)(r_{,i}n_j - r_{,j}n_i)\right) \tag{2.145}$$

The $\boldsymbol{g}$ tensor is exactly the same as before (see below equation 2.96) with the range of indices 1, 2.

The rest of the equations can be derived in a manner analogous to the three-dimensional case. Closed-form expressions exist for the corner tensor C_{ij} for two-dimensional problems [see, for example, Mukherjee (1982)].

2.2.6 Plane Stress

Now, $\sigma_{13} = \sigma_{23} = \sigma_{33} = 0$. It is useful to assume also $v_{1,3} = v_{2,3} = v_{3,1} = v_{3,2} = 0$, so that $\omega_{13} = \omega_{23} = 0$. Using the definition of Jaumann rates (equation 2.90), $\hat{\sigma}_{13} = \hat{\sigma}_{23} = \hat{\sigma}_{33} = 0$. Thus, the left-hand side of equation (2.100) needs to be summed for $i, j = 1, 2$. The right-hand side, of course, needs to be summed for $i, j = 1, 2$ and, here, $\varepsilon_{33}^{(R)}$ can be eliminated in favor of $\varepsilon_{11}^{(R)} + \varepsilon_{22}^{(R)}$ using Hooke's law with $\sigma_{33}^{(R)} = 0$. The final equation, again with the range of indices 1, 2, is

$$v_j(p) = \int_{\partial B} [U_{ij}(p, Q)\tau_i^{(s)}(Q) - T_{ij}(p, Q)v_i(Q)]\, ds(Q)$$
$$+ \int_B \bar{\lambda} U_{ij,i}(p, q)\, d_{kk}^{(n)}(q)\, da(q) + \int_B 2GU_{ij,k}(p, q)\, d_{ik}^{(n)}(q)\, da(q)$$
$$+ \int_B U_{ij,m}(p, q)[g_{mi}(q) - \sigma_{mi}(q)\, d_{33}(q)]\, da(q) \qquad (2.146)$$

The plane stress kernels U_{ij} and T_{ij} must be used in the above equations. These are obtained from equations (2.144) and (2.145) by replacing ν by $\bar{\nu} = \nu/(1 + \nu)$. Also, the plane stress Lamé constant is

$$\bar{\lambda} = \frac{2G\bar{\nu}}{1 - 2\bar{\nu}} = \frac{2G\nu}{1 - \nu} \qquad (2.147)$$

and the form of $\mathbf{g}$ is the same as that for plane strain.

It is convenient to eliminate d_{33} in terms of d_{11} and d_{22}. Using equation (2.89) with $i = j = 3$ and $\overset{*}{\sigma}_{33} = 0$, one gets

$$d_{33} = -\frac{\nu}{1 - \nu}(d_{11} + d_{22}) + \frac{\nu}{1 - \nu}(d_{11}^{(n)} + d_{22}^{(n)}) + d_{33}^{(n)} \qquad (2.148)$$

The presence of $d_{33}^{(n)}$ in the above equation cannot be avoided. If, however, the deformation is incompressible, $d_{11} + d_{22} + d_{33} = 0$ and equation (2.146) can be expressed purely in terms of tensor components in the 1 and 2 directions.

2.2.7 *Axisymmetric Problems*

A common assumption for metal deformation is that the nonelastic deformation is incompressible, that is, $d_{11}^{(n)} + d_{22}^{(n)} + d_{33}^{(n)} = 0$. For metal forming problems, the total strains are typically large and comparable to the nonelastic strains. Hence, sometimes one also assumes that the total deformation is incompressible, that is, $d_{11} + d_{22} + d_{33} = 0$. Both these assumptions are made in this section. The contents of this section have been published by Rajiyah and Mukherjee (1987, 1989), and the reader is referred to those papers for further details. A discussion of small-strain axisymmetric elasto-viscoplastic problems appears in Sarihan and Mukherjee (1982) and Mukherjee (1982).

The starting point here is the three-dimensional equation (2.117). In view of the above assumptions, the second integral in equation (2.117) vanishes and g_{mi} is replaced by g'_{mi} where (see equation 2.96)

$$\boldsymbol{g}' = \boldsymbol{\sigma} \cdot \boldsymbol{\omega} + \boldsymbol{d} \cdot \boldsymbol{\sigma} \tag{2.149}$$

Also, now (see equation 2.129)

$$\boldsymbol{\tau}^{(s)} = \boldsymbol{\tau}^{(c)} - \boldsymbol{n} \cdot \boldsymbol{g}' \tag{2.150}$$

An axisymmetric body with axisymmetric loading is considered in this section. Using cylindrical polar coordinates R, θ, and Z, the nonzero components of displacements, stresses, and strains are u_R, u_Z; ε_{RR}, $\varepsilon_{\theta\theta}$, ε_{ZZ}, $\varepsilon_{RZ}(=\varepsilon_{ZR})$; σ_{RR}, $\sigma_{\theta\theta}$, σ_{ZZ}, and $\sigma_{RZ}(=\sigma_{ZR})$. All dependent variables are functions of R, Z, and t. Some torsion problems can be independent of θ, but these are not considered in this formulation.

The notation used here is shown in figure 2.5. The source point is denoted

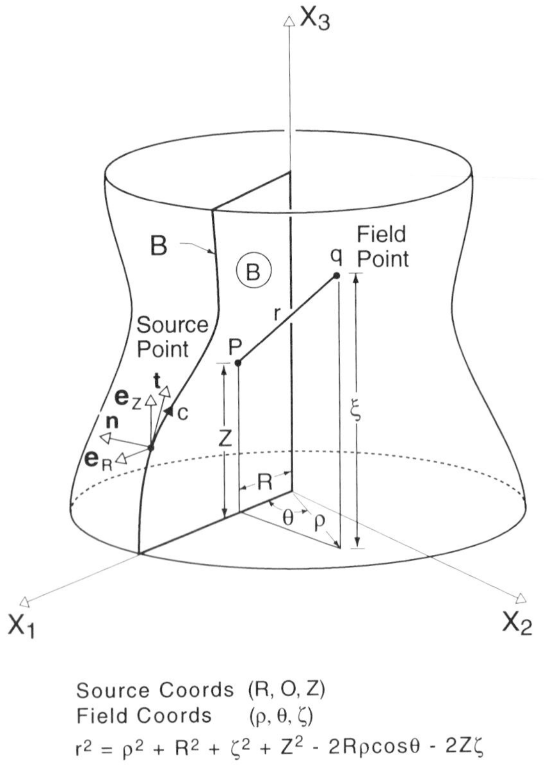

Figure 2.5. Geometry of the axisymmetric problem.

by (R, O, Z) and the field point by (ρ, θ, ζ). Since the problem is axisymmetric, it is sufficient to choose the source point in the x_1–x_3 plane.

The axisymmetric BEM equation for the velocity is derived from equation (2.117) by integrating the kernels U_{ij}, T_{ij}, etc., for the field point moving around a ring, keeping the source point fixed. Integrating θ from 0 to 2π in equation (2.117) results in ($j = 1$ and 3, no sum over ρ or ζ)

$$\begin{aligned}
v_j = &\int_{\partial B} [U_{\rho j}\tau_\rho^{(s)} + U_{\zeta j}\tau_\zeta^{(s)} - T_{\rho j}v_\rho - T_{\zeta j}v_\zeta]\rho\, dc \\
&+ 2G\int_B \Bigg[U_{\rho j,\rho}d_{\rho\rho}^{(n)} + U_{\rho j,\zeta}d_{\rho\zeta}^{(n)} + U_{\zeta j,\rho}d_{\zeta\rho}^{(n)} \\
&+ U_{\zeta j,\zeta}d_{\zeta\zeta}^{(n)} + \frac{U_{\rho j}d_{\theta\theta}^{(n)}}{\rho}\Bigg]\rho\, d\rho\, d\zeta + \int_B \Big[U_{\rho j,\rho}[\sigma_{\rho\rho}d_{\rho\rho} + \sigma_{\rho\zeta}(d_{\rho\zeta} - \omega_{\rho\zeta})] \\
&+ U_{\rho j,\zeta}[\sigma_{\rho\rho}d_{\rho\zeta} + \sigma_{\rho\zeta}d_{\zeta\zeta} - \sigma_{\zeta\zeta}\omega_{\rho\zeta}] \\
&+ U_{\zeta j,\rho}[+\sigma_{\rho\rho}\omega_{\rho\zeta} + \sigma_{\rho\zeta}d_{\rho\rho} + \sigma_{\zeta\zeta}d_{\rho\zeta}] \\
&+ U_{\zeta j,\zeta}[\sigma_{\rho\zeta}(d_{\rho\zeta} + \omega_{\rho\zeta}) + \sigma_{\zeta\zeta}d_{\zeta\zeta}] + \frac{U_{\rho j}\sigma_{\theta\theta}d_{\theta\theta}}{\rho}\Big]\rho\, d\rho\, d\zeta
\end{aligned} \tag{2.151}$$

where, because of axisymmetry, $\dot{u}_R = \dot{u}_1$, $\dot{u}_z = \dot{u}_3$, and $dc = \sqrt{d\rho^2 + d\zeta^2}$ is an element on the boundary of the $\rho - \zeta$ plane. The domain B and boundary ∂B in the above equation now refer to a generator plane of the axisymmetric solid and its boundary (excluding the portion on the x_3-axis), respectively (figure 2.5) so that the three-dimensional problem is effectively reduced to a two-dimensional one. Also, equation (2.151) is valid for the velocity of an internal source point. The corresponding boundary integral equation is obtained, as usual, by taking the limit $p \to P$, which introduces the corner tensor C_{ij} (see equation 2.118).

The kernels for the case where p does not lie on the axis of symmetry are (Mukherjee 1982)

$$\begin{aligned}
U_{\rho 1}(p, q) = &\frac{1}{16\pi(1-\nu)G}\frac{k}{\sqrt{R\rho}}\Bigg\{\left[2(3-4\nu)\gamma + \frac{(Z-\zeta)^2}{R\rho}\right]K(k) \\
&- \left[2(3-4\nu)(1+\gamma) + \left(\frac{\gamma}{\gamma-1}\right)\frac{(Z-\zeta)^2}{R\rho}\right]E(k)\Bigg\}
\end{aligned} \tag{2.152}$$

$$U_{\rho 3}(p, q) = \frac{(\zeta - Z)k}{16\pi(1-\nu)GR\rho\sqrt{R\rho}}\left(RK(k) + \frac{\rho - R\gamma}{\gamma - 1}E(k)\right) \tag{2.153}$$

$$U_{\zeta 1}(p, q) = \frac{(Z-\zeta)k}{16\pi G(1-\nu)R\rho\sqrt{R\rho}}\left(\rho K(k) - \frac{\rho\gamma - R}{\gamma - 1}E(k)\right) \tag{2.154}$$

$$U_{\zeta 3}(p, q) = \frac{1}{8\pi(1-\nu)G}\frac{k}{\sqrt{R\rho}}\left((3-4\nu)K(k) + \frac{(Z-\zeta)^2}{2R\rho(\gamma-1)}E(k)\right) \tag{2.155}$$

where $K(k)$ and $E(k)$ are complete elliptic integrals of the first and second kind, respectively, and $k = \sqrt{2/(1+\gamma)}$. Further, $\gamma = 1 + \{(Z-\zeta)^2 + (R-\rho)^2\}/2R\rho$.

The traction kernels are given in terms of derivatives of $U_{\rho j}$, etc., by the equations ($j = 1$ and 3)

$$\frac{1}{2G}T_{\rho j}(p, Q) = \left[\frac{1-\nu}{1-2\nu}\frac{\partial U_{\rho j}}{\partial \rho} + \frac{\nu}{1-2\nu}\left(\frac{1}{\rho}U_{\rho j} + \frac{\partial U_{\zeta j}}{\partial \zeta}\right)\right]n_\rho + \frac{1}{2}\left(\frac{\partial U_{\rho j}}{\partial \zeta} + \frac{\partial U_{\zeta j}}{\partial \rho}\right)n\zeta \tag{2.156}$$

$$\frac{1}{2G}T_{\zeta j}(p, Q) = \left[\frac{1-\nu}{1-2\nu}\frac{\partial U_{\zeta j}}{\partial \zeta} + \frac{\nu}{1-2\nu}\left(\frac{1}{\rho}U_{\rho j} + \frac{\partial U_{\rho j}}{\partial \rho}\right)\right]n_\zeta + \frac{1}{2}\left(\frac{\partial U_{\rho j}}{\partial \zeta} + \frac{\partial U_{\zeta j}}{\partial \rho}\right)n_\rho \tag{2.157}$$

where n_ρ and n_ζ are the components of the outward unit normal at the field point on the boundary in the ρ and ζ directions.

The derivatives of the displacement kernels, which occur in the above equations as well as in the area integrals of equation (2.151), are given in Sarihan (1982).

If p lies on the axis of symmetry, as must be considered for solid bodies without axial holes in them, one gets

$$U_{\rho 1} = 0, \qquad U_{\rho 3} = \frac{(\zeta - Z)\rho}{8(1-\nu)Gr^3}$$
$$U_{\zeta 1} = 0, \qquad U_{\zeta 3} = \frac{1}{8(1-\nu)G}\left(\frac{3-4\nu}{r} + \frac{(\zeta-Z)^2}{r^3}\right) \tag{2.158}$$

$$T_{\rho 1} = 0, \qquad T_{\rho 3} = -\frac{1}{4(1-\nu)r^2}\left[\frac{3\rho(\zeta-Z)}{r^2}\frac{\partial r}{\partial n} + (1-2\nu)\left(\frac{\rho n_\zeta}{r} - \frac{\zeta - Z}{r}n_\rho\right)\right]$$
$$T_{\zeta 1} = 0, \qquad T_{\zeta 3} = -\frac{\partial r/\partial n}{4(1-\nu)r^2}\left((1-2\nu) + \frac{3(\zeta-Z)^2}{r^2}\right) \tag{2.159}$$

where

$$\frac{\partial r}{\partial n} = \frac{\zeta - Z}{r} n_\zeta + \frac{\rho n_\rho}{r}$$

It is interesting to note that even though, for example, $U_{\rho 1} = 0$ in equation (2.158), kernels like $U_{\rho 1,\rho}$ for p on the axis of symmetry are nonzero. These are determined by considering the appropriate components of the three-dimensional $U_{ij,k}$ kernel, setting $R = 0$ and then integrating θ from 0 to 2π.

The scaled Lagrangian traction rates are related to the Cauchy traction rates as follows (see equation 2.150):

$$\begin{aligned}\tau_\rho^{(s)} &= \tau_\rho^{(c)} - n_\rho[\sigma_{\rho\rho} d_{\rho\rho} + \sigma_{\rho\zeta}(d_{\rho\zeta} - \omega_{\rho\zeta})] \\ &\quad - n_\zeta[\sigma_{\rho\rho} d_{\rho\zeta} + \sigma_{\rho\zeta} d_{\zeta\zeta} - \sigma_{\zeta\zeta}\omega_{\rho\zeta}] \\ \tau_\zeta^{(s)} &= \tau_\zeta^{(c)} - n_\rho[+\sigma_{\rho\rho}\omega_{\rho\zeta} + \sigma_{\rho\zeta} d_{\rho\rho} + \sigma_{\zeta\zeta} d_{\rho\zeta}] \\ &\quad - n_\zeta[\sigma_{\rho\zeta}(d_{\rho\zeta} + \omega_{\rho\zeta}) + \sigma_{\zeta\zeta} d_{\zeta\zeta}]\end{aligned} \tag{2.160}$$

Velocity gradients at an internal point are obtained, as usual, by differentiating equation (2.151) at an internal source point p. Referring to equation (2.151), the term on the left-hand side becomes $v_{j,\bar{I}}$ and the boundary integral on the right-hand side becomes

$$\int_{\partial B} [U_{\rho j,\bar{I}}\, \tau_\rho^{(s)} + U_{\zeta j,\bar{I}}\, \tau_\zeta^{(s)} - T_{\rho j,\bar{I}}\, v_\rho - T_{\zeta j,\bar{I}}\, v_\zeta] \rho\, dc$$

The derivatives can immediately be moved under the integral sign in the above expression since p is an internal point, Q is a boundary point, and the above integral is regular. Such, however, is not the case for the domain integrals in equation (2.151), which are, in general, $1/r$ singular. As has been discussed before (see section 2.2.4), the best method for evaluating derivatives of the domain integrals in equation (2.151) appears to be a regularization procedure. In the work on axisymmetric problems by Rajiyah and Mukherjee (1987, 1989), however, a different method was used. The integral was decomposed into one on the region $B - B_\eta(p)$ and the others on $B_\eta(p)$, where $B_\eta(p)$ is a small circle in the generator plane of the solid (figure 2.5), centered at p and of radius η. Convected differentiation of the integral over $B - B_\eta(p)$ was carried out using the method outlined by Bui (1978). Free terms result from this process. The appropriate free terms from the various derivatives of the displacement kernels, for the cases where p

Table 2.4. Free terms in the velocity gradient equations for the case $p \notin$ (the axis of symmetry)

Kernel	Free terms from radial differentiation ($\bar{l} = 1$)	Free terms from axial differentiation ($\bar{l} = 3$)
$U_{\rho 1,\rho}$	$\frac{5-8\nu}{16(1-\nu)G}$	0
$U_{\rho 1,\zeta}$	0	$\frac{7-8\nu}{16(1-\nu)G}$
$U_{\zeta 1,\rho}$	0	$-\frac{1}{16(1-\nu)G}$
$U_{\zeta 1,\zeta}$	$-\frac{1}{16(1-\nu)G}$	0
$U_{\rho 3,\rho}$	0	$-\frac{1}{16(1-\nu)G}$
$U_{\rho 3,\zeta}$	$-\frac{1}{16(1-\nu)G}$	0
$U_{\zeta 3,\rho}$	$\frac{7-8\nu}{16(1-\nu)G}$	0
$U_{\zeta 3,\zeta}$	0	$\frac{5-8\nu}{16(1-\nu)G}$

does not lie on the axis of symmetry and p lies on the axis of symmetry, are listed in tables 2.4 and 2.5, respectively. It is very interesting to note from table 2.5 that free terms result from some first derivatives of $U_{\rho 1}$ and $U_{\zeta 1}$ if p lies on the axis of symmetry, even though $U_{\rho 1} = U_{\zeta 1} = 0$ in this case.

Table 2.5. Nonzero free terms in the velocity gradient equations for the case $p \in$ (the axis of symmetry)

Kernel	Free terms from radial differentiation ($\bar{l} = 1$)	Kernel	Free terms from axial differentiation ($\bar{l} = 3$)
$U_{\rho 1,\rho}$	$\frac{17-20\nu}{60(1-\nu)G}$	$U_{\rho 3,\rho}$	$\frac{1}{60(1-\nu)G}$
$U_{\zeta 1,\zeta}$	$-\frac{1}{30(1-\nu)G}$	$U_{\zeta 3,\zeta}$	$\frac{7-10\nu}{30(1-\nu)G}$
$U_{\rho 1}$	$-\frac{1}{12(1-\nu)G}$	$U_{\rho 3}$	$-\frac{1}{12(1-\nu)G}$

The actual free terms in an equation for $v_{j,\bar{l}}$ are obtained by multiplying the appropriate terms in table 2.4 or 2.5 by the corresponding terms from equation (2.151). Thus, for example, the explicit form of the equation for $v_{1,\bar{1}}$ is (no sum over ρ, or ζ; $p \notin x_3$-axis)

$$
\begin{aligned}
v_{1,\bar{1}} = & \int_{\partial B} [U_{\rho 1,\bar{1}} \tau_\rho^{(s)} + U_{\zeta 1,\bar{1}} \tau_\zeta^{(s)} - T_{\rho 1,\bar{1}} v_\rho - T_{\zeta 1,\bar{1}} v(\zeta)] \rho \, dc \\
& + 2G \int_{B - B_\eta(p)} \Bigg[U_{\rho 1,\rho\bar{1}} d_{\rho\rho}^{(n)} + U_{\rho 1,\zeta\bar{1}} d_{\rho\zeta}^{(n)} + U_{\zeta 1,\rho\bar{1}} d_{\zeta\rho}^{(n)} \\
& + U_{\zeta 1,\zeta\bar{1}} d_{\zeta\zeta}^{(n)} + U_{\rho 1,\bar{1}} \frac{d_{\theta\theta}^{(n)}}{\rho} \Bigg] \rho \, d\rho \, d\zeta + \frac{5 - 8\nu}{8(1-\nu)} d_{\rho\rho}^{(n)}(p) \\
& - \frac{1}{8(1-\nu)} d_{\zeta\zeta}^{(n)}(p) + \int_{B - B_\eta(p)} \Bigg[U_{\rho 1,\rho\bar{1}} [\sigma_{\rho\rho} d_{\rho\rho} + \sigma_{\rho\zeta}(d_{\rho\zeta} - \omega_{\rho\zeta})] \\
& + U_{\rho 1,\zeta\bar{1}} [\sigma_{\rho\rho} d_{\rho\zeta} + \sigma_{\rho\zeta} d_{\zeta\zeta} - \sigma_{\zeta\zeta} \omega_{\rho\zeta}] \\
& + U_{\zeta 1,\rho\bar{1}} [+\sigma_{\rho\rho} \omega_{\rho\zeta} + \sigma_{\rho\zeta} d_{\rho\rho} + \sigma_{\zeta\zeta} d_{\rho\zeta}] \\
& + U_{\zeta 1,\zeta\bar{1}} [\sigma_{\rho\zeta}(d_{\rho\zeta} + \omega_{\rho\zeta}) + \sigma_{\zeta\zeta} d_{\zeta\zeta}] + \frac{U_{\rho 1,\bar{1}} \sigma_{\theta\theta} d_{\theta\theta}}{\rho} \Bigg] \rho \, d\rho \, d\zeta \\
& + \frac{5 - 8\nu}{16G(1-\nu)} \{\sigma_{\rho\rho}(p) \, d_{\rho\rho}(p) + \sigma_{\rho\zeta}(p)[d_{\rho\zeta}(p) - \omega_{\rho\zeta}(p)]\} \\
& - \frac{1}{16G(1-\nu)} \{\sigma_{\rho\zeta}(p)[d_{\rho\zeta}(p) + \omega_{\rho\zeta}(p)] + \sigma_{\zeta\zeta}(p) \, d_{\zeta\zeta}(p)\}
\end{aligned}
\tag{2.161}
$$

The Cauchy principal values of the integrals on $B - B_\eta(p)$, as $\eta \to 0$, must be determined accurately in a successful numerical implementation of the problem. Some details of this implementation are given later in this chapter. The explicit forms of the kernels $U_{ij,kl}$ are given in Rajiyah (1987).

As mentioned before, it is a simple matter to obtain the Jaumann and then the material rates of the Cauchy stress at an internal point once the velocity gradients have been determined at that point. For this purpose, it is useful to record the relationship between the material and Jaumann rates of components of the Cauchy stress in cylindrical polar coordinates. These equations take the form

$$\overset{\wedge}{\sigma}_{RR} = \dot{\sigma}_{RR} - 2\sigma_{RZ}\omega_{RZ}$$

$$\overset{\wedge}{\sigma}_{\theta\theta} = \dot{\sigma}_{\theta\theta}$$

$$\overset{\wedge}{\sigma}_{RZ} = \dot{\sigma}_{RZ} + \sigma_{RR}\omega_{RZ} - \sigma_{ZZ}\omega_{RZ}$$

$$\overset{\wedge}{\sigma}_{ZZ} = \dot{\sigma}_{ZZ} + 2\sigma_{RZ}\omega_{RZ} \tag{2.162}$$

The boundary stress rates, at any time, are best obtained from a boundary algorithm (see Mukherjee, 1982, p.19) rather than by trying to take the limit of an equation like (2.161) as $p \to P$. This approach, which requires tangential differentiation of the displacement (or velocity in a rate formulation) components at a boundary point, was first suggested for linear elastic problems by Rizzo and Shippy (1968). This idea has been generalized to nonelastic problems with small strains and rotations by Sarihan and Mukherjee (1982).

The method used here for large-strain large-rotation problems is very similar to that outlined by Mukherjee (1982). Once the iterative calculations are completed at the end of a time step, the boundary values of v_i, $\tau_i^{(s)}$, and $v_{i,j}$ are determined at each boundary node. It is then a simple matter to determine $\tau_i^{(c)}$ through the use of equations (2.160).

At this stage, it is convenient to transform $\tau_R^{(c)}$ and $\tau_Z^{(c)}$ into local coordinates normal and tangential to the boundary at a boundary point P. Now, from equation (2.132)

$$\overset{\wedge}{\sigma}_{nn} = \tau_n^{(c)}, \qquad \overset{\wedge}{\sigma}_{nc} = \tau_c^{(c)} \tag{2.163}$$

where $\overset{\wedge}{\sigma}_{nn}$ and $\overset{\wedge}{\sigma}_{nc}$ are the normal and shearing components of the Jaumann rates for the Cauchy stress at P. The rest of the algorithm follows that outlined by Mukherjee (1982) with the constitutive model (equation 2.89) written as

$$\frac{\partial v_c}{\partial c} = d_{cc} = \frac{1}{E}[\overset{\wedge}{\sigma}_{cc} - \nu(\overset{\wedge}{\sigma}_{nn} + \overset{\wedge}{\sigma}_{\theta\theta})] + d_{cc}^{(n)} \tag{2.164}$$

$$\frac{v_R}{R} = d_{\theta\theta} = \frac{1}{E}[\overset{\wedge}{\sigma}_{\theta\theta} - \nu(\overset{\wedge}{\sigma}_{nn} + \overset{\wedge}{\sigma}_{cc})] + d_{\theta\theta}^{(n)} \tag{2.165}$$

These equations are solved for the unknown stress rates $\overset{\wedge}{\sigma}_{cc}$ and $\overset{\wedge}{\sigma}_{\theta\theta}$, and then these, together with $\overset{\wedge}{\sigma}_{nn}$ and $\overset{\wedge}{\sigma}_{nc}$, are transformed back to global coordinates. Finally, the material rates $\dot{\sigma}_{RR}$, etc., in global coordinates, are obtained from equation (2.162).

2.2.8 Derivative Boundary Integral Equations (DBEM) for Plane Strain Problems

A derivative BEM (DBEM) formulation (Ghosh et al. 1986), for planar problems, has been used to carry out sensitivity analyses for large-deformation problems. The primary reason for this is that the DBEM delivers very accurate stresses and stress rates on the boundary of a body. These quantities are crucial for shape optimization studies. Details of this work are given in chapter 4.

The derivative boundary element formulation uses derivatives of displacements (or velocities), together with tractions (or traction rates), as primary boundary variables. For two-dimensional problems, partial integration is carried out on the term $T_{ij}(p, Q)v_i(Q)$ in equation (2.117) so that one gets $\delta_i = \partial v_i/\partial s$ (where s is the distance measured along ∂B) as the boundary variable, together with a kernel W_{ij}. When the distance between a source and a field point (denoted by r) tends to zero, this kernel has a singularity of the order of $\ln r$. This is weaker compared to the $1/r$ singularity of T_{ij}. Details are given by Ghosh et al. (1986). Unless otherwise indicated, the range of indices in this section is 1, 2.

The kernel W_{ij} has the form

$$W_{ij} = \frac{1}{4\pi(1-\nu)}[2(1-\nu)\psi\delta_{ij} + \gamma_{ik}r_{,j}r_{,k} + (1-2\nu)\gamma_{ji}\ln r] \qquad (2.166)$$

Here, ψ is the angle between the vector $\boldsymbol{r}(P, Q)$ and a reference direction. Also, $\gamma_{11} = \gamma_{22} = 0$ and $\gamma_{12} = -\gamma_{21} = 1$.

The DBEM equation below is quite similar to equation (2.143) for $p \to P$. It is written below for the case $d_{kk}^{(n)} = d_{11}^{(n)} + d_{22}^{(n)} + d_{33}^{(n)} = 0$. Note that plane strain kernels must be used in this equation. Also, the equation below is only valid in a simply connected domain. For multiply connected domains, please see Ghosh et al. (1986).

$$\begin{aligned} 0 = &\int_{\partial B} [U_{ij}(\boldsymbol{b}, P, Q)\tau_i^{(s)}(\boldsymbol{b}, Q) - W_{ij}(\boldsymbol{b}, P, Q)\delta_i(\boldsymbol{b}, Q)]\, ds(\boldsymbol{b}, Q) \\ &+ \int_B [2GU_{ij,k}(\boldsymbol{b}, P, q)\, d_{ik}^{(n)}(\boldsymbol{b}, q)]\, da(\boldsymbol{b}, q) \\ &+ \int_B U_{ij,m}(\boldsymbol{b}, P, q) g_{mi}(\boldsymbol{b}, q)\, da(\boldsymbol{b}, q) \end{aligned} \qquad (2.167)$$

The explicit inclusion of the shape design vector $\boldsymbol{b}$ in the above equation is for shape sensitivity analysis, which is discussed in chapter 4. The presence of $\boldsymbol{b}$ in the above can be ignored for the time being.

It should be noted that for some problems with prescribed velocity on a portion ∂B_1 of ∂B, prescription of $\boldsymbol{\delta}$ on ∂B_1 may lead to loss of information on the velocity itself (Zhang et al. 1992b). This may lead to loss of uniqueness of the solution obtained from this formulation. In such cases, this difficulty can be overcome by appending constraint equations of the type

$$\int_A^B \boldsymbol{\delta}\, ds = \boldsymbol{v}(B) - \boldsymbol{v}(A) \tag{2.168}$$

where A and B are suitably chosen points on the boundary ∂B where the velocities are known.

An algebraic boundary algorithm (section 2.2.3) is best suited for the determination of stress rates and velocity gradients at a regular point on the boundary. The equations (2.131 through 2.134), however, can be explicitly solved for planar problems [see Sladek and Sladek (1986) and Cruse and Vanburen (1971) for the linear elastic case]. The result is

$$\hat{\sigma}_{ij} = A_{ijk}\tau_k^{(c)} + B_{ijk}\delta_k + C_{ijkl}d_{kl}^{(n)} + D_{ij}d_{kk}^{(n)} \tag{2.169}$$

$$h_{ij} = E_{ijk}\tau_k^{(c)} + F_{ijk}\delta_k + G_{ijkl}d_{kl}^{(n)} \tag{2.170}$$

where

$$A_{ijk} = (n_i n_j + c_1 t_i t_j)n_k + (n_i t_j + n_j t_i)t_k$$

$$B_{ijk} = c_2 t_i t_j t_k$$

$$C_{ijkl} = -\, c_2 t_i t_j t_k t_l$$

$$D_{ij} = \nu c_2 t_i t_j$$

$$E_{ijk} = \frac{c_3}{2G} n_i n_j n_k + \frac{1}{G} t_i n_j t_k$$

$$F_{ijk} = \gamma_{ij} n_k + t_i t_j t_k - c_1 n_i n_j t_k$$

$$G_{ijkl} = c_3 n_i n_j n_k n_l + 2 t_i n_j n_k t_l$$

with $c_1 = \nu/(1-\nu)$, $c_2 = 2G/(1-\nu)$, $c_3 = (1-2\nu)/(1-\nu)$, and t_i are the components of the unit (anticlockwise) tangent vector to ∂B at a point on it.

The spin $\boldsymbol{\omega}$ can be obtained from h_{ij} by using equation (2.20) and the material rate of $\boldsymbol{\sigma}$ can be obtained from its Jaumann rate from equation (2.90).

The velocity gradient $\boldsymbol{h}$, at an internal point, is quite analogous to equation (2.141). This is

$$\begin{aligned}
h_{j\bar{l}}(\boldsymbol{b},p) &= v_{j,\bar{l}}(p) \\
&= \int_{\partial B} [U_{ij,\bar{l}}(\boldsymbol{b},p,Q)\tau_i^{(s)}(\boldsymbol{b},Q) - W_{ij,\bar{l}}(\boldsymbol{b},Q)\delta_i(\boldsymbol{b},Q)]\, ds(\boldsymbol{b},Q) \\
&\quad - 2Gd_{ik}^{(n)}(\boldsymbol{b},p)\int_{\partial B} U_{ij,k}(\boldsymbol{b},p,Q)n_l(\boldsymbol{b},Q)\, ds(\boldsymbol{b},Q) \\
&\quad - g_{mi}(\boldsymbol{b},p)\int_{\partial B} U_{ij,m}(\boldsymbol{b},p,Q)n_l(\boldsymbol{b},Q)\, ds(\boldsymbol{b},Q) \\
&\quad + \int_B 2GU_{ij,k\bar{l}}(\boldsymbol{b},p,q)[d_{ik}^{(n)}(\boldsymbol{b},q) - d_{ik}^{(n)}(\boldsymbol{b},p)]\, da(\boldsymbol{b},q) \\
&\quad + \int_B U_{ij,m\bar{l}}(\boldsymbol{b},p,q)[g_{mi}(\boldsymbol{b},q) - g_{mi}(\boldsymbol{b},p)]\, da(\boldsymbol{b},q) \qquad (2.171)
\end{aligned}$$

where $h_{j\bar{l}} = \partial v_j/\partial x_{\bar{l}}(p)$.

2.2.9 *Derivative Boundary Integral Equations (DBEM) for Plane Stress Problems*

This time, the DBEM equation at a boundary point P is of the form (see equation 2.146)

$$\begin{aligned}
0 &= \int_{\partial B} [U_{ij}(\boldsymbol{b},P,Q)\tau_i^{(s)}(\boldsymbol{b},Q) - W_{ij}(\boldsymbol{b},P,Q)\delta_i(\boldsymbol{b},Q)]\, ds(Q) \\
&\quad + \int_B \bar{\lambda}U_{ij,i}(\boldsymbol{b},P,Q)\, d_{kk}^{(n)}(\boldsymbol{b},q)\, da(\boldsymbol{b},q) \\
&\quad + \int_B 2GU_{ij,k}(\boldsymbol{b},P,Q)\, d_{ik}^{(n)}(\boldsymbol{b},q)\, da(\boldsymbol{b},q) \\
&\quad + \int_B U_{ij,m}(\boldsymbol{b},p,q)[g_{mi}(\boldsymbol{b},p,q) - \sigma_{mi}(\boldsymbol{b},q)\, d_{33}(\boldsymbol{b},q)]\, da(\boldsymbol{b},q) \qquad (2.172)
\end{aligned}$$

where, of course, the plane stress kernels (see section 2.2.6) must be used. The plane stress version of W_{ij} is obtained from equation (2.166) by replacing ν with $\bar{\nu} = \nu/(1 + \nu)$. This time, with $d_{11}^{(n)} + d_{22}^{(n)} + d_{33}^{(n)} = 0$, one gets (see equation 2.148)

$$d_{33} = -\frac{\nu}{1-\nu}d_{kk} + \frac{2\nu - 1}{1-\nu}d_{kk}^{(n)} \qquad (2.173)$$

The stress rate and velocity gradient equations at a boundary point, analogous to equations (2.169) and (2.170), respectively, now take the form (Zhang 1991; Zhang et al. 1992a)

$$\hat{\sigma}_{ij} = A_{ijk}\tau_k^{(c)} + B_{ijk}\delta_k + C_{ijkl}d_{kl}^{(n)} \tag{2.174}$$

$$h_{ij} = E_{ijk}\tau_k^{(c)} + F_{ijk}\delta_k + H_{ijkl}d_{kl}^{(n)} \tag{2.175}$$

where A_{ijk}, etc., have the same forms as given in equations (2.169) and (2.170) with ν replaced by $\bar{\nu} = \nu/(1 + \nu)$ and

$$H_{ijkl} = n_i n_j(n_k n_l + \nu t_k t_l) + 2t_i n_j n_k t_l \tag{2.176}$$

The equation for $h_{\bar{j}\bar{l}}$ at an internal point p can be derived from equation (2.172) in a manner analogous to the derivation of equation (2.171) from (2.167).

2.2.10 Sharp Corners for Planar Problems

The modeling of sharp corners has been discussed in detail by Zhang and Mukherjee (1991) for the elastic case and by Zhang et al. (1992a) for elastoplastic problems. Two kinds of situations can arise: (a) special corners across which the Cauchy stress tensor $\boldsymbol{\sigma}$ remains continuous throughout a deformation process, and (b) general corners across which $\boldsymbol{\sigma}$ can suffer a jump discontinuity. It is also possible for $\boldsymbol{\sigma}$ to become unbounded at a corner, but such situations are not considered in this book. The plane strain problem is considered below.

In certain special situations, the Cauchy stress tensor (and therefore its material rate) remains continuous at a corner throughout the deformation history. An example is a right-angled corner in which the corner angle remains a right angle throughout a deformation process. Another example is a corner that arises from using symmetry in a problem where the point was originally regular. In such cases, one can write corner equations in a manner analogous to the cases that have been considered before by Zhang and Mukherjee (1991) and Zhang et al. (1992a):

$$\dot{\sigma}_{ij}^{-} = \dot{\sigma}_{ij}^{+} \tag{2.177}$$

where $\dot{\sigma}_{ij}$, on either side of a corner, can be obtained from equations (2.90), (2.169), and (2.170) (for the plane strain case), together with equation (2.20), as functions of the components of $\boldsymbol{\tau}^{(c)}$, $\boldsymbol{\delta}$, $\boldsymbol{d}^{(n)}$, $\boldsymbol{n}$, and $\boldsymbol{t}$.

A possible option for the general case is to invoke continuity of the velocity $\mathbf{v}$ at a corner. This leads to integral constraints as shown below. Suppose that AC and CB are two smooth segments that meet at corner C. Then, continuity of $\mathbf{v}$ at C demands that

$$v_i(P) + \int_P^C \delta_i^- \, ds = v_i(Q) - \int_C^Q \delta_i^+ \, ds \tag{2.178}$$

where P within AC and Q within CB are points at which the velocities are known. Of course, if velocities are not known at any point within smooth segments contiguous to a corner, then velocity information from points farther away must be used and equation (2.178) must be suitably modified. Knowledge of v at any one point on ∂B is sufficient for this idea to work.

Equation (2.178) gives two extra equations at each corner and this extra information is sufficient, in many cases, for solving the problem. A word of caution here. Sometimes, as discussed in detail by Zhang and Mukherjee (1991), δ_n and associated rotations can become singular at a corner. In such situations, care must be exercised in using equation (2.178) in the general case. Suitable shape functions for δ_n must be employed to reflect this singular behavior.

2.3 Finite Element Formulations

The focus of this book, of course, is boundary element analysis. Finite element analysis of large deformation problems in solid mechanics is very well developed. There are review articles such as that of Cheng and Kikuchi (1985). Numerous papers have been published on the subject. A sampling (which is, by no means, exhaustive) of recent papers includes: Bohatier and Chenot (1989), Healy and Dodds (1992), Maniatty et al. (1991), Nemat-Nasser and Li (1992), Nishiguchi et al. (1990), Perić et al. (1992), Saran and Runesson (1992), Saran and Wagoner (1991), Simo and Miehe (1992), Weber and Anand (1990), and Zabaras and Arif (1992).

The purpose of this section is not to present a detailed review of FEM applications. A hypoelastic FEM model, however, has been used to generate numerical results for some problems, for comparison with BEM results. Some of these numerical results are discussed in section 2.4. A brief review of a hypoelastic FEM model is presented below. It should be mentioned here that the current trend in FEM research is the use of hyperelastic models. Such an approach is needed for elastoplastic problems in which the elastic strains are large. For problems with small elastic strains, results from hypo- and hyperelastic models are essentially the same.

2.3.1 A Three-Dimensional FEM in an Updated Lagrangian Formulation

An updated Lagrangian formulation is used here in which the configuration of the body at time t is used as the reference configuration for the time step t to $t + \Delta t$. In this formulation, for example,

$$\boldsymbol{F}_t^{t+\Delta t} = \frac{\partial \boldsymbol{x}(t + \Delta t)}{\partial \boldsymbol{x}(t)} \cong \boldsymbol{I}$$

Following Chandra and Mukherjee (1984b), a Galerkin-type weighted residual formulation of the rate of equilibrium equation (2.41), with a virtual velocity field δv_j as the weighting function, is (the range of indices in this section is $1, 2, 3$)

$$\int_{B_0} (\dot{S}_{ij,i}) \delta v_j \, dV = 0 \tag{2.179}$$

Using the divergence theorem, equation (2.179) becomes

$$\int_{B_0} \dot{S}_{ij} \delta(v_{j,i}) \, dV - \int_{\partial B_{0\tau}} \tau_j^{(s)} \, \delta v_j \, dS = 0 \tag{2.180}$$

where, in the updated Lagrangian formulation (see equation 2.49)

$$\tau_j^{(s)} = N_i \dot{S}_{ij} \tag{2.181}$$

and $\partial B_0 \tau$ is that part of ∂B on which tractions are prescribed.

The next steps are the use of equation (2.96) relating $\dot{\boldsymbol{S}}$ to $\hat{\boldsymbol{\sigma}}$ in an updated Lagrangian framework, together with the assumption that the body is nearly incompressible ($d_{kk} = 0$). Consequences of these assumptions are

$$\dot{\boldsymbol{S}} = \hat{\boldsymbol{\sigma}} - \boldsymbol{\sigma} \cdot \boldsymbol{\omega} - \boldsymbol{d} \cdot \boldsymbol{\sigma} \tag{2.182}$$

$$\hat{\boldsymbol{K}} = \hat{\sigma} \tag{2.183}$$

The last equation is a consequence of equations (2.30), (2.94), and (2.90). A differentiated form of equation (2.30) is

$$\dot{\boldsymbol{K}} = J\dot{\boldsymbol{\sigma}} + \dot{J}\boldsymbol{\sigma} \tag{2.184}$$

This equation, with $\dot{J} = 0$ [equation (2.94) with $d_{kk} = 0$] and $J = 1$ (updated Lagrangian formulation), simply becomes $\dot{\boldsymbol{K}} = \dot{\boldsymbol{\sigma}}$. Now, equation (2.90) with $\boldsymbol{\sigma} = \boldsymbol{K}$ [equation (2.30) with $J = 1$] gives equation (2.183). It is important to note here that rates of quantities in the updated Lagrangian formulation must be handled with care. Thus, for example, one gets $\boldsymbol{F} = \boldsymbol{I}$ but (see equation 2.18)

$$\dot{\boldsymbol{F}} = \boldsymbol{d} + \boldsymbol{\omega} \tag{2.185}$$

Use of equations (2.182) and (2.183), together with the definitions of $\boldsymbol{d}$ and $\boldsymbol{\omega}$ in terms of the velocity gradients (equations 2.19 and 2.20), in the Galerkin formulation (2.180) results in the equation

$$\int_{B_0} [\hat{K}_{ij}\delta(d_{ij}) - \tfrac{1}{2}\sigma_{ij}\delta(2d_{ik}d_{kj} - v_{k,i}v_{k,j})]\,dV - \int_{\partial B_{0T}} \tau_j^{(s)}\,\delta v_j\,dS = 0 \tag{2.186}$$

This is the basic equation for the FEM formulation.

A discretized version of equation (2.186) is obtained by writing

$$\boldsymbol{v} = [N]\{v\} \tag{2.187}$$

where $[N]$ is the shape function matrix and $\{v\}$ contains nodal velocities. From equations (2.187) and (2.19)

$$\{d\} = [B]\{v\} \tag{2.188}$$

At this point, the kinematic assumption (2.88) and the hypoelastic law (2.89) are invoked. In matrix form, equation (2.89) is

$$\{\hat{\sigma}\} = [C](\{d\} - \{d^{(n)}\}) \tag{2.189}$$

Finally, using equations (2.187), (2.188), (2.189), and (2.183) in (2.186) results in the discretized equations for the components of the nodal velocities at the element level. These are then assembled in the usual way to yield the discretized global equations

$$([K] + [K_G] + [K_C])\{v\} = \{P^{(n)}\} + \{P^{(B)}\} \tag{2.190}$$

where $[K]$ is the elastic stiffness matrix, $[K_G]$ is the geometric stiffness matrix, $[K_C]$ is the load correction matrix, and $\{P^{(n)}\}$ and $\{P^{(B)}\}$ are load vectors.

The correspondence of the terms in equations (2.186) and (2.190) is discussed next. The first integral in equation (2.186) gives the elastic stiffness matrix. Such an elastic element stiffness matrix may be expressed as

$$[k] = \int_{V_e} [B]^{\mathrm{T}}[C][B]\,dV \tag{2.191}$$

and the nonelastic load vector

$$\{p^{(n)}\} = \int_{V_e} [B]^{\mathrm{T}}[C]\{d^{(n)}\}\, dV \tag{2.192}$$

where equations (2.191) and (2.192) are written at the element level and V_e is the volume of a finite element. These terms are quite analogous to corresponding finite element formulations for small deformation (Mukherjee 1982). For the discretized equations, lower-case letters denote matrices and load vectors at the element level, while upper case letters represent assembled matrices and load vectors.

The second stiffness matrix in equation (2.190), the geometric stiffness matrix $[k_G]$, originates from the second term in equation (2.186). This is a stress-dependent symmetric matrix, which comes about as a consequence of including rotation effects in the stress rate equation (2.182).

The surface integral term in equation (2.186) includes the traction rate $\boldsymbol{\tau}^{(s)}$, which has been discussed before. This surface integral gives rise to terms that are velocity dependent and others that are not. The velocity-dependent terms give rise to the load correction matrix $[k_C]$, while the rest are grouped together in the load vector $\{p^{(B)}\}$.

The geometric stiffness matrix. Using equation (2.18), one can write

$$\{h\} = [B_h]\{v\} \tag{2.193}$$

Substituting the above in the second term of equation (2.186) results in the definition of the geometric correction matrix at the element level as

$$[k_G] = \int_{V_e} [B_h]^{\mathrm{T}}[D_G^*][B_h]\, dV \tag{2.194}$$

where $[D_G^*]$ is a stress-dependent symmetric matrix given in detail by Yamada and Hirakawa (1978).

The load correction matrix. The load correction matrix, as stated before, arises from the surface integral in equation (2.186). The traction $\boldsymbol{\tau}^{(s)}$, which appears in the surface integral, must be written in terms of prescribed boundary variables. The dead load and follower load situations were discussed in section 2.2.2. Boundary conditions for other situations that arise in metal forming problems are discussed later.

2.4 Numerical Implementation and Results

First, numerical implementation of the BEM equations presented in section 2.2 are discussed. Next, numerical results for sample planar (from Chandra and Mukherjee 1983) and axisymmetric (from Rajiyah and Mukherjee 1987) problems are presented. Further results for planar and axisymmetric metal forming processes are presented in subsequent chapters.

2.4.1 Viscoplastic Constitutive Models

Constitutive assumptions regarding additive decomposition of $\boldsymbol{d}$ (equation 2.88) and the hypoelastic law (equation 2.89) have been presented before. The nonelastic stretch $\boldsymbol{d}^{(n)}$ must now be determined from an appropriate plastic or viscoplastic law.

A general form of such constitutive models (see equations 2.91) may be written as

$$d_{ij}^{(n)} = f_{ij}(\sigma_{ij}, \hat{\sigma}_{ij}, q_{ij}^{(k)}) \tag{2.195}$$

$$\hat{q}_{ij}^{(k)} = g_{ij}^{(k)}(\sigma_{ij}, \hat{\sigma}_{ij}, q_{ij}^{(k)}) \tag{2.196}$$

Here, $q_{ij}^{(k)}$ are state (or internal) variables that can be scalars or tensors. These are often physically motivated by micromechanical considerations like dislocation density, resistance to dislocation motion, and so on. These models, however, are phenomenological in nature, and the state variables, in mechanics terminology, are analogous to quantities like current yield stress, kinematic hardening parameters, or accumulated plastic work. A small number of these state variables are assumed to completely characterize the present deformation state of a material, and the history dependence of the nonelastic stretch, up to the current time, is assumed to be completely represented by their current values. For the viscoplastic models used to obtain the numerical results presented in this book, $\hat{\sigma}_{ij}$ is absent in the right-hand sides of the above equations, and $\boldsymbol{d}^{(n)}$ and $\hat{\boldsymbol{q}}^{(k)}$, at any time, are assumed to depend only on the current values of the stress, state variables, and temperature in an explicit manner. Further details are available in chapter 2 of Mukherjee (1982).

Anand's model. A typical viscoplastic constitutive model, suitable for metal deformation at elevated temperatures, is due to Anand (1982). It has one scalar internal variable, s. The model, adapted to the present multiaxial large-deformation situation, is described by the equations (Rajiyah and Mukherjee 1987)

$$d_{ij}^{(n)} = \frac{3}{2}\frac{d^{(n)}}{\sigma}\sigma'_{ij} \tag{2.197}$$

where σ'_{ij} are the components of the deviatoric part of the Cauchy stress and σ is the stress invariant defined as

$$\sigma = \sqrt{\tfrac{3}{2}\sigma'_{ij}\sigma'_{ij}}$$

The invariant $d^{(n)}$ is given by

$$d^{(n)} = Ae^{-Q/kT}\left(\frac{\sigma}{s}\right)^{1/m}, \qquad \sigma < s \tag{2.198}$$

together with the evolution equation

$$\dot{s} = h_0\left(1 - \frac{s}{s_c}\right)d^{(n)} \tag{2.199}$$

with

$$s_c = \tilde{s}\left(\frac{d^{(n)}}{A}e^{Q/kT}\right)^n \tag{2.200}$$

Here, T is the temperature in kelvins, Q is the activation energy, and k is Boltzmann's constant. Also, A, h_0, $\tilde{s}$, m, and n are material constants, of which m and n are, in general, temperature dependent. The particular parameters used here are representative of Fe–0.05 carbon steel in a temperature range of 1173 to 1573 K and strain rate range of 1.4×10^{-4} to $2.3 \times 10^{-2}\,\text{s}^{-1}$. These parameters have been used for all the isothermal (at $T = 1173$ K) simulations for axisymmetric problems reported in this section. They are (Anand 1982)

$$A = 10^{11}\,\text{s}^{-1}, \qquad h_0 = 1329.22\ \text{MPa}$$

$$\tilde{s} = 147.6\ \text{MPa} \qquad m = 0.147, \qquad n = 0.03$$

$$Q/k = 3.248 \times 10^4\ \text{K}$$

together with the elastic constants (at 1173 K)

$$E = 5.88 \times 10^3\ \text{MPa} \qquad \upsilon = 0.3$$

Also, the initial value of s is taken to be 47.11 MPa.

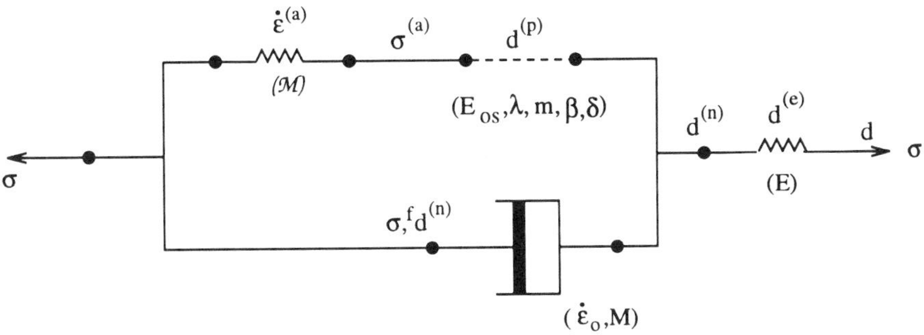

Figure 2.6. Hart's (1976) model in tension.

Hart's model. Hart's model (Hart 1976) has been used to describe material behavior for the planar numerical examples discussed in this section. This model, for small-strain elasto-viscoplastic problems, is discussed by Mukherjee (1982) and an extension for large-strain problems is presented by Mukherjee and Chandra (1987).

The model for uniaxial tension is shown in figure 2.6. It has two state variables, a scalar H (the hardness), and a tensor $\varepsilon_{ij}^{(a)}$ (the anelastic strain). Material response in dilation is assumed to be elastic. According to this model (extended to large deformation),

$$\overset{\wedge}{\varepsilon}_{ij}^{(a)} = d_{ij}^{(n)} - d_{ij}^{(p)} \tag{2.201}$$

that is, the nonelastic stretch is decomposed into a plastic and an anelastic part. Since $\boldsymbol{\varepsilon}^{(a)}$ is a proper state variable, which reflects the magnitude and direction of the prior deformation history, its Jaumann rate is used in equation (2.201). The tensor $\boldsymbol{d}^{(p)}$ is the completely irrecoverable and path-dependent permanent strain. The other state variable, H, called hardness, is analogous to the current yield stress in an elastoplastic material.

The deviatoric component of the Cauchy stress, σ'_{ij}, is decomposed into two auxiliary tensors

$$\sigma'_{ij} = \sigma'^{(a)}_{ij} + \sigma'^{(f)}_{ij} \tag{2.202}$$

The flow rules for kinematic quantities are of the form

$$\varepsilon_{ij}^{(a)} = \frac{3}{2}\frac{\varepsilon^{(a)}}{\sigma^{(a)}}\sigma'^{(a)}_{ij} \tag{2.203}$$

$$d_{ij}^{(p)} = \frac{3}{2}\frac{d^{(p)}}{\sigma^{(a)}}\sigma'^{(a)}_{ij} \tag{2.204}$$

$$d_{ij}^{(n)} = \frac{3}{2}\frac{d^{(n)}}{\sigma^{(f)}}\sigma'^{(f)}_{ij} \tag{2.205}$$

where $d^{(n)}$, $d^{(p)}$, $\varepsilon^{(a)}$; σ, $\sigma^{(a)}$, and $\sigma^{(f)}$ are scalar invariants of the corresponding tensors, defined as

$$d^{(n)} = \left(\frac{2}{3} d_{ij}^{(n)} d_{ij}^{(n)}\right)^{1/2}, \qquad d^{(p)} = \left(\frac{2}{3} d_{ij}^{(p)} d_{ij}^{(p)}\right)^{1/2}$$

$$\varepsilon^{(a)} = \left(\frac{2}{3} \varepsilon_{ij}^{(a)} \varepsilon_{ij}^{(a)}\right)^{1/2}, \qquad \sigma = \left(\frac{3}{2} \sigma'_{ij} \sigma'_{ij}\right)^{1/2}$$

$$\sigma^{(a)} = \left(\frac{3}{2} \sigma_{ij}^{\prime(a)} \sigma_{ij}^{\prime(a)}\right)^{1/2}, \qquad \sigma^{(f)} = \left(\frac{3}{2} \sigma_{ij}^{\prime(f)} \sigma_{ij}^{\prime(f)}\right)^{1/2} \tag{2.206}$$

The scalar invariants are related to each other through the uniaxial equations (figure 2.6)

$$\sigma^{(a)} = \mathcal{M} \varepsilon^{(a)} \tag{2.207}$$

$$d^{(n)} = \dot{\varepsilon}_0 \left(\frac{\sigma^{(f)}}{\sigma_0}\right)^{M} \tag{2.208}$$

$$d^{(p)} = E_0 \left[\ln\left(\frac{H}{\sigma^{(a)}}\right)\right]^{-1/\lambda} \tag{2.209}$$

$$E_0 = E_{0s} \left(\frac{H}{H_s}\right)^{m} \tag{2.210}$$

$$\dot{H} = d^{(p)} H \Gamma(H, \sigma^{(a)}) \tag{2.211}$$

$$\Gamma(H, \sigma^{(a)}) = \left(\frac{\beta}{H}\right)^{\delta} \left(\frac{\sigma^{(a)}}{H}\right)^{\beta/H} \tag{2.212}$$

Equation (2.207) represents a linear anelastic element, (2.208) a nonlinear dashpot, (2.209) and (2.210) a "plastic" element, and, finally, (2.211) and (2.212) describe strain hardening. The flow parameters are $\mathcal{M}$, M, m, λ, $\dot{\varepsilon}_0$ (at a reference stress level σ_0), and E_{0s} (at a reference hardness level H_s); and β and δ are strain-hardening parameters. The above equations are written for a uniform temperature throughout the body. These material parameters, for a given material or alloy, must be determined from relaxation and tension tests (Hart 1976, Mukherjee 1982).

2.4.2 *Planar Problems*

Discretization of integral equations. Boundary value problems with irregular geometry must, in general, be solved by numerical methods. The first step is the discretization of a planar body (plane strain or plane stress) into surface elements and internal cells. A discretized version of the plane strain boundary integral equation related to equation (2.143), considering the limit as $p \rightarrow P$ (also see the three-dimensional equation 2.118) is

$$C_{ij}(P_M)v_i(P_M) = \sum_{N_s} \int_{\Delta s_N} [U_{ij}(P_M, Q)\tau_i^{(s)}(Q) - T_{ij}(P_M, Q)v_i(Q)]\, ds(Q)$$

$$+ \sum_{n_i} \int_{\Delta A_n} [\lambda U_{ij,i}(P_M, q)\eta^{(n)}(q) + 2GU_{ij,k}(P_M, q)\, d_{ik}^{(n)}(q)]\, da(q)$$

$$+ \sum_{n_i} \int_{\Delta A_n} U_{ij,m}(P_M, q)g_{mi}(q)\, da(q) \tag{2.213}$$

where the boundary of the body in the configuration ∂B is divided into N_s boundary segments and the interior into n_i internal cells and $v_i(P_M)$ are the components of velocities at a point P that coincides with node M.

Suitable shape functions must now be chosen for the variation of tractions and velocities on the surface element Δs_N and the variation of the nonelastic strain rates and velocities over an internal cell ΔA_n. One gets integrals of the type (writing $V_{ijk} = U_{ij,k}$)

$$\Delta \bar{U}_{ij}^{M,N} = \int_{\Delta s_N} U_{ij}(P_M, Q)N_\alpha J_b\, d\xi \tag{2.214}$$

$$\Delta \bar{T}_{ij}^{M,N} = \int_{\Delta s_N} T_{ij}(P_M, Q)N_\alpha J_b\, d\xi \tag{2.215}$$

$$\Delta \bar{V}_{ijk}^{M,N} = \int_{\Delta A_n} V_{ijk}(P_M, q)M_\alpha J_a\, d\xi\, d\eta \tag{2.216}$$

where N_α are the shape functions for velocities and tractions on a boundary element and M_α are the shape functions for $\boldsymbol{d}^{(n)}$ over an internal cell. Of course, similar integrals must be evaluated for $\eta^{(n)}$ and $\boldsymbol{g}$. The boundary and domain Jacobians, used to map the integrals into standard straight lines and squares, are denoted by J_b and J_a, respectively, and ξ and η are intrinsic coordinates. It is usual to adopt an isoparametric representation for shapes of the boundary elements and the shape functions for $\boldsymbol{v}$ and $\boldsymbol{\tau}^{(s)}$.

When a surface element or internal cell is sufficiently far away from a source point, it is easy to evaluate the above integrals by Gaussian quadrature. Singular integrals, which occur when a boundary element includes the source point P_M, need special care. One approach is analytic integration. This is generally possible for simple situations, such as straight boundary elements and linear shape functions. This approach has been used to obtain the numerical results presented later in this section. Another approach is the use of logarithmically weighted Gaussian integration for equation (2.214) and the use of rigid-body translation modes (Lachat 1975; also see Mukherjee 1982) for equation (2.215). These integrals occur in BEM applications in linear elasticity and are discussed in numerous papers and several books.

Another "generic" issue in the BEM is the modeling of sharp corners on ∂B. In order to obtain the numerical results presented in this section, sharp corners have been taken care of by putting double field points at corners (conforming elements). Nonconforming elements, in which the source points (collocation points) are placed inside boundary elements, can also be used. Use of conforming elements is a more elegant approach and leads to fewer equations in the discretized system. Finally, the corner tensor $\boldsymbol{C}$ is known in closed form (Mukherjee 1982, Guiggiani and Casalini 1987, Guiggiani and Gigante 1990). Alternatively, its calculation can be avoided by using rigid-body translation modes.

Equation (2.213) is eventually reduced to an algebraic system of the type

$$[C]\{v\} = [\Delta U]\{\tau^{(s)}\} + [\Delta T]\{v\} + \{b\} \tag{2.217}$$

where the vector $\{b\}$ contains the contributions from the second and third integrals on the right-hand side of equation (2.213). It should be noted that $\boldsymbol{\tau}^{(s)}$ and $\boldsymbol{b}$, in general, contain unknown velocity gradients, as well [see equations (2.96) and (2.98)]. Thus, iterations become necessary to solve the above equation. This issue is addressed in the next subsection (Solution Strategy).

Prescribed boundary conditions are input into equation (2.217) at this stage. For example, in a dead load problem (see section 2.2.2), half the velocities and traction rates $\boldsymbol{\tau}^{(s)}$ are given as functions of time. A switching of the columns of $[\Delta U]$ and $[\Delta T]$ is carried out and the unknown components x_i of the boundary values of $\boldsymbol{v}$ and $\boldsymbol{\tau}^{(s)}$ are written as

$$[A]\{x\} = [B]\{y\} + \{b\} \tag{2.218}$$

where $\boldsymbol{y}$ contains the known components of $\boldsymbol{v}$ and $\boldsymbol{\tau}^{(s)}$ on ∂B.

The velocity equation at an internal point p and the corresponding equation for the velocity gradient at an internal point are discretized in similar fashion. A discretized version of the velocity gradient equation has the form

$$h_{j\bar{l}}(p_r) = \sum_{N_S} \int_{\partial S_N} [U_{ij,\bar{l}}(p_r, Q)\tau_i^{(s)}(Q) - T_{ij,\bar{l}}(p_r, Q)v_i(Q)]\, ds(Q)$$

$$+ \sum_{n_i} \frac{\partial}{\partial x_{\bar{l}}} \int_{\Delta A_n} \lambda U_{ij,i}(p_r, q)\, d_{kk}^{(n)}(q)\, da(q)$$

$$+ \sum_{n_i} \frac{\partial}{\partial x_{\bar{l}}} \int_{\Delta A_n} 2GU_{ij,k}(p_r, q)\, d_{ik}^{(n)}(q)\, da(q)$$

$$+ \sum_{n_i} \frac{\partial}{\partial x_{\bar{l}}} \int_{\Delta A_n} U_{ij,m}(p_r, q) g_{mi}(q)\, da(q) \tag{2.219}$$

This equation is also written in matrix form analogous to equation (2.218).

One now faces the issue of evaluation of singular domain integrals when $p_r \in \Delta A_n$. Several approaches are possible. One choice is to first integrate those expressions containing products of the singular kernels, shape functions, and Jacobians [analogous to equations (2.214) through (2.216)] in closed form, and then to differentiate the resulting expressions with respect to the coordinates of a source point. This has been done, for simple choices of geometry of internal cells and shape functions, to obtain the numerical results presented in this section.

A second choice is to differentiate the expressions under the integral sign, to get $O(1/r^2)$ integrals for two-dimensional problems, together with free terms. This approach has been employed to obtain the axisymmetric results presented in section 2.4.3, and some details of integrating the $O(1/r^2)$ kernels are given there.

A third choice is regularization to weaken the Cauchy singular integrals to weakly singular ones of $O(1/r)$. This approach is described in detail in section 2.2.4. The weakly singular integrals can be numerically evaluated using one of several methods [e.g., the polar mapping method of Mukherjee (1982) or other mapping methods such as the one by Nagarajan and Mukherjee (1993)]. This third choice should be pursued for general situations where the first approach of analytic integration is not possible.

Solution strategy. The basic strategy is to integrate stepwise in time with iterations at each time step. The boundary unknowns $\boldsymbol{v}$ and $\boldsymbol{\tau}^{(s)}$ are treated as "primary" unknowns in this approach and the internal unknowns (here, the velocity gradients) are "secondary." Okada et al. (1990) treated all unknowns as "primary."

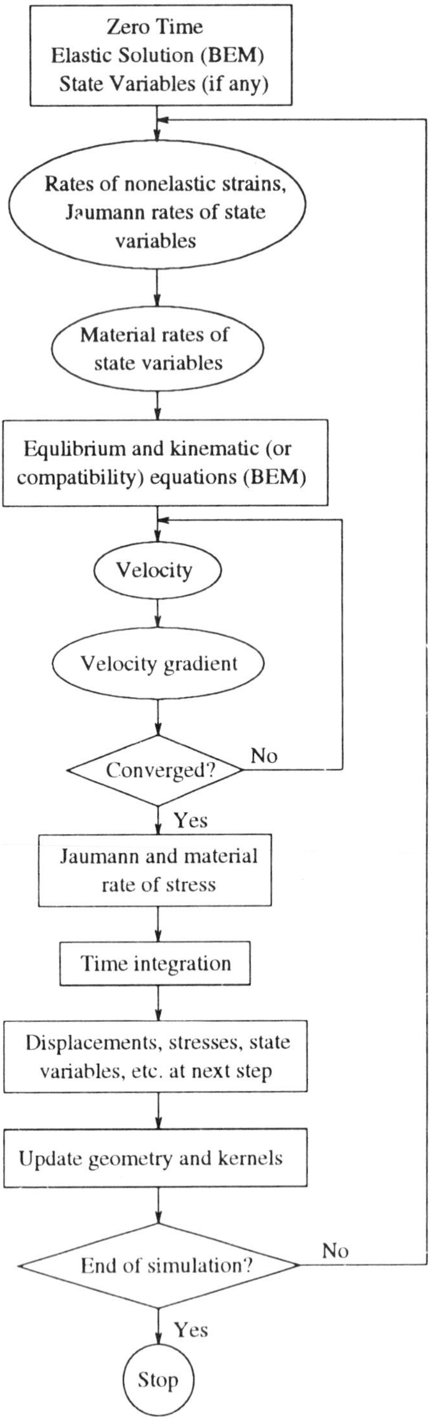

Figure 2.7. Flow chart for the solution of large-strain elastic-viscoplastic problems (from Chandra and Mukherjee 1983).

The solution strategy is best explained by referring to figure 2.7. The steps followed by Chandra and Mukherjee (1983) are described below.

1. The elasticity problem is first solved to obtain initial displacements. For this step, equation (2.218) is solved with $\{b\} = 0$ for the unknown components of $\boldsymbol{u}$ and $\boldsymbol{\tau}$ in terms of the prescribed ones.
2. The internal values of the displacement gradients are obtained from a truncated version of equation (2.219) with $d_{ik}^{(n)} = 0$ and $v_{k,n} = 0$, and $\boldsymbol{v}$ and $\boldsymbol{\tau}^{(s)}$ replaced by $\boldsymbol{u}$ and $\boldsymbol{\tau}$ in the rest of the equation. The initial stresses are determined from strains through Hooke's law. Also, initial rotations are determined from the displacement gradients.
3. The tensor $d_{ik}^{(n)}$ at time zero is obtained from the constitutive equation (2.195).
4. The first approximation to $v_j(P)$ and $\tau_j^{(s)}(P)$ at $t = 0$ is calculated from equation (2.213) with $v_{k,l}$ set to zero. It must be noted that the velocity gradients occur in one of the domain integrals, as well as in the expression for $\tau_j^{(s)}$.
5. The first approximation to $v_{j,\bar{l}}(p) = h_{j\bar{l}}(p)$ at $t = 0$ is obtained from equation (2.219) with all the velocity gradients in the integrals on the right-hand side of this equation set to zero. The $v_{j,l}(p)$ values are extrapolated to get $v_{j,\bar{l}}(P)$.
6. The velocity gradients are inserted into equation (2.213) and the full equation is solved to determine a second approximation to $v_j(P)$ and $\tau_j^{(s)}(P)$. These, and the first approximations of velocity gradients, are then used in equation (2.219) to obtain a second approximation for $v_{j,l}(p)$.
7. Step (6) is repeated until the required convergence is achieved and $v_j(P)$ is determined at time zero.
8. The quantities $v_j(p)$, $v_{j,\bar{l}}(p)$, and then $\hat{\sigma}_{ij}$ are calculated at time zero. The material rate of the Cauchy stress is obtained from the Jaumann rate using equation (2.90).
9. The relevant quantities are calculated at time Δt from the values and rates at $t = 0$. Steps (3) through (8) are repeated to obtain the rates at $t = \Delta t$, and so on. The time histories of the various quantities are thus obtained by marching forward in time and suitable updating of the geometry and the kernels.

It should be noted here that a better way to calculate $\boldsymbol{h}(P)$ (see step 5) and $\hat{\sigma}_{ij}(P)$ (step 8) is to use the boundary algorithm [see equations (2.169) and (2.170)].

Ideally, the geometry and the kernels should be updated at each time step. In practice, updating is done as often as is necessary during the integration process. It should be noted that the time marching process must be carried out with the material derivatives of the tensors in order to obtain their integrated values in the global coordinate frame as functions of $\boldsymbol{x}$ and time.

Numerical results for sample problems. The numerical results presented here are taken from Chandra and Mukherjee (1983). The generalized Hart (1976) model is used to describe the viscoplastic deformation of metallic solids. The BEM program uses straight boundary elements and polygonal internal cells. The velocity and tractions $\boldsymbol{v}$ and $\boldsymbol{\tau}^{(s)}$ are taken to be piecewise linear on the boundary elements and $\boldsymbol{d}^{(n)}$ and $\boldsymbol{g}$ are taken to be piecewise constant on the internal cells. All integrations are carried out analytically. The domain integrals in equation (2.216) are evaluated by first performing the integrations for an arbitrary source point p_r and then differentiating the integrals at p_r. Results for large deformation of a slab with a hole in plane strain are shown in figures 2.8 through 2.10. The material properties used here for Hart's constitutive model are representative of 304 stainless steel at 400°C:

$$\lambda = 0.15, \qquad M = 7.8, \qquad m = 5$$

$$.\,// = 0.132 \times 10^{8}\ \text{psi}$$

$$E = 0.244 \times 10^{8}\ \text{psi}, \qquad \nu = 0.298$$

$$\dot{\varepsilon}_0 = 3.15\ \text{s}^{-1} \quad \text{at} \quad \sigma_0 = 10\ \text{ksi}$$

$$\dot{\varepsilon}^*_{sT} = 1.269 \times 10^{-24}\ \text{s}^{-1} \quad \text{at} \quad \sigma^*_s = 10\ \text{ksi}, \qquad T_B = 673\ \text{K}$$

$$\beta = 0.179 \times 10^{6}\ \text{psi}, \qquad \delta = 1.33$$

Only a quarter of the slab is modeled because of symmetry. The BEM mesh, with 37 boundary nodes and 20 internal cells, and the FEM mesh, with 102 nodes and 41 elements, are available in Chandra and Mukherjee (1983). The FEM formulation is described in section 2.3 of this book [see, also, Chandra and Mukherjee (1984b)]. It uses a piecewise quadratic description of velocities over triangular finite elements.

Displacements and stresses for the problem are shown in figures 2.8 and 2.9. The locations of the ends of the horizontal and vertical diameters of the hole, points $(a_0, 0)$ and $(0, a_0)$, as functions of u_2^∞ / L_0 (with $u_2^\infty = v_2^\infty t$) are shown in figure 2.8. The horizontal diameter of the hole is first seen to decrease due to the Poisson effect and then increase with deformation. The redistribution of stress along the line AB in figure 2.9 shows the expected drop in normalized stress concentration from approximately 3.16 to 2.35. The stress becomes more evenly distributed with deformation. In contrast to the small-strain problem discussed in Mukherjee (1982, p.71), a special feature

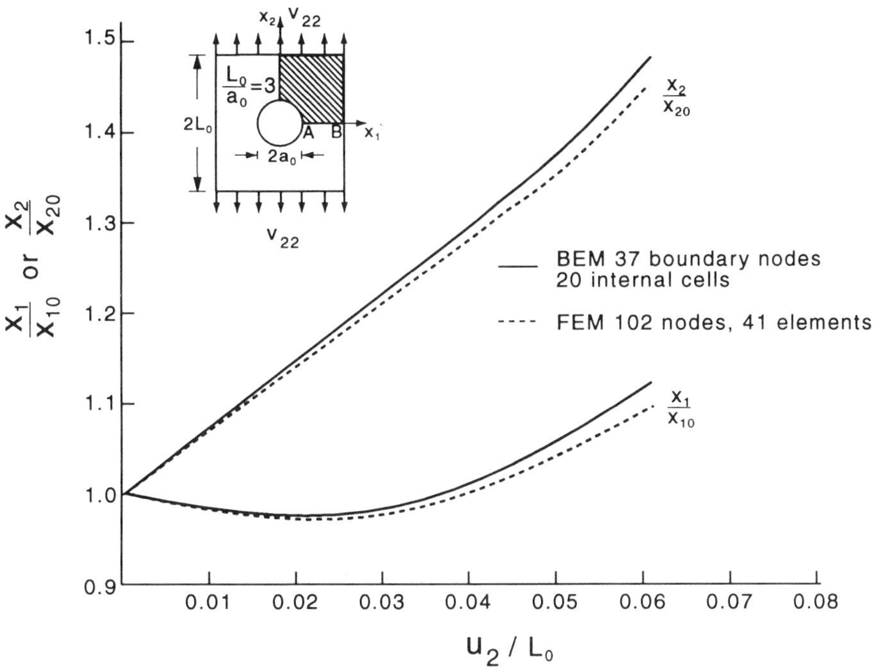

Figure 2.8. Comparison of BEM and FEM solutions for hole growth in a square block in plane strain extension at constant velocity: $v_2^\infty = 10^{-3}$ inch/s; 304 SS at 400°C (from Chandra and Mukherjee 1983).

of the large-deformation problem is that the stress concentration at a remote strain of 6 percent increases near the free edge of the slab. This is a consequence of "caving-in" of the free edge of the slab, as shown in figure 2.10. A vertical equilibrium check for the stress distribution at a remote strain of 6 percent in figure 2.9 shows that the BEM result underpredicts the remote stress resultant by about 6 percent. This is considered quite good in view of the fact that the slab has undergone a substantial amount of deformation at this stage.

Researchers at U.S. Steel (Appleby et al. 1982, Bourcier et al. 1984) have carried out some FEM analyses and large-deformation experiments on a slab with a hole. The calculations described in this section are able to predict the important qualitative features of the experiments in a very satisfactory fashion. Bourcier et al. (1984), for example, measured an initial decrease of the horizontal diameter of the hole in a slab, followed by an increase of this diameter with deformation. They also observed a caving-in of the free edge of the slab. The magnitudes of the calculated displacements are quite

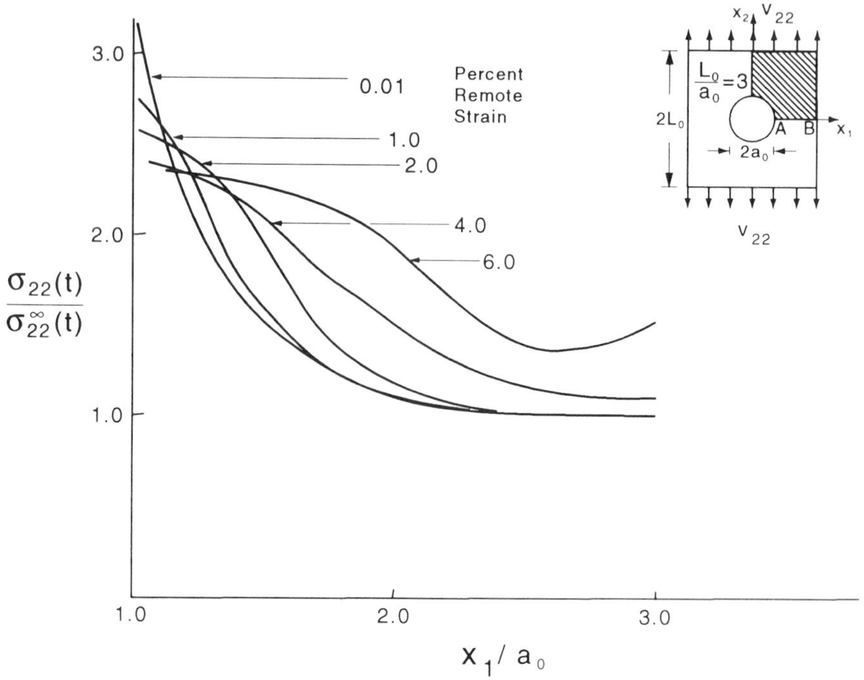

Figure 2.9. Stress redistribution along AB at various stages of plane strain hole growth in a square block in plane strain extension at constant velocity: $v_2^{\infty} = 10^{-3}$ inch/s; 304 SS at 400°C (from Chandra and Mukherjee 1983).

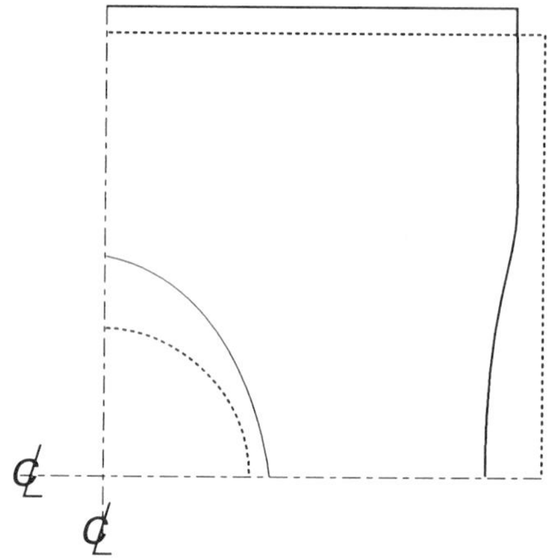

Figure 2.10. Deformation of a block with a circular hole for the problem described in figures 2.8 and 2.9: $\varepsilon_{22}^{\infty}= 0.06$ (from Chandra and Mukherjee 1983).

close to the measured ones. Detailed quantitative comparisons of theory and experiment have not been carried out because Hart model parameters for the materials used by Bourcier et al. (1984) are not known.

2.4.3 *Axisymmetric Problems*

Discretization of integral equations. Discretization of integral equations for the axisymmetric problem, for $\boldsymbol{v}(P)$ (equation 2.151), $\boldsymbol{v}(p)$, and $\boldsymbol{h}(p)$ [see equation (2.161)] is carried out in a manner entirely analogous to the planar case [see equation (2.213)]. These equations will not be repeated here, but some of them are given by Rajiyah and Mukherjee (1987). Integrations of the kernels that contain elliptic functions over boundary elements and over internal cells, however, must be carried out numerically since analytic integration is not possible.

Of crucial importance in the BEM is the accurate numerical evaluation of the integrals of products of singular kernels and shape functions of the unknowns, over boundary elements, as well as internal cells. It is pointed out, again, that the kernels U_{ij} are $\ln r$ singular, while T_{ij} are $1/r$ singular over the boundary. Also, first as well as second spatial derivatives of U_{ij}, the latter being $1/r^2$ singular, must be integrated over internal cells. All these kernels contain elliptic functions, so analytical/numerical methods must be developed for this purpose. Also, these elliptic integrals are highly sensitive and great care must be exercised in integrating them.

Special methods have been described by Mukherjee (1982) for the integration of the kernels U_{ij} and T_{ij} over boundary elements and for the integration of $U_{ij,k}$ over internal cells. In short, a simple transformation $r = e^2$ is very useful for the first integration, while a special mapping technique has been used successfully for the domain integration of kernels like $U_{\rho j,\rho}$. The T_{ij} integrals have been obtained indirectly through the use of a rigid-body translation mode in the Z-direction and an inflation mode in the R-direction. Domain integration of second derivatives of U_{ij} was avoided by Sarihan and Mukherjee (1982) by use of the so-called "strain rate gradient method."

The "strain rate gradient method" is impractical for these large-deformation problems since such a method would introduce spatial derivatives of terms like $\sigma_{\rho\rho} d_{\rho\rho}$ in the domain integrals in an equation like (2.161). Brebbia et al. (1984) suggested an elegant method for the integration of $U_{ij,kl}$ over internal cells. This method, however, relies on the separation of the radial and the angular dependence of the $U_{ij,kl}$ kernels. This is possible for two-dimensional but not for axisymmetric problems where complicated elliptic functions of R, Z, ρ, and ζ are present in the kernels. An accurate method for calculating these integrals for large-deformation axisymmetric problems is outlined below (Rajiyah and Mukherjee 1987).

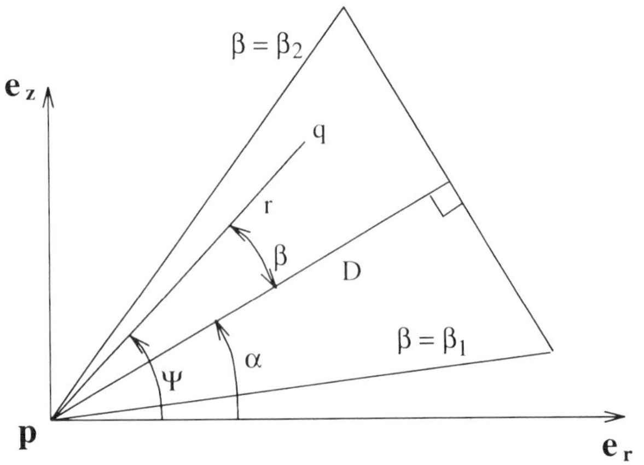

Figure 2.11. Notation used for the evaluation of domain integrals.

Singular integration of $U_{ik,kl}$ over internal cells. It is easiest to explain this procedure in terms of local polar coordinates (r, ψ) centered at the source point (R, Z) and lying in the R–Z plane. Here, r is the distance between the source and the field point and ψ is the angle made by the line pq to the R-axis in the R–Z plane (figure 2.11).

The kernels $U_{\rho 1, \rho \bar{1}}$, etc., in equation (2.161) contain terms with the strongest singularity, $1/r^2$, and the next strongest, $1/r$. A typical term containing $U_{\rho 1, \rho \bar{1}}$ in equation (2.161) can be written as

$$\int_B U_{\rho 1, \rho \bar{1}}(p, q) g(q)\, dA_q = \int_B \left[U_{\rho 1, \rho \bar{1}}(p, q) g(q) - \frac{f(\psi)}{r^2} g(p) \right] dA_q + g(p) \int_B \frac{f(\psi)}{r^2}\, dA_q \tag{2.220}$$

where $f(\psi)/r^2$ is the term with the strongest singularity in $U_{\rho 1, \rho \bar{1}}$ and $g(q)$ is a term multiplying this kernel. The functions $f(\psi)$, for the various kernels $U_{ij,kl}$, have been obtained using the symbolic computer package MACSYMA (Rand 1984) and are listed in table 2.6. It should be noted here that $U_{ij,kl}$, when p lies on the axis of symmetry, can be put directly into the form $f(\psi)/r^2$.

The first domain integral in equation (2.220) is now only $1/r$ singular and can be obtained by mapping followed by Gaussian integration as described by Mukherjee (1982). The second integral on a triangular cell, using the notation given in figure 2.11, becomes

$$g(p) \int_{\beta_1}^{\beta_2} d\psi \int_{\varepsilon}^{D/\cos\beta} \frac{f(\psi)}{r}\, dr = g(p) \int_{\beta_1}^{\beta_2} f(\beta) \ln\left(\frac{D}{\cos\beta} \right) d\beta \tag{2.221}$$

Table 2.6. Singular functions from $U_{ij,kl}$

$\{c_1 = 1/[8\pi(1-\nu)G],\ c_2 = (3-4\nu)\}$

Kernel	$f(\psi)/c_1$
$U_{\rho 1,\rho\bar{1}}$	$(\cos 2\psi - \cos 4\psi - c_2 \cos 2\psi)$
$U_{\rho 1,\zeta\bar{1}}$	$-(\sin 4\psi + c_2 \sin 2\psi)$
$U_{\zeta 1,\rho\bar{1}}$	$(\sin 2\psi - \sin 4\psi)$
$U_{\zeta 1,\zeta\bar{1}}$	$\cos 4\psi$
$U_{\rho 1,\rho\bar{2}}$	$-(\sin 4\psi + c_2 \sin 2\psi)$
$U_{\rho 1,\zeta\bar{2}}$	$(\cos 2\psi + \cos 4\psi + c_2 \cos 2\psi)$
$U_{\zeta 1,\rho\bar{2}}$	$\cos 4\psi$
$U_{\zeta 1,\zeta\bar{2}}$	$(\cos 2\psi + \cos 4\psi + c_2 \cos 2\psi)$
$U_{\rho 2,\rho\bar{1}}$	$(\sin 2\psi - \sin 4\psi)$
$U_{\rho 2,\zeta\bar{1}}$	$\cos 4\psi$
$U_{\zeta 2,\rho\bar{1}}$	$(\cos 4\psi - \cos 2\psi - c_2 \cos 2\psi)$
$U_{\zeta 2,\zeta\bar{1}}$	$(\sin 4\psi - c_2 \sin 2\psi)$
$U_{\rho 2,\rho\bar{2}}$	$\cos 4\psi$
$U_{\rho 2,\zeta\bar{2}}$	$(\sin 2\psi + \sin 4\psi)$
$U_{\zeta 2,\rho\bar{2}}$	$(\sin 4\psi - c_2 \sin 2\psi)$
$U_{\zeta 2,\zeta\bar{2}}$	$-(\cos 2\psi + \cos 4\psi - c_2 \cos 2\psi)$

The value of the integral at the lower limit $\varepsilon \to 0$ vanishes when all contiguous elements around p are taken into account since, due to the nature of the kernels,

$$\int_0^{2\pi} f(\psi)\, d\psi = 0 \qquad \text{if } p \notin \text{(axis of symmetry)}$$

and

$$\int_0^{\pi} f(\psi)\, d\psi = 0 \qquad \text{if } p \in \text{(axis of symmetry)}$$

It should be noted here that the above integral between 0 and π, for the case where p lies on the axis of symmetry, vanishes for the combination $U_{\rho j,\rho\bar{1}} + U_{\rho j,\bar{1}}/\rho$ rather than for each term separately. Here, $d_{\rho\rho} = d_{\theta\theta}$ and $\sigma_{\rho\rho} = \sigma_{\theta\theta}$.

Thus, the integral in equation (2.221) is now completely regular and can be evaluated by the usual Gaussian quadrature. Of course, the complete integral over the domain B is obtained by summing over triangular cells as usual. Also, the above procedure is only used for accurate evaluation of the singular integrals. If the domain cell is far from a source point, regular Gaussian integration is sufficient. "Nearly singular" cases, where the source point is outside but very near the domain of integration, require special care, usually perhaps the use of more Gauss points in these regions. Such special treatment for "nearly singular" cases has not been necessary in this work.

It should be mentioned here that the above procedure was inspired by the method outlined by Brebbia et al. (1984) for the evaluation of similar singular integrals in two dimensions. The method described above, however, is completely general and can be applied to very complicated singular kernels, and allows the prescription of arbitrary shape functions for the functions $g(q)$. It can be generalized even further by adding and subtracting the $1/r$ singular portions also from $U_{\rho 1,\rho\bar{1}}g(q)$. That would then lead to regular integrals that could, in principle, be evaluated directly by Gaussian quadrature without the need of any shape functions for $g(q)$. Shape functions for $g(q)$ have been used to obtain the numerical results presented in this work because the unknown velocity gradients occur in the "geometric correction" integrals in equations such as (2.151) and (2.161).

Finally, attention should be drawn to the method of Huang and Du (1989), which has been described in detail in section 2.2.4. This approach could also be applied to axisymmetric problems, perhaps more easily than the method outlined above. This is an interesting topic for further research.

Constitutive model and shape functions. For the numerical examples that follow, the constitutive model due to Anand (1982) [see equations (2.197) through (2.200)] has been used to describe viscoplastic behavior. The boundary ∂B is discretized either into straight or into curved quadratic elements in this study. The traction rates and velocities are assumed to be piecewise quadratic on the boundary elements. Region B is discretized into triangular cells. Terms multiplying the kernels in the domain integrals in equation (2.151), for example, $d_{\rho\rho}^{(n)}$ or $\sigma_{\rho\rho}d_{\rho\rho} + \sigma_{\rho\zeta}(d_{\zeta} - \omega_{\rho\zeta})$, are assumed to be piecewise linear on the internal cells, with the nodes lying on the vertices of the triangles. Tangential derivatives of velocities are obtained carefully on the boundary by differentiating the boundary shape functions and taking account of the curvature (when present) of the boundary elements. Gaussian quadrature for integration of nonsingular kernels typically uses ten Gauss points on a boundary element and seven Gauss points on a triangular internal cell.

Numerical results. The problem of expansion of a sphere under a time-varying pressure loading at the inner surface is presented here. This problem is chosen because the BEM results can be compared with a direct solution in this case. Other numerical results from this formulation are available in Rajiyah and Mukherjee (1987, 1989) and also for axisymmetric extrusion problems later in this book.

Generating direct solutions for large expansion of a sphere made of a compressible material is not an easy task. However, the corresponding problem for a sphere made of an incompressible material ($\nu = 0.5$) can be solved fairly easily; this solution is briefly outlined by Rajiyah and Mukherjee (1987). The BEM formulation presented here does not admit the incompressible case ($\nu = 0.5$) since the Lamé parameter λ, which occurs in the hypoelastic equation (2.89) as well as in Hooke's law for the reference field, blows up in this limit. It is possible to generate an alternative BEM formulation for the incompressible case by suitable modification of the appropriate equations, as is done for compressible linear elasticity. Such has not been done in this work. Instead, in order to validate the present code, a value close to $\nu = 0.5$ has been chosen and the BEM results are compared with direct solutions for the expansion of a sphere made of an incompressible material. The value of ν closest to 0.5 that gives stable results from the BEM

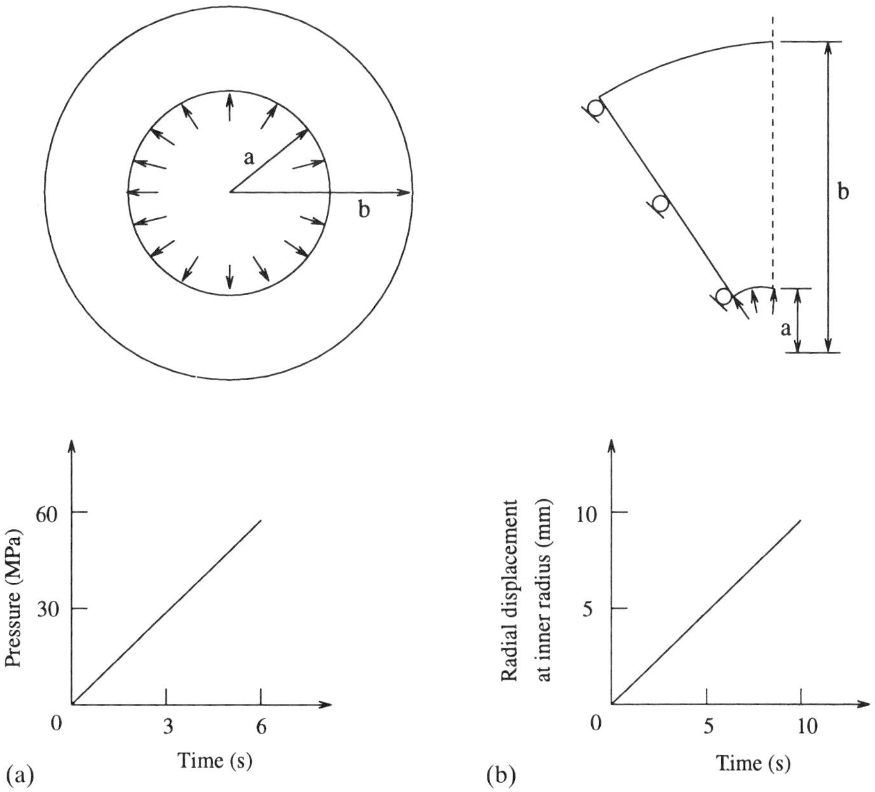

Figure 2.12. Inflation of a sphere: (a) prescribed pressure rate at the inner radius; (b) prescribed velocity at the inner radius ($a = 0.1$ m, $b = 0.15$ m, $\dot{p} = 10$ MPa/s, $v = 1$ mm/s).

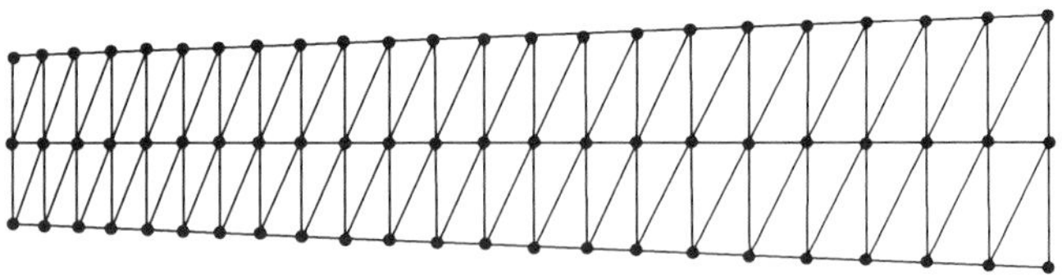

Figure 2.13. Boundary element mesh for inflation of a sphere: 27 boundary nodes, 13 quadratic boundary elements, and 88 internal cells.

formulation is 0.488. This value has therefore been used. Since the direct solution has been verified to be insensitive to the value of ν in the neighborhood of 0.5, this comparison of the BEM and direct results is considered valid and useful.

A 5-degree section of a sphere is modeled as shown in figure 2.12, and the BEM mesh for this problem is given in figure 2.13. Only quadratic boundary interpolation is considered here. The numerical results for the prescribed rate of pressure problem are given in figures 2.14 through 2.16.

The agreement between the direct and BEM solutions is very good, even for stresses inside the sphere. This is expected from the BEM since internal stress rates are obtained through exact differentiation of the velocity equations. This procedure eliminates jumps in stresses across interelement

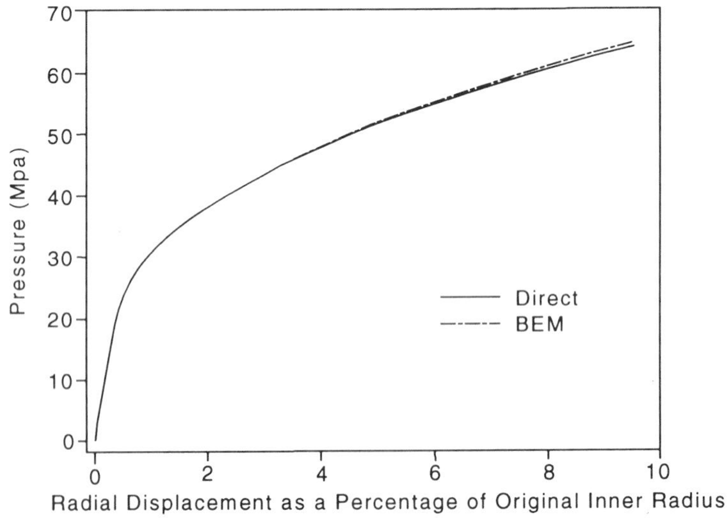

Figure 2.14. Pressure as a function of radial displacement of the inside surface of a sphere—$\dot{p}$ prescribed on inside surface (from Rajiyah and Mukherjee 1987).

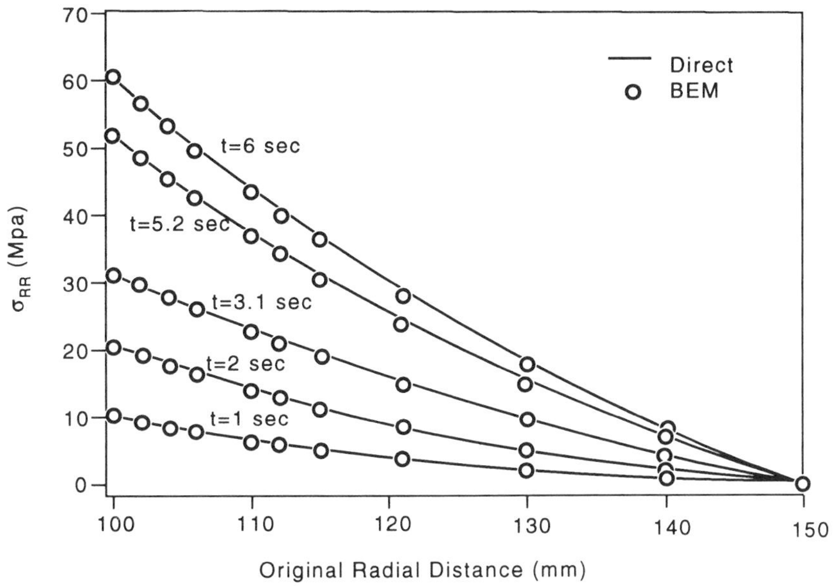

Figure 2.15. Same situation as in figure 2.14—comparison of BEM and direct solutions for the redistribution of radial stress (from Rajiyah and Mukherjee 1987).

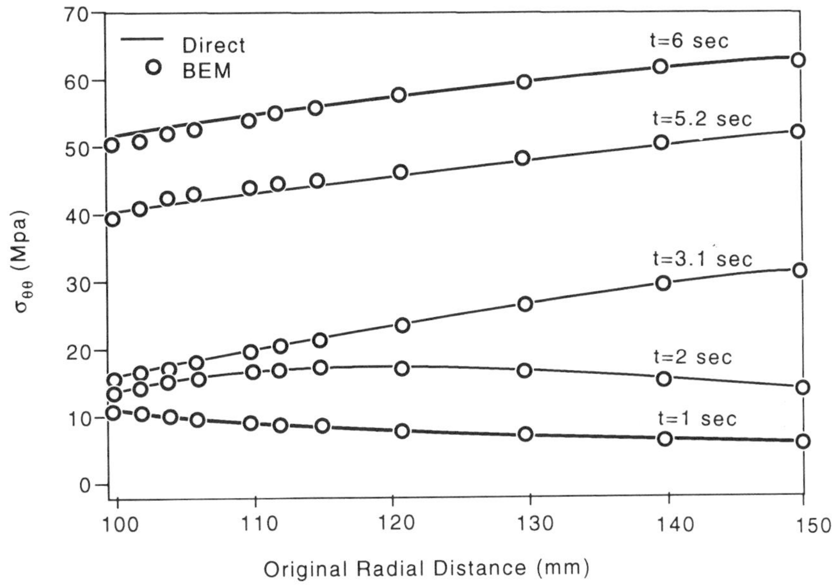

Figure 2.16. Same situation as in figure 2.14—comparison of BEM and direct solutions for the redistribution of tangential stress (from Rajiyah and Mukherjee 1987).

boundaries—as is typical in the finite element model (FEM)—and greatly contributes to accurate determination of the stresses, as well as the nonelastic strains, which are driven by the stresses.

The CPU time for the above problem, with 27 boundary nodes and 42 internal nodes, was 4918.7 seconds on an IBM 3090/400 computer at Cornell University.

References

Anand, L. (1982). "Constitutive Equations for the Rate-Dependent Deformation of Metals at Elevated Temperatures," *J. Eng. Mat. Technol., ASME*, **104**, 12–17.

Appleby, E. J., Devenpeck, M. L., Richmond, O., and Whitmore, R. W. (1982). "Stretching of Notched Steel Sheets: a Theoretical–Experimental Comparison," *Numerical Methods in Industrial Forming Processes* (ed. J. F. T. Pittman, R. D. Wood, J. M. Alexander, and O. C. Zienkiewicz), pp. 723–732. Pineridge Press, Swansea, U.K.

Atluri, S. N. (1984). "On Constitutive Relations at Finite Strain: Hypo-elasticity and Elasto-plasticity with Isotropic or Kinematic Hardening," *Computer Meth. Appl. Mech. Eng.*, **43**, 137–171.

Betti, E. (1872). "Teoria Dell Elasticita," *Il Nuovo Ciemento, Series 2*, 7–10.

Bohatier, C. and Chenot, J.-L. (1989). "Finite Element Formulation for Non-Steady-State Large Deformations with Sliding or Evolving Contact Boundary Conditions," *Int. J. Num. Meth. Eng.*, **28**, 753–768.

Bourcier, R. J., Smelser, R. E., Richmond, O., and Koss, D. A. (1984). "Deformation and Failure at Isolated Holes in Plane-Strain Tension," *Int. J. Fracture*, **24**, 289–297.

Brebbia, C. A., Telles, J. C. F., and Wrobel, L. C. (1984). *Boundary Element Techniques—Theory and Applications in Engineering*. Springer-Verlag, Berlin.

Bui, H. D. (1978). "Some Remarks About the Formulation of Three-Dimensional Thermoelastoplastic Problems by Integral Equations," *Int. J. Solids Structures*, **14**, 935–939.

Chandra, A. and Mukherjee, S. (1983). "Applications of the Boundary Element Method to Large Strain Large Deformation Problems of Viscoplasticity," *J. Strain Anal.*, **18**, 261–270.

Chandra, A. and Mukherjee, S. (1984a). "Boundary Element Formulations for Large Strain-Large Deformation Problems of Viscoplasticity," *Int. J. Solids Structures*, **20**, 41–53.

Chandra, A. and Mukherjee, S. (1984b). "A Finite Element Analysis of Metal-Forming Problems with an Elasto-Viscoplastic Material Model," *Int. J. Num. Meth. Eng.*, **20**, 1613–1628.

Chandra, A. and Mukherjee, S. (1986). "An Analysis of Large Strain Viscoplasticity Problems Including the Effects of Induced Material Anisotropy," *J. Appl. Mech., ASME*, **53**, 77–82.

Chen, Z. Q. and Ji, X. (1990). "A New Approach to Finite Deformation Problems of Elastoplasticity–Boundary Element Analysis Method," *Computer Meth. Appl. Mech. Eng.*, **78**, 1–18.

Cheng, J.-H. and Kikuchi, N. (1985). "An Analysis of Metal Forming Processes Using Large Deformation Elastic-Plastic Formulations," *Computer Meth. Appl. Mech. Eng.*, **49**, 71–108.

Cruse, T. A. and Vanburen, W. (1971). "Three-Dimensional Elastic Stress Analysis of a Fracture Specimen with an Edge Crack," *Int. J. Fracture*, **7**, 1–15.

Fung, Y. C. (1965). *Foundations of Solid Mechanics*. Prentice-Hall, Englewood Cliffs, N.J.

Ghosh, N., Rajiyah, H., Ghosh, S., and Mukherjee, S. (1986). "A New Boundary Element Method Formulation for Linear Elasticity," *J. Appl. Mech., ASME*, **53**, 69–76.

Guiggiani, M. and Casalini, P. (1987). "Direct Computation of Cauchy Principal Value Integrals in Advanced Boundary Elements," *Int. J. Num. Meth. Eng.*, **24**, 1711–1720.

Guiggiani, M. and Gigante, A. (1990). "A General Algorithm for Multidimensional Cauchy Principal Value Integrals in the Boundary Element Method," *J. Appl. Mech., ASME*, **57**, 906–915.

Gurtin, M. E. (1981). *An Introduction to Continuum Mechanics*. Academic Press, New York.

Hart, E. W. (1976). "Constitutive Relations for the Nonelastic Deformation of Metals," *J. Eng. Mater. Technol., ASME*, **98**, 193–202.

Healy, B. E. and Dodds Jr., R. H. (1992). "A Large Stain Plasticity Model for Implicit Finite Element Analyses," *Comput. Mech.*, **9**, 95–112.

Hill, R. (1978). "Aspects of Invariance in Solid Mechanics," *Advances in Applied Mechanics*, vol. 18 (ed. C.-S. Yih), pp. 1–75. Academic Press, New York.

Huang, Q. and Du, Q. (1988). "An Improved Formulation for Domain Stress Evaluation by Boundary Element Method in Elastoplastic Problems" [in English], *Acta Mech. Solida Sinica*, **2**, 19–24.

Jin, H., Runesson, K. and Mattiasson, K. (1989). "Boundary Element Formulation in Finite Deformation Plasticity Using Implicit Integration," *Computers Structures*, **31**, 25–34.

Lachat, J. C. (1975). "A Further Development of the Boundary-Integral Technique for Elastostatics," Ph.D. thesis, University of Southampton, U.K.

Maniatty, A. M., Dawson, P. R. and Weber, G. G. (1991). "An Eulerian Elasto-Viscoplastic Formulation for Steady-State Forming Processes," *Int. J. Mech. Sci.*, **33**(5), 361–377.

Mukherjee, S. (1982). *Boundary Element Methods in Creep and Fracture*. Elsevier, London.

Mukherjee, S. and Chandra, A. (1984). "Boundary Element Formulations for Large Strain-Large Deformation Problems of Plasticity and Viscoplasticity," *Developments in Boundary Element Methods—3*, (ed. P. K. Banerjee and S. Mukherjee), chap. 2, pp. 27–58. Elsevier, Barking, Essex, U.K.

Mukherjee, S. and Chandra, A. (1987). "Nonlinear Solid Mechanics," *Boundary Element Methods in Mechanics*, vol. 3, *Computational Methods in Mechanics* (ed. D. E. Beskos), chap. 6, pp. 285–331. Elsevier, Amsterdam.

Mukherjee, S. and Chandra, A. (1991). "A Boundary Element Formulation for Design Sensitivities in Problems Involving Both Geometric and Material Nonlinearities," *Math. Computer Modelling*, **15**, 245–255.

Nagarajan, A. and Mukherjee, S. (1993). "A Mapping Method for Numerical Evaluation of Two-Dimensional Integrals with $1/r$ Singularity," *Comput. Mech.*, **12**, 19–26.

Nanson, E. J. (1877–78). "Note on Hydrodynamics," *The Messenger of Mathematics*, **7**, 182–183.

Nemat-Nasser, S. and Li, Y.-F. (1992). "A New Explicit Algorithm for Finite-Deformation Elastoplasticity and Elastoviscoplasticity: Performance Evaluation," *Computers Structures*, **44**(5), 937–963.

Nishiguchi, I., Sham, T.-L. and Krempl, E. (1990). "A Finite Deformation Theory of Viscoplasticity Based on Overstress: Part II—Finite Element Implementation and Numerical Experiments," *J. Appl. Mech*, *ASME*, **57**, 553–561.

Ogden, R. W. (1984). *Non-Linear Elastic Deformations*. Ellis Horwood, Chichester, U.K.

Okada, H., Rajiyah, H., and Atluri, S. N. (1990). "A Full Tangent Stiffness Field-Boundary Element Formulation for Geometric and Material Nonlinear Problems of Solid Mechanics," *Int. J. Num. Meth. Eng.*, **29**, 15–35.

Perić, D., Owen, D. R. J. and Honnor, M. E. (1992). "A Model for Finite Strain Elasto-Plasticity Based on Logarithmic Strains: Computational Issues," *Computer Meth. Appl. Mech. Eng.*, **94**, 35–61.

Rajiyah, H. (1987). "An Analysis of Large Deformation Axisymmetric Inelastic Problems by the Finite Element and Boundary Element Methods," Ph.D. thesis, Cornell University, Ithaca, N.Y.

Rajiyah, H. and Mukherjee, S. (1987). "Boundary Element Analysis of Inelastic Axisymmetric Problems with Large Strains and Rotations," *Int. J. Solids Structures*, **23**, 1679–1698.

Rajiyah, H. and Mukherjee, S. (1989). "A Note on the Efficiency of the Boundary Element Method for Inelastic Axisymmetric Problems with Large Strains," *J. Appl. Mech.*, *ASME*, **56**, 721–724.

Rand, R. H. (1984). *Computer Algebra in Applied Mathematics—An Introduction to MACSYMA*. Pitman, Boston.

Rizzo, F. J. and Shippy, D. J. (1968). "A formulation and solution procedure for the general nonhomogeneous elastic inclusion problem," *Int. J. Solids Structures*, **4**, 1161–1179.

Rubinstein, R. and Atluri, S. N. (1983). "Objectivity of Incremental Constitutive Relations over Finite Time Steps in Computational Finite Deformation Analysis," *Computer Meth. Appl. Mech. Eng.*, **36**, 277–290.

Saran, M. J. and Runesson, K. (1992). "A Generalized Closest-Point-Projection Method for Deformation-Neutralized Formulation in Finite-Strain Plasticity," *Eng. Comput.*, **9**, 359–370.

Saran, M. J. and Wagoner, R. H. (1991). "A Consistent Implicit Formulation for Nonlinear Finite Element Modeling with Contact and Friction: Part I—Theory," *J. Appl. Mech.*, *ASME*, **58**, 499–506.

Sarihan, V. (1982). "Axisymmetric Viscoplastic Deformation by the Boundary Element Method," Ph.D. thesis, Cornell University, Ithaca, N.Y.

Sarihan, V. and Mukherjee, S. (1982). "Axisymmetric Viscoplastic Deformation by the Boundary Element Method," *Int. J. Solids Structures*, **18**, 1113–1128.

Simo, J. C. and Miehe, C. (1992). "Associative Coupled Thermoplasticity at Finite Strains: Formulation, Numerical Analysis and Implementation," *Computer Meth. Appl. Mech. Eng.*, **98**, 41–104.

Sladek, J. and Sladek, V. (1986). "Computation of Stresses by BEM in 2D Elastostatics," *Acta Technica CSAV*, **31**, 523–531.

Sokolnikoff, I. S. (1964). *Tensor Analysis—Theory and Applications to Geometry and Mechanics of Continua*, 2d ed. John Wiley, New York.

Thomson, W. [later Lord Kelvin] (1848). "Note on the Integration of the Equations of Equilibrium of an Elastic Solid," *Cambridge Dublin Math. J.*, **3**, 87–89.

Weber, G. and Anand, L. (1990). "Finite Deformation Constitutive Equations and a Time Integration Procedure for Isotropic, Hyperelastic-Viscoplastic Solids," *Computer Meth. Appl. Mech. Eng.*, **79**, 173–202.

Yamada, Y. and Hirakawa, H. (1978). "Large Deformation and Instability Analysis in Metal Forming Process," *Applications of Numerical Methods to Forming Processes*, AMD-28 (ed. A. Armen and R. F. Jones), pp. 27–38. American Society of Mechanical Engineers, New York.

Zabaras, N. and Arif, A. F. M. (1992). "A Family of Integration Algorithms for Constitutive Equations in Finite Deformation Elasto-Viscoplasticity," *Int. J. Num. Meth. Eng.*, **33**, 59–84.

Zhang, Q. (1991). "Shape Design Sensitivity Analysis by the Boundary Element Method," Ph.D. thesis, Cornell University, Ithaca, N.Y.

Zhang, Q. and Mukherjee, S. (1991). "Design Sensitivity Coefficients for Linear Elastic Bodies with Zones and Corners by the Derivative Boundary Element Method," *Int. J. Solids Structures*, **27**, 983–998.

Zhang, Q., Mukherjee, S., and Chandra, A. (1992a). "Design Sensitivity Coefficients for Elasto-Viscoplastic Problems by Boundary Element Methods," *Int. J. Num. Meth. Eng.*, **34**, 947–966.

Zhang, Q., Mukherjee, S., and Chandra, A. (1992b). "Shape Design Sensitivity Analysis for Geometrically and Materially Nonlinear Problems by the Boundary Element Method," *Int. J. Solids Structures*, **29**, 2503–2525.

3

Thermal Problems

3.1 Steady-State Conduction

3.1.1 Direct Formulation

One of the most popular and earlier applications of the boundary element method is found in the steady-state heat conduction or potential flow problems governed by the Laplace equation. The heat transfer process in many manufacturing processes, at least as a first attempt, may also be modelled as steady-state conduction.

The steady-state heat conduction equation with no internal heat generation in a domain Ω bounded by a boundary Γ can be written as

$$\frac{\partial^2 T}{\partial x_i^2} = 0 \qquad \text{in } \Omega \tag{3.1}$$

subject to the boundary conditions

$$T = \overline{T} \qquad \text{on } \Gamma_T \tag{3.2}$$

and

$$k \frac{\partial T}{\partial x_i} n_i = \overline{q} \qquad \text{on } \Gamma_q \tag{3.3}$$

where the surface $\Gamma \equiv \Gamma_T \cup \Gamma_q$ and k is the thermal conductivity. The surface flux $\overline{q}$ may also include contributions from convective cooling, which may be written as

$$q^{(c)} = h(T - T_\infty) \tag{3.4}$$

The Laplace operator being self-adjoint, an adjoint equation may be written as

$$\frac{\partial^2 G}{\partial x_i^2} = \delta(x_i(q) - x_i(p)) \tag{3.5}$$

in an infinite domain with homogeneous boundary conditions. Here, p is the source point and q is the field point in the domain. The fundamental solution $G(p, q)$ of the adjoint equation subject to homogeneous boundary conditions can be expressed (Jaswon and Symn 1977) as

$$G(p, q) = \frac{1}{4\pi r} \qquad \text{in 3D} \tag{3.6}$$

and

$$G(p, q) = -\frac{1}{2\pi} \ln(r) \qquad \text{in 2D} \tag{3.7}$$

Here, r is the Euclidian distance between the source point p and the field point q. At this point, one may pursue a direct approach (e.g., Banerjee and Butterfield 1981, Mukherjee 1982) or an indirect approach (e.g., Beskos 1987, Jaswon and Symn 1977). The direct approach lends itself to easier physical interpretations and is pursued here.

Taking the inner product of equation (3.1) with $G(p, q)$ and the inner product of equation (3.5) with $T(q)$, subtracting one from the other and applying the divergence theorem, an integral representation of the governing equation (3.1) may be obtained as

$$T(p) = \int_\Gamma \left(G(p, Q) \frac{\partial T(Q)}{\partial n_Q} - \frac{\partial G(p, Q)}{\partial n_Q} T(Q) \right) d\Gamma \tag{3.8}$$

where, p and q are source and field points, respectively, in the domain and P and Q are source and field points on the boundary. The internal equation (3.8) relates the temperature at an internal point to the temperature and flux fields at the boundary. It is important to note here, that T and $\partial T/\partial n_Q$ at every boundary field point Q need to be known for determination of $T(p)$. For a well-posed problem, however, only half of the set of T and $\partial T/\partial n_Q$ on the boundary is known. This brings us to the most important step in BEM and its point of departure from the traditional Green's function approach. At this time, we take the limit as the internal source point p

approaches a point P on the boundary. The resulting boundary equation may be written as

$$C(P)T(P) = \int_{\Gamma} \left(G(P,Q)\frac{\partial T(Q)}{\partial n_Q} - \frac{\partial G(P,Q)}{\partial n_Q} T(Q) \right) d\Gamma \tag{3.9}$$

The fundamental solution $G(p, q)$ for the steady-state heat conduction problem contains a singularity of order $(1/r)$ for a three-dimensional problem and a singularity of order $\ln(r)$ for two-dimensional applications as the distance r between a source point and a field point tends to zero. Accordingly, as $r \rightarrow 0$, the kernel $\partial G/\partial n_Q$ contains a singularity of order $(1/r^2)$ in 3D and a singularity of order $(1/r)$ in 2D applications. The coefficient C in equation (3.9) in general depends on the local geometry at P and is a residual term arising from integration of singular kernels. It is well known (e.g., Mukherjee 1982) that $C = \frac{1}{2}$ if the boundary is locally smooth at P. It may be evaluated analytically for corners in two dimensions. For all two-dimensional and three-dimensional cases (with or without corners), it is also possible to evaluate indirectly appropriate combinations of $C(P)$ and integrals of $\partial G(P, P)/\partial n_Q$ by requiring that the BEM formulation must satisfy the equipotential solution. Imposing a constant temperature with zero flux, and without any loss of generality choosing the magnitude of the constant temperature to be unity everywhere, we get

$$C(P) = -\int_{\Gamma} \frac{\partial G(P,Q)}{\partial n_Q} d\Gamma \tag{3.10}$$

Thus, equation (3.9) may also be written as

$$0 = \int_{\Gamma} \left(G(P,Q)\frac{\partial T(Q)}{\partial n_Q} - \frac{\partial G(P,Q)}{\partial n_Q} \{T(Q) - T(P)\} \right) d\Gamma \tag{3.11}$$

Equation (3.11) may now be solved for a well-posed problem by discretizing and solving the unknown half of the set of temperatures and fluxes on the boundary in terms of the known half.

3.1.2 *Alternative Complex Variable Approach*

Noting that the well-known Cauchy integral formula for complex analytic function is a generalized expression of the boundary integral equation for the real potential in a complex domain, various researchers (e.g., Hrodmadka and Lai 1987, Kwak and Choi 1987) have developed the system of equations for potential problems directly from the Cauchy integral formula. Such an approach is called the complex variable boundary element method (CVBEM), and all the calculations are also carried out in the complex arithmetic.

Let $\phi(Z)$ be an analytic function in a simply connected complex domain Ω with a simple closed boundary Γ. Then

$$\phi(Z) = T(x, y) + i\bar{T}(x, y), \qquad Z \in \Omega \tag{3.12}$$

where $Z = x + iy$, $i = \sqrt{-1}$, and T and $\bar{T}$ are two real-variable functions. T is the sought-after potential function and $\bar{T}$ is its conjugate, which is usually called the stream function and satisfies the Cauchy–Riemann conditions

$$\frac{\partial T}{\partial x} = \frac{\partial \bar{T}}{\partial y} \quad \text{and} \quad \frac{\partial T}{\partial y} = -\frac{\partial \bar{T}}{\partial x} \tag{3.13}$$

Each real function is harmonic and satisfies the Laplace equation

$$\frac{\partial^2 T}{\partial x^2} + \frac{\partial^2 T}{\partial y^2} = 0 \quad \text{and} \quad \frac{\partial^2 \bar{T}}{\partial x^2} + \frac{\partial^2 \bar{T}}{\partial y^2} = 0 \tag{3.14}$$

Hence, the complex variable approach is suitable only for problems whose governing equations are reducible to Laplace equations.

The Cauchy integral formula then states that (Carrier et al. 1966)

$$\frac{1}{2\pi i}\int_\Gamma \frac{\phi(Z)}{Z - Z_0}\,dZ = \begin{cases} \phi(Z_0), & Z_0 \in \Omega \\ c\phi(Z_0), & Z_0 \in \Gamma \\ 0, & Z_0 \text{ outside } \Gamma \end{cases} \tag{3.15}$$

where c denotes the interior angle at Z_0 on the boundary divided by 2π. Hence, $c = \frac{1}{2}$ for a locally smooth boundary at Z_0. Here, Z_0 denotes a source point P on the boundary.

The complex expressions in equation (3.15) may now be written in terms of their real and imaginary parts. Denoting the unit tangent and normal vectors at x (or a field point Q) on the boundary as s and n, respectively, we get,

$$dZ = \frac{dZ}{ds}\,ds \tag{3.16}$$

and

$$(\bar{Z} - \bar{Z}_0)\frac{dZ}{ds} = (x - x_0)(s + in) \tag{3.17}$$

Here $(x - x_0)$ is the same as vector $\boldsymbol{r}$ from the source point x_0 (or P) to the field point x (or Q). Using equations (3.16) and (3.17) in equation (3.15) and separating the real and imaginary parts, a pair of integral equations for T and $\bar{T}$ may be obtained as

$$C(P)T(P) = \frac{1}{2\pi} \int_\Gamma \left(\bar{T}(Q) \frac{\partial(\ln r)}{\partial s_Q} + T(Q) \frac{\partial(\ln r)}{\partial n_Q} \right) d\Gamma \tag{3.18a}$$

and

$$C(P)\bar{T}(P) = \frac{1}{2\pi} \int_\Gamma \left(\bar{T}(Q) \frac{\partial(\ln r)}{\partial n_Q} - T(Q) \frac{\partial(\ln r)}{\partial s_Q} \right) d\Gamma \tag{3.18b}$$

The equations (3.18*a* and *b*) are coupled for T and its conjugate $\bar{T}$. Hence, the two equations must be solved simultaneously for both T and $\bar{T}$. The Cauchy–Riemann conditions (3.13) provide a set of constraints between T and $\bar{T}$. Utilizing these and integrating equation (3.18*a*) by parts, we get,

$$C(P)T(P) = \int_\Gamma \left(G(P,Q) \frac{\partial T(Q)}{\partial n_Q} - \frac{\partial G(P,Q)}{\partial n_Q} T(Q) \right) d\Gamma \tag{3.19}$$

which is exactly the same as equation (3.9) obtained before through direct formulation for two-dimensional problems.

3.1.3 A Derivative BEM (DBEM) Formulation

Generally, the unknowns in the BEM consist of two kinds of variables: the potential function on the Neumann part of the boundary and its normal derivative on the Dirichlet part of the boundary. In conventional BEM formulations, both the potential and its normal derivative are approximated by the same interpolation function even though they are different in smoothness. The solution thus obtained on the boundary, however, is known to be usually more accurate (Kwak and Choi 1987) than those obtained by classical FEM with similar meshes. The BEM solves directly for the flux on the boundary, while in FEM it is calculated by numerical differentiation of the potential and extrapolation to the boundary, which may be a source of poor accuracy.

There also exists a very important class of problems in manufacturing where the tangential derivative of the potential on the boundary is as important as its normal derivatives (e.g., Liggett and Liu 1983, Mukherjee and Chandra 1991). In the conventional BEM, the tangential derivative of the potential is obtained by numerical differentiation of the boundary data.

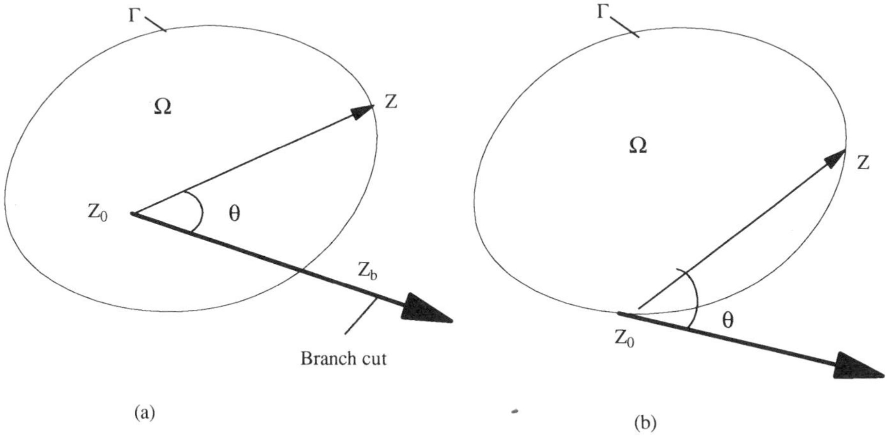

Figure 3.1. Multivaluedness of complex logarithm function: (a) when the point z_0 is within the domain, (b) when the point z_0 is on the boundary.

This degrades the accuracy of the final solution. To alleviate this problem, BEM formulations may be developed with both tangential and normal derivatives of the potential as primary variables. A complex-variable approach may be used for this purpose.

Considering a simply connected domain, a derivative BEM formulation for steady-state heat conduction may be obtained by integration by parts of Cauchy's integral formula. This may be written as,

$$\frac{1}{2\pi i}\int_{\Gamma} \ln(Z - Z_0)\frac{d\phi(Z)}{dZ}\,dZ = \begin{cases} \phi(Z_b) - \phi(Z_0), & Z_0 \in \Omega \\ 0, & Z_0 \in \Gamma \text{ and outside } \Gamma \end{cases} \tag{3.20}$$

Here Z_b is a point of intersection on Γ by a branch cut originating at the source point Z_0 in Ω and passing to infinity (as shown in figure 3.1*a*). This is due to the multivaluedness of the complex logarithmic function, which has appeared as a result of the integration by parts. A special case should be taken when Z_0 is on Γ, such that the branch cut should not cross any point on the boundary except the point Z_0 itself (as shown in figure 3.1*b*).

The complex expressions in equation (3.20) may now be rewritten in terms of their real and imaginary parts:

$$\ln(Z - Z_0) = \ln|x - x_0| + i\theta(x - x_0) \tag{3.21}$$

and

$$\frac{d\phi}{dZ} = \frac{\partial\phi}{\partial s}\, ds = \left(\frac{\partial T}{\partial s} + i\frac{\partial \bar{T}}{\partial s}\right) ds = \left(\frac{\partial T}{\partial s} + i\frac{\partial T}{\partial n}\right) ds \tag{3.22}$$

where θ denotes the angle between the vector $x - x_0$ (or $\boldsymbol{r}$) from a source point to a field point and a reference direction corresponding to that of the branch cut. This reference direction can be taken arbitrarily as long as continuity and single-valuedness of the potential are ensured. Substituting equations (3.21) and (3.22) into equation (3.20) and writing the real and imaginary parts separately, we get

$$\int_\Gamma \left(G(P,Q)\frac{\partial T(Q)}{\partial n_Q} + H(P,Q)\frac{\partial T(Q)}{\partial s} \right) d\Gamma = 0 \tag{3.23a}$$

and

$$\int_\Gamma \left(H(P,Q)\frac{\partial T(Q)}{\partial n_Q} - G(P,Q)\frac{\partial T(Q)}{\partial s} \right) d\Gamma = 0 \tag{3.23b}$$

where a field point at x on the boundary is denoted by Q and a source point at x_0 on the boundary is denoted by P. Here, the kernel function $H(P,Q)$ denotes the conjugate of the fundamental solution $G(P,Q)$ and is given as

$$H(P,Q) = -\frac{1}{2\pi}\,\theta(r) \tag{3.23c}$$

It is interesting to note that both of equations (3.23*a*,*b*) are expressed in terms of the derivatives of potential T only. The conjugate function $\bar{T}$ does not appear in equations (3.23*a*,*b*). This implies that the two real integral equations in (3.23) hold simultaneously for the derivatives of T at each point of the boundary Γ. Thus, one can use either equation (3.23*a*) or (3.23*b*) at any source point P, provided that continuity and single-valuedness conditions for the potential are satisfied. Integration by parts of the conventional BEM equation (3.9) for steady-state heat conduction would have resulted in equation (3.23*a*) only. It is also observed that the kernel H is regular, while $\partial G(P,Q)/\partial n_Q$ was $(1/r)$ singular as $r \to 0$. This obviates difficulties in numerical implementation relating to the singularities of the kernels. As observed by Kwak and Choi (1987), however, the regular nature of H makes all of its elements similar in magnitude. This may lead to ill-conditioning of the coefficient matrix. This problem can easily be avoided by suitably selecting equations out of (3.23*a*) and (3.23*b*) such that the G-term occurs on the diagonal of the coefficient matrix. This can be achieved by using equation (3.23*a*) on the Dirichlet part (where T and hence $\partial T/\partial s$ is specified) and equation (3.23*b*) on the Neumann part (where $\partial T/\partial n$ is specified) of the boundary Γ.

Hence, for numerical implementation, we may use

$$\int_\Gamma G(P,Q)\frac{\partial T(Q)}{\partial n_Q}\,d\Gamma = -\int_\Gamma H(P,Q)\frac{\partial T(Q)}{\partial s}\,d\Gamma \qquad \text{on } \Gamma_T \tag{3.24a}$$

and

$$\int_\Gamma G(P,Q)\frac{\partial T(Q)}{\partial s}\,d\Gamma = \int_\Gamma H(P,Q)\frac{\partial T(Q)}{\partial n}\,d\Gamma \qquad \text{on } \Gamma_q \tag{3.24b}$$

The tangential derivative $\partial T/\partial s$ is specified over Γ_T and the normal derivative $\partial T/\partial n$ is specified over Γ_q.

For derivative BEM, the primary variables are $\partial T/\partial s$ and $\partial T/\partial n$. This makes it prone to degeneration. For a typical conduction problem with zero flux over one part and constant temperature (i.e., $\partial T/\partial s = 0$) over the rest of the boundary, the boundary conditions in terms of $\partial T/\partial s$ and $\partial T/\partial n$ become homogeneous. Unless special care is taken, such a system will not be able to use the information about the magnitude of the constant temperature and will falsely result in a trivial solution.

To ensure a nontrivial unique solution in such cases, some auxiliary conditions relating the magnitude of the potential T need to be imposed on the system. Assuming that the potential is continuous over the entire boundary, the definition of $\partial T/\partial s$ is utilized for this purpose. If T_1 and T_2 are the temperatures at the starting and ending of the Neumann boundary Γ_q (where $\partial T/\partial s$ is unknown), we may write

$$\int_{\Gamma_q} \frac{\partial T}{\partial s}\,ds = \Delta T = T_2 - T_1 \tag{3.25}$$

and such a condition is imposed for every segment of Γ_q.

Now, equation (3.24) should be solved together with equation (3.25). This can be done using various techniques. The classical Lagrange multiplier technique can be used efficiently for this purpose.

A discretized form of equation (3.24) may eventually be written as

$$\begin{bmatrix} G_{ss} & H_{sn} \\ -H_{ns} & G_{nn} \end{bmatrix} \begin{Bmatrix} \dfrac{\partial T^{(u)}}{\partial n} \\ \dfrac{\partial T^{(u)}}{\partial s} \end{Bmatrix} = \begin{bmatrix} -H_{ss} & -G_{sn} \\ -G_{ns} & H_{nn} \end{bmatrix} \begin{Bmatrix} \dfrac{\partial T^{(k)}}{\partial s} \\ \dfrac{\partial T^{(k)}}{\partial n} \end{Bmatrix} = \{b\} \tag{3.26}$$

where a superscript (u) denotes the unknowns and a superscript (k) denotes the known quantities. It should be noted here that, unlike conventional BEM, both rows and columns are exchanged in equation (3.26) rather than just the columns. This ensures that terms involving G always appear on the diagonal and that the coefficient matrix is well conditioned.

After discretization, the constraint (3.25) becomes

$$C^{\mathrm{T}} \frac{\partial T^{(u)}}{\partial s} = d \tag{3.27}$$

Introducing the Lagrange multiplier λ for constraint (3.25), the complete system of equations may be obtained as

$$\begin{bmatrix} G_{ss} & H_{sn} & 0 \\ -H_{ns} & G_{nn} & C \\ 0 & C^T & 0 \end{bmatrix} \begin{Bmatrix} \dfrac{\partial T^{(u)}}{\partial n} \\ \dfrac{\partial T^{(u)}}{\partial s} \\ \lambda \end{Bmatrix} = \begin{Bmatrix} b \\ ---- \\ d \end{Bmatrix} \tag{3.28}$$

Once the tangential derivatives $\partial T/\partial s$ and the normal derivatives $\partial T/\partial n$ are obtained everywhere on the boundary by solving equation (3.28), the potential T is calculated by integrating $\partial T/\partial s$ along the boundary.

3.2 Steady-State Conduction–Convection

3.2.1 Formulation

The heat transfer in many manufacturing processes, such as single point machining, continuous casting, thin film deposition for electronic packaging, may be modeled as a steady-state conduction–convection phenomenon with constant or variable velocity fields.

The basic steady-state conduction–convection equation with no internal heat generation in a three-dimensional domain Ω bounded by a surface Γ can be described as

$$\rho c V_i \frac{\partial T}{\partial x_i} = k \frac{\partial^2 T}{\partial x_i^2} \qquad \text{in } \Omega \tag{3.29}$$

Subject to the boundary conditions

$$T = \bar{T} \qquad \text{on } \Gamma_T \tag{3.30}$$

and

$$k\frac{\partial T}{\partial x_i}n_i = \overline{q} \qquad \text{on } \Gamma_q \tag{3.31}$$

where the surface Γ is $\Gamma_T \cup \Gamma_q$, k is the thermal conductivity, ρ is the density, and c is the heat capacity. The surface flux $\overline{q}$ can also include contributions from convective cooling, which may be expressed as

$$q^{(c)} = h(T - T_\infty) \tag{3.32}$$

where h is the convective heat transfer coefficient. Equation (3.29) applies to an Eulerian reference frame that remains fixed while material flows through it. The convective term represents the energy transported by the material as it moves through the reference frame.

The velocity field V_i may be decomposed as

$$V_i = V_i^{(c)} + V_i^{(v)} \tag{3.33}$$

into a constant part and a variable part. An adjoint equation based on the constant velocity part may now be written as

$$-\rho c V_i^{(c)}\frac{\partial G}{\partial x_i} = k\frac{\partial^2 G}{\partial x_i^2} + \delta(x_i(q) - x_i(p)) \tag{3.34}$$

in an infinite domain with homogeneous boundary conditions. Here, p is the source point and q is the field point in the domain. The fundamental solution $G(p, q)$ of the adjoint equation subject to homogeneous boundary conditions can be expressed (Carslaw and Jaeger 1986) as

$$G(p, q) = \frac{1}{4\pi k r}\exp\left(\frac{V^{(c)}r - V_i^{(c)}[x_i(q) - x_i(p)]}{2\kappa}\right) \qquad \text{in 3D} \tag{3.35}$$

and

$$G(p, q) = \frac{1}{2\pi k}\exp\left(\frac{-V_i^{(c)}[x_i(q) - x_i(p)]}{2\kappa}\right)K_0\left(\frac{V^{(c)}r}{2\kappa}\right) \qquad \text{in 2D} \tag{3.36}$$

Here r is the distance between the source point p and the field point q, $V^{(c)}$ is the magnitude of the constant part of the velocity field, $\kappa = k/\rho c$ is the thermal diffusivity, and K_0 is the modified Bessel function of the second kind of order zero.

Applying the divergence theorem, an integral representation of the governing equation (3.29) may be obtained as

$$
\begin{aligned}
T(p) = &-k\int_{\Gamma}\left(\frac{\partial G(p,Q)}{\partial n_Q}T(Q)-G(p,Q)\frac{\partial T(Q)}{\partial n_Q}\right)d\Gamma \\
&-\rho c\int_{\Gamma} G(p,Q)(V_i^{(c)}(q) + V_i^{(v)}(q))n_i(Q)T(Q)\,d\Gamma \\
&+\rho c\int_{\Omega} G(p,Q)V_{i,i}^{(v)}(q)T(q)\,d\Omega \\
&+\rho c\int_{\Omega}\frac{\partial G}{\partial x_i}V_i^{(v)}(q)T(q)\,d\Omega
\end{aligned}
\tag{3.37}
$$

where p and q are source and field points, respectively, in the domain; P and Q are source and field points, respectively, on the boundary. Equation (3.37) is also called the internal equation since it relates the temperature at an internal point to the temperature and flux fields at the boundary. It is important to note here that T and $\partial T/\partial n_Q$ at every field point Q need to be known for determination of $T(p)$. For a well-posed problem, however, only half of the set of T and $\partial T/\partial n_Q$ on the boundary is known. In addition, cases involving variable velocities require evaluation of the two domain integrals.

This brings us to the point of departure of the boundary element formulation from the traditional Green's function approach. At this time, we take the limit as the internal source point p approaches a point P on the boundary. The resulting boundary equation may be expressed as

$$
\begin{aligned}
C(P)T(P) = &-k\int_{\Gamma}\left(\frac{\partial G(P,Q)}{\partial n_Q}T(Q)-G(P,Q)\frac{\partial T(Q)}{\partial n_Q}\right)d\Gamma \\
&-\rho c\int_{\Gamma} G(P,Q)(V_i^{(c)}(Q) + V_i^{(v)}(Q))n_i(Q)T(Q)\,d\Gamma \\
&+\rho c\int_{\Omega} G(P,Q)V_{i,i}^{(v)}(q)T(q)\,d\Omega \\
&+\rho c\int_{\Omega}\frac{\partial G(P,q)}{\partial x_i}V_i^{(v)}(q)T(q)\,d\Omega
\end{aligned}
\tag{3.38}
$$

Several features of equations (3.37) and (3.38) should be discussed at this point. The fundamental solution $G(p,q)$ for the steady-state conduction–convection problem contains a singularity of the order $(1/r)$ for a three-dimensional problem and a singularity of order $\ln(r)$ (contained in the Bessel function) for a two-dimensional application as the distance r between a source point and a field point tends to zero. The kernels $G(p, Q)$, $\partial G(p, Q)/\partial n_Q$ never become singular since Q is a boundary point and p is strictly an interior point. Thus, the distance between p and Q never becomes zero. Such, however, is not the case for $G(P, Q)$, $\partial G(P, Q)/\partial n_Q$ or $G(p, q)$, $\partial G(p, q)/\partial x_i$. The kernels $G(P, Q)$ and $\partial G(P, Q)/\partial n_Q$ appear in boundary integrals, while the kernels $G(p, q)$ and $\partial G/\partial x_i$ appear in domain integrals. The coefficient C in equation (3.38), in general, depends on the local geometry at P and is a residual term arising from integration of singular kernels. Here, $C = k/2$ if the boundary is locally smooth at P. It may be evaluated analytically for corners in two dimensions. It is also possible to evaluate indirectly appropriate combinations of $C(P)$ and integrals involving $\partial G(P, P)/\partial n_Q$ by requiring that the BEM formulation should satisfy the equipotential field with zero flux.

Imposing the equipotential condition on the entire domain (i.e., constant temperature and zero flux) and, without loss of generality, choosing the magnitude of the constant temperature field to be unity everywhere, we get from equation (3.38),

$$\begin{aligned} C(P) = &- k \int_\Gamma \frac{\partial G(P, Q)}{\partial n_Q} \, d\Gamma \\ &- \rho c \int_\Gamma G(P, Q)(V_i^{(c)}(Q) + V_i^{(v)}(Q)) n_i(Q) \, d\Gamma \\ &+ \rho c \int_\Omega G(P, q) V_{i,i}^{(v)}(q) \, d\Omega + \rho c \int_\Omega \frac{\partial G(P, q)}{\partial x_i} V_i^{(v)} \, d\Omega \end{aligned} \tag{3.39}$$

Thus, equation (3.38) may also be written as

$$\begin{aligned} 0 = &- k \int_\Gamma \left(\frac{\partial G(P, Q)}{\partial n_Q} (T(Q) - T(P)) - G(P, Q) \frac{\partial T(Q)}{\partial n_Q} \right) d\Gamma \\ &- \rho c \int_\Gamma G(P, Q)(V_i^{(c)}(Q) + V_i^{(v)}(Q))[T(Q) - T(P)] \, d\Gamma \\ &+ \rho c \int_\Omega G(P, q) \, V_{i,i}^{(v)}(q)[T(q) - T(P)] \, d\Omega \\ &+ \rho c \int_\Omega \frac{\partial G(P, q)}{\partial x_i} V_i^{(v)}[T(q) - T(P)] \, d\Omega \end{aligned} \tag{3.40}$$

3.2.2 Numerical Implementation

Numerical implementation of the BEM equations for the conduction–convection problem with variable velocity field is discussed in this section.

The boundary Γ is discretized into N_s boundary segments and the domain Ω is divided into N_d internal cells. Polynomial shape functions may now be assumed over these boundary segments and internal cells. The boundary segments may be straight or curved and the internal cells may be polygonal or with curved sides. In the following development, linear shape functions over straight-sided boundary elements as well as linear shape functions over triangular internal cells are considered.

A discretized version of the boundary equation (3.38) may now be written as

$$
\begin{aligned}
C(P^M)T(P_M) = &-k\sum_{i=1}^{N_s}\int_{\Delta s_i}\left(\frac{\partial G(P_M,Q)}{\partial n_Q}T(Q) - G(P_M,Q)\frac{\partial T(Q)}{\partial n_Q}\right)ds_Q \\
&-\rho c\sum_{i=1}^{N_s}\int_{\Delta s_i} G(P_M,Q)(V_l^{(c)}(Q)+V_l^{(v)}(Q))n_l(Q)T(Q)\,ds_Q \\
&+\rho c\sum_{j=1}^{N}\int_{\Delta\Omega_j}\left(G(P_M,q)V_{l,l}^{(v)}(q)+\frac{\partial G(P_M,q)}{\partial x_l}V_l^{(v)}(q)\right)T(q)\,d\Omega
\end{aligned}
\tag{3.41}
$$

where $T(P_M)$ represents the temperature at a point P on the boundary that coincides with the boundary node M. Assuming both temperature and normal flux to be linear over each boundary element and the temperature distribution to be linear over each internal cell, we get

$$
\begin{aligned}
C(P^M)T(P_M) = &-k\sum_{i=1}^{N_s}\int_{\Delta s_i}\left(k\frac{\partial G(P_M,Q)}{\partial n_Q}\right. \\
&\left.+\rho c G(P_M,Q)\big(V_i^{(v)}(Q)+V_i^{(v)}(Q)\big)\right)(\Psi_1T_1+\Psi_2T_2)\,ds_Q \\
&+k\sum_{i=1}^{N_s}\int_{\Delta s_i} T(P_M,Q)(\Psi_1q_1^{(n)}+\Psi_2q_2^{(n)})\,ds_Q \\
&+\rho c\sum_{j=1}^{N_d}\int_{\Delta\Omega_j}\left(G(P_M,q)V_{l,l}^{(v)}(q)+\frac{\partial G(P_M,q)}{\partial x_l}v_l^{(v)}(q)\right) \\
&\times(\Phi_1T_{j1}+\Phi_2T_{j2}+\Phi_3T_{j3})\,d\Omega
\end{aligned}
\tag{3.42}
$$

where the shape functions for the boundary elements are

$$\Psi_1 = \tfrac{1}{2}(1 - \eta) \tag{3.43a}$$

and

$$\Psi_2 = \tfrac{1}{2}(1 + \eta) \tag{3.43b}$$

Here η is a dimensionless local coordinate over individual boundary segments. T_1, T_2, $q_1^{(n)}$, and $q_2^{(n)}$ are nodal quantities of the ith boundary element.

The shape functions over the triangular elements are

$$\Phi_1 = \frac{\dfrac{1.5}{A} - \dfrac{A\mu}{2.0} - 1.5\nu}{1.5A} \tag{3.44a}$$

$$\Phi_2 = \frac{\dfrac{1.5}{A} + A\mu}{1.5A} \tag{3.44b}$$

and

$$\Phi_3 = \frac{\dfrac{1.5}{A} - \dfrac{A\mu}{2.0} + 1.5\nu}{1.5A} \tag{3.44c}$$

where $A = \sqrt{3.0}$; μ and ν are dimensionless local coordinates over individual triangular elements; and T_{j1}, T_{j2}, and T_{j3} are nodal temperatures of the jth triangular internal cell. Defining coefficient matrices at the element level as

$$\begin{aligned} a^{\gamma}_{M,j} = C_M \delta_{M,j} + \int_{\Delta s_j} \Psi_\gamma \bigg(k \frac{\partial G(P_M, Q)}{\partial n_Q} \\ + \rho c G(P_M, Q)(V_l^{(c)}(Q) + V_l^{(v)}(Q)) n_l \bigg) \, ds_Q \end{aligned} \tag{3.45}$$

$$b^{\gamma}_{M,j} = -k \int_{\Delta s_j} \Psi_\gamma G(P_M, Q) \, ds_Q \tag{3.46}$$

and

$$d^{\gamma}_{M,j} = \rho c \int_{\Delta \Omega_i} \Phi_\gamma \bigg(G(P_M, Q) V_{l,l}^{(v)} + \frac{\partial G(P_M, q)}{\partial x_l} V_l^{(v)} \bigg) \, d\Omega \tag{3.47}$$

equation (3.42) may now be expressed as

$$\sum_{j=1}^{N_b} A_{ij} T_j = \sum_{j=1}^{N_b} B_{ij} q_j^{(n)} + \sum_{l=1}^{N_t} D_{il} T_l \tag{3.48a}$$

where N_b is the total number of boundary nodes and N_t is the total of boundary and internal nodes. Each nodal coefficient A_{ij} is equal to the sum of a_{ij}^2 of element $(j-1)$ and $a_{ij}^{\prime}$ of element (j) for an anticlockwise node numbering system. The same procedure also applies to B_{ij}. For D_{ij}, contributions from all internal cells are added at a common internal point.

Equation (3.48*a*) may also be written as

$$\sum_{j=1}^{N_b} (A_{ij} - D_{ij}) T_j = \sum_{j=1}^{N_b} B_{ij} q_j^{(n)} + \sum_{l=N_b+1}^{N_t} D_{il} T_l \tag{3.48b}$$

or

$$\sum_{j=1}^{N_b} \widetilde{A}_{ij} T_j = \sum_{j=1}^{N_b} B_{ij} q_j^{(n)} + \sum_{l=N_b+1}^{N_t} \widetilde{D}_{il} T_l \tag{3.48c}$$

For the case involving only constant convective velocity, the domain integrals and the resulting $D_{il} T_l$ term vanish. Accordingly, a pure boundary representation

$$\sum_{j=1}^{N_b} A_{ij} T_j = \sum_{j=1}^{N_b} B_{ij} a_j^{(n)} \tag{3.49}$$

may be obtained. At each location over the entire boundary of the domain, either T or $q^{(n)}$ or a combination of T and $q^{(n)}$ is prescribed for a well-posed problem. Accordingly, equation (3.49) may be rearranged as

$$\sum_{j=1}^{N_b} \widetilde{A}_{ij} Y_j^{(u)} = \sum_{j=1}^{N_b} \widetilde{B}_{ij} Y_j^{(k)} \tag{3.50}$$

where the matrix coefficients $\widetilde{A}_{ij}$ and $\widetilde{B}_{ij}$ are

$$\widetilde{A}_{ij} = \begin{cases} A_{ij}, & \text{for } q_j^{(n)} \text{ specified} \\ B_{ij}, & \text{for } T_j \text{ specified} \\ A_{ij} + B_{ij} h, & \text{for convective heat loss} \end{cases} \tag{3.51}$$

and

$$\widetilde{B}_{ij} = \begin{cases} B_{ij}, & \text{for } q_j^{(n)} \text{ specified} \\ A_{ij}, & \text{for } T_j \text{ specified} \\ -B_{ij}h, & \text{for convective heat loss} \end{cases} \tag{3.52}$$

and the column vectors $Y_j^{(u)}$ and $Y_j^{(k)}$ are

$$Y_j^{(u)} = \begin{cases} T_j, & \text{for } q_j^{(n)} \text{ specified or convective heat loss} \\ q_j^{(n)}, & \text{for } T_j \text{ specified} \end{cases} \tag{3.53}$$

$$Y_j^{(k)} = \begin{cases} q_j^{(n)}, & \text{for } q_j^{(n)} \text{ specified} \\ T_j, & \text{for } T_j \text{ specified} \\ T_\infty, & \text{for convective heat loss} \end{cases} \tag{3.54}$$

Equation (3.50) can now be solved for $Y_j^{(u)}$ to obtain the unknown temperature and fluxes at the appropriate nodes. Once T and $q^{(n)}$ have been obtained over the entire boundary, the internal equation (3.37) may be used as an algebraic expression to determine temperature and flux at any desired internal point. It is important to note here that determination of temperature and flux at an internal point requires only algebraic evaluation once temperature and flux are known at every boundary node. For conduction–convection problems involving variable velocity fields, the existence of the domain integrals involving temperature at internal points prohibits direct solution of equation (3.50). For a well-posed problem, either T or $q^{(n)}$ or a combination of T and $q^{(n)}$ is known at every node on the boundary. Accordingly, equation (3.50) for the variable velocity case may also be rearranged as

$$\sum_{j=1}^{N_b} \widetilde{A}_{ij} Y_j^{(u)} = \sum_{j=1}^{N_b} \widetilde{B}_{ij} Y_j^{(k)} + \sum_{l=N_b+1}^{N_t} \widetilde{D}_{mi} T_i \tag{3.55}$$

after appropriate regrouping of the known and unknown quantities on the boundary has been carried out. Equation (3.55), however, involves $Y_j^{(u)}$ and the unknown temperatures T_l at internal points. This may be solved using an iterative scheme that places degrees of freedom (d.o.f.) on the boundary nodes only or a noniterative direct solution scheme where the temperatures at internal points are included as d.o.f.

Using the noniterative scheme, we may write

$$\sum_{j=1}^{N_b} \widetilde{A}_{mj} Y_j^{(u)} = \sum_{j=1}^{N_b} \widetilde{B}_{mj} Y_j^{(k)} + \sum_{l=N_b+1}^{N_t} \widetilde{D}_{ml} T_l \tag{3.56}$$

where $m = N_b + 1$ to N_t, by placing source points on the internal points. Equations (3.55) and (3.56) are now solved simultaneously to obtain $Y_j^{(u)}$ and internal temperature T_l. The noniterative scheme increases the number of d.o.f. Thus, it requires handling of a bigger coefficient matrix.

Alternatively, one can use an iterative scheme. A guess value of T_l is assumed and equation (3.55) is solved with that guess value to obtain an estimate for $Y_j^{(u)}$. One T and $q^{(n)}$ are obtained everywhere on the boundary, and the internal equation (3.37) is used to calculate T_l at interior points. The newly found values of T_l are now used in equation (3.55) to obtain a new estimate for $Y_j^{(u)}$, which in turn is used again in the internal equation to update the values of T_l. The process is repeated until the values of temperatures and fluxes at all the nodal points converge. In the iterative scheme, the d.o.f. are limited to boundary nodes only, so the boundary nature of the problem along with its associated advantages are retained. However, iterations are introduced and the rate of convergence of these iterations needs to be investigated.

3.2.3 *Evaluation of Singular Integrals*

Integrals of kernels over linear boundary elements and triangular internal cells in equation (3.48) need to be evaluated very carefully (Cristescu and Lougignac 1978, Pima and Fernandes 1981). For two-dimensional problems, $G(P_M, Q)$ is $\log(r)$ singular and may be evaluated analytically or numerically through log-weighted Gaussian quadrature. The kernel $\partial G(P_M, Q)/\partial n_Q$ is $(1/r)$ singular. Using an indirect approach and requiring that the governing equation satisfy isothermal boundary conditions over the entire boundary, we get

$$\sum_{j=1}^{N_b} A_{lj} = \sum_{i=1}^{N_t} D_{li} \tag{3.57}$$

where, without any loss of generality, the resulting constant-temperature solution is taken to be unity. Equation (3.57) may be rearranged as

$$A_{ll} - D_{ll} = \sum_{\substack{i=1 \\ i \neq l}}^{N_t} D_{li} - \sum_{\substack{j=1 \\ j \neq l}}^{N_b} A_{lj} \tag{3.58}$$

and $(A_{ll} - D_{ll})$ representing the desired combination of the diagonal elements of A_{M_j} (including $C(P_M)$) and D_{M_i} can be evaluated in terms of the off-diagonal elements.

3.2.4 Numerical Results and Verification

Numerical results for example problems involving spatially varying convective velocities are presented in this section.

An entrance length problem is considered here. A schematic diagram of the problem geometry along with boundary conditions is shown in figure 3.2. Three different mesh sizes (Gupta et al. 1994) are considered. Mesh A (as shown in figure 3.2a) is a spatially uniform coarse grid. For this problem, steep temperature gradients are expected near $x = 1$. Accordingly, meshes B and C (figures 3.2b,c) are progressively refined near $x = 1$.

Various researchers (e.g., Skerget and Brebbia 1983, Tanaka et al. 1986) have attempted to solve conduction–convection problems using the fundamental solution for the Laplace equation. As observed by Gupta et al. (1994), such formulations produce significant inaccuracies with the increase in convective velocity. To remedy this problem, Gupta et al. (1994) have developed an alternate BEM formulation based on the fundamental solutions for a constant-velocity moving heat source problem, and such an approach is pursued here.

Effect of different fundamental solutions. Accordingly, comparisons between the solutions obtained from these different approaches are investigated first. Two cases, one with constant velocity and the other with a variable velocity are considered. For both cases, two different mesh sizes A and C (as shown in figures 3.2a and 3.2c) are considered to investigate the mesh dependency of the solution. Tables 3.1a and 3.1b show, respectively, the temperatures at $y = 0$ for constant velocity cases at $Pe = 20$ and $Pe = 60$. There also exists an analytical solution (Chan and Chandra 1991a,b,c) for this problem with constant convective velocity. It may be observed from tables 3.1a and 3.1b that BEM results obtained with the moving heat source fundamental solution agree very well with the analytical solutions. The maximum discrepancy between them is 0.32% for the coarse mesh and it reduces to 0.07% for the fine mesh. On the other hand, the BEM results using the fundamental solution corresponding to the Laplace equation do not fare so well. This is evident in relatively larger errors compared to the analytical solution. Further, spurious negative temperatures in the BEM results may be observed at various locations. For a given mesh, the discrepancy from the analytical solution also increases with increasing Peclet number.

A spatially varying velocity profile is considered next. The spatial variation of the convective velocity in the x direction is represented as

$$V_1(y) = 6y(1 - y) \tag{3.59}$$

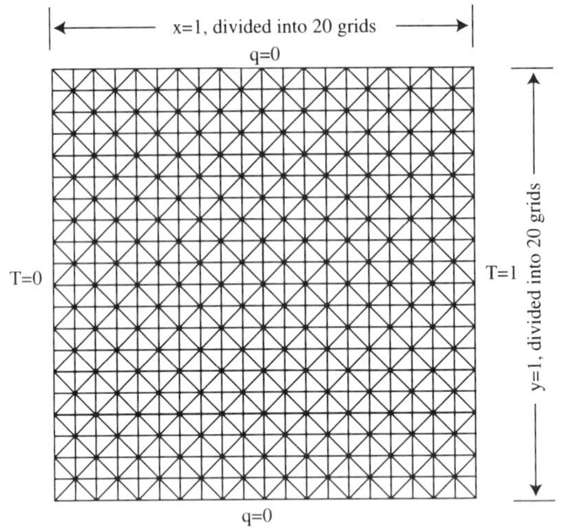

(a)

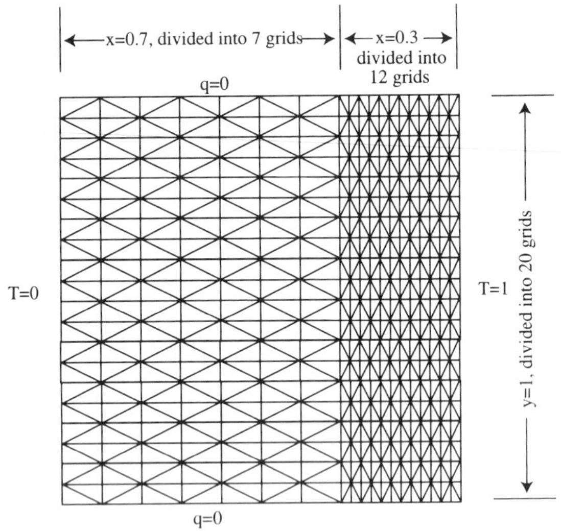

(b)

Figure 3.2. (a) Mesh A (coarse mesh), (b) Mesh B (intermediate mesh), (c) Mesh C (fine mesh).

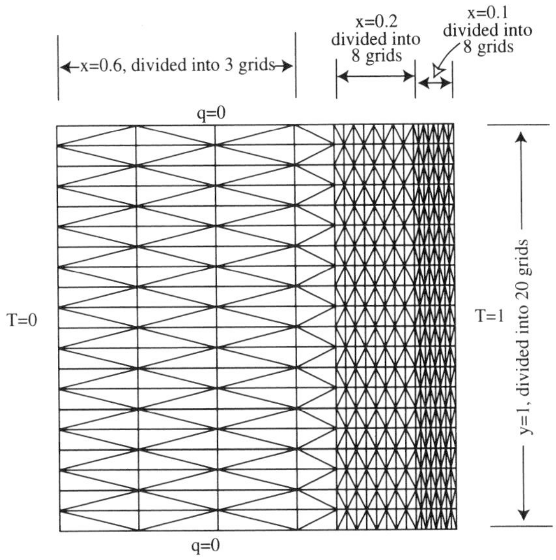

(c)

Figure 3.2—*continued*

Table 3.1a. Comparisons of BEM results with analytical solutions for $Pe = 20$

	Moving heat source		$y = 0.0$ Laplace equation		Analytical
x:	Mesh size A	Mesh size C	Mesh size A	Mesh size C	
0.00	0.0	0.0	0.0	0.0	0.0
0.10	$0.131\,07 \times 10^{-7}$		$0.128\,08 \times 10^{-3}$		$0.131\,69 \times 10^{-7}$
0.20	$0.110\,28 \times 10^{-6}$	$0.109\,38 \times 10^{-6}$	$0.129\,32 \times 10^{-3}$	$0.180\,43 \times 10^{-3}$	$0.220\,40 \times 10^{-6}$
0.30	$0.828\,17 \times 10^{-6}$		$0.974\,60 \times 10^{-4}$		$0.829\,40 \times 10^{-6}$
0.40	$0.613\,11 \times 10^{-5}$	$0.614\,65 \times 10^{-5}$	$0.306\,68 \times 10^{-4}$	$0.975\,62 \times 10^{-4}$	$0.614\,22 \times 10^{-5}$
0.50	$0.453\,10 \times 10^{-4}$		$-0.843\,35 \times 10^{-4}$		$0.453\,98 \times 10^{-4}$
0.60	$0.334\,12 \times 10^{-3}$	$0.335\,72 \times 10^{-3}$	$-0.209\,42 \times 10^{-4}$	-0.53746×10^{-3}	$0.335\,46 \times 10^{-3}$
0.70	$0.247\,25 \times 10^{-2}$	$0.248\,06 \times 10^{-2}$	$0.431\,28 \times 10^{-3}$	$0.211\,40 \times 10^{-2}$	$0.247\,88 \times 10^{-2}$
0.75	$0.671\,90 \times 10^{-2}$	$0.674\,30 \times 10^{-2}$	$0.263\,99 \times 10^{-2}$	$0.596\,70 \times 10^{-2}$	$0.673\,79 \times 10^{-2}$
0.80	$0.182\,61 \times 10^{-1}$	$0.183\,29 \times 10^{-1}$	$0.100\,66 \times 10^{-1}$	$0.167\,93 \times 10^{-1}$	$0.183\,16 \times 10^{-1}$
0.85	$0.496\,26 \times 10^{-1}$	$0.498\,23 \times 10^{-1}$	$0.336\,32 \times 10^{-1}$	$0.470\,57 \times 10^{-1}$	$0.497\,87 \times 10^{-1}$
0.90	0.134 87	0.135 43	0.106 34	0.132 22	0.135 34
0.9125		0.173 90		0.170 56	0.173 77
0.9250		0.223 29		0.219 91	0.223 13
0.9375		0.286 71		0.283 37	0.286 50
0.9500	0.366 70	0.368 15	0.327 63	0.365 08	0.367 88
0.9625		0.472 73		0.470 20	0.472 37
0.9750		0.607 02		0.605 10	0.606 53
0.9875		0.779 46		0.777 36	0.778 80
1.0	1.0	1.0	1.0	1.0	1.0

Table 3.1b. Comparisons of BEM results with analytical solutions for $Pe = 60$

$y = 0.0$	Moving heat source		Laplace equation		Analytical
x:	Mesh size A	Mesh size C	Mesh size A	Mesh size C	
0.00	0.0	0.0	0.0	0.0	0.0
0.10	-0.223×10^{-13}		0.29853×10^{-3}		0.3532×10^{-23}
0.20	0.1526×10^{-13}	-0.147×10^{-13}	0.31626×10^{-3}	0.38674×10^{-3}	0.1425×10^{-20}
0.30	-0.311×10^{-13}		0.27699×10^{-3}		0.5749×10^{-18}
0.40	-0.849×10^{-14}	-0.161×10^{-13}	0.19094×10^{-3}	0.41932×10^{-3}	0.2319×10^{-14}
0.50	0.7244×10^{-13}		0.34802×10^{-4}		0.9357×10^{-13}
0.60	0.3751×10^{-10}	0.3776×10^{-10}	-0.2472×10^{-3}	0.33998×10^{-3}	0.3775×10^{-10}
0.70	0.15120×10^{-7}	0.15237×10^{-7}	-0.7554×10^{-3}	0.20554×10^{-3}	0.15230×10^{-7}
0.75	0.30355×10^{-6}	0.30604×10^{-6}	-0.2661×10^{-2}	0.96979×10^{-4}	0.30590×10^{-6}
0.80	0.60939×10^{-5}	0.61466×10^{-5}	-0.6094×10^{-3}	-0.8765×10^{-4}	0.61442×10^{-5}
0.85	0.12234×10^{-3}	0.12346×10^{-3}	-0.1195×10^{-1}	-0.4217×10^{-3}	0.12341×10^{-3}
0.90	0.24567×10^{-2}	0.24796×10^{-2}	-0.3317×10^{-1}	-0.6934×10^{-4}	0.24788×10^{-2}
0.9125		0.52494×10^{-2}		0.12786×10^{-2}	0.52475×10^{-2}
0.9250		0.11113×10^{-1}		0.49655×10^{-2}	0.11109×10^{-1}
0.9375		0.23527×10^{-1}		0.14250×10^{-1}	0.23518×10^{-1}
0.9500	0.49399×10^{-1}	0.49808×10^{-1}	-0.22345	0.36677×10^{-1}	0.49787×10^{-1}
0.9625		0.10545		0.88886×10^{-1}	0.10540
0.9750		0.22326		0.20529	0.22313
0.9875		0.47268		0.45104	0.47237
1.0	1.0	1.0	1.0	1.0	1.0

while the convective velocity in the y direction is considered to be zero. The above velocity field can be decomposed into a constant part of magnitude 1.5 and a variable part $V_1^{(v)} = 6y(1 - y) - 1.5$. The BEM results are presented in table 3.2. It is observed again that, compared to the moving heat source BEM results, the BEM formulation based on the Green's function for the Laplace equation produces large errors and the errors increase with increasing Peclet number. Both iterative and noniterative schemes produced very similar results when moving heat source fundamental solutions are used. For the BEM formulation using moving heat source fundamental solution, the iterative scheme converged in fewer than 32 iterations without any relaxation. However, when the fundamental solution from the Laplace equation was used, the iterative scheme did not converge without any relaxation. Even with a relaxation factor of 0.05, the solution did not converge in 500 iterations. Consequently, the results presented in table 3.2 for the Laplace equation Green's function are obtained by using the noniterative method.

Comparison between iterative and noniterative schemes. In the present work both iterative and noniterative schemes are used to obtain solutions for

Table 3.2. Comparisons of BEM results for variable velocity cases

$x = 0.925$	Moving heat source		Laplace equation	
y:	$Pe = 60$	$Pe = 100$	$Pe = 60$	$Pe = 100$
0.00	0.272 20	0.138 47	0.307 00	0.204 82
0.05	0.212 38	$0.890\,98 \times 10^{-1}$	0.236 55	0.127 36
0.10	0.121 41	$0.357\,49 \times 10^{-1}$	0.126 54	$0.376\,87 \times 10^{-1}$
0.15	$0.569\,44 \times 10^{-1}$	$0.104\,87 \times 10^{-1}$	$0.518\,76 \times 10^{-1}$	$0.457\,98 \times 10^{-2}$
0.20	$0.241\,23 \times 10^{-1}$	$0.244\,04 \times 10^{-2}$	$0.167\,22 \times 10^{-1}$	-0.4950×10^{-3}
0.25	$0.101\,39 \times 10^{-1}$	$0.571\,25 \times 10^{-3}$	$0.437\,65 \times 10^{-2}$	-0.4797×10^{-3}
0.30	$0.467\,09 \times 10^{-2}$	$0.145\,99 \times 10^{-3}$	$0.841\,43 \times 10^{-3}$	-0.1642×10^{-3}
0.35	$0.256\,36 \times 10^{-2}$	$0.528\,45 \times 10^{-4}$	-0.1938×10^{-3}	-0.1301×10^{-3}
0.40	$0.168\,33 \times 10^{-2}$	$0.247\,33 \times 10^{-4}$	-0.4319×10^{-3}	-0.5294×10^{-3}
0.45	$0.131\,84 \times 10^{-1}$	$0.158\,11 \times 10^{-4}$	-0.5461×10^{-3}	-0.6157×10^{-4}
0.50	$0.121\,57 \times 10^{-2}$	$0.135\,38 \times 10^{-4}$	-0.5435×10^{-3}	-0.2895×10^{-4}
0.55	$0.131\,81 \times 10^{-2}$	$0.158\,10 \times 10^{-4}$	-0.5508×10^{-3}	-0.6033×10^{-4}
0.60	$0.168\,33 \times 10^{-2}$	$0.247\,33 \times 10^{-4}$	-0.4249×10^{-3}	-0.5437×10^{-4}
0.65	$0.256\,36 \times 10^{-2}$	$0.528\,44 \times 10^{-2}$	-0.1872×10^{-3}	-0.1313×10^{-3}
0.70	$0.467\,08 \times 10^{-2}$	$0.145\,99 \times 10^{-3}$	$0.835\,43 \times 10^{-3}$	-0.1637×10^{-3}
0.75	$0.101\,39 \times 10^{-1}$	$0.571\,25 \times 10^{-3}$	$0.437\,42 \times 10^{-2}$	-0.4795×10^{-3}
0.80	$0.241\,23 \times 10^{-1}$	$0.244\,04 \times 10^{-2}$	$0.167\,22 \times 10^{-1}$	-0.4950×10^{-3}
0.85	$0.569\,44 \times 10^{-1}$	$0.104\,87 \times 10^{-1}$	$0.518\,76 \times 10^{-1}$	$0.457\,98 \times 10^{-2}$
0.90	0.121 419	$0.357\,40 \times 10^{-1}$	0.126 54	$0.376\,87 \times 10^{-2}$
0.95	0.212 38	$0.890\,97 \times 10^{-1}$	0.236 55	0.127 36
1.0	0.272 20	0.138 47	0.307 00	0.204 82

steady-state conduction–convection equations with variable velocities. A comparison of the results obtained from both schemes is presented here. The results obtained for the boundary as well as internal nodes are compared. It has been observed that both iterative and noniterative schemes produce very similar results. Temperatures for the boundary nodes and a few representative internal nodes (for Peclet number 60 and 100) are presented, respectively, in tables 3.3a and 3.3b. No relaxation is used in the iterative scheme and it can be observed that results compare extremely well for both the methods. Both methods took about the same computational time for the above-mentioned problems.

Effect of mesh size. Temperatures on the lower boundary, $y = 0$, are presented for all three different meshes (as shown in figures 3.2a,b,c) for Peclet numbers of 20, 40, 60, and 100 in tables (3.4a–d), respectively, and are presented graphically in figures (3.3a–d), respectively. The temperatures at $x = 0.95$ for the above Peclet numbers are presented in tables (3.5a–d) and are graphically presented in figures (3.4a–d) respectively. As can be seen, a finer mesh is required near $x = 1$ to capture the steep temperature gradients,

Table 3.3a. Comparison between iterative and noniterative schemes for boundary nodes at $y = 0$

$y = 0.0$	Iterative scheme		Noniterative scheme	
x:	$Pe = 40$	$Pe = 100$	$Pe = 40$	$Pe = 100$
0.0	0.0	0.0	0.0	0.0
0.40	$0.670\,44 \times 10^{-3}$	$0.128\,10 \times 10^{-4}$	$0.670\,48 \times 10^{-3}$	$0.128\,87 \times 10^{-4}$
0.70	$0.123\,13 \times 10^{-1}$	$0.624\,15 \times 10^{-3}$	$0.123\,13 \times 10^{-1}$	$0.623\,84 \times 10^{-3}$
0.750	$0.276\,72 \times 10^{-1}$	$0.212\,04 \times 10^{-2}$	$0.276\,72 \times 10^{-1}$	$0.211\,96 \times 10^{-2}$
0.80	$0.595\,25 \times 10^{-1}$	$0.683\,52 \times 10^{-2}$	$0.595\,25 \times 10^{-1}$	$0.683\,42 \times 10^{-2}$
0.850	0.126 03	$0.218\,01 \times 10^{-1}$	0.126 03	$0.218\,00 \times 10^{-1}$
0.90	0.260 77	$0.688\,82 \times 10^{-1}$	0.260 77	$0.688\,81 \times 10^{-1}$
0.9500	0.538 79	0.285 00	0.538 79	0.285 00
1.0	1.0	1.0	1.0	1.0

Table 3.3b. Comparison between iterative and noniterative schemes for internal nodes at $x = 0.975$

$x = 0.975$	Iterative scheme		Noniterative scheme	
y:	$Pe = 40$	$Pe = 100$	$Pe = 40$	$Pe = 100$
0.00	0.750 20	0.561 74	0.750 20	0.561 74
0.15	0.493 51	0.168 69	0.493 51	0.168 69
0.30	0.292 29	$0.478\,86 \times 10^{-1}$	0.292 29	$0.478\,86 \times 10^{-1}$
0.45	0.228 50	$0.248\,48 \times 10^{-1}$	0.228 51	$0.248\,48 \times 10^{-1}$
0.60	0.239 89	$0.478\,86 \times 10^{-1}$	0.239 86	$0.285\,54 \times 10^{-1}$
0.75	0.339 59	0.686 63	0.339 62	$0.686\,63 \times 10^{-1}$
0.90	0.596 64	0.283 79	0.596 64	0.283 79
1.0	0.750 19	0.561 73	0.750 20	0.561 73

particularly for high Peclet numbers. It is also observed that the mesh can be made coarser near the left-hand boundary without any effect on the temperature predictions. This reduces the computation time considerably. Also, it was observed during the course of present study that mesh refinement in the y-direction did not have any considerable effect on the solutions. This may be due to the fact that the temperature variation in the y-direction is significantly smaller than that in the x-direction.

Effect of decomposition level. Due to spatially varying convective velocity, the temperature appears in the domain integral. This requires special care. In the present work, an arbitrary velocity field is decomposed into a constant part and a variable part. The effect of the constant convective velocity is introduced in the BEM formulation through the moving heat source

Table 3.4. Effect of grid size at boundary nodes ($y = 0$) for $Pe = 20$, 40, 60, and 100

x:	Mesh A	Mesh B	Mesh C
		(a) $Pe = 20$	
0.0000	0.0	0.0	0.0
0.2000	$0.357\,95 \times 10^{-3}$	$0.498\,87 \times 10^{-3}$	$0.100\,76 \times 10^{-2}$
0.4000	$0.292\,01 \times 10^{-2}$	$0.341\,98 \times 10^{-2}$	$0.475\,68 \times 10^{-2}$
0.6000	$0.224\,74 \times 10^{-1}$	$0.219\,25 \times 10^{-1}$	$0.207\,18 \times 10^{-1}$
0.7000	$0.618\,87 \times 10^{-1}$	$0.551\,78 \times 10^{-1}$	$0.537\,85 \times 10^{-1}$
0.7250	–	$0.724\,91 \times 10^{-1}$	$0.709\,64 \times 10^{-1}$
0.7500	0.102 26	$0.943\,98 \times 10^{-1}$	$0.926\,69 \times 10^{-1}$
0.7750	–	0.122 57	0.120 56
0.8000	0.168 02	0.158 68	0.156 28
0.8250	–	0.204 75	0.201 84
0.8500	0.273 33	0.263 17	0.259 61
0.8750	–	0.336 66	0.332 24
0.9000	0.437 08	0.428 04	0.422 03
0.9125	–	–	0.475 03
0.9250	–	0.539 92	0.533 97
0.9375	–	–	0.598 81
0.9500	0.678 62	0.673 90	0.669 43
0.9625	–	–	0.745 62
0.9750	–	0.829 45	0.827 02
0.9875	–	–	0.912 86
1.0	1.0	1.0	1.0
		(b) $Pe = 40$	
0.0000	0.0	0.0	0.0
0.2000	$0.875\,41 \times 10^{-5}$	$0.363\,11 \times 10^{-4}$	$0.117\,03 \times 10^{-3}$
0.4000	$0.306\,73 \times 10^{-3}$	$0.394\,50 \times 10^{-3}$	$0.670\,44 \times 10^{-3}$
0.6000	$0.499\,02 \times 10^{-2}$	$0.413\,63 \times 10^{-2}$	$0.362\,14 \times 10^{-2}$
0.7000	$0.201\,00 \times 10^{-1}$	$0.134\,33 \times 10^{-1}$	$0.123\,13 \times 10^{-1}$
0.7250	–	$0.203\,43 \times 10^{-1}$	$0.187\,47 \times 10^{-1}$
0.7500	$0.402\,34 \times 10^{-1}$	$0.299\,35 \times 10^{-1}$	$0.276\,72 \times 10^{-1}$
0.7750	–	0.4918×10^{-1}	$0.406\,77 \times 10^{-1}$
0.8000	$0.802\,42 \times 10^{-1}$	$0.641\,87 \times 10^{-1}$	$0.595\,25 \times 10^{-1}$
0.8250	–	$0.934\,93 \times 10^{-1}$	$0.867\,83 \times 10^{-1}$
0.8500	0.158 61	0.135 67	0.126 03
0.8750	–	0.195 91	0.182 10
0.9000	0.307 89	0.280 91	0.260 77
0.9125	–	–	0.313 44
0.9250	–	0.398 62	0.376 87
0.9375	–	–	0.451 79
0.9500	0.576 01	0.556 94	0.538 79
0.9625	–	–	0.638 32
0.9750	–	0.760 32	0.750 20
0.9875	–	–	0.873 11
1.0	1.0	1.0	1.0

Table 3.4—*continued*

x:	Mesh A	Mesh B	Mesh C
		(c) *Pe* = 60	
0.0000	0.0	0.0	0.0
0.2000	$0.287\ 24 \times 10^{-5}$	$0.628\ 67 \times 10^{-5}$	$0.222\ 52 \times 10^{-4}$
0.4000	$0.779\ 79 \times 10^{-4}$	$0.907\ 75 \times 10^{-4}$	$0.150\ 61 \times 10^{-3}$
0.6000	$0.200\ 32 \times 10^{-2}$	$0.126\ 55 \times 10^{-2}$	$0.965\ 13 \times 10^{-1}$
0.7000	$0.101\ 33 \times 10^{-1}$	$0.475\ 56 \times 10^{-2}$	$0.386\ 54 \times 10^{-1}$
0.7250	–	$0.802\ 37 \times 10^{-2}$	$0.654\ 89 \times 10^{-1}$
0.7500	$0.227\ 58 \times 10^{-1}$	$0.128\ 33 \times 10^{-1}$	$0.104\ 98 \times 10^{-1}$
0.7750	–	$0.205\ 46 \times 10^{-1}$	$0.168\ 32 \times 10^{-1}$
0.8000	$0.509\ 36 \times 10^{-1}$	$0.327\ 54 \times 10^{-1}$	$0.268\ 59 \times 10^{-1}$
0.8250	–	$0.520\ 46 \times 10^{-1}$	$0.427\ 16 \times 10^{-1}$
0.8500	0.113 11	$0.824\ 31 \times 10^{-1}$	$0.677\ 19 \times 10^{-1}$
0.8750	–	0.129 99	0.106 92
0.9000	0.246 79	0.203 62	0.167 45
0.9125	–	–	0.213 01
0.9250	–	0.315 53	0.272 20
0.9375	–	–	0.347 09
0.9500	0.517 94	0.480 34	0.439 74
0.9625	–	–	0.552 16
0.9750	–	0.710 42	0.685 42
0.9875	–	–	0.838 37
1.0	1.0	1.0	1.0
		(d) *Pe* = 100	
0.0000	0.0	0.0	0.0
0.2000	$0.381\ 55 \times 10^{-6}$	$0.535\ 44 \times 10^{-6}$	$0.135\ 05 \times 10^{-5}$
0.4000	$0.174\ 77 \times 10^{-4}$	$0.118\ 44 \times 10^{-4}$	$0.128\ 10 \times 10^{-4}$
0.6000	$0.742\ 18 \times 10^{-3}$	$0.252\ 10 \times 10^{-3}$	$0.115\ 42 \times 10^{-3}$
0.7000	$0.480\ 12 \times 10^{-2}$	$0.116\ 67 \times 10^{-2}$	$0.624\ 15 \times 10^{-3}$
0.7250	–	$0.226\ 71 \times 10^{-2}$	$0.121\ 00 \times 10^{-2}$
0.7500	$0.121\ 99 \times 10^{-1}$	$0.397\ 50 \times 10^{-2}$	$0.212\ 04 \times 10^{-2}$
0.7750	–	$0.715\ 47 \times 10^{-2}$	$0.381\ 50 \times 10^{-2}$
0.8000	$0.309\ 16 \times 10^{-1}$	$0.128\ 19 \times 10^{-1}$	$0.683\ 52 \times 10^{-2}$
0.8250	–	$0.229\ 05 \times 10^{-1}$	$0.122\ 19 \times 10^{-1}$
0.8500	$0.778\ 48 \times 10^{-1}$	$0.408\ 17 \times 10^{-1}$	$0.218\ 01 \times 10^{-1}$
0.8750	–	$0.724\ 93 \times 10^{-1}$	$0.388\ 22 \times 10^{-1}$
0.9000	0.192 91	0.128 06	$0.688\ 82 \times 10^{-1}$
0.9125	–	–	$0.966\ 07 \times 10^{-1}$
0.9250	–	0.224 01	0.138 47
0.9375	–	–	0.199 23
0.9500	0.459 81	0.384 89	0.285 00
0.9625	–	–	0.403 29
0.9750	–	0.639 82	0.561 74
0.9875	–	–	0.764 97
1.0	1.0	1.0	1.0

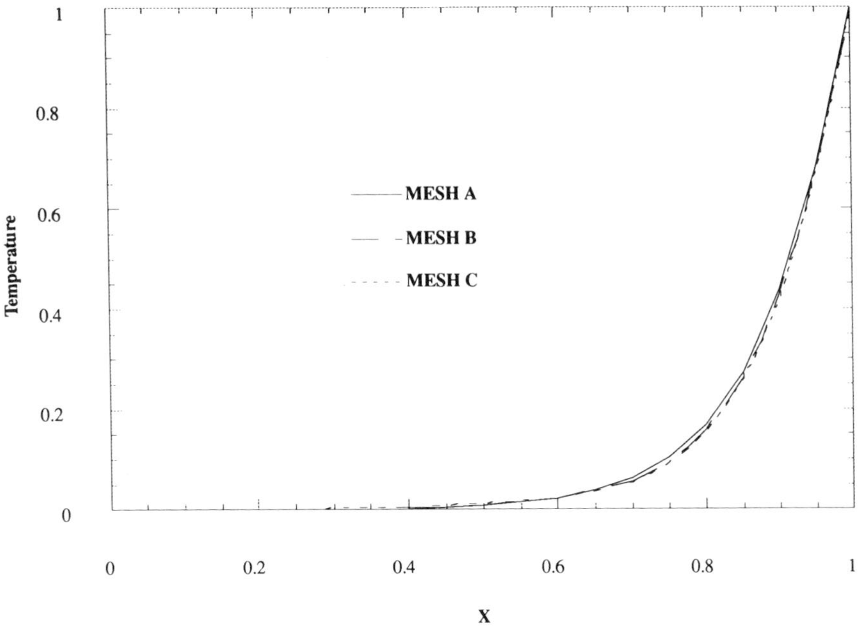

(a)

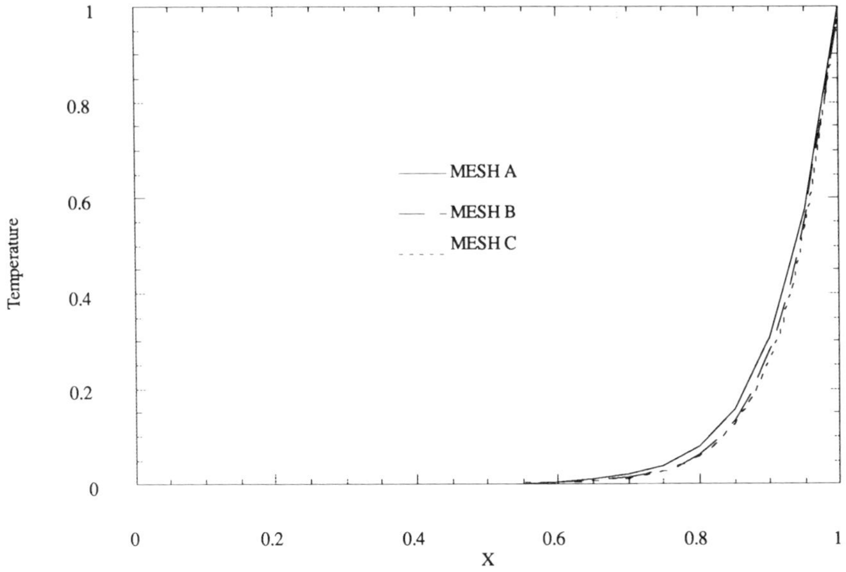

(b)

Figure 3.3. Effect of grid size at boundary nodes ($y = 0$): (a) Peclet number = 20, (b) Peclet number = 40, (c) Peclet number = 60, (d) Peclet number = 100.

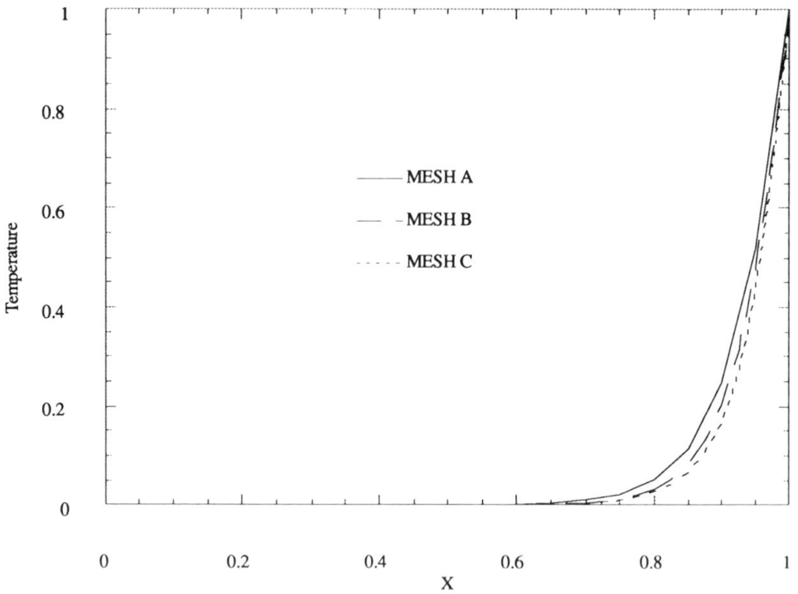

(c)

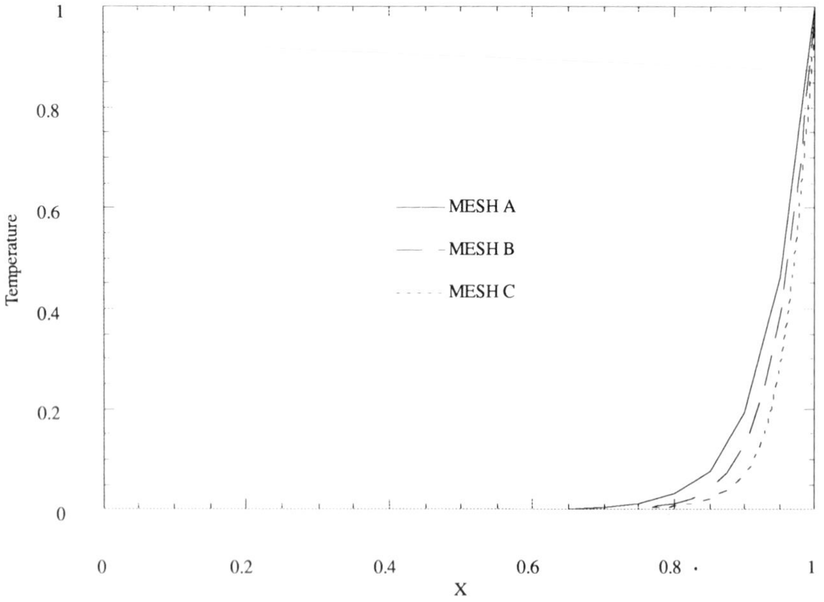

(d)

Figure 3.3—*continued*

Table 3.5. Effect of grid size at internal nodes ($x = 0.95$) for $Pe = 20$, 40, 60, and 100

y:	Mesh A	Mesh B	Mesh C
		(a) $Pe = 20$	
0.00	0.67862	0.67390	0.66943
0.05	0.65020	0.64253	0.63723
0.10	0.58395	0.57469	0.56970
0.15	0.50782	0.49734	0.49288
0.20	0.43472	0.42403	0.42000
0.25	0.37034	0.36092	0.35794
0.30	0.31890	0.31149	0.30929
0.35	0.28007	0.27512	0.27385
0.40	0.25372	0.25088	0.25022
0.45	0.23841	0.23703	0.23680
0.50	0.23343	0.23256	0.23247
0.55	0.23841	0.23703	0.23680
0.60	0.25372	0.25088	0.25022
0.65	0.28007	0.27512	0.27385
0.70	0.31890	0.31149	0.30929
0.75	0.37034	0.36092	0.35794
0.80	0.43472	0.42403	0.42000
0.85	0.50782	0.49734	0.49288
0.90	0.58395	0.57469	0.56970
0.95	0.65020	0.64253	0.63723
1.00	0.67862	0.67390	0.66943
		(b) $Pe = 40$	
0.00	0.57601	0.55694	0.53879
0.05	0.52607	0.49717	0.47682
0.10	0.41623	0.38207	0.36305
0.15	0.30922	0.27102	0.25539
0.20	0.22173	0.18733	0.17492
0.25	0.15752	0.12946	0.12143
0.30	0.11304	0.93840×10^{-1}	0.88679×10^{-1}
0.35	0.83730	0.71981×10^{-1}	0.69329×10^{-1}
0.40	0.65104	0.59531×10^{-1}	0.58286×10^{-1}
0.45	0.54890	0.52990×10^{-1}	0.52611×10^{-1}
0.50	0.51606	0.50982×10^{-1}	0.50860×10^{-1}
0.55	0.54890	0.52990×10^{-1}	0.52611×10^{-1}
0.60	0.65104	0.59531×10^{-1}	0.58286×10^{-1}
0.65	0.83730	0.71891×10^{-1}	0.69329×10^{-1}
0.70	0.11304	0.93840×10^{-1}	0.88679×10^{-1}
0.75	0.15752	0.12946	0.12143
0.80	0.22173	0.18733	0.17492
0.85	0.30922	0.27102	0.25539
0.90	0.41623	0.38207	0.36305
0.95	0.52607	0.49717	0.47682
1.00	0.57601	0.55694	0.53879

Table 3.5—*continued*

y:	Mesh A	Mesh B	Mesh C
		(c) $Pe = 60$	
0.00	0.517 94	0.483 04	0.439 74
0.05	0.456 67	0.401 54	0.358 98
0.10	0.325 65	0.266 90	0.231 18
0.15	0.218 23	0.156 35	0.130 92
0.20	0.137 84	$0.888\ 90 \times 10^{-1}$	$0.719\ 44 \times 10^{-1}$
0.25	$0.885\ 40 \times 10^{-1}$	$0.505\ 45 \times 10^{-1}$	$0.410\ 55 \times 10^{-1}$
0.30	$0.545\ 60 \times 10^{-1}$	$0.309\ 57 \times 10^{-1}$	$0.256\ 75 \times 10^{-1}$
0.35	$0.345\ 89 \times 10^{-1}$	$0.203\ 51 \times 10^{-1}$	$0.179\ 15 \times 10^{-1}$
0.40	$0.210\ 82 \times 10^{-1}$	$0.149\ 05 \times 10^{-1}$	$0.138\ 71 \times 10^{-1}$
0.45	$0.140\ 71 \times 10^{-1}$	$0.121\ 80 \times 10^{-1}$	$0.119\ 07 \times 10^{-1}$
0.50	$0.117\ 17 \times 10^{-1}$	$0.113\ 73 \times 10^{-1}$	$0.113\ 15 \times 10^{-1}$
0.55	$0.140\ 71 \times 10^{-1}$	$0.121\ 80 \times 10^{-1}$	$0.119\ 07 \times 10^{-1}$
0.60	$0.210\ 82 \times 10^{-1}$	$0.149\ 05 \times 10^{-1}$	$0.138\ 71 \times 10^{-1}$
0.65	$0.345\ 89 \times 10^{-1}$	$0.203\ 51 \times 10^{-1}$	$0.179\ 15 \times 10^{-1}$
0.70	$0.545\ 60 \times 10^{-1}$	$0.309\ 57 \times 10^{-1}$	$0.256\ 75 \times 10^{-1}$
0.75	$0.885\ 40 \times 10^{-1}$	$0.505\ 45 \times 10^{-1}$	$0.410\ 55 \times 10^{-1}$
0.80	0.137 84	$0.888\ 90 \times 10^{-1}$	$0.719\ 44 \times 10^{-1}$
0.85	0.218 23	0.156 35	0.130 93
0.90	0.325 65	0.266 90	0.231 18
0.95	0.456 67	0.401 54	0.358 98
1.00	0.517 94	0.480 34	0.437 94
		(d) $Pe = 100$	
0.00	0.459 81	0.384 89	0.285 00
0.05	0.387 01	0.286 50	0.195 85
0.10	0.237 67	0.150 86	$0.926\ 44 \times 10^{-1}$
0.15	0.146 20	$0.667\ 48 \times 10^{-1}$	$0.354\ 91 \times 10^{-1}$
0.20	$0.811\ 05 \times 10^{-1}$	$0.283\ 04 \times 10^{-1}$	$0.129\ 69 \times 10^{-1}$
0.25	$0.510\ 80 \times 10^{-1}$	$0.121\ 95 \times 10^{-1}$	$0.524\ 28 \times 10^{-2}$
0.30	$0.273\ 32 \times 10^{-1}$	$0.542\ 78 \times 10^{-2}$	$0.240\ 93 \times 10^{-2}$
0.35	$0.157\ 28 \times 10^{-1}$	$0.247\ 44 \times 10^{-2}$	$0.131\ 25 \times 10^{-2}$
0.40	$0.652\ 40 \times 10^{-2}$	$0.121\ 79 \times 10^{-2}$	$0.827\ 90 \times 10^{-3}$
0.45	$0.223\ 84 \times 10^{-2}$	$0.704\ 99 \times 10^{-3}$	$0.622\ 25 \times 10^{-3}$
0.50	$0.682\ 16 \times 10^{-3}$	$0.573\ 56 \times 10^{-3}$	$0.564\ 28 \times 10^{-3}$
0.55	$0.223\ 84 \times 10^{-2}$	$0.704\ 99 \times 10^{-3}$	$0.622\ 25 \times 10^{-3}$
0.60	$0.652\ 40 \times 10^{-2}$	$0.121\ 79 \times 10^{-2}$	$0.827\ 90 \times 10^{-3}$
0.65	$0.157\ 28 \times 10^{-1}$	$0.247\ 44 \times 10^{-2}$	$0.131\ 25 \times 10^{-2}$
0.70	$0.273\ 32 \times 10^{-1}$	$0.542\ 78 \times 10^{-2}$	$0.240\ 93 \times 10^{-2}$
0.75	$0.510\ 80 \times 10^{-1}$	$0.121\ 95 \times 10^{-1}$	$0.524\ 28 \times 10^{-2}$
0.80	$0.811\ 05 \times 10^{-1}$	$0.283\ 04 \times 10^{-1}$	$0.129\ 69 \times 10^{-1}$
0.85	0.146 20	$0.667\ 47 \times 10^{-1}$	$0.354\ 91 \times 10^{-1}$
0.90	0.237 67	0.150 866	$0.926\ 43 \times 10^{-1}$
0.95	0.387 01	0.286 50	0.195 84
1.00	0.459 81	0.384 89	0.285 00

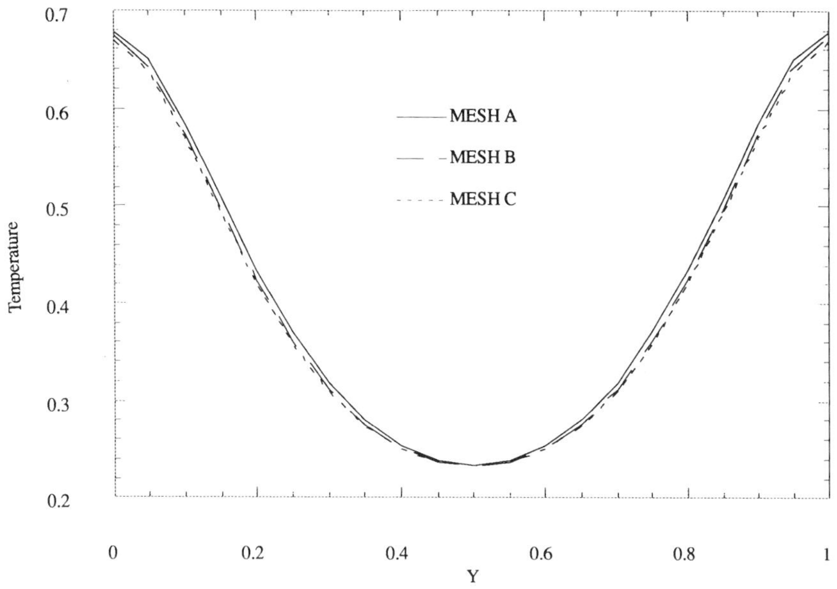

(a)

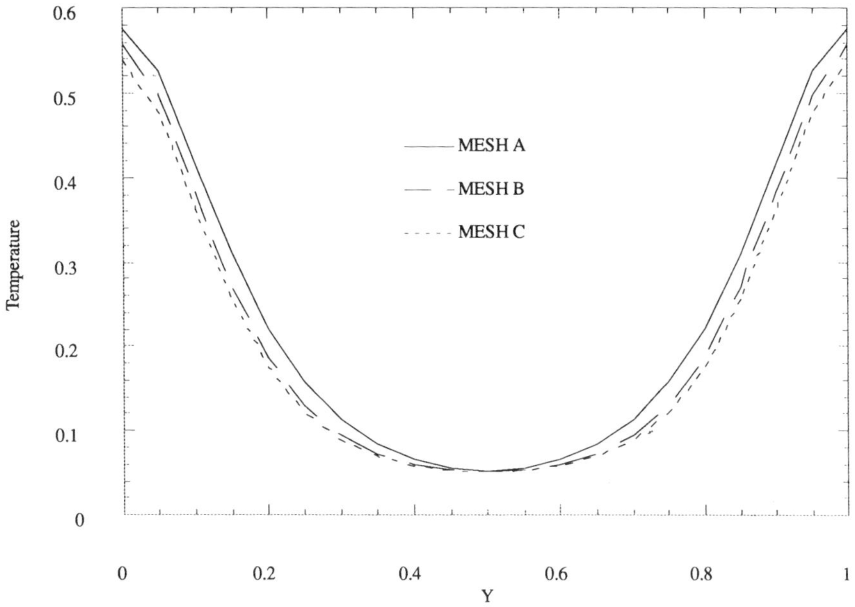

(b)

Figure 3.4. Effect of grid size at internal nodes ($x = 0.95$): (a) Peclet number = 20, (b) Peclet number = 40, (c) Peclet number = 60, (d) Peclet number = 100.

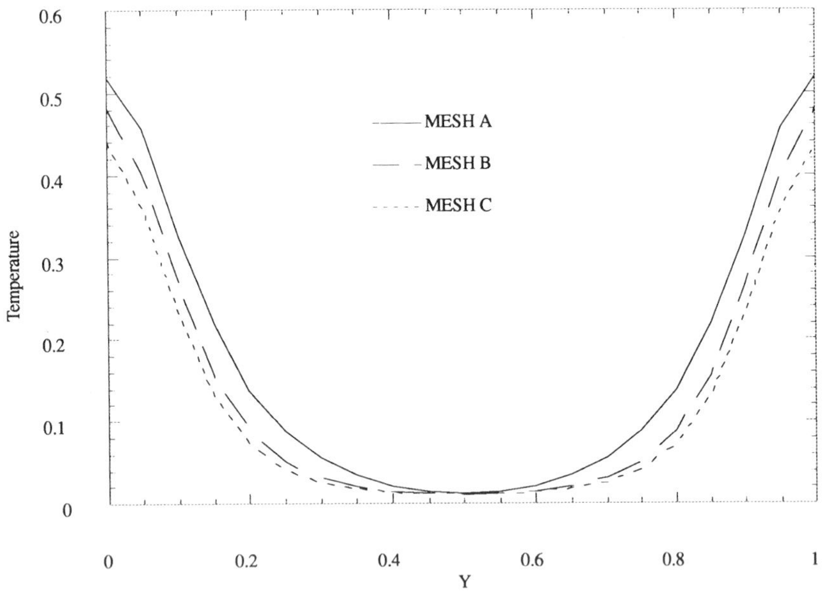

(c)

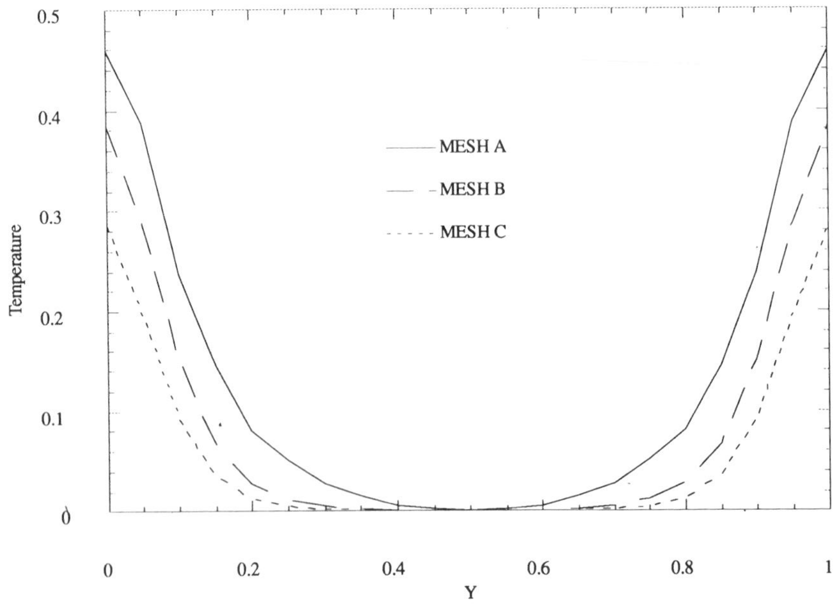

(d)

Figure 3.4—*continued*

Table 3.6. Effect of decomposition level for a constant velocity case at $Pe = 60$

$y = 0$ x:	Analytical	CVEL* = 60.0 VVEL† = 0.0	CVEL = 55.0 VVEL = 5.0	CVEL = 65.0 VVEL = −5.0	CVEL = 10.0 VVEL = 50.0
0.00	0.0	0.0	0.0	0.0	0.0
0.20	0.1425×10^{-20}	-0.472×10^{-13}	0.8215×10^{-14}	0.2936×10^{-13}	-0.1615×10^{-7}
0.40	0.2319×10^{-15}	-0.161×10^{-10}	0.8439×10^{-12}	0.1786×10^{-11}	-0.1230×10^{-6}
0.60	0.3775×10^{-10}	0.3775×10^{-10}	-0.1409×10^{-9}	0.26825×10^{-9}	-0.1912×10^{-5}
0.70	0.15230×10^{-7}	0.15237×10^{-7}	0.11508×10^{-7}	0.19225×10^{-7}	-0.1432×10^{-4}
0.7250	0.68256×10^{-7}	0.68287×10^{-7}	0.53705×10^{-7}	0.83932×10^{-7}	-0.2086×10^{-4}
0.7500	0.30590×10^{-6}	0.30604×10^{-6}	0.24941×10^{-6}	0.36612×10^{-6}	-0.3070×10^{-4}
0.7750	0.13720×10^{-5}	0.13715×10^{-5}	0.11538×10^{-5}	0.15986×10^{-5}	-0.4642×10^{-4}
0.8000	0.61442×10^{-5}	0.61466×10^{-5}	0.53289×10^{-5}	0.69816×10^{-5}	-0.7150×10^{-4}
0.8250	0.27536×10^{-4}	0.27547×10^{-4}	0.24584×10^{-4}	0.30495×10^{-4}	-0.1106×10^{-3}
0.8500	0.12341×10^{-3}	0.12346×10^{-3}	0.11335×10^{-3}	0.13320×10^{-3}	-0.1595×10^{-3}
0.8750	0.55308×10^{-3}	0.55328×10^{-3}	0.52241×10^{-3}	0.58178×10^{-3}	-0.1317×10^{-3}
0.9000	0.24788×10^{-2}	0.24796×10^{-2}	0.24165×10^{-2}	0.25337×10^{-2}	0.92509×10^{-3}
0.9125	0.52475×10^{-2}	0.52494×10^{-2}	0.51380×10^{-2}	0.53441×10^{-2}	0.27016×10^{-2}
0.9250	0.11109×10^{-1}	0.11113×10^{-1}	0.10922×10^{-1}	0.11275×10^{-1}	0.70199×10^{-2}
0.9375	0.23518×10^{-1}	0.23527×10^{-1}	0.22307×10^{-1}	0.23800×10^{-1}	0.17167×10^{-1}
0.9500	0.49787×10^{-1}	0.49808×10^{-1}	0.49304×10^{-1}	0.50243×10^{-1}	0.40605×10^{-1}
0.9625	0.10540	0.10545	0.10473	0.10608	0.93678×10^{-1}
0.9750	0.22313	0.22326	0.22236	0.22407	0.21052
0.9875	0.47237	0.47268	0.47148	0.47378	0.45775
1.00	1.0	1.0	1.0	1.0	1.0

*CVEL, constant velocity; †VVEL, variable velocity.

fundamental solution, while the effect of the variable convective velocity is incorporated through the domain integral. The sensitivity of the results to results to this decomposition level is investigated next. Since, analytical solutions are only available for constant-velocity cases, in our investigation a constant-velocity field is artificially decomposed into a constant part and a variable part. The constant part is absorbed in the boundary integral, while the variable part is included in the domain integral. Table 3.6 presents the results for various decomposition levels for $Pe = 60$. The results are also compared with analytical solutions. It is observed that the magnitude of error decreases as a smaller fraction of the total velocity is assumed to be the variable part and included in the domain integral. This is expected since the moving heat source fundamental solution with $V_i^{(c)}$ approaching zero behaves like the fundamental solution for the Laplace equation. Finally, best comparisons are obtained when the total velocity is treated as constant and included in a boundary integral, obviating the need for the domain term.

Tables 3.7 and 3.8 show the effect of decomposition level for a spatially varying velocity case. Four decomposition levels with constant velocity equal to $0.5Pe$, $1.0Pe$, $1.5Pe$ and $3.0Pe$ are considered. The average velocity for the above velocity field is $1.0Pe$. The temperatures at $y = 0$ for $Pe = 20$, 40, 60, and 100 are presented in tables 3.7a–d, respectively. Figures 3.5a–d graphically represent the temperatures for these four decomposition levels for $Pe = 20$, 40, 60, and 100 respectively. The temperatures at $x = 0.9875$ for the above Peclet numbers are presented in tables 3.8a–d. The same result is also graphically presented in figures 3.6a–d. The temperature varies for different decomposition levels. In general, the temperature decreases with the increase in the constant velocity part. This variation is observed to increase with the increase in Peclet number. However, no closed form analytical solution is available for the spatially varying convective velocity. Accordingly, the numerical results could not be compared against such analytical solutions.

In order to understand the effect of decomposition level, an example problem with spatially varying but piecewise constant velocity profile is also investigated. In each constant velocity zone, an accurate BEM solution based on the Green's function for the constant velocity case is obtained. The complete solution is then synthesized by matching the solutions across the interfaces (Chan and Chandra 1991a). The matched solution is then compared to the BEM solution obtained from the present scheme using five different decomposition levels. The temperature differences between the matched solution and those obtained from the present scheme are presented in tables 3.9(a,b) and 39(c,d) for two different Peclet numbers of 10 and 30. The ordinate locations $y = 0$ and 1.0 represent the bottom and top boundaries, while $y = 0.3$ and 0.7 represent the two interfaces. At each of

Table 3.7. Effect of decomposition level for variable velocity case at boundary nodes ($y = 0$) for $Pe = 20$, 40, 60, and 100

x:	CVEL = 0.5Pe	CVEL = 1.0Pe	CVEL = 1.5Pe	CVEL = 3.0Pe
		(a) $Pe = 20$		
0.0	0.0	0.0	0.0	0.0
0.20	$0.390\,95 \times 10^{-3}$	$0.769\,48 \times 10^{-3}$	$0.100\,76 \times 10^{-2}$	$0.132\,92 \times 10^{-2}$
0.40	$0.288\,95 \times 10^{-2}$	$0.410\,94 \times 10^{-2}$	$0.475\,68 \times 10^{-2}$	$0.550\,47 \times 10^{-2}$
0.60	$0.196\,42 \times 10^{-1}$	$0.203\,13 \times 10^{-1}$	$0.207\,18 \times 10^{-1}$	$0.207\,91 \times 10^{-1}$
0.70	$0.554\,90 \times 10^{-1}$	$0.546\,21 \times 10^{-1}$	$0.537\,85 \times 10^{-1}$	$0.513\,39 \times 10^{-1}$
0.725	$0.724\,81 \times 10^{-1}$	$0.717\,28 \times 10^{-1}$	$0.709\,64 \times 10^{-1}$	$0.684\,26 \times 10^{-1}$
0.750	$0.942\,42 \times 10^{-1}$	$0.934\,95 \times 10^{-1}$	$0.926\,69 \times 10^{-1}$	$0.895\,35 \times 10^{-1}$
0.775	0.122 21	0.121 47	0.120 56	0.116 76
0.80	0.158 07	0.157 30	0.156 28	0.151 59
0.825	0.203 85	0.203 03	0.201 84	0.195 99
0.850	0.261 96	0.261 03	0.259 61	0.252 25
0.875	0.335 08	0.333 99	0.332 24	0.322 96
0.90	0.425 82	0.424 31	0.422 03	0.410 35
0.9125	0.478 89	0.477 36	0.475 03	0.462 93
0.9250	0.537 49	0.536 11	0.533 97	0.522 59
0.9375	0.601 76	0.600 62	0.598 81	0.588 78
0.9500	0.671 73	0.670 86	0.669 43	0.661 15
0.9625	0.747 26	0.764 65	0.745 62	0.739 38
0.9750	0.827 96	0.827 62	0.827 02	0.822 98
0.9875	0.913 01	0.913 00	0.912 86	0.911 15
1.0	1.0	1.0	1.0	1.0
		(b) $Pe = 40$		
0.0	0.0	0.0	0.0	0.0
0.20	$0.382\,79 \times 10^{-4}$	$0.907\,78 \times 10^{-4}$	$0.117\,03 \times 10^{-3}$	$0.101\,85 \times 10^{-3}$
0.40	$0.359\,27 \times 10^{-3}$	$0.583\,46 \times 10^{-3}$	$0.670\,44 \times 10^{-3}$	$0.592\,40 \times 10^{-3}$
0.60	$0.324\,52 \times 10^{-2}$	$0.356\,00 \times 10^{-2}$	$0.362\,14 \times 10^{-2}$	$0.316\,60 \times 10^{-2}$
0.70	$0.133\,36 \times 10^{-1}$	$0.128\,28 \times 10^{-1}$	$0.123\,13 \times 10^{-1}$	$0.105\,74 \times 10^{-1}$
0.725	$0.198\,58 \times 10^{-1}$	$0.193\,51 \times 10^{-1}$	$0.187\,47 \times 10^{-1}$	$0.161\,78 \times 10^{-1}$
0.750	$0.292\,02 \times 10^{-1}$	$0.285\,47 \times 10^{-1}$	$0.276\,72 \times 10^{-1}$	$0.235\,15 \times 10^{-1}$
0.775	$0.427\,88 \times 10^{-1}$	$0.419\,18 \times 10^{-1}$	$0.406\,77 \times 10^{-1}$	$0.344\,96 \times 10^{-1}$
0.80	$0.625\,23 \times 10^{-1}$	$0.613\,19 \times 10^{-1}$	$0.595\,25 \times 10^{-1}$	$0.503\,78 \times 10^{-1}$
0.825	$0.911\,19 \times 10^{-1}$	$0.894\,02 \times 10^{-1}$	$0.867\,83 \times 10^{-1}$	$0.732\,99 \times 10^{-1}$
0.850	0.132 36	0.129 87	0.126 03	0.106 24
0.875	0.191 39	0.187 72	0.182 10	0.153 31
0.90	0.274 44	0.268 91	0.260 77	0.219 73
0.9125	0.328 17	0.322 27	0.313 44	0.267 27
0.9250	0.391 43	0.385 67	0.376 87	0.328 84
0.9375	0.465 23	0.459 99	0.451 79	0.404 55
0.9500	0.550 37	0.545 93	0.538 79	0.495 21
0.9625	0.647 39	0.643 95	0.638 32	0.601 49
0.9750	0.756 18	0.753 95	0.750 20	0.723 35
0.9875	0.875 48	0.874 67	0.873 11	0.859 20
1.0	1.0	1.0	1.0	1.0

Table 3.7—*continued*

x:	CVEL = 0.5Pe	CVEL = 1.0Pe	CVEL = 1.5Pe	CVEL = 3.0Pe

(c) Pe = 60

0.0	0.0	0.0	0.0	0.0
0.20	0.932 71 × 10^{-5}	0.201 84 × 10^{-4}	0.222 52 × 10^{-4}	0.947 98 × 10^{-5}
0.40	0.921 93 × 10^{-4}	0.144 14 × 10^{-3}	0.150 61 × 10^{-3}	0.778 49 × 10^{-4}
0.60	0.884 26 × 10^{-3}	0.984 06 × 10^{-3}	0.965 13 × 10^{-3}	0.590 64 × 10^{-3}
0.70	0.442 79 × 10^{-2}	0.415 67 × 10^{-2}	0.386 54 × 10^{-2}	0.274 57 × 10^{-2}
0.725	0.730 96 × 10^{-2}	0.698 65 × 10^{-2}	0.654 89 × 10^{-2}	0.453 04 × 10^{-2}
0.750	0.117 83 × 10^{-1}	0.112 66 × 10^{-1}	0.104 98 × 10^{-1}	0.686 58 × 10^{-2}
0.775	0.189 10 × 10^{-1}	0.180 87 × 10^{-1}	0.168 32 × 10^{-1}	0.108 85 × 10^{-1}
0.80	0.302 60 × 10^{-1}	0.289 19 × 10^{-1}	0.268 59 × 10^{-1}	0.171 86 × 10^{-1}
0.825	0.483 14 × 10^{-1}	0.460 99 × 10^{-1}	0.427 16 × 10^{-1}	0.271 03 × 10^{-1}
0.850	0.769 39 × 10^{-1}	0.732 63 × 10^{-1}	0.677 19 × 10^{-1}	0.426 29 × 10^{-1}
0.875	0.122 04	0.115 94	0.106 92	0.668 73 × 10^{-1}
0.90	0.191 86	0.181 82	0.167 45	0.104 58
0.9125	0.241 00	0.229 70	0.213 03	0.136 28
0.9250	0.301 96	0.290 17	0.272 20	0.184 19
0.9375	0.376 84	0.365 31	0.347 09	0.250 75
0.9500	0.467 59	0.457 05	0.439 74	0.340 40
0.9625	0.575 93	0.567 16	0.552 16	0.457 90
0.9750	0.702 71	0.696 54	0.685 42	0.607 72
0.9875	0.846 80	0.844 06	0.838 37	0.791 82
1.0	1.0	1.0	1.0	1.0

(d) Pe = 100

0.0	0.0	0.0	0.0	0.0
0.20	0.122 38 × 10^{-5}	0.183 68 × 10^{-5}	0.135 05 × 10^{-5}	0.166 46 × 10^{-6}
0.40	0.127 51 × 10^{-4}	0.162 51 × 10^{-4}	0.128 10 × 10^{-4}	0.223 89 × 10^{-5}
0.60	0.129 35 × 10^{-3}	0.138 15 × 10^{-3}	0.115 42 × 10^{-3}	0.289 51 × 10^{-4}
0.70	0.785 26 × 10^{-3}	0.718 01 × 10^{-3}	0.624 15 × 10^{-3}	0.224 86 × 10^{-3}
0.725	0.152 47 × 10^{-2}	0.141 17 × 10^{-2}	0.121 00 × 10^{-2}	0.376 91 × 10^{-3}
0.750	0.281 55 × 10^{-2}	0.265 01 × 10^{-2}	0.212 04 × 10^{-2}	0.578 30 × 10^{-3}
0.775	0.517 64 × 10^{-2}	0.465 93 × 10^{-2}	0.381 50 × 10^{-2}	0.102 12 × 10^{-2}
0.80	0.948 85 × 10^{-2}	0.844 52 × 10^{-2}	0.683 52 × 10^{-2}	0.179 01 × 10^{-2}
0.825	0.173 63 × 10^{-1}	0.152 71 × 10^{-1}	0.122 19 × 10^{-1}	0.315 64 × 10^{-2}
0.850	0.317af2222 × 10^{-1}	0.275 61 × 10^{-1}	0.218 01 × 10^{-1}	0.556 22 × 10^{-2}
0.875	0.578 07 × 10^{-1}	0.496 11 × 10^{-1}	0.388 22 × 10^{-1}	0.979 59 × 10^{-2}
0.90	0.104 44	0.887 21 × 10^{-1}	0.688 82 × 10^{-1}	0.172 64 × 10^{-1}
0.9125	0.142 31	0.122 64	0.966 07 × 10^{-1}	0.251 40 × 10^{-1}
0.9250	0.193 54	0.170 50	0.138 47	0.418 48 × 10^{-1}
0.9375	0.262 25	0.236 76	0.199 23	0.718 85 × 10^{-1}
0.9500	0.352 96	0.326 45	0.285 00	0.124 87
0.9625	0.470 58	0.445 38	0.403 29	0.216 24
0.9750	0.619 28	0.598 79	0.561 74	0.369 93
0.9875	0.799 98	0.788 54	0.764 97	0.619 82
1.0	1.0	1.0	1.0	1.0

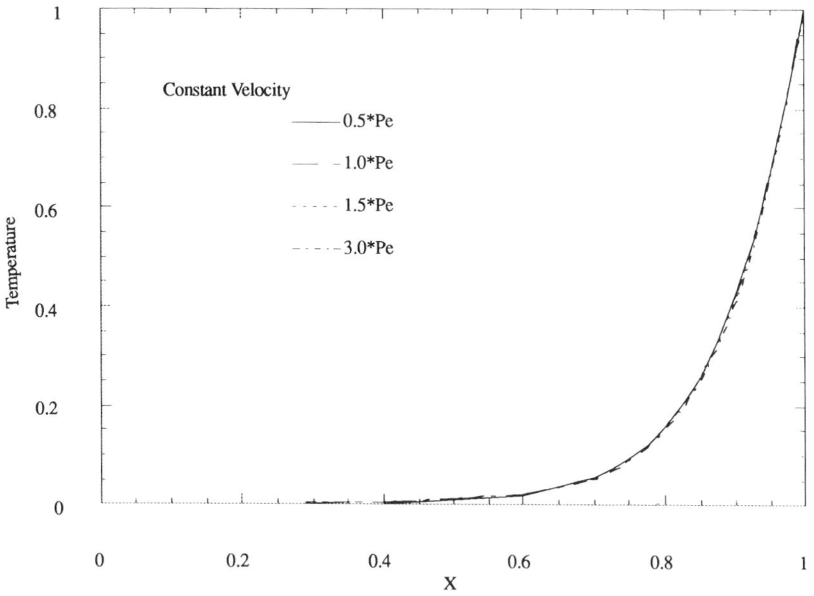

(a)

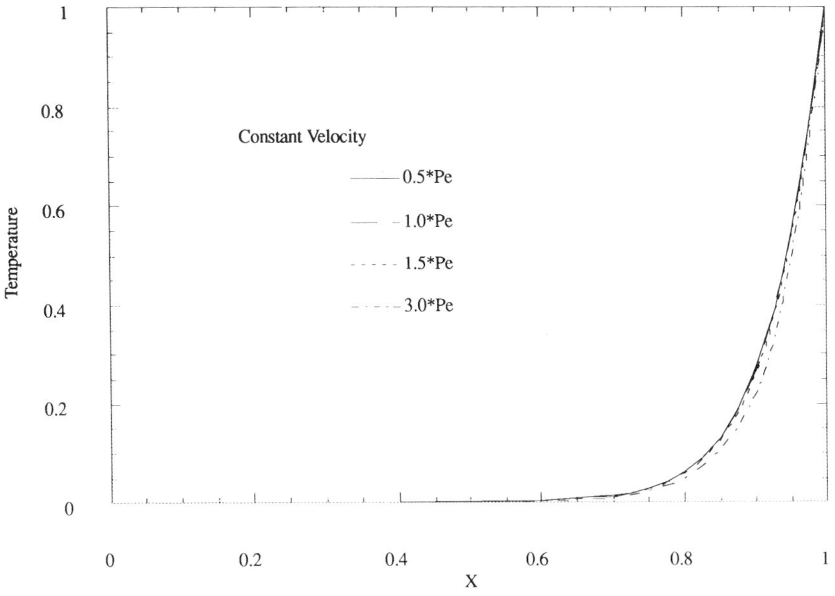

(b)

Figure 3.5. Effect of decomposition level at boundary nodes ($y = 0$): (a) Peclet number = 20, (b) Peclet number = 40, (c) Peclet number = 60, (d) Peclet number = 100.

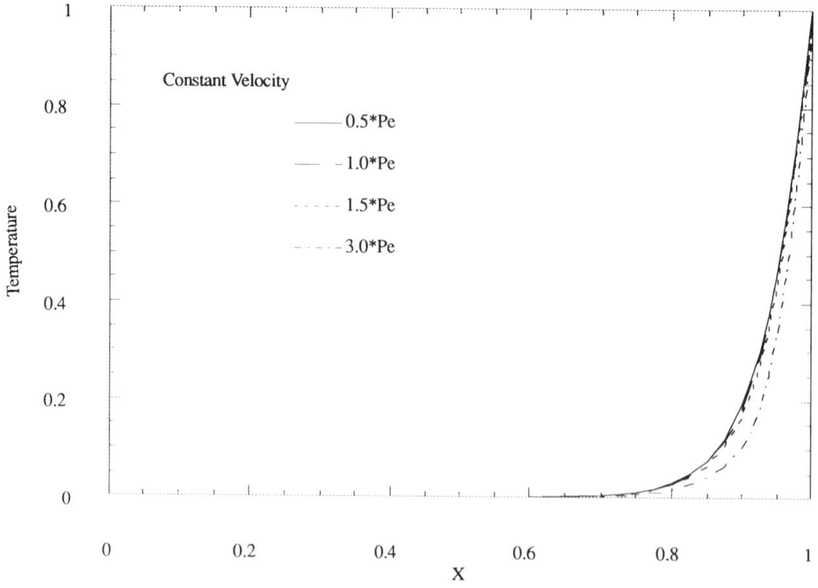

(c)

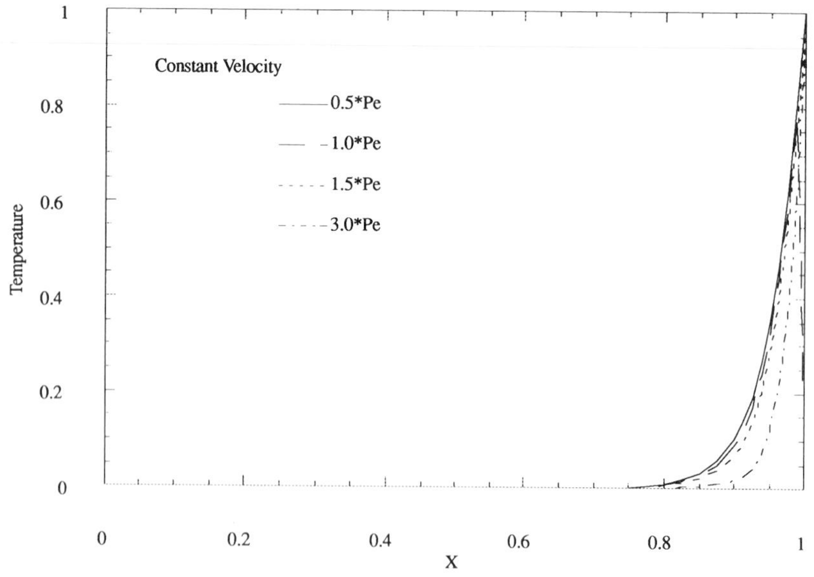

(d)

Figure 3.5—*continued*

Table 3.8. Effect of decomposition level for variable velocity case at internal nodes ($x = 0.9875$) for $Pe = 20$, 40, 60, and 100

y:	CVEL = 0.5Pe	CVEL = 1.0Pe	CVEL = 1.5Pe	CVEL = 3.0Pe
		(a) Pe = 20		
0.0	0.913 01	0.913 00	0.912 86	0.911 15
0.05	0.898 75	0.898 63	0.898 35	0.896 03
0.10	0.870 14	0.869 99	0.869 67	0.867 14
0.15	0.835 75	0.835 73	0.835 46	0.832 58
0.20	0.800 55	0.800 68	0.800 56	0.798 25
0.25	0.767 78	0.768 26	0.768 36	0.766 27
0.30	0.740 17	0.740 84	0.741 06	0.739 82
0.35	0.718 28	0.719 39	0.719 68	0.718 72
0.40	0.703 30	0.704 61	0.704 54	0.704 17
0.45	0.694 28	0.696 23	0.695 55	0.695 25
0.50	0.691 84	0.693 92	0.692 59	0.692 52
0.55	0.693 40	0.694 60	0.695 55	0.695 25
0.60	0.700 32	0.698 87	0.704 54	0.704 17
0.65	0.715 58	0.714 19	0.719 68	0.718 72
0.70	0.740 09	0.740 79	0.741 06	0.739 82
0.75	0.768 81	0.770 24	0.768 36	0.766 27
0.80	0.801 31	0.802 10	0.800 56	0.798 25
0.85	0.836 16	0.836 48	0.835 46	0.832 58
0.90	0.870 34	0.870 37	0.869 67	0.867 14
0.95	0.898 86	0.898 85	0.898 35	0.896 03
1.0	0.913 09	0.913 17	0.918 26	0.911 15
		(b) Pe = 40		
0.0	0.875 48	0.874 67	0.873 11	0.859 20
0.05	0.840 72	0.839 71	0.837 85	0.821 37
0.10	0.774 43	0.773 50	0.771 93	0.757 06
0.15	0.700 25	0.700 44	0.699 75	0.685 26
0.20	0.632 10	0.633 49	0.634 46	0.626 84
0.25	0.574 94	0.578 12	0.580 59	0.575 18
0.30	0.532 33	0.535 98	0.539 48	0.539 84
0.35	0.499 68	0.504 93	0.509 64	0.511 13
0.40	0.479 80	0.484 62	0.489 45	0.494 32
0.45	0.465 86	0.472 07	0.477 79	0.482 55
0.50	0.463 60	0.468 82	0.473 98	0.480 26
0.55	0.465 87	0.472 09	0.477 79	0.482 55
0.60	0.479 69	0.484 41	0.489 45	0.494 32
0.65	0.499 55	0.504 70	0.509 64	0.511 13
0.70	0.532 35	0.536 03	0.539 48	0.539 84
0.75	0.575 00	0.578 23	0.580 59	0.575 18
0.80	0.632 13	0.633 54	0.634 46	0.626 84
0.85	0.700 26	0.700 45	0.699 75	0.685 26
0.90	0.774 43	0.773 50	0.771 93	0.757 06
0.95	0.840 73	0.839 72	0.837 85	0.821 37
1.0	0.875 48	0.874 67	0.873 11	0.859 20

Table 3.8—*continued*

y:	CVEL = 0.5Pe	CVEL = 1.0Pe	CVEL = 1.5Pe	CVEL = 3.0Pe
		(c) Pe = 60		
0.0	0.864 80	0.844 06	0.838 37	0.791 82
0.05	0.789 96	0.787 34	0.781 24	0.729 00
0.10	0.686 89	0.685 81	0.682 05	0.643 55
0.15	0.580 26	0.583 05	0.582 40	0.549 96
0.20	0.491 80	0.498 05	0.502 05	0.489 95
0.25	0.422 51	0.433 03	0.440 40	0.433 44
0.30	0.374 95	0.386 10	0.395 56	0.402 02
0.35	0.336 60	0.351 58	0.363 71	0.371 73
0.40	0.316 29	0.329 91	0.342 14	0.357 95
0.45	0.298 34	0.315 40	0.329 74	0.344 57
0.50	0.298 14	0.312 57	0.325 67	0.344 38
0.55	0.298 34	0.315 41	0.329 74	0.344 57
0.60	0.316 29	0.329 90	0.342 14	0.357 95
0.65	0.336 60	0.351 57	0.363 71	0.371 73
0.70	0.374 95	0.386 11	0.395 56	0.402 02
0.75	0.422 51	0.433 04	0.440 40	0.433 44
0.80	0.491 80	0.498 05	0.502 05	0.489 95
0.85	0.580 26	0.583 05	0.582 40	0.549 96
0.90	0.686 89	0.685 81	0.682 05	0.643 55
0.95	0.789 96	0.787 34	0.781 24	0.728 99
1.0	0.846 80	0.844 06	0.838 37	0.791 82
		(d) Pe = 100		
0.0	0.799 98	0.788 54	0.764 97	0.619 82
0.05	0.698 81	0.690 10	0.666 92	0.516 27
0.10	0.533 59	0.536 07	0.527 97	0.447 42
0.15	0.386 99	0.402 89	0.405 68	0.351 69
0.20	0.283 33	0.307 87	0.322 71	0.317 02
0.25	0.205 27	0.240 36	0.262 30	0.264 68
0.30	0.158 22	0.193 40	0.219 61	0.248 81
0.35	0.112 71	0.157 79	0.189 64	0.219 45
0.40	0.096 99	0.137 58	0.169 28	0.214 52
0.45	0.071 87	0.121 48	0.157 73	0.199 37
0.50	0.078 29	0.120 55	0.153 93	0.203 95
0.55	0.071 87	0.121 48	0.157 73	0.199 37
0.60	0.096 99	0.137 58	0.169 28	0.214 52
0.65	0.112 71	0.157 79	0.189 64	0.219 45
0.70	0.158 22	0.193 40	0.219 61	0.248 81
0.75	0.205 27	0.240 36	0.262 30	0.264 68
0.80	0.283 33	0.307 87	0.322 71	0.317 02
0.85	0.386 99	0.402 89	0.405 68	0.351 69
0.90	0.533 59	0.536 07	0.527 97	0.447 42
0.95	0.698 81	0.690 10	0.666 92	0.516 26
1.0	0.799 98	0.788 54	0.764 97	0.619 82

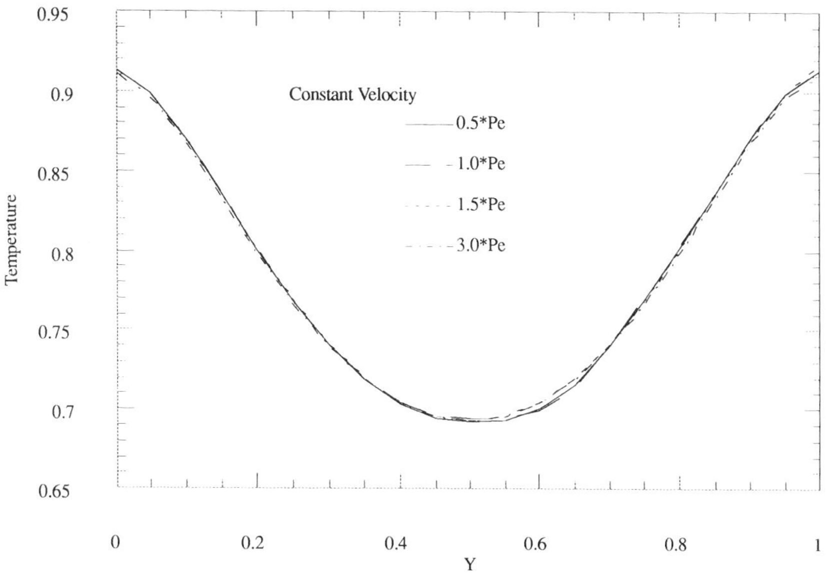

(a)

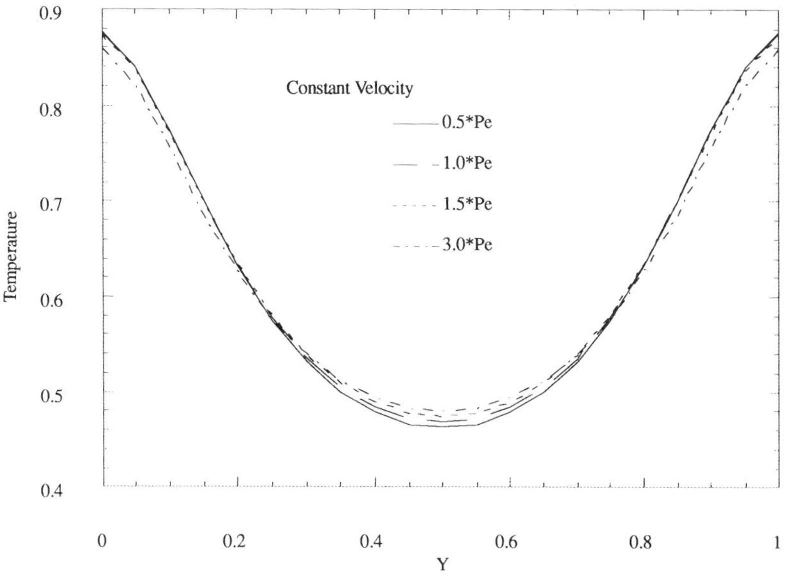

(b)

Figure 3.6. Effect of decomposition level at internal nodes ($x = 0.9875$): (a) Peclet number = 20, (b) Peclet number = 40, (c) Peclet number = 60, (d) Peclet number = 100.

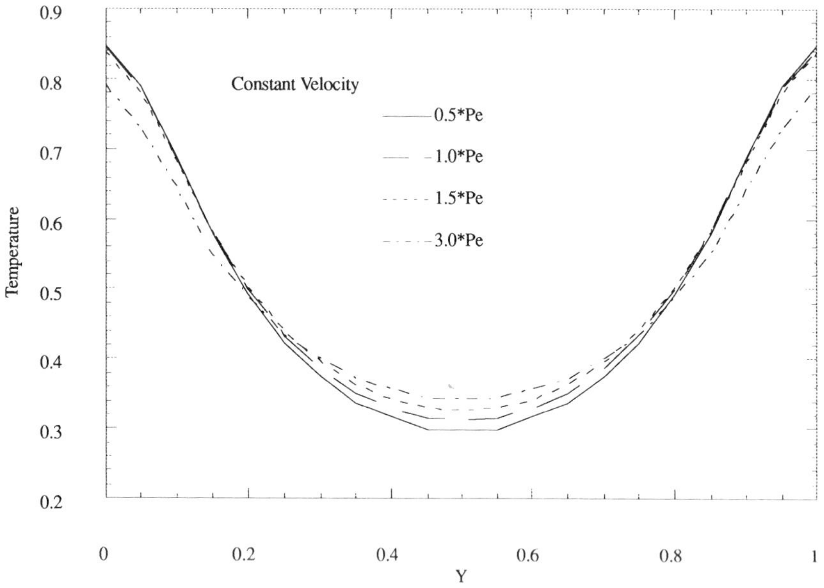

(c)

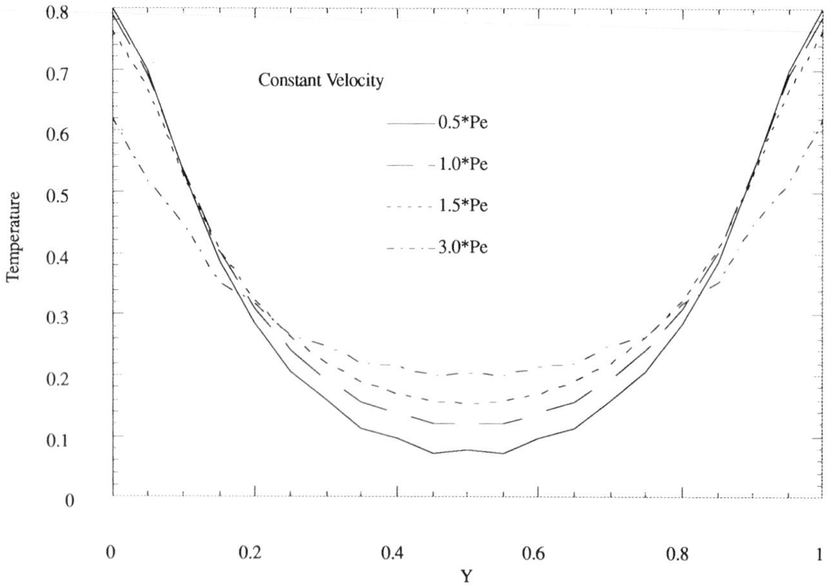

(d)

Figure 3.6—*continued*

Table 3.9a. Effect of decomposition level for a piecewise constant velocity in y direction for $Pe = 10$ at $y = 0.0$ and $y = 0.3$

x:	CVEL = 0.5	CVEL = 0.75	CVEL = 1.0	CVEL = 1.25	CVEL = 1.5
			$y = \mathbf{0.0}$		
0.00	0.0	0.0	0.0	0.0	0.0
0.10	3.4900×10^{-5}	-6.0000×10^{-7}	-2.8400×10^{-5}	-7.2600×10^{-5}	-1.0150×10^{-4}
0.20	1.1160×10^{-4}	2.3900×10^{-5}	-4.5700×10^{-5}	-1.5780×10^{-4}	-2.3090×10^{-4}
0.30	2.4600×10^{-4}	7.7999×10^{-5}	-5.8000×10^{-5}	-2.7500×10^{-4}	-4.1800×10^{-4}
0.40	4.7100×10^{-4}	1.7900×10^{-4}	-6.3000×10^{-5}	-4.3500×10^{-4}	-6.8800×10^{-4}
0.50	8.2400×10^{-4}	3.5100×10^{-4}	-5.0999×10^{-5}	-6.3999×10^{-4}	-1.0560×10^{-3}
0.60	1.3280×10^{-3}	6.1299×10^{-4}	-1.4998×10^{-5}	-8.7299×10^{-4}	-1.5160×10^{-4}
0.70	1.9099×10^{-3}	9.1999×10^{-4}	-1.99974×10^{-5}	-1.1200×10^{-3}	-2.0200×10^{-3}
0.80	2.2299×10^{-3}	1.0600×10^{-3}	-3.99947×10^{-5}	-1.3499×10^{-3}	-2.4499×10^{-3}
0.90	1.6499×10^{-3}	6.4998×10^{-4}	-3.1000×10^{-4}	-1.4200×10^{-3}	-2.3800×10^{-3}
1.00	0.0	0.0	0.0	0.0	0.0
			$y = \mathbf{0.3}$		
0.00	0.0	0.0	0.0	0.0	0.0
0.10	2.4599×10^{-5}	7.6488×10^{-6}	-4.4780×10^{-6}	-3.3821×10^{-5}	-4.8806×10^{-5}
0.20	8.3901×10^{-5}	3.8422×10^{-5}	7.6158×10^{-6}	-7.6167×10^{-5}	-1.1574×10^{-4}
0.30	1.9084×10^{-4}	9.8215×10^{-5}	3.9639×10^{-5}	-1.4392×10^{-4}	-2.2333×10^{-4}
0.40	3.7985×10^{-4}	2.0955×10^{-4}	1.07390×10^{-4}	-2.4850×10^{-4}	-3.9318×10^{-4}
0.50	7.2017×10^{-4}	4.2035×10^{-4}	2.3317×10^{-4}	-3.7802×10^{-4}	-6.3596×10^{-4}
0.60	1.3219×10^{-3}	8.0654×10^{-4}	4.3068×10^{-4}	-4.7566×10^{-4}	-9.3741×10^{-4}
0.70	2.2196×10^{-3}	1.3749×10^{-3}	6.5489×10^{-4}	-5.4220×10^{-4}	-1.3333×10^{-3}
0.80	3.0380×10^{-3}	1.7980×10^{-3}	6.4440×10^{-4}	-8.3370×10^{-4}	-2.0262×10^{-3}
0.90	2.8400×10^{-3}	1.4616×10^{-3}	1.3130×10^{-4}	-1.3638×10^{-3}	-2.6998×10^{-3}
1.00	0.0	0.0	0.0	0.0	0.0

Table 3.9b. Effect of decomposition level for a piecewise constant velocity in y direction for $Pe = 10$ at $y = 0.7$ and $y = 1.0$

x:	CVEL = 0.5	CVEL = 0.75	CVEL = 1.0	CVEL = 1.25	CVEL = 1.5
			$y = \mathbf{0.7}$		
0.00	0.0	0.0	0.0	0.0	0.0
0.10	-8.3722×10^{-4}	-8.5417×10^{-4}	-8.6629×10^{-4}	-8.9564×10^{-4}	-9.1062×10^{-4}
0.20	-2.3974×10^{-3}	-2.4429×10^{-3}	-2.4737×10^{-3}	-2.5575×10^{-3}	-2.5971×10^{-3}
0.30	-5.3173×10^{-3}	-5.4099×10^{-3}	-5.4685×10^{-3}	-5.6521×10^{-3}	-5.7315×10^{-3}
0.40	-1.0798×10^{-2}	-1.0968×10^{-2}	-1.1070×10^{-2}	-1.1426×10^{-3}	-1.1571×10^{-2}
0.50	-2.0994×10^{-2}	-2.1294×10^{-2}	-2.1481×10^{-2}	-2.2092×10^{-2}	-2.2350×10^{-2}
0.60	-3.9464×10^{-2}	-3.9979×10^{-2}	-4.0355×10^{-2}	-4.1261×10^{-2}	-4.1732×10^{-2}
0.70	-7.09784×10^{-2}	-7.1823×10^{-2}	-7.2543×10^{-2}	-7.3740×10^{-4}	-7.4531×10^{-2}
0.80	-0.2177×10^{-2}	-0.2189×10^{-2}	-0.2200×10^{-2}	-0.2215×10^{-2}	-0.2227×10^{-2}
0.90	-0.1490×10^{-2}	-0.15038×10^{-2}	-0.1517×10^{-2}	-0.1532×10^{-2}	-0.1545×10^{-2}
1.00	0.0	0.0	0.0	0.0	0.0
			$y = \mathbf{1.0}$		
0.00	0.0	0.0	0.0	0.0	0.0
0.10	1.5770×10^{-2}	1.2530×10^{-2}	9.4600×10^{-3}	6.0900×10^{-3}	3.1099×10^{-3}
0.20	8.7360×10^{-3}	7.1700×10^{-3}	5.8020×10^{-3}	3.9870×10^{-3}	2.6440×10^{-3}
0.30	3.7660×10^{-3}	3.1680×10^{-3}	2.7580×10^{-3}	1.8510×10^{-3}	4.4200×10^{-3}
0.40	1.4996×10^{-3}	1.2824×10^{-3}	1.2096×10^{-3}	7.2810×10^{-4}	6.2230×10^{-4}
0.50	5.9310×10^{-4}	5.1070×10^{-4}	5.1480×10^{-4}	2.7960×10^{-4}	2.5770×10^{-4}
0.60	2.4091×10^{-4}	2.0636×10^{-4}	2.1371×10^{-4}	1.1391×10^{-4}	1.1056×10^{-4}
0.70	9.7690×10^{-5}	8.2110×10^{-5}	8.4400×10^{-5}	4.8570×10^{-5}	4.7240×10^{-5}
0.80	3.6464×10^{-5}	2.9516×10^{-5}	2.9641×10^{-5}	1.8548×10^{-5}	1.7087×10^{-5}
0.90	1.0513×10^{-5}	7.8300×10^{-6}	7.6170×10^{-5}	5.0050×10^{-6}	4.7220×10^{-6}
1.00	0.0	0.0	0.0	0.0	0.0

Table 3.9c. Effect of decomposition level for a piecewise constant velocity in y direction for $Pe = 30$ at $y = 0.0$ and $y = 0.3$

x:	CVEL = 0.5	CVEL = 0.75	CVEL = 1.0	CVEL = 1.25	CVEL = 1.5
			$y = 0.0$		
0.00	0.0	0.0	0.0	0.0	0.0
0.10	2.47200×10^{-8}	-1.9358×10^{-7}	-4.4669×10^{-7}	-7.4796×10^{-7}	-1.1015×10^{-6}
0.20	-2.6261×10^{-6}	-3.7667×10^{-6}	-5.0569×10^{-6}	-6.5515×10^{-6}	-8.2652×10^{-6}
0.30	8.4200×10^{-7}	-4.2180×10^{-6}	-9.7740×10^{-6}	-1.6017×10^{-5}	-2.2983×10^{-5}
0.40	3.9940×10^{-6}	-1.7381×10^{-5}	-4.0133×10^{-5}	-6.4883×10^{-5}	-9.1733×10^{-5}
0.50	1.7720×10^{-5}	-6.9270×10^{-5}	-1.5905×10^{-4}	-2.5352×10^{-4}	-3.5302×10^{-4}
0.60	7.3799×10^{-5}	-2.6320×10^{-4}	-6.0070×10^{-4}	-9.4410×10^{-4}	-1.2951×10^{-3}
0.70	2.6929×10^{-4}	-9.3950×10^{-4}	-2.1145×10^{-3}	-3.2715×10^{-3}	-4.4185×10^{-3}
0.80	7.5300×10^{-4}	-3.0200×10^{-3}	-6.5950×10^{-3}	-9.9930×10^{-3}	-1.3267×10^{-2}
0.90	1.1300×10^{-3}	-7.4700×10^{-3}	-1.5450×10^{-2}	-2.2780×10^{-2}	-2.9650×10^{-2}
1.00	0.0	0.0	0.0	0.0	0.0
			$y = 0.3$		
0.00	0.0	0.0	0.0	0.0	0.0
0.10	1.6691×10^{-8}	-3.6404×10^{-8}	-1.0206×10^{-7}	-1.8264×10^{-7}	-2.7947×10^{-7}
0.20	1.05634×10^{-7}	-1.9000×10^{-7}	-5.4606×10^{-7}	-9.7187×10^{-7}	-1.4713×10^{-6}
0.30	5.4025×10^{-6}	-8.1006×10^{-7}	-2.3923×10^{-6}	-4.2352×10^{-6}	-6.3387×10^{-6}
0.40	2.5337×10^{-6}	-3.3277×10^{-6}	-1.0010×10^{-5}	-1.7594×10^{-5}	-2.5998×10^{-5}
0.50	1.2203×10^{-5}	-1.2721×10^{-5}	-4.0402×10^{-5}	-7.1022×10^{-5}	-1.0399×10^{-4}
0.60	5.6164×10^{-5}	-4.8837×10^{-5}	-1.6265×10^{-5}	-2.8522×10^{-4}	-4.1425×10^{-4}
0.70	3.2211×10^{-4}	-1.2315×10^{-4}	-5.9497×10^{-4}	-1.0888×10^{-3}	-1.5989×10^{-3}
0.80	1.5023×10^{-3}	-4.5203×10^{-4}	-2.4627×10^{-3}	-4.4936×10^{-3}	-6.5344×10^{-3}
0.90	1.2564×10^{-2}	4.1629×10^{-3}	-3.9665×10^{-3}	-1.1676×10^{-2}	-1.9031×10^{-2}
1.00	0.0	0.0	0.0	0.0	0.0

Table3.9d. Effect of decomposition level for a piecewise constant velocity in y direction for $Pe = 30$ at $y = 0.7$ and $y = 1.0$

x:	CVEL = 0.5	CVEL = 0.75	CVEL = 1.0	CVEL = 1.25	CVEL = 1.5
			$y = 0.0$		
0.00	0.0	0.0	0.0	0.0	0.0
0.10	-1.2388×10^{-7}	-1.7698×10^{-7}	-2.4263×10^{-7}	-3.2322×10^{-7}	-4.2005×10^{-7}
0.20	-7.7487×10^{-7}	-1.0705×10^{-6}	-1.4265×10^{-6}	-1.8523×10^{-6}	-2.3518×10^{-6}
0.30	-4.0750×10^{-6}	-5.4253×10^{-6}	-7.00756×10^{-6}	-8.8504×10^{-6}	-1.0954×10^{-5}
0.40	-2.0823×10^{-5}	-2.6684×10^{-5}	3.3367×10^{-5}	-4.0951×10^{-5}	-4.9355×10^{-5}
0.50	-1.0638×10^{-4}	-1.3131×10^{-4}	-1.5899×10^{-4}	-1.8961×10^{-4}	-2.2258×10^{-4}
0.60	-5.5006×10^{-5}	-6.5506×10^{-4}	-7.6887×10^{-4}	-8.9144×10^{-4}	-1.0204×10^{-3}
0.70	-2.8938×10^{-3}	-3.3391×10^{-3}	-3.8109×10^{-3}	-4.3048×10^{-3}	-4.8149×10^{-3}
0.80	-1.5748×10^{-2}	-1.7702×10^{-2}	-1.8713×10^{-2}	-2.1744×10^{-2}	-2.3784×10^{-2}
0.90	-7.8672×10^{-3}	-8.7074×10^{-2}	-9.5203×10^{-2}	-1.0291×10^{-1}	-1.10268×10^{-1}
1.00	0.0	0.0	0.0	0.0	0.0
			$y = 1.0$		
0.00	0.0	0.0	0.0	0.0	0.0
0.10	8.6703×10^{-3}	1.1014×10^{-2}	8.01860×10^{-3}	4.2513×10^{-3}	5.300×10^{-5}
0.20	-9.6350×10^{-5}	5.7080×10^{-5}	1.1679×10^{-4}	8.3031×10^{-5}	2.400×10^{-6}
0.30	-2.3451×10^{-5}	-2.1457×10^{-6}	1.0511×10^{-6}	1.0044×10^{-6}	5.6000×10^{-8}
0.40	-3.4768×10^{-6}	-2.0940×10^{-7}	1.9400×10^{-8}	-5.9298×10^{-8}	-1.1058×10^{-8}
0.50	-5.4547×10^{-7}	-1.5471×10^{-8}	2.1242×10^{-9}	-8.3707×10^{-9}	-2.1519×10^{-9}
0.60	-8.1234×10^{-8}	-1.0684×10^{-9}	1.6535×10^{-10}	-5.11598×10^{-1}	-2.6331×10^{-10}
0.70	-1.5965×10^{-8}	-6.4340×10^{-11}	2.1029×10^{-11}	-2.1944×10^{-11}	-3.3282×10^{-11}
0.80	-2.8681×10^{-9}	-3.3633×10^{-12}	2.9218×10^{-12}	-4.8660×10^{-13}	-4.0303×10^{-12}
0.90	-4.9641×10^{-10}	-2.6651×10^{-13}	3.2432×10^{-13}	2.3740×10^{-14}	-3.0602×10^{-13}
1.00	0.0	0.0	0.0	0.0	0.0

these ordinate locations, the temperature difference is presented at various values of x. It may be observed that the best comparison is obtained for the case where the average velocity is used as the constant part that enters the fundamental solution. However, more investigation is needed to derive more conclusive insights.

3.3 Transient Conduction with Moving Boundaries and Phase Changes

3.3.1 Formulation

Many engineering applications require solution of heat conduction equations in the presence of moving boundaries and phase transformations. Applications of this kind include, for example, the production and melting of ice, solidification of castings, and the aerodynamic heating of missiles. All of these problems share the characteristic of an interface boundary which moves into the solid (melting) or into the liquid (solidification) region in accordance with the relative magnitudes of the temperature gradients on either side of it. The temperature gradient (flux) discontinuity at any point of the interface is related to its normal velocity by the equation balancing the rate of heat flow with the energy rate required to create a fresh amount of solid (or liquid) per unit time. In all of these problems, the thermal behavior of the media on either side of the interface is assumed to be governed by the well-known partial differential of heat conduction. Even though this differential equation is linear, the above problems are nonlinear due to the presence of the moving interface. Problems of this kind are well known as "Stefan" problems. The pervasiveness of this sort of problem, particularly when considered in generalized form, has prompted a surge of interest in this area in recent years.

The heat transfer equation to be treated here descends from the general equation

$$c\frac{\partial T}{\partial t} - k\nabla^2 T = LS \tag{3.60}$$

where T is temperature (°C), t is time (s), c is the heat capacity (cal/cm^3), k is the thermal conductivity [cal/(cm − s − °C], L is the latent heat content per unit volume of solid (cal/cm^3) and S is the volumetric rate of solid production due to phase change. Thermal diffusivity κ may be defined as $\kappa = k/\rho c$. Typically, the right-hand side of equation (3.60) needs to be evaluated as part of the solution. The parameters c and k are assumed to be constant within each constituent phase.

Assuming that the phase constitution is a function of temperature only and denoting the solid volume fraction as h, one may write

$$S = \frac{\partial h}{\partial T}\frac{\partial T}{\partial t} \tag{3.61}$$

and

$$\left(c - L\frac{\partial h}{\partial T}\right)\frac{\partial T}{\partial t} - k\nabla^2 T = 0 \tag{3.62}$$

thus expressing the inherent nonlinearity of the problem through the coefficients of the governing equation. This often presents severe computational problems, because the magnitude of the latent heat term may be enormous compared to other terms. For many practical problems, the phase change occurs only in a range of temperature and space that is quite narrow compared to the overall scales of the problem. Thus, the problem features a zone of crucial, intense activity, usually more concentrated than the mesh spacing, whose movements are also not known a priori. This makes it difficult to solve. Various investigators have proposed ways to solve equation (3.62) by moderating the sharpness of the $L\ \partial h/\partial T$ term, artificially extending the range of phase change temperature and diffusing the latent heat effects. Rolph and Bathe (1982) present a scheme designed to handle either finite or infinitesimal width phase change zones. They discretize the latent heat effects by using an iterative scheme which keeps track of a latent heat budget node by node, integrating over an appropriate contributory volume.

Alternatively, one may choose not to resist the localization of phase change processes (Curran et al. 1986), and in fact may exaggerate it, to treat this as a moving boundary problem. O'Neill (1983) has developed such an approach, where the phase change is confined conceptually to an infinitesimally thin surface across which the discontinuity of flux of sensible heat and the latent heat must balance. Phase change across the surface is regarded as instantaneous and complete, and the two contiguous phases occupy mutually exclusive zones. In this approach, the governing equation becomes

$$c\frac{\partial T}{\partial t} - k\,\nabla^2 T = 0 \tag{3.63}$$

within each phase Ω_l and Ω_s, with the condition

$$L\frac{ds}{dt} = \left(k\frac{\partial T}{\partial n}\boldsymbol{n}\right)^S - \left(k\frac{\partial T}{\partial n}\boldsymbol{n}\right)^L \tag{3.64}$$

On the interphase boundary Γ_{SL}. Here s is the location of a point on the interphase boundary and $\boldsymbol{n}^{\alpha}$ is a unit vector pointing normally outward from the α phase. Γ_{SL}, however, is generally an interior boundary. Thus, much ingenuity has been expended on accommodating a moving phase interface on a fixed grid as well as on transforming the problem to a moving and deforming numerical coordinate system.

The BEM is a good candidate for addressing Stefan-type problems because it offers complete and efficient multidimensional solutions while focusing primarily on system boundaries.

The adjoint equation to the transient heat conduction equation (3.63) may be written as

$$c\frac{\partial G}{\partial t} + k\,\nabla^2 G = \Delta(p, q) \tag{3.65}$$

subject to homogeneous boundary conditions. The fundamental solution to equation (3.65) may be expressed as

$$G(p, t_F; q, \tau) = \frac{\exp\left(-\dfrac{r^2}{4\kappa(t_F - \tau)}\right)}{4\pi\kappa(t_F - \tau)^{m/2}} \tag{3.66}$$

where r is the Euclidian distance between a source point p and a field point q, and m equals the number of spatial dimensions of the problem. It is also assumed here that the material properties remain constant within each phase, although they may vary from one phase to the other. Thus, they may be taken out of the equation by scaling the coordinates.

An integral equation representation of the governing equation may now be obtained by taking the inner product of equation (3.63) with G and subtracting from it the inner product of equation (3.65) with T.

The inner product of (3.63) with G may be expressed as

$$I_1 = \int_{r=0}^{t_F} d\tau \int_{\Omega_\alpha(\tau)} G\left(\nabla^2 T - \frac{\partial T}{\partial_\tau}\right) d\Omega \tag{3.67}$$

or

$$I_1 = \int_{\tau=0}^{t_F} d\tau\left\{\int_{\Gamma_\alpha(\tau)}\left(G\frac{\partial T}{\partial n} - T\frac{\partial G}{\partial n}\right) d\Gamma + \int_{\Omega_\alpha(\tau)} T\nabla^2 G\, d\Omega\right\}$$
$$-\int_{\tau=0}^{t_F} d\tau \int_{\Omega_\alpha(\tau)}\left(\frac{\partial(GT)}{\partial\tau} - T\frac{\partial G}{\partial\tau}\right) d\Omega \tag{3.68}$$

where n is the normal direction to the phase boundary Γ_α. It should be emphasized here that the domain $\Omega_\alpha(\tau)$ and its boundary $\Gamma_\alpha(\tau)$ of the α phase are functions of time and are not known a priori. It is this fact that distinguishes this class of Stefan problems. In contrast to the usual treatments of fixed domain problems, the order of time differentiation and spatial integration cannot be interchanged here without special measures. The difficulty stems from the fact that the limits of spatial integrations are functions of time.

Let us focus now on an integral of the form

$$\frac{\partial}{\partial \tau} \int_{\Omega_\alpha(\tau)} g \, d\Omega$$

when g is any integrable and differentiable function of space and time. The above integral may be transformed as,

$$\frac{\partial}{\partial \tau} \int_{\Omega_\alpha(\tau)} g(\boldsymbol{x}, \tau) \, d\Omega = \frac{\partial}{\partial \tau} \int_{\Omega_0} g(\boldsymbol{x}_0, \tau) J(\boldsymbol{x}_0, \tau) \, d\Omega_0 \tag{3.69}$$

where J is the determinant of the transformation between a reference or initial configuration Ω_0 and the current configuration $\Omega_\alpha(\tau)$. The integrand on the right-hand side of equation (3.69) is written entirely in the Ω_0 system. Thus, the differentiation on the right-hand side of equation (3.69) is performed holding $\boldsymbol{x}_0$ (not $\boldsymbol{x}$) constant. Since Ω_0 is independent of time, we may now write

$$\frac{d}{d\tau} \int_{\Omega_0} gJ \, d\Omega_0 = \int_{\Omega_0} \left(J \frac{dg}{d\tau} + g \frac{dJ}{d\tau} \right) d\Omega_0 \tag{3.70}$$

The derivative of J may also be expressed as

$$\frac{dJ}{d\tau} = J(\boldsymbol{\nabla} \cdot \boldsymbol{V}) \tag{3.71}$$

and

$$\frac{dg}{d\tau} = \left(\frac{\partial g}{\partial \tau} \right)_r + \boldsymbol{V} \cdot \boldsymbol{\nabla} g \tag{3.72}$$

where $\boldsymbol{V}$ is the velocity of a point in the changing volume of integration or on its boundary. Hence,

$$\boldsymbol{V} \equiv \frac{d\boldsymbol{r}}{d\tau} \tag{3.73}$$

Accordingly, we may write

$$\frac{d}{d\tau}\int_{\Omega_0} gJ\,d\Omega_0 = \int_{\Omega_0}\left[\left(\frac{\partial g}{\partial \tau}\right)_{\boldsymbol{x}} + \boldsymbol{\nabla}\cdot \boldsymbol{V}g\right]J\,d\Omega_0$$

$$= \int_{\Omega_\alpha(\tau)}\left[\left(\frac{\partial g}{\partial \tau}\right)_{\boldsymbol{x}} + \boldsymbol{\nabla}\cdot \boldsymbol{V}g\right]d\Omega \tag{3.74}$$

Application of the divergence theorem now yields

$$\frac{\partial}{\partial\tau}\int_{\Omega_\alpha(\tau)} g\,d\Omega = \int_{\Omega_\alpha(\tau)}\frac{\partial g}{\partial\tau}d\Omega + \int_{\Gamma_\alpha(\tau)} \boldsymbol{n}\cdot\boldsymbol{V}g\,d\Gamma \tag{3.75}$$

where it is understood now that the differentiations are performed holding $\boldsymbol{x}$ constant. This may be recognized as a generalized Reynolds transport theorem, where the factor $\boldsymbol{n}\cdot\boldsymbol{V}$ is the normal component of the velocity of the boundary Γ. For melting or solidification problems, the notion of $\Gamma_\alpha(\tau)$ is due only to phase change and not to movement of material. Thus $\boldsymbol{n}\cdot\boldsymbol{V}$ may be designated as $ds/d\tau$, the rate of progress of the phase change front. By implication, $\boldsymbol{V}$ in the interior as used above is arbitrary, in the sense that it is any velocity field compatible with $ds/d\tau$ on the surface, as dictated by the progress of the phase change.

Substituting the results of equation (3.75) into equation (3.68) and substituting GT for g, we get

$$I_1 = \int_0^{t_F} d\tau\int_{\Gamma_\alpha(\tau)}\left(G\frac{\partial T}{\partial n} - T\frac{\partial G}{\partial n} + GT\frac{ds}{d\tau}\right)d\Gamma + \int_{\Omega_\alpha(0)} GT\,d\Omega$$

$$+ \int_0^{t_F} d\tau\int_{\Omega_\alpha(\tau)} T\left(\frac{\partial G}{\partial\tau} + \boldsymbol{\nabla}^2 G\right)d\Omega \tag{3.76}$$

where it is realized that G equals zero at $\tau = t_F$ and $\int_{\Omega_\alpha(t_F)} GT\,d\Omega$ vanishes as τ approaches t_F. Considering again the inner product of equation (3.65) with T, we get,

$$I_2 = \int_0^{t_F} d\tau\int_{\Omega_\alpha(\tau)} T\left(\frac{\partial G}{\partial\tau} + \boldsymbol{\nabla}^2 G\right)d\Omega - T(p, t_F) \tag{3.77}$$

Finally, the governing equation for heat transfer in moving boundary phase change problems may be obtained by subtracting equation (3.77) from (3.76).

This yields (in a physical coordinate system)

$$T(p,t_F) = \kappa \int_0^{t_F} d\tau \int_{\Gamma_\alpha(\tau)} \Bigg(G(p,t_F;Q,\tau)\frac{\partial T}{\partial n_Q}(Q,\tau)$$
$$- \frac{\partial G(p,t_F;Q,\tau)}{\partial n_Q} T(Q,\tau) + \frac{1}{\kappa} G(p,t_F;Q,\tau)\ T(Q,\tau)V_n(Q)\Bigg)\, d\Gamma$$
$$+ \int_{\Omega_\alpha(0)} G(p,t_F;q,0)T(q,0)\, d\Omega \tag{3.78}$$

where n_Q is the outward normal at a field point Q to the boundary $\Gamma_\alpha\ (\tau)$ and $V_n(Q)$ is the outward normal component of the velocity of this field point.

Equation (3.78) gives the temperature, $T(p,t_F)$ at the source point p in the domain $\Omega_\alpha(t_F)$ at the final time t_F in terms of boundary temperatures and normal fluxes at every point on the boundary. In addition, it depends on the domain integral involving the initial temperature $T(q,0)$. A very important simplification arises when the initial temperature has the uniform value T_i throughout the body. In that case, one can eliminate the domain integral in equation (3.78) by writing it in terms of a modified temperature

$$\hat{T}(q,t) = T(q,t) - T_{\text{init}} \tag{3.79}$$

with this modification, no integrations over the domain $B(t)$ are required.

Taking the usual limit as the interior source point p approaches a boundary source point P, we get

$$C(P)\hat{T}(P,t_F) = \kappa \int_0^{t_F} d\tau \int_{\Gamma_\alpha(\tau)} \Bigg(G(P,t_F;Q,\tau)\frac{\partial \hat{T}}{\partial n_Q}(Q,\tau) - \frac{\partial G(P,t_F;Q,\tau)}{\partial n_Q}\hat{T}(Q,\tau)$$
$$+ \frac{1}{\kappa} G(P,t_F;Q,\tau)\hat{T}(Q,\tau)V_n(Q)\Bigg)\, d\Gamma \tag{3.80}$$

where $C(P)$ is the usual corner tensor whose value depends on the local geometry of the boundary $\Gamma_\alpha(t_F)$ at P. If the boundary $\Gamma_\alpha(t_F)$ at P is smooth, $C(P){=}k/2$. Otherwise, $C(P) = k\theta$ where θ is the included angle at P in radians. It should also be noted here that equations (3.78 and 3.80) contain boundary integrals only when $\Omega_\alpha(0) = 0$. It may also happen in a given problem that part of the boundary $\Gamma_\alpha(\tau)$ is moving while the rest is not. In that case equation (3.80) must be suitably applied with proper conditions on the moving and stationary parts of $\Gamma_\alpha(\tau)$.

A detailed description of numerical implementation of equation (3.80) for solidification problems is given later in chapter 7.

3.4 Transient Conduction–Convection

3.4.1 Formulation

The heat transfer in many manufacturing processes requires a transient conduction–convection model. This is particularly true when one tries to investigate features of a process at a shorter time scale. While the heat transfer in a turning operation may be modeled as a steady-state conduction–convection phenomenon, a transient model is needed to understand chip curling. There are also various other processes, such as mining, grinding, welding, where the process is either intermittent (chip formation in mining or grinding) or involves transient effects (moving boundary and phase change in welding) that are crucial for developing an accurate model. Many problems of this class may be treated as transient conduction–convection problems. In the following development, as a first attempt, the convective velocity is assumed to remain constant.

The BEM has been applied to transient conduction–convection problems by various researchers. Curran et al. (1980) used a time-dependent fundamental solution defined only for the diffusive terms. This requires a domain integral. Aral and Tang (1989) used a secondary reduction BEM (SR-BEM) algorithm to reduce the BEM finite difference computations for parabolic partial differential equations to a boundary-only form. Their algorithm uses the fundamental solution for the steady-state diffusion operator and requires internal discretization. The SR-BEM algorithm is essentially a procedure for efficiently eliminating the internal degrees of freedom. Accordingly, the temporal features of a transient problem and its associated time integration issues are not addressed. Typically, very small time steps are needed when a transient problem is attempted by employing a steady-state fundamental solution (Brebbia et al. 1984). Moreover, such an algorithm is susceptible to inherent numerical oscillations and false diffusion. Recently, Lim et al. (1994) have developed a transient BEM formulation based on the time-dependent fundamental solution of the transient convection–diffusion operator. This approach, which reduces the governing parabolic partial differential equation to a boundary-only form, is pursued here.

The governing equation for the transient conduction–convection problem may be expressed as

$$\left(\frac{\partial T(x,t)}{\partial t} + v_i \frac{\partial T(x,t)}{\partial x_i} \right) - \kappa \frac{\partial^2 T(x,t)}{\partial x_i^2} = 0 \qquad \text{in } \Omega \tag{3.81}$$

Here v_i is the convective velocity and $\kappa = k/\rho c$ is the thermal diffusivity. It is assumed that thermal conductivity, specific heat, and density remain constant. For situations involving spatial variations of κ, the domain B may be divided into several zones and the thermal diffusivity κ may be assumed to be piecewise constant over each of these zones. A BEM formulation for planar problems is presented in this section. A similar technique may also be used to obtain a three-dimensional BEM formulation.

In the present applications, the convective velocity is assumed to be constant in time and space. A boundary-only form of the BEM formulation is developed first. Cases involving spatially nonuniform convection velocities may also be analyzed using zoning, such that the velocity is constant in each zone, or by introducing a domain integral incorporating the velocity gradients. The zoning approach has been used by Chan and Chandra (1991b, 1991c) for steady-state machining processes.

For constant velocities of convection, we may introduce a coordinate system with its axes oriented in parallel and normal directions to the velocity vector. Without any loss of generality, the governing equation may be expressed as

$$\frac{\partial T(x,t)}{\partial t} + Pe\frac{\partial T(x,t)}{\partial x_1} = \nabla^2 T(x,t) \qquad \text{in } \Omega \tag{3.82}$$

where the x_1 direction is along the direction of the resultant convection velocity. Pe is the Peclet number ($Pe = VL/\kappa$). Here V is the convective velocity in the x_1 direction, and L is a characteristic length. The boundary conditions are

$$T(x,t) = \overline{T}(x,t) \qquad \text{on } \Gamma_T \tag{3.83}$$

and

$$\frac{\partial T(x,t)}{\partial x_i} n_i = \overline{q}(x,t) \qquad \text{on } \Gamma_q \tag{3.84}$$

Equation (3.82) applies to an Eulerian reference frame that remains spatially fixed while the material flows through it. The convective term represents the energy transported by the material as it moves through the reference frame. The surface flux $\overline{q}$ includes a contribution from convective cooling losses, $q^{(c)}$, which may be written as

$$q^{(c)}(x,t) = Nu(T(x,t) - T_\infty) \tag{3.85}$$

T_∞ represents the ambient temperature and Nu is the Nusselt number (hL/k), where h is the convective heat transfer coefficient.

Let us also consider the adjoint equation,

$$-\frac{\partial G(x,t)}{\partial t} - Pe\frac{\partial G(x,t)}{\partial x_1} = \nabla^2 G(x,t) + \delta(x(q)-x(p))\delta(t-\tau) \tag{3.86}$$

where p and q are a source point and field point, respectively, in the domain (P and Q represent a source point and field point, respectively, on the boundary), τ is the final time, and t is the time variable.

Applying the divergence theorem and following the procedure of Tanaka et al. (1986) and Chan and Chandra (1991b,c), an integral representation of the governing equation may be obtained as

$$\begin{aligned} T(p,\tau;Q,t) = \int_0^\tau dt \int_{\Gamma(t)} \biggl(& G(p,\tau;Q,t)\frac{\partial T}{\partial n_Q}(Q,t) \\ & - \frac{\partial G}{\partial n_Q}(p,\tau;Q,t)T(Q,t)\biggr) d\Gamma \\ & - Pe\int_0^\tau dt\int_{\Gamma(t)} G(p,\tau;Q,t)T(Q,t)n_1(Q,t)\, d\Gamma \end{aligned} \tag{3.87}$$

The normal flux at the boundary $q(Q,t)$ is equal to $\partial T(Q,t)/\partial n_Q$. In deriving equation (3.87), it is assumed that

$$T(x,0) = 0 \tag{3.88}$$

Thus, $T(x,t)$ represents the temperature rise from a uniform initial temperature in the body. A domain integral appears in equation (3.87) if the initial temperature distribution is not uniform (Zabaras and Mukherjee 1987). However, in many practical problems, the initial temperature distribution remains uniform and, for simplicity, this is assumed here.

The fundamental solution $G(p,\tau;q,t)$ (Carslaw and Jaeger 1986) is given as

$$\begin{aligned} G(p,\tau;q,t) = & \frac{1}{4\pi(\tau-t)} \\ & \times \exp\left(\frac{-\{[(x_1(q)-x_1(p)) + Pe(\tau-t)]^2 + (x_2(q)-x_2\cdot(p))^2\}}{4(\tau-t)}\right) \end{aligned} \tag{3.89}$$

for planar problems. This may also be written as

$$\begin{aligned} G(p,\tau;q,t) = & \frac{1}{4\pi(\tau-t)}\exp\left(-\frac{r^2}{4(\tau-t)}\right)\exp\left(-\frac{Pe}{2}(x_1(q)-x_1(p))\right) \\ & \times \exp\left(-\frac{Pe^2}{4}(\tau-t)\right) \end{aligned} \tag{3.90}$$

Here r is the Euclidean distance from the source point to the field point. In equation (3.90), the first exponent represents the conduction effects, while the second and third exponents represent the effects due to convection and to transient convection, respectively.

A boundary integral equation for the transient conduction–convection problem may now be obtained by taking the limit as the source point p inside the domain approaches the point P on the boundary ($p \to P$). This gives

$$C(P)T(P,\tau) = \int_0^\tau dt \int_{\Gamma(t)} \left(G(P,\tau;Q,t)\frac{\partial T}{\partial n_Q}(Q,t) - \frac{\partial G}{\partial n_Q}(P,\tau;Q,t)T(Q,t) \right) d\Gamma$$
$$- Pe \int_0^\tau dt \int_{\Gamma(t)} G(P,\tau;Q,t)T(Q,t)n_1(Q,t)\, d\Gamma \qquad (3.91)$$

The coefficient $C(P)$, in general, depends on the local geometry at P. If the boundary is locally smooth at P, $C = 1/2$ for two-dimensional problems. Otherwise, it may be evaluated indirectly (Banerjee and Butterfield 1981, Mukherjee 1982, Brebbia et al. 1984). In the present work, the coefficient $C(P)$ at a geometric corner is evaluated directly (Brebbia et al. 1984) in a Cauchy principal value sense.

3.4.2 Numerical Implementation

3.4.2.1 Discretization

Numerical implementation of the BEM equations (3.87, 3.91) for the transient conduction–convection problem is discussed in this section. The first step is the discretization of the boundary of the two-dimensional domain into boundary elements. In this case, discretizations in both time and space are needed. Accordingly, the boundary is discretized into N spatial boundary elements and the time dimension is subdivided into F time steps. A linear interpolation in space and time is used here for temperature and heat flux:

$$T(x,t) = \sum_{j=1}^{N} \sum_{k=1}^{F} \psi_j \phi^k T_j^k \qquad (3.92)$$

and

$$q(x,t) = \sum_{j=1}^{N} \sum_{k=1}^{F} \psi_j \phi^k q_j^k \qquad (3.93)$$

where ψ and ϕ are interpolation functions in space and time, respectively.

A discretized version of the boundary integral equation (3.91) may now be written as

$$C(P_M)T(P_M, t_F; Q, t) =$$

$$\sum_{k=1}^{F} \int_{T_{k-1}}^{t_k} \phi^{k-1}\, dt \Bigg[\sum_{j=1}^{N} \Bigg(\int_{\Delta\Gamma_j(t)} G(P_M, t_F; Q, t)(\psi_j q_j^{k-1} + \psi_{j+1} q_{j+1}^{k-1})\, d\Gamma_j$$

$$- \int_{\Delta\Gamma_j(t)} \frac{\partial G}{\partial n_Q}(P_M, t_F; Q, t)(\psi_j T_j^{k-1} + \psi_{j+1} T_{j+1}^{k-1})\, d\Gamma_j$$

$$- \int_{\Delta\Gamma_j(t)} G(P_M, t_F; Q, t) Pe\, n_1(Q)(\psi_j T_j^{k-1} + \psi_{j+1} T_{j+1}^{k-1})\, d\Gamma_j \Bigg) \Bigg]$$

$$+ \sum_{k=1}^{F} \int_{t_{k-1}}^{t_k} \phi^k\, dt \Bigg[\sum_{j=1}^{N} \Bigg(\int_{\Delta\Gamma_j(t)} G(P_M, t_F; Q, t)(\psi_j q_j^k + \psi_{j+1} q_{j+1}^k)\, d\Gamma_j$$

$$- \int_{\Delta\Gamma_j(t)} \frac{\partial G}{\partial n_Q}(P_M, t_F; Q, t)(\psi_j T_j^k + \psi_{j+1} T_{j+1}^k)\, d\Gamma_j$$

$$- \int_{\Delta\Gamma_j(t)} G(P_M, t_F; Q, t) Pe\, n_1(Q)(\psi_j T_j^k + \psi_{j+1} T_{j+1}^k)\, d\Gamma_j \Bigg) \Bigg] \quad (3.94)$$

In equation (3.94), $T(P_M)$ represents the temperature at a point P that coincides with node M. T_j^k, T_{j+1}^k, q_j^k, and q_{j+1}^k are nodal values of temperature and flux at time t^k.

Defining ($q = 1, 2$),

$$a_\gamma^k(P_M, \Delta\Gamma_j) = \int_{t_{k-1}}^{t_k} \phi^k\, dt \int_{\Delta\Gamma_j(t_k)} \psi_\gamma \Bigg(\frac{\partial G}{\partial n_Q}(P_M, t_F; Q, t)$$

$$+ Pe\, n_1(Q) G(P_M, t_F; Q, t) \Bigg) d\Gamma_j \quad (3.95)$$

and

$$b_\gamma^k(P_M, \Delta\Gamma_j) = \int_{t_{k-1}}^{t_k} \phi^k dt_k \int_{\Delta\Gamma_j(t_k)} \psi_\gamma [G(P_M, t_F; Q, t)]\, d\Gamma_j \quad (3.96)$$

as the coefficient matrices relating the Mth source point with the jth boundary element, we may express equation (3.94) in a matrix form:

$$\sum_{k=1}^{F} \sum_{j=1}^{N} (-A_{ij}^{k-1} T_j^{k-1} + B_{ij}^{k-1} q_j^{k-1} - A_{ij}^k T_j^k + B_{ij}^k q_j^k) = 0 \quad (3.97)$$

Assembled nodal coefficients A_{ij}^k and B_{ij}^k are formed as

$$A_{ij}^k = a_{(\gamma=2)}^k(i, \Delta\Gamma_{j-1} + a_{(\gamma=1)}^k(i, \Delta\Gamma_j) + C(i)\Delta(F, k)\Delta(i, j) \tag{3.98}$$

and

$$B_{ij}^k = b_{(\gamma=2)}^k(i, \Delta\Gamma_{j-1} + b_{\gamma=1)}^k(i, \Delta\Gamma_j) \tag{3.99}$$

Here, $\gamma = 1$ or 2 represents the first or second node of the spatial discretization for the corresponding boundary element. The associated time integration scheme is discussed in detail later.

Integrals of kernels over the elements in equations (3.94–3.96) must be obtained carefully. Details of the integration procedure are discussed in a later section.

At each time step and at each location over the entire boundary of the domain, either T or q (or a combination of T and q) is prescribed for a well-posed problem. Also, at each time step, T and q are known at every location on the boundary for all previous time steps. Equation (3.97) may then be rearranged as

$$\sum_{j=1}^{N} \tilde{A}_{ij}^F Y_j^F = F_i + \sum_{j=1}^{N} (A_{ij}^{F-1} T_j^{F-1} - B_{ij}^{F-1} q_j^{F-1})$$
$$- \sum_{k=1}^{F-1} \sum_{j=1}^{N} (-A_{ij}^{k-1} T_j^{k-1} + B_{ij}^{k-1} q_j^{k-1} - A_{ij}^k T_j^k + B_{ij}^k q_j^k) \tag{3.100}$$

where F_i is the vector formed by multiplying the known variables by the appropriate coefficient matrix at the final time. Equation (3.100) may now be used to solve for the unknown variables Y_j^F at the final time.

3.4.2.2 Integration of kernels in time and space

As observed from equation (3.90), as t approaches the final time τ, the quantity $(\tau - t)$ approaches zero. This quantity, $(\tau - t)$, appears in the denominator of the expression for $G(P, \tau; Q, t)$ and $\partial G/\partial n_Q\ (P, \tau; Q, t)$, as well as in the denominator of one of the exponents. So, $G(P, \tau; Q, t)$ or $\partial G/\partial n_Q(P, \tau; Q, t)$ is regular as t approaches τ for $r \neq 0$. However, a singularity occurs if $r \to 0$ at the same time as $t \to \tau$. This is true for the diagonal terms of the G or $\partial G/\partial n_Q$ kernel (when P *equiv* Q) at the final time.

In the present work, the regular integrands (for all terms when $t < \tau$ and for off-diagonal terms when $t = \tau$) are integrated numerically. The time integrations are performed by a quadrature scheme for improper integrals (Press et al. 1986), while the spatial integrations are done by Gauss quadrature.

It is important to note here that the order of the singularity for the diagonal term at the final time is not known a priori. In the present work, the term $\exp[-Pe^2/4(\tau-t)]$ appearing in both G and $\partial G/\partial n_Q$ is first expanded in a Taylor series in $(\tau-t)$. We obtain

$$G(P,\tau;Q,t)=\frac{1}{4\pi(\tau-t)}\exp\left(-\frac{r^2}{4(\tau-t)}\right)$$
$$\times\exp\left(-\frac{Pe}{2}(x_1(Q)-x_1(P))\right)\left(1+\sum_{n=1}^{\infty}\frac{(-Pe^2/4)^n}{n!}(\tau-t)^n\right) \tag{3.101}$$

and

$$\frac{\partial G}{\partial n_Q}(P,\tau;Q,t)=\frac{-1}{8\pi(\tau-t)^2}\exp\left(-\frac{r^2}{4(\tau-t)}\right)$$
$$\times\exp\left(-\frac{Pe}{2}(x_1(Q)-x_1(P))\right)[(x_1(Q)-x_1(P)+Pe(\tau-t))n_1$$
$$+(x_2(Q)-x_2(P))n_2]\left(1+\sum_{n=1}^{\infty}\frac{(-Pe^2/4)^n}{n!}(\tau-t)^n\right) \tag{3.102}$$

Integrating $\phi^k G(k=F-1,F)$ in time with $t_k=t_F$, we get

$$\int_{t_{F-1}}^{t_F}\phi^{F-1}G(P,t_F;Q,t)\,dt_F=\frac{1}{4\pi\Delta t_F}\exp\left(-\frac{Pe}{2}(x_1(Q)-x_1(P))\right)$$
$$\times\left(I_1+\sum_{n=1}^{\infty}\frac{(-Pe^2/4)^n}{n!}I_2\right) \tag{3.103a}$$

$$\int_{t_{F-1}}^{t_F}\phi^{F}G(P,t_F;Q,t)\,dt_F=\frac{1}{4\pi}\exp\left(-\frac{Pe}{2}(x_1(Q)-x_1(P))\right)$$
$$\times\left(I_3+\sum_{n=1}^{\infty}\frac{(-Pe^2/4)^n}{n!}I_4\right)-\int_{t_{F-1}}^{t_F}\phi^{F-1}G(P,t_F;Q,t)\,dt_F \tag{3.103b}$$

where

$$I_1=\frac{r^2}{4}\left(\frac{e^{-z_{F-1}}}{z_{F-1}}-E_1(z_{F-1})\right) \tag{3.103c}$$

with

$$z_{F-1}=\frac{r^2}{4\Delta t_{F-1}}\qquad\text{and}\qquad E_1(z_{F-1})=\int_{z_{F-1}}^{\infty}\frac{e^{-\xi}}{\xi}\,d\xi \tag{3.103d}$$

Also,

$$I_2 = \left(\frac{r^2}{4}\right)^{n+1}\left(- \sum_{m=1}^{n+1} A^{-1}(m, n+2)(z_{F-I}^{m-(n+2)} e^{-z_{F-1}}) + B^{-1}(n+2)E_1(z_{k-1})\right) \tag{3.103e}$$

with

$$A(m,n) = \prod_{l=1}^{m} (l-n) \qquad \text{and} \qquad B(n) = \prod_{l=1}^{n-1} (l-n) \tag{3.103f}$$

and

$$I_3 = E_1(z_{F-1}) \tag{3.103g}$$

$$I_4 = \left(\frac{r^2}{4}\right)^{n}\left(- \sum_{m=1}^{n} A^{-1}(m, n+1)(z_{F-1}^{m-(n+1)} e^{-z_{F-1}}) + B^{-1}(n+1)E_1(z_{F-1})\right) \tag{3.103h}$$

It should be noted here that as z_k approaches zero, $E_1(z_k)$ approaches $\ln(z_k)$. Accordingly, the order of singularity in G is of $\ln(r)$ in spatial dimensions and this occurs at the final time.

Similarly, time integration of $\phi^k \; \partial G/\partial n_Q$ gives

$$\int_{t_{F-1}}^{t_F} \phi^{F-1} \frac{\partial G}{\partial n_Q}(P, t_F; Q, t)\, dt_F = \frac{d}{8\pi \Delta t_k} \exp\left(- \frac{Pe}{2}(x_1(Q) - x_1(P))\right) \times \left(I_3 + \sum_{n=1}^{\infty} \frac{(-Pe^2/4)^n}{n!} I_4 \right) - \frac{Pe\, n_1(Q)}{2} \int_{t_{F-1}}^{t_F} \phi^{F-1} G(P, t_F; Q, t)\, dt_F \tag{3.104a}$$

$$\begin{aligned} \int_{t_{F-1}}^{t_F} \phi^{F} \frac{\partial G}{\partial n_Q}(P, t_F; Q, t)\, dt_F &= \frac{d}{8\pi} \exp\left(- \frac{Pe}{2}(x_1(Q) - x_1(P))\right) \\ &\times \left[J_1 + \left(- \frac{Pe^2}{4}\right) I_3 + \sum_{n=2}^{\infty} \frac{(-Pe^2/4)^n}{n!} J_2 \right] \\ &- \frac{Pe\, n_1(Q)}{8\pi} \exp\left(- \frac{Pe}{2}(x_1(Q) - x_1(P))\right)\left(I_3 + \sum_{n=1}^{\infty} \frac{(-Pe^2/4)^n}{n!} I_4 \right) \\ &- \int_{t_{F-1}}^{t_F} \phi^{F-1} \frac{\partial G}{\partial n_Q}(P, t_F; Q, t)\, dt_F \end{aligned} \tag{3.104b}$$

Here,

$$d = (x_i(Q) - x_i(P))n_i(Q) \qquad \text{when } i = 1, 2 \tag{3.104c}$$

$$J_1 = \frac{4}{r^2} e^{-z_{F-1}} \tag{3.104d}$$

$$J_2 = \left(\frac{r^2}{4}\right)^{n-1} \left(- \sum_{m=1}^{n-1} A^{-1}(m,n)(z_{F-1}^{m-n} e^{-z_{F-1}}) + B^{-1}(n) I_3 \right) \tag{3.104e}$$

Observations of the terms involved in J_1 also reveal that the order of singularity in $\partial G/\partial n_Q$ is $1/r$ in spatial dimensions, and this also occurs at the final time.

After the orders of spatial singularities (at $t = \tau$) in G and $\partial G/\partial n_Q$ are determined, appropriate integration schemes may be used for the diagonal terms at the final time step. Since the kernel G contains a $\ln r$ singularity, a numerical scheme for integrating improper integrals (Press et al. 1986) utilizing Romberg integration as an open interval is used for spatial integration of the singular terms of G. The kernel $\partial G/\partial n_Q$, however, contains stronger singularities of order $1/r$. The diagonal terms of $\partial G/\partial n_Q$ are integrated analytically using the methods of Banerjee and Butterfield (1981), Mukherjee (1982), and Brebbia et al. (1984).

3.4.3 Example Problems and Numerical Results

Figure 3.7 shows a schematic diagram of the test problem. The boundary conditions are

$$T = 1, \qquad x_1 = 0, \qquad 0 \le x_2 \le 1 \tag{3.105a}$$

$$T = 0, \qquad x_1 = 1, \qquad 0 \le x_2 \le 1 \tag{3.105b}$$

$$q = 0, \qquad x_2 = 0, \qquad 0 \le x_1 \le 1 \tag{3.105c}$$

$$q = 0, \qquad x_2 = 1, \qquad 0 \le x_1 \le 1 \tag{3.105d}$$

and the initial condition is

$$T = 0, \qquad t = 0 \tag{3.105e}$$

The boundary conditions specified above represent a one-dimensional problem. A closed-form analytical solution is obtained using the method of separation of variables. This may be expressed as

$$T(x,t) = 2 \sum_{n=1}^{\infty} \frac{n\pi}{\lambda_n} (1 - \exp(-\lambda_n t) \exp\left(\frac{(Pe)x}{2}\right) \sin(n\pi x) \tag{3.106}$$

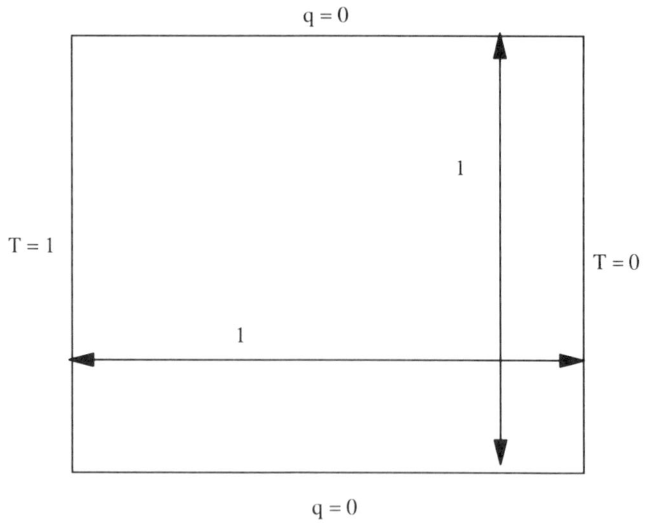

Figure 3.7. Schematic diagram of the test problem (uniform grid and a total of 40 nodes are used; time increment depends on the Peclet number).

where

$$\lambda_n = \left(\frac{Pe}{2}\right)^2 + (n\pi)^2$$

A standard successive-over-relaxation finite difference method (SOR) is also used to solve equation (3.82) subject to boundary and initial conditions in equations (3.105*a*–*e*). In this finite difference solution, the conduction term is approximated by a central difference, the convective term is modeled by an upwind difference, and the transient part is modeled by the Crank–Nicolson technique.

Figure 3.8 shows the transient fields for $Pe = 20$ for five different times. The BEM results are obtained using a uniform mesh size of 0.1 and a constant time increment of 0.01. The SOR results were obtained using the same uniform mesh size and time increment. It can be observed that the BEM results closely match the analytical results. Further, the BEM performs much better in predicting the temperature profiles. In SOR, the false diffusion caused by the upwind difference of the convective term introduces large errors in the temperature field. The results from SOR using a uniform mesh size of 0.01 and time increment of $\Delta t = 10^{-4}$ are plotted in figure 3.9. It is clear that the SOR results can be improved by reducing the grid size. However, the computation time required to obtain these fine-mesh results is considerably longer because of the decrease in Δx and Δt. Moreover, these refinements make the computational time prohibitively large for higher Peclet numbers. In the BEM formulation, the Green's function used is the transient moving heat source solution (Carslaw and Jaeger 1986), which gives a better approximation of the convective term. Consequently, false diffusion is minimized (Tanaka et al. 1986, Chan and Chandra 1991b).

To illustrate the capability of the BEM formulation in handling high Peclet number convection, the transient fields for $Pe = 2000$ for five different times are plotted in figure 3.10. The results were obtained for a uniform mesh size of 0.1 and a constant time increment of 10^{-5} for the BEM calculation and a uniform mesh size of 0.01 and a constant time increment of 10^{-5} for the SOR calculation. It can be observed that there is a very good match between the BEM results and those of the asymptotic solutions. The analytical solution given in equation (3.106) fails to converge for $Pe = 2000$. Consequently, we have to compare it with the asymptotic solution. In the limit of large Pe numbers, the convection term dominates and the problem can be solved using asymptotic expansion. The outer region is dominated by convection, whose solution is a sharp temperature front moving downstream with the wave velocity. Around the wave front, the inner region, scaled as $Pe^{-1/2}$, is dominated by diffusion. The solution can be expressed as

$$T(x_1, t) = \begin{cases} \dfrac{1}{2}\left[2 - \operatorname{erfc}\left(-\dfrac{(x_1 - t)Pe}{2t^{1/2}}\right)\right], & x_1 \le Pe \\ \dfrac{1}{2}\operatorname{erfc}\left(\dfrac{(x_1 - t)Pe}{2t^{1/2}}\right), & x_1 > Pe \end{cases} \tag{3.107}$$

Figure 3.11 shows that the phenomenon is accurately captured by the BEM; however, the SOR is totally inadequate. It should be pointed out that the time increment was reduced from 10^{-2} ($Pe = 20$) to 10^{-5} ($Pe = 2000$) in the BEM calculations. The BEM formulation is unconditionally stable. It was observed that the term $(Pe^2/4)^n\,(\tau - t)^n$ in equations (3.101) and (3.102) has a tendency to create an overflow in the computer for high Peclet numbers. This may be avoided by choosing a time increment so that the above term is less than the overflow capacity of the computer. Table 3.10 shows the maximum Peclet numbers for a fixed time increment. The table was generated by fixing the time increment and increasing the Peclet number until the program terminated with an overflow error message. Double precision was used in these calculations. In the case of SOR, the time increment is limited by the stability criterion of the Courant number being less than 1 (Anderson et al. 1989). The Courant number may be defined as $Pe(\Delta t/\Delta x)$. As Pe increases, the grid size has to be decreased to capture the high gradients.

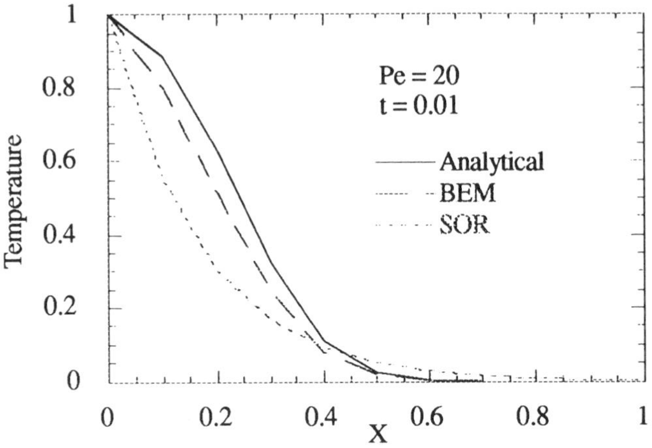
Pe = 20
t = 0.01
Analytical
BEM
SOR
Temperature
X
0
0.2
0.4
0.6
0.8
1

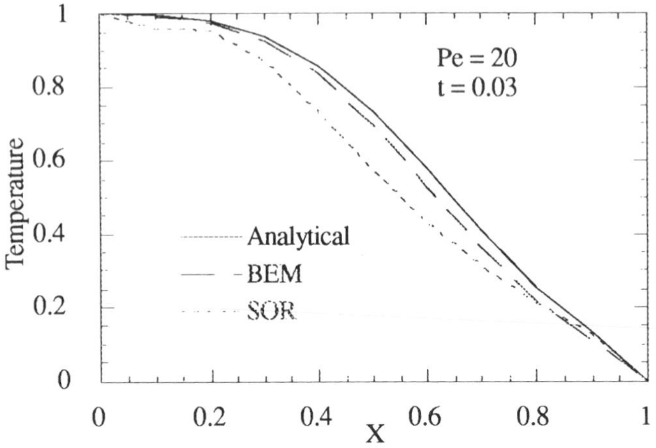
Pe = 20
t = 0.03
Analytical
BEM
SOR
Temperature
X
0
0.2
0.4
0.6
0.8
1

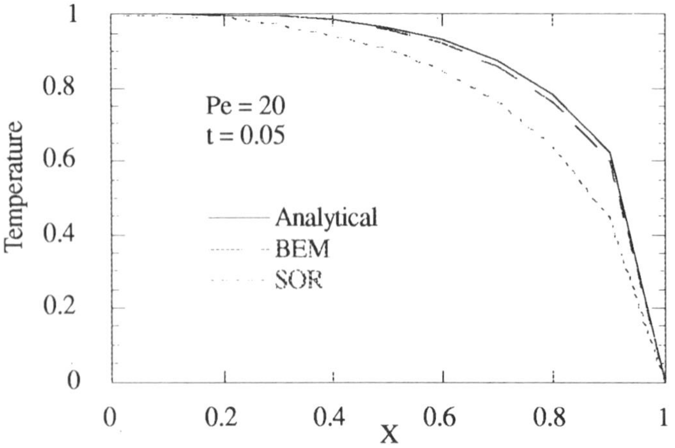
Pe = 20
t = 0.05
Analytical
BEM
SOR
Temperature
X
0
0.2
0.4
0.6
0.8
1

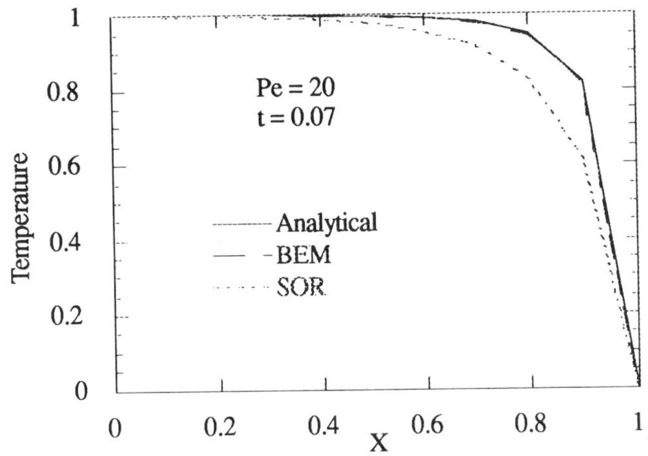

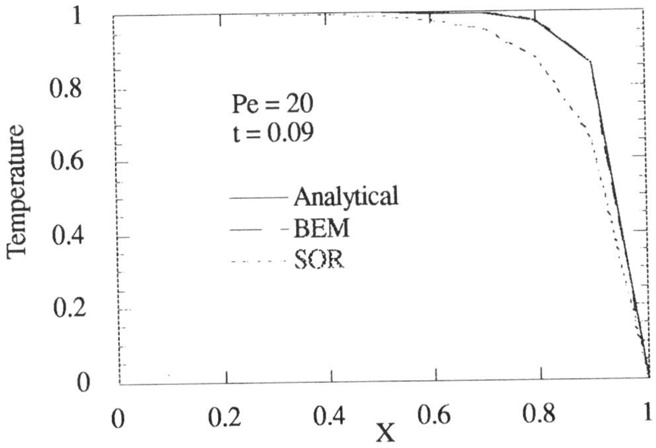

Figure 3.8. Comparison of the BEM solution with the analytical and SOR solutions at $Pe = 20$ and $\Delta x = 0.1$ for $t = 0.01$, $t = 0.03$, $t = 0.05$, $t = 0.07$, and $t = 0.09$ as indicated.

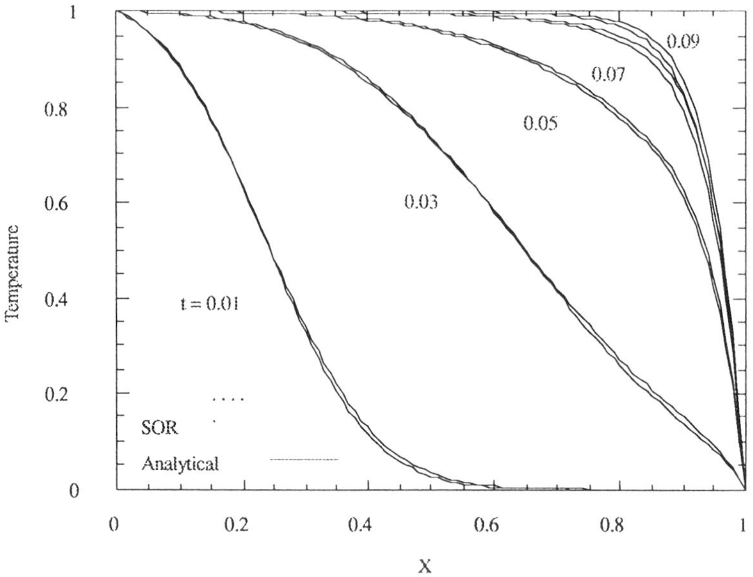

Figure 3.9. Comparison of the SOR solution with the analytical solution at $Pe = 20$, $\Delta x = 0.1$, $t = 0.01$ through 0.09 and $\Delta t = 10 \times 10^{-5}$.

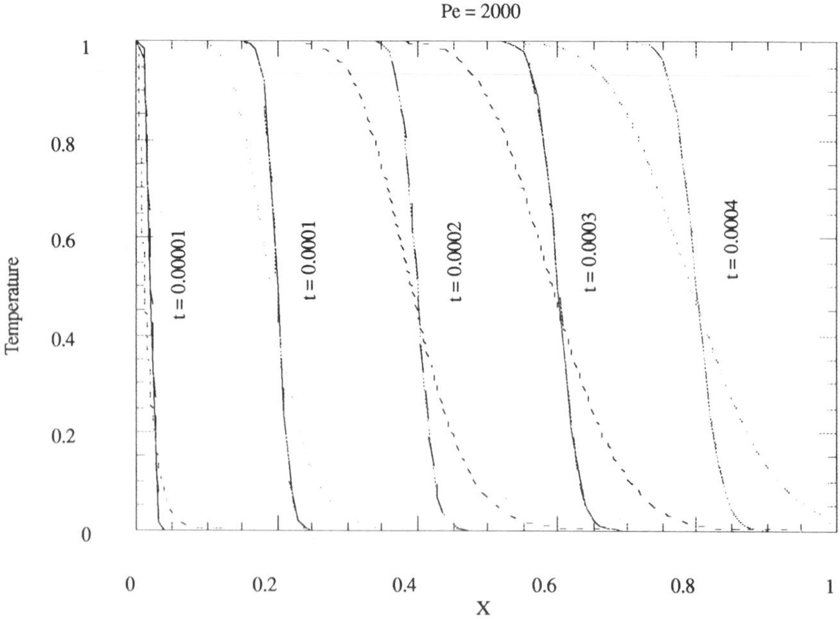

Figure 3.10. Comparison of the BEM solution with the asymptotic and SOR solutions at $Pe = 2000$ and $\Delta x = 0.1$ (dashed lines = SOR results; solid lines = BEM results and analytical results).

Table 3.10. Relationship between Δt and maximum *Pe*

	BEM		SOR	
Δt	Maximum *Pe*	ΔX	Maximum *Pe*	ΔX
1×10^{-2}	100	0.1	10	0.1
1×10^{-3}	300	0.1	30	0.03
1×10^{-4}	1000	0.1	100	0.01
1×10^{-5}	2100	0.1	300	0.003

This can be observed from the results of $Pe = 20$ given in figures 3.9 and 3.10. To capture the steep gradient, the grid size should be 0.01 and the time increment reduced to 10^{-4}. As a result, the time increment should be restricted to $\Delta t \sim Pe^{-2}$.

To compare the accuracies of the BEM and SOR methods, we define the following errors. The error at each time step is defined as

$$E = \left(\sum_{i=1}^{n} \frac{(T - T_i)^2}{N} \right)^{1/2} \tag{3.108}$$

where T is the analytical solution, T_i is the approximate solution, and N is the number of nodes. The errors of the centerline temperature ($y = 0.5$) for both BEM and SOR results are tabulated in table 3.11. It can be observed that, for the same mesh size, BEM results have smaller errors than the SOR, except for the case of $Pe = 5.0$. In an attempt to determine the reason, temperature fields at various time steps were plotted (figure 3.11). It can be observed that the SOR prediction gives a lower temperature at the beginning of the domain and a higher temperature at the end, resulting in a smaller error (equation 3.108).

3.5 Thermal Stresses and Thermomechanical Aspects

The stress fields associated with spatial and temporal variations of temperatures greatly influence the performance of various manufacturing processes both in terms of product quality and process efficiency. The stress fields associated with solidification processes play crucial roles in determining the quality of the final casting. In machining processes, tool wear on the rake face is significantly affected by chip curling, which in turn is primarily governed by the associated residual stresses in the chip.

Thermomechanical coupling also plays significant roles in numerous manufacturing processes. Forming (cold and hot), machining, welding,

Table 3.11. RMS error of BEM and SOR results

		RMS Error	
Pe	*t*	BEM	SOR
1.0	0.01	4.77801×10^{-2}	7.78740×10^{-2}
	0.03	1.38802×10^{-2}	2.08699×10^{-2}
	0.05	9.01710×10^{-3}	1.14195×10^{-2}
	0.07	7.04800×10^{-3}	7.01487×10^{-3}
	0.09	5.77009×10^{-3}	4.44575×10^{-3}
5.0	0.01	5.28524×10^{-2}	8.86331×10^{-2}
	0.03	1.90902×10^{-2}	1.92125×10^{-2}
	0.05	1.42936×10^{-2}	1.04627×10^{-2}
	0.07	1.17259×10^{-2}	8.56085×10^{-2}
	0.09	9.41962×10^{-3}	8.38559×10^{-3}
10.0	0.01	5.66623×10^{-2}	0.109351
	0.03	2.52636×10^{-2}	3.55529×10^{-2}
	0.05	1.97946×10^{-2}	2.92629×10^{-2}
	0.07	1.39803×10^{-2}	3.64733×10^{-2}
	0.09	8.56032×10^{-3}	4.92900×10^{-3}
20.0	0.01	5.51768×10^{-2}	0.164496
	0.03	3.23170×10^{-2}	9.54526×10^{-2}
	0.05	1.44366×10^{-2}	9.56546×10^{-2}
	0.07	3.44543×10^{-3}	8.66238×10^{-3}
	0.09	1.55643×10^{-3}	7.89693×10^{-3}

solidification, and so on, are only a few examples of processes influenced by thermomechanical coupling. Various recently developed processes such as sinter-forming, hot isostatic pressing (HIP), and thin-film deposition are primarily governed by thermomechanical coupling. In many of these processes, the strain rate is relatively low ($\sim 10^{-3}$–10^{-2}/s) and a quasi-coupled approach, where the temperature field influences the deformation field but the deformation field does not influence the temperature field, becomes feasible. For other processes involving high strain rates (e.g., machining), fully coupled governing equations must be solved simultaneously.

3.5.1 *Constitutive Laws*

Assuming small strains and small rotations, the total strain rate tensor $\dot{\varepsilon}_{ij}$ in a general problem may be additively decomposed as

$$\dot{\varepsilon}_{ij} = \dot{\varepsilon}_{ij}^{(e)} + \dot{\varepsilon}_{ij}^{(n)} + \dot{\varepsilon}_{ij}^{(T)} \tag{3.109}$$

where $\dot{\varepsilon}_{ij}^{(e)}$, $\dot{\varepsilon}_{ij}^{(n)}$, and $\dot{\varepsilon}_{ij}^{(T)}$ are elastic, nonelastic, and thermal strain rates, respectively. The range of indices is 1, 2 for two-dimensional and 1, 2, 3 for three-dimensional problems. The total strain rates are related to the velocity gradients as

$$\dot{\varepsilon}_{ij} = \tfrac{1}{2}(V_{i,j} + V_{j,i}) \tag{3.110}$$

The elastic strain rate is related to the stress rate through the usual Hooke's law

$$\dot{\sigma}_{ij} = \lambda \dot{\varepsilon}_{kk}^{(e)} \delta_{ij} + 2G\dot{\varepsilon}_{ij}^{(e)} \tag{3.111}$$

and the associated stress field satisfies the equilibrium equation.

The nonelastic strain rate and the thermal strain rates need to be characterized next. The nonelastic strain rates are obtained from a suitable constitutive model for plasticity or viscoplasticity. Details of such models are discussed in chapter 2.

The nonelastic strain rate must be obtained from a suitable plastic or viscoplastic constitutive law. For J_2 flow theory of plasticity, we may write

$$\dot{\varepsilon}_{ij}^{(n)} = f_{ij}(\sigma_{ij}, \dot{\sigma}_{ij}) \tag{3.112}$$

Alternatively, a viscoplastic model in terms of the state variables may be expressed as

$$\dot{\varepsilon}_{ij}^{(n)} = f_{ij}(\sigma_{ij}, q_{ij}^{(k)}, T) \tag{3.113}$$

and

$$q_{ij}^{(k)} = g_{ij}^{(k)}(\sigma_{ij}, q_{ij}^{(l)}, T) \tag{3.114}$$

where $q_{ij}^{(k)}$ are suitably defined internal of state variables and a superposed dot denotes a time derivative. Examples of such constitutive models are the hyperbolic sine law (Tien and Richmond 1982), and the viscoplastic models due to Hart (1976) and Anand (1982). It is important to note here that the nonelastic strain is, in general, history dependent, and usually volume conserving.

The thermal strains are typically assumed to be dilatational while the nonelastic strains are deviatoric. Thus,

$$\varepsilon_{ij}^{(T)} = \delta_{ij} \int_{T_R}^{T} \alpha(\theta)\, d\theta = f(T)\, \delta_{ij} \tag{3.115}$$

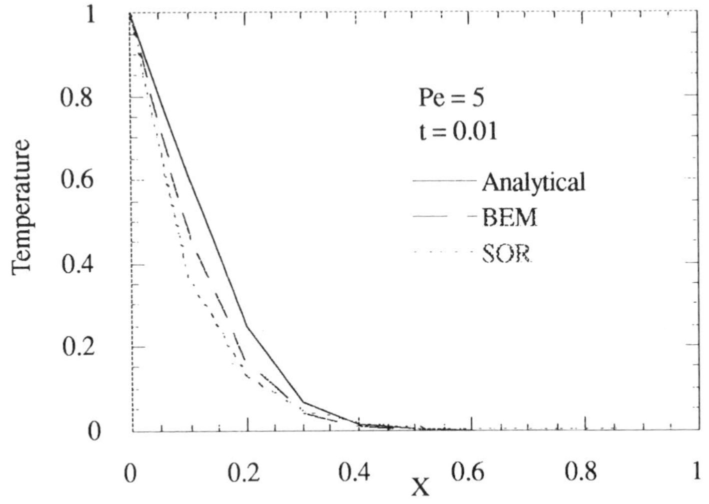
Pe = 5
t = 0.01
Analytical
BEM
SOR
Temperature
X
0
0.2
0.4
0.6
0.8
1

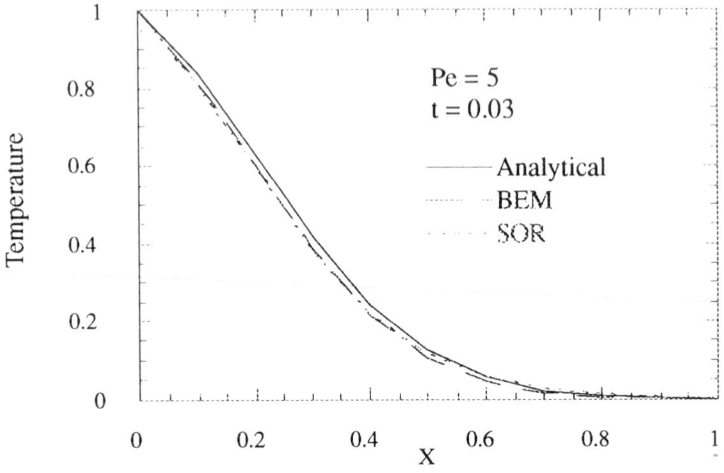
Pe = 5
t = 0.03
Analytical
BEM
SOR
Temperature
X
0
0.2
0.4
0.6
0.8
1

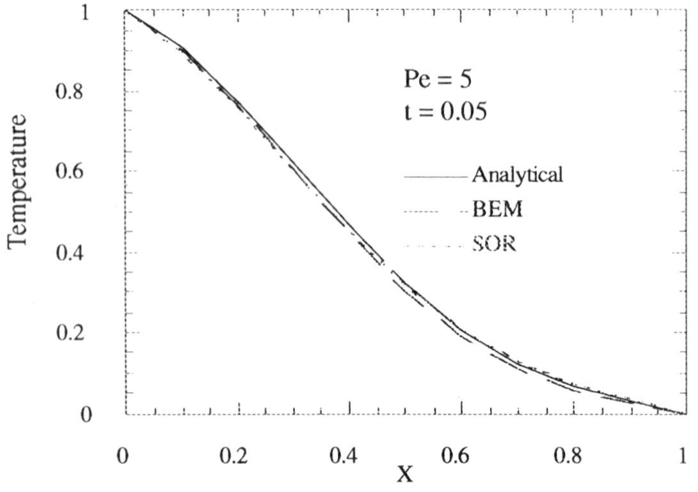
Pe = 5
t = 0.05
Analytical
BEM
SOR
Temperature
X
0
0.2
0.4
0.6
0.8
1

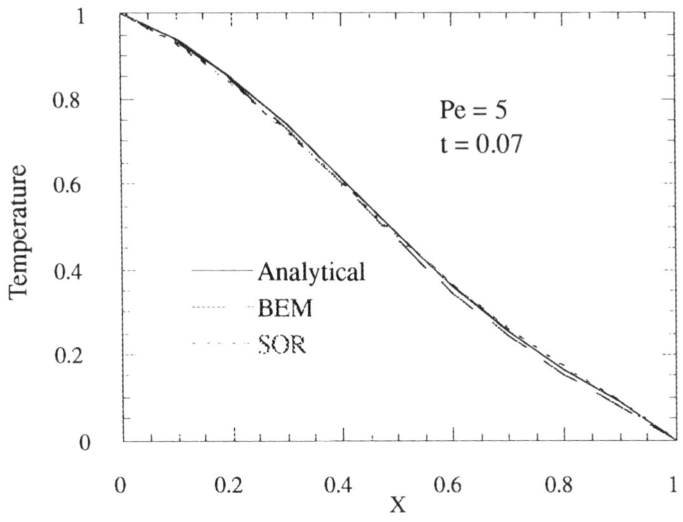

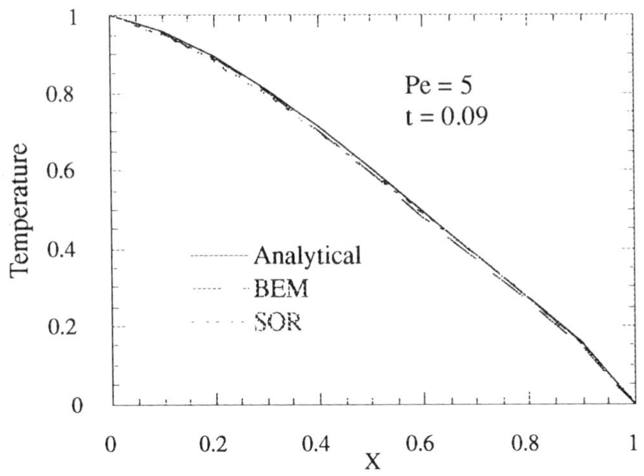

Figure 3.11. Comparison of the BEM solution with the analytical solution and SOR solutions at $Pe = 5$, and $\Delta x = 0.1$.

while

$$\dot{\varepsilon}_{kk}^{(n)} = 0 \tag{3.116}$$

where α is the temperature-dependent coefficient of linear thermal expansion, δ_{ij} is the kronecker delta, and T_R is a reference temperature at which the thermal strains are zero. For solidification problems, T_R is usually taken to be the melting temperature T_m. Now, depending on the particular natures of $\varepsilon_{ij}^{(n)}$ and $\varepsilon_{ij}^{(T)}$, a hierarchy of different thermoelastic and thermoplastic models may be constructed.

3.5.2 Stationary Thermoelasticity in Nonhomogeneous Media

In many practical problems, the nonelastic strains may be neglected and the heat transfer may be modeled as a steady-state heat conduction phenomenon. Hence,

$$\nabla^2 T = 0 \tag{3.117}$$

Applying the Laplace differential operator to the functional $f(T)$ defined in equation (3.115), we get

$$\nabla^2 f(T(x)) = \alpha(T)\nabla^2 T(x) + \frac{\partial \alpha'(T)}{\partial T} T_{,i}(x) T_{,i}(x) \tag{3.118}$$

In the context of linearized theory, that is, assuming the heat fluxes to be sufficiently small, the second term of the right-hand side of equation (3.118) may be neglected in comparison to the first term. This yields

$$\nabla^2 f(T) = 0 \tag{3.119}$$

Equation (3.119) implies that a potential function F may be constructed such that

$$f = -F_{,i} \qquad \text{with} \qquad F_{,ii} = \text{constant} \tag{3.120}$$

The condition (3.120) is called the *stationary condition* and allows a boundary integral only formulation for homogeneous media.

The BEM formulation for stationary thermoelasticity in non-homogeneous media may now be initiated by writing the equation of equilibrium

$$\sigma_{ji,j} = 0 \tag{3.121}$$

in the absence of any body forces and substituting Hooke's law of elasticity (3.111), the strain rate decomposition equation (3.109), and the kinematic equation (3.110) into equation (3.121). This yields

$$[\lambda u_{k,k}\delta_{ij} + G(u_{i,j} + u_{j,i})]_{,j} = [(3\lambda + 2G)f(T)]_{,i} \tag{3.122}$$

The Lamé parameters λ and G are related to Young's modulus E and Poisson's ratio ν as

$$\lambda = \frac{\nu E}{(1+\nu)(1-2\nu)} \quad \text{and} \quad G = \frac{E}{2(1+\nu)} \tag{3.123}$$

The components of the traction vector at a point on the boundary of the domain may be expressed as

$$\tau_i = [\lambda u_{k,k}\delta_{ij} + G(u_{i,j} + u_{j,i})]n_j - (3\lambda + 2G)f(T)n_i \tag{3.124}$$

where n_i denotes the components of the outward normal at a point on the boundary.

In many applications, the elastic parameters may be assumed to remain constant over the entire domain. In reality, however, these parameters depend on the temperature. The changes in temperatures may be large enough for many practical situations to require considerations of temperature-dependent material parameters. In the following development, spatial variations in elastic parameters (due to temperature variations) are considered. The classical stationary thermoelasticity problem in homogeneous media is then deduced as a special case.

Using equation (3.123), the governing Navier equation (3.122) may be written (Ghosh and Mukherjee 1984, Fleuries and Predeleanu 1987, Sladek et al 1990) as

$$\begin{aligned} Gu_{i,jj} &+ \frac{G}{1-2\nu}u_{k,ki} + \lambda_{,i}u_{k,k} + G_{,j}(u_{i,j} + u_{j,i}) \\ &= 2G\left(\frac{1+\nu}{1-2\nu}\right)f(T)_{,i} + f(T)(3\lambda_{,i} + 2G_{,i}) \end{aligned} \tag{3.125}$$

and the traction vector τ_i may be expressed as

$$\tau_i = G\left[\left((u_{i,j} + u_{j,i})n_j + \frac{2\nu}{1-2\nu}u_{k,k}n_i\right) - \frac{2(1+\nu)}{(1-2\nu)}f(T)n_i\right] \tag{3.126}$$

Assuming G_0 and ν_0 to be the nominal base properties for the corresponding auxiliary problem in a homogeneous medium, the auxiliary equation may be written as

$$u_{i,jj}^{(R)} + \frac{1}{1-2\nu_0} u_{k,ki}^{(R)} = -\frac{\Delta(p,q)}{G_0}\delta_{ij}e_j \tag{3.127}$$

and the auxiliary solution $u_i^{(R)}$ may be expressed as

$$u_i^{(R)} = U_{ij}e_j \tag{3.128}$$

and

$$\tau_i^{(R)} = G_0\left((u_{i,j}^{(R)} + u_{j,i}^{(R)})n_j + \frac{2\nu_0}{1-2\nu_0}u_{k,k}^{(R)}n_i\right) \tag{3.129}$$

The BEM formulation for a stationary thermoelasticity problem may now be obtained through a weighted residual approach where the auxiliary solution in equation (3.128) is used as the weighting function. This gives

$$\int_B u_i^{(R)}\left[Gu_{i,jj} + \frac{G}{1-2\nu}u_{k,ki} + \lambda_{,i}u_{k,k} + G_{,j}(u_{i,j}+u_{j,i})\right]dA$$
$$= \int_B u_i^{(R)}\left(2G\frac{(1+\nu)}{(1-2\nu)}f_{,i}(T) + f(T)(3\lambda_{,i} + 2G_{,i})\right)dA \tag{3.130}$$

Using the definitions of tractions in equations (3.126) and (3.129) and the divergence theorem, equation (3.130) may be written as

$$\frac{G(p)}{G_0}u_j(p) = \int_{dB}\left(U_{ij}(p,Q)\tau_i(Q) - \frac{G(Q)}{G_0}T_{ij}(p,Q)u_i(Q)\right)ds_Q$$
$$+ \int_B [U_{ij,k}(p,q) + U_{kj,i}(p,q)G_{,k}(q) + U_{kj,k}(p,q)\lambda_{,i}(q)]u_i(q)\,dA_q$$
$$+ \int_B \left(2U_{ij}(p,q)G(q)\frac{1+\nu(q)}{1-2\nu(q)}f_{,i}(T;q)\right)dA_q$$
$$- \int_B U_{kj,k}(p,q)\left[G(q)\left(\frac{1}{1-2\nu(q)} - \frac{1}{1-2\nu_0}\right)u_i(q)\right]_{,i}dA_q \tag{3.131}$$

where the two point kernels U_{ij} and T_{ij} are the usual ones for homogeneous isotropic elasticity. For three-dimensional cases,

$$U_{ij} = \frac{1}{16\pi(1-\nu)Gr}[(3-4\nu)\delta_{ij} + r_{,i}r_{,j}] \tag{3.132a}$$

and

$$T_{ij} = -\frac{1}{8\pi(1-\nu)r^2}\left([(1-2\nu)\,\delta_{ij} + 3r_{,i}r_{,j}\frac{\partial r}{\partial n} - (1-2\nu)(r_{,i}n_j - r_{,j}n_i)\right) \tag{3.132b}$$

In two dimensions,

$$U_{ij} = -\frac{1}{8\pi(1-\nu)G}[(3-4\nu)\ln r\,\delta_{ij} - r_{,i}r_{,j}] \tag{3.133a}$$

and

$$T_{ij} = -\frac{1}{4\pi(1-\nu)r}\left([(1-2\nu)\,\delta_{ij} + 2r_{,i}r_{,j}]\frac{\partial r}{\partial n} - (1-2\nu)(r_{,i}n_j - r_{,j}n_i)\right) \tag{3.133b}$$

in plane strain. Expressions for U_{ij} and T_{ij} in plane stress may be obtained by replacing ν with $\overline{\nu} = \nu/(1+\nu)$ in equations (3.133). Here, lower-case letters p and q denote points inside the domain and capital letters P and Q denote points on the boundary ∂B. A comma denotes a derivative with respect to a field point and r denotes the distance between a source point p and a field point q.

The Poisson's ratio ν is typically a weak function of temperature. Hence, in many problems we may assume ν to be constant (ν_0). This reduces equation (3.131) to

$$\begin{aligned}\frac{G(p)}{G_0}u_j(p) = {}&\int_{\partial B}\left(U_{ij}(p,Q)\tau_i(Q) - \frac{G(Q)}{G_0}T_{ij}(p,Q)u_i(Q)\right)ds_Q\\ &+\int_B\left((U_{ij,k}(p,q) + U_{kj,i}(p,q))G_{,k}(q)\right.\\ &\left.+\frac{2\nu}{1-2\nu}U_{kj,k}(p,q)G_{,i}(q)\right)u_i(q)\,dA_q\\ &+\int_B\frac{2(1+\nu)}{(1-2\nu)}U_{ij,i}(p,q)G(q)f(T;q)\,dA_q\end{aligned} \tag{3.134}$$

It should be noted that, for homogeneous media, the first domain integral on the right-hand side of the internal equation (3.134) becomes zero since $G_{,i} = 0$ for such cases. Hence, the internal equation for thermoelastic deformation in a homogeneous medium requires only one domain integral involving $f(T)$, while for nonhomogeneous media an additional domain integral involving spatial gradients of elastic moduli and the displacement field is involved. The appearance of the displacement field in the domain integral requires an iterative approach if degrees of freedom are put on the boundary only.

To obtain a boundary equation from the internal equation (3.134), we take the limit $p \to P$ as the internal source point approaches the boundary. This yields,

$$\frac{G(P)}{G_0} C_{ij}(P) u_i(P) = \int_{\partial B} \left(U_{ij}(P, Q) \tau_i(Q) - \frac{G(Q)}{G_0} T_{ij}(P, Q) u_i(Q) \right) ds_Q$$

$$+ \int_B \left(\{ U_{ij,k}(P, q) + U_{kj,i}(P, q) G_{,k}(q) \right.$$

$$\left. + \frac{2\nu}{1 - 2\nu} U_{kj,k}(P, q) G_{,i}(q) \right) u_i(q) \, dA_q$$

$$+ \int_B \frac{2(1 + \nu)}{1 - 2\nu} U_{ij,i}(P, q) G(q) f(T; q) \, dA_q \tag{3.135}$$

where, $C_{ij}(P)$ is the usual corner tensor that depends on the local geometry at (P). As discussed earlier, it may be evaluated directly or indirectly by requiring that the solution satisfy rigid-body modes. It should also be noted that equation (3.135) is independent of the particular choice of G_0. This becomes obvious when the explicit forms of U_{ij} and T_{ij} shown in equations (3.132) and (3.133) are considered.

Once the displacement field has been obtained throughout the body, the displacement gradients must be obtained by differentiating the displacement field. The stresses can then be obtained from Hooke's law:

$$\sigma_{ij} = G \left(u_{i,j} + u_{j,i} + \frac{2\nu}{1 - 2\nu} u_{k,k} \, \delta_{ij} - \frac{2(1 + \nu)}{1 - 2\nu} f(T) \, \delta_{ij} \right) \tag{3.136}$$

The displacement gradients can be obtained by one of two methods.

Analytical differentiation. In this approach, equation (3.134) is differentiated at an internal source point to give

$$\frac{G(p)}{G_0} u_{j,L}(p) = -\frac{G_{,L}(p)}{G_0} u_j(p) + \int_{\partial B} U_{ij,L}(p,Q)\tau_i(Q)\,ds_Q$$

$$-\int_{\partial B} \frac{G(Q)}{G_0} T_{ij,L}(p,Q)u_i(Q)\,ds_Q$$

$$+\frac{\partial}{\partial x_L}\int_B \bigg(\{U_{ij,k}(p,q)+U_{kj,i}(p,q)\}G_{,k}(q)$$

$$+\frac{2\nu}{1-2\nu}U_{kj,k}(p,q)G_{,i}(q)\bigg)u_i(q)\,dA_q$$

$$+\frac{\partial}{\partial x_L}\int_B \frac{2(1+\nu)}{(1-2\nu)} U_{ij,i}(p,q)G(q)f(T;q)\,dA_q \tag{3.137}$$

This approach has the advantage of delivering a continuous strain and stress field inside B. The difficulty in implementing this method arises from the fact that, since $U_{ij,k}$ has a $1/r$ singularity, the last two integrals must first be integrated analytically for an arbitrary source point p and then differentiated at p. This approach has been implemented by several researchers for problems where integrals of type $\int_B U_{ij,k}(p,q)\,dA_q$ are involved. In this problem. However, the domain integrals also contain the shear modulus, unknown displacement field, temperature, and so on. The complicated nature of the domain integrals renders the analytical differentiation method difficult to implement in this problem.

Numerical differentiation. In this approach, the displacements are interpolated within each internal cell by suitable shape functions and then these shape functions are differentiated element-wise to give the strains. This strategy, therefore, uses the BEM to determine displacements throughout the body and then a method analogous to the FEM to obtain the strains. This method can be easily implemented but has the disadvantages of allowing discontinuities in stresses at internal nodes and across intercell boundaries.

Numerical implementations. For numerical implementations, the boundary of the body ∂B is discretized into N_s segments and the interior is discretized into n_i internal cells. A discretized version of equation (3.135) may then be obtained as

$$
\begin{aligned}
\frac{G(P_M)}{G_0} C_{ij}(P_M)u_i(P_M) &= \sum_{N_s} \int_{\Delta c_N} U_{ij}(P_M, Q)\tau_i(Q)\, ds_Q \\
&- \sum_{N_s} \int_{\Delta c_N} T_{ij}(P_M, Q)\frac{G(Q)}{G_0} u_i(Q)\, ds_Q \\
&+ \sum_{n_i} \int_{\Delta A_n} (U_{ij,k}(P_M, q) + U_{kj,i}(P_M, q)G_{,k}(q)u_i(q)\, dA_q \\
&+ \sum_{n_i} \int_{\Delta A_n} \frac{2\nu}{1-2\nu} U_{kj,k}(P_M, q)G_{,i}(q)u_i(q)\, dA_q \\
&+ \sum_{n_i} \int_{\Delta A_n} \frac{2(1+\nu)}{(1-2\nu)} U_{ij,i}(P_M, q)G(q)f(T; q)\, dA_q
\end{aligned}
\tag{3.138}
$$

where P_M is a boundary point P that coincides with a node M.

The temperature field is first obtained on ∂B and in B by solving the steady heat equation by the BEM. Suitable shape functions must now be chosen for the tractions, displacements, and temperatures on ∂B and in B. The traction and displacement components and the temperature are assumed to be piecewise linear on the boundary segments Δc_N. Double nodes are placed at corners to allow for jumps in normals or traction components across them.

The shear modulus is assumed to be a function of temperature (e.g., linear)

$$G = G_0(k_1 + k_2 T) \tag{3.139}$$

Experimental results show that this is a fairly good assumption unless the temperature is very high. Of course, this assumption can be relaxed and G can be made a nonlinear function of temperature.

The displacements and temperature in the present development are assumed to be piecewise linear on the triangular internal cells ΔA_n with the sampling points placed at the vertices of each triangle. By virtue of equation (3.139), the shear modulus becomes piecewise linear and the gradients of G become piecewise constant on the internal cells. The gradients of G may also be evaluated pointwise internally from the equation

$$G_{,i} = \frac{\partial G}{\partial T} T_{,i} \tag{3.140}$$

In such a case, shape functions of temperatures on internal cells would become unnecessary. This may improve the accuracy but will be computationally more expensive. The numerical procedure will also involve evaluations of boundary and domain integrals containing singularities of orders $\ln r$ and $(1/r)$ in two-dimensional applications and $(1/r)$ and $(1/r^2)$ in three-dimensional applications.

The numerical discretization transforms equation (3.135) into an algebraic system of the type

$$[A]\{u\} + [B]\{\tau\} = \{C\} + \{D\} \tag{3.141}$$

where the coefficient matrices $[A]$ and $[B]$ contain boundary integrals of the kernels, U_{ij}, and so on. $\{C\}$ represents the area integrals containing components of the unknown displacement field and $\{D\}$ is obtained from the last (known) integral in equation (3.138). The presence of u_i in $\{C\}$ requires iterations in order to solve equation (3.141). This procedure is described below.

1. The starting values of the internal displacements are taken to be zero. Equation (3.141) with $\{C\} = 0$ is solved for the unknown boundary components of the displacements and tractions in terms of the prescribed ones and the known $\{D\}$. The matrix $\{D\}$ is only calculated once and remains invariant throughout the iterative process.
2. With $\{C\}$ still equal to zero, the boundary values of $\{u\}$ and $\{\tau\}$ are used in a discretized version of equation (3.134) to yield the first approximation to the internal displacement field.
3. A first approximation to $\{C\}$ is obtained with the internal displacement field from step (2), and equation (3.141) is solved again to obtain the second approximation to the boundary values of $\{u\}$ and $\{\tau\}$. All these quantities are then used in the discretized version of the internal equation to yield the second approximation to $u_i(p)$.
4. Step (3) is repeated as many times as necessary until convergence is achieved and the displacement field is known everywhere and the traction field is known on ∂B.

Evaluation of internal stresses. A finite element type approach has been used here to obtain the displacement gradients from the displacement field. The piecewise linear internal displacement field is differentiated to yield piecewise constant displacement gradients on the internal cells. The internal stresses are then obtained from equation (3.136).

Evaluation of boundary stresses. The boundary stresses can be determined accurately from operations only on the boundary. Once the displacements and tractions are evaluated on the boundary, the tangential derivatives of the displacements, $\partial u_i/\partial s$, are obtained by differentiating the boundary shape functions. All the components of boundary stresses can now be obtained from the constitutive and stress-traction equations.

3.5.2.1 Special case for homogeneous media

For stationary thermoelasticity in homogeneous media, we may now invoke equations (3.119) and (3.120). Furthermore, the first domain integral on the right-hand side of equation (3.135) becomes zero for homogeneous media. This finally yields a boundary-only formulation for stationary thermoelasticity problems in homogeneous media. For homogeneous media equation (3.135) reduces to

$$C_{ij}(P)u_i(P) = \int_{\partial B} (U_{ij}(P,Q)\tau_i(Q) - T_{ij}(P,Q)u_i(Q))\,ds_Q + \int_B \frac{2G(1+\nu)}{1-2\nu} U_{ij,i}(P,Q)f(T;q)\,dA_q \tag{3.142}$$

By virtue of the stationary conditions (3.119) and (3.120), the domain integral on the right-hand side of equation (3.142) may be written as

$$\int_B \frac{2G(1+\nu)}{1-2\nu} U_{ij,i}(P,Q)f(T;q)\,dA_q = \int_B f(T;q)\ \nabla^2 U_j(P,Q)\,dA_q \tag{3.143}$$

or

$$\int_B \frac{2G(1+\nu)}{1-2\nu} U_{ij,i}(P,Q)f(T;q)\,dA_q = \int_B U_j(P,Q)\nabla^2 f(T;q)\,dA_q - \int_{\partial B} [f(T;q)Z_j(P,Q) - \alpha(T)q^{(n)}(Q)U_j(P,Q)]\,ds_Q \tag{3.144}$$

By virtue of equation (3.119), the domain integral on the right-hand side of equation (3.144) vanishes. Hence, for stationary thermoelasticity in homogeneous media (Rizzo and Shippy 1977, Sladek and Sladek 1984, Sladek et al. 1990)

$$C_{ij}(P)u_i(P) = \int_{\partial B} [U_{ij,i}(P,Q)\tau_i(Q) - T_{ij}(P,Q)u_i(Q)]\,ds_Q + \int_{\partial B} [f(T;q)Z_j(P,Q) - \alpha(T;Q)q^{(n)}(Q)U_j(P,Q)]\,ds_Q \tag{3.145}$$

involving boundary integrals only. The kernels in equations (3.143–3.145) may be written as

$$U_j(P, Q) = \frac{B}{4\pi(d-1)}\left(\frac{1+\nu}{1-\nu}\right) r^{(3-d)} r_{,j} \tag{3.146}$$

$$Z_j(P, Q) = n_i(Q) U_{j,i}(P, Q)$$

$$= \frac{1}{4\pi(d-1) r^{(d-2)}}\left(\frac{1+\nu}{1-\nu}\right)(B\delta_{ij} - r_{,i} r_{,k}) n_i(Q) \tag{3.147}$$

where, for three-dimensional problems,

$$A = 1, \qquad B = 1, \qquad C = 2 \qquad \text{and} \qquad d = 3 \tag{3.148a}$$

and for two-dimensional plane strain problems,

$$A = -\ln r, \qquad B = -0.5(1 + 2\ln r)$$

$$C = -(1 + 2\nu + 2\ln r) \qquad \text{and} \qquad d = 2 \tag{3.148b}$$

For two-dimensional plane stress problems, ν should be replaced by $\bar{\nu} = \nu/(1+\nu)$.

Once the temperature and its associated flux field is obtained, the last boundary integral in equation (3.145) is known completely. Hence, equation (3.145) after discretization may be written as

$$[A]\{u\} + [B]\{\tau\} = \{f\} \tag{3.149}$$

The right-hand side of equation (3.149) is known a priori and requires integrations over boundary only. Hence, it is very similar to discretized BEM equations for linear elasticity in homogeneous media. Solution of equation (3.149) yields the displacements and tractions everywhere on the boundary. Using equation (3.136), the stresses at an internal source paint may be obtained as

$$\begin{aligned}
\sigma_{lj}(p) = &-\frac{2G(1+\nu)}{1-2\nu} f(T;p)\,\delta_{lj} \\
&+ \int_{\partial B} D_{lji}(P, Q)\left(\tau_i(Q) + \frac{2G(1+\nu)}{1-2\nu} n_i(Q) f(T;Q)\right) ds_Q \\
&+ \int_{\partial B} T_{ljim}(P, Q) \hat{D}_m u_i(Q)\, ds_Q \\
&+ \int_{\partial B} [\alpha(T;Q) q^{(n)}(Q) F_{ij}(P, Q) - f(T;Q) H_{ij}(P, Q)]\, ds_Q
\end{aligned} \tag{3.150}$$

where the differential operator $\hat{D}_m$ is defined as

$$\hat{D}_m = \delta_{m3}\frac{\partial}{\partial t(Q)} \quad \text{in two dimensions} \tag{3.151}$$

and

$$\hat{D}_m = \left(\rho_m(Q)\frac{\partial}{\partial t(Q)} - t_m(Q)\frac{\partial}{\partial \rho(Q)}\right) \quad \text{in three dimensions} \tag{3.152}$$

Here, $t(Q)$ is the unit tangent vector to the boundary at Q and the other tangent vector (in three dimensions) is defined as

$$\rho_i = \varepsilon_{ijk} n_j t_k \tag{3.153}$$

where $n_j(Q)$ is the unit outward normal vector to the boundary surface ∂B.

The kernels in equation (3.150) may be defined as

$$D_{lji}(P, Q) = \frac{1}{4\pi(1-\nu)(d-1)r^{d-1}}[(1-2\nu)(\delta_{li}r_{,j} + \delta_{ji}r_{,l} - \delta_{lj}r_{,i}) + dr_{,i}r_{,j}r_{,l}] \tag{3.154}$$

$$\begin{aligned} T_{ljim}(P, Q) = \frac{G}{4\pi(1-\nu)(d-1)r^{d-1}} & \{4\nu\delta_{ij}r_{,k}\varepsilon_{kim} \\ & + (1-2\nu)[r_{,l}\varepsilon_{jim} + r_{,j}\varepsilon_{lim} - r_{,k}(\delta_{il}\varepsilon_{jkm}) + \delta_{ij}\varepsilon_{lkm})] \\ & - dr_{,i}r_{,k}(r_{,l}\varepsilon_{jkm} + \mathrm{r}_{,j}\,\varepsilon_{lkm})\} \end{aligned} \tag{3.155}$$

$$F_{lj}(P, Q) = \frac{G}{4\pi(d-1)r^{d-2}}\left(\frac{1+\nu}{1-\nu}\right)\left(\frac{\delta_{jl}}{(1-2\nu)}C - 2r_{,j}r_{,i}\right) \tag{3.156}$$

and

$$H_{lj}(P, Q) = \frac{G}{\pi(d-1)r^{d-1}}\left(\frac{1+\nu}{1-\nu}\right)(\delta_{jl} - dr_{,j}r_{,l})r_{,i}n_i(Q) \tag{3.157}$$

where the quantities C, d, etc., are given in equations (3.148) and (3.149).

Finally, the stress tensor at a boundary point can be computed from the boundary values of temperature, tractions and tangential derivatives of displacements. For a BEM formulation with displacements and tractions as fundamental variables, the tangential derivatives of displacements need to be evaluated numerically. The calculation involves four vectors: the traction $\boldsymbol{\tau}$, the tangential derivative of displacement $\partial \boldsymbol{u}/\partial s$, and the unit normal and tangent vectors $\boldsymbol{n}$ and $\boldsymbol{t}$; two tensors are the stress $\boldsymbol{\sigma}$ and displacement gradient $\partial \boldsymbol{u}/\partial \boldsymbol{x}$. The four vectors (involving 12 scalar components for a two-dimensional problem) are known at P. For a two-dimensional problem, the 12 unknown tensor components σ_{ij} and $u_{i,j}$ can now be determined from the following set of three equations:

$$\sigma_{ij} = G\left(u_{i,j} + u_{j,i} + \frac{2\nu}{1-2\nu} u_{k,k}\,\delta_{ij} - \frac{2(1+\nu)}{1-2\nu} f(T)\,\delta_{ij}\right) \tag{3.158}$$

$$\tau_j = \sigma_{ij} n_i \tag{3.159}$$

$$\frac{\partial u_i}{\partial s} = u_{i,j} t_j \tag{3.160}$$

The stress at a boundary point P may be expressed (in two dimensions) as

$$\sigma_{ij}(P) = A_{ijk}(P)\left(\tau_k(P) + \frac{2G(1+\nu)}{(1-2\nu)} n_k(P) f(T;P)\right) + B_{ijk}(P)\frac{\partial u}{\partial t(P)} - \frac{2G(1+\nu)}{(1-2\nu)}\delta_{ij} f(T;P) \tag{3.161}$$

where

$$A_{ijk} = \left(n_i n_j + \frac{\nu}{1-\nu} t_i t_j\right) n_k + (n_i t_j + n_j t_i) t_k \tag{3.162}$$

and

$$B_{ijk} = \frac{2G}{1-\nu} t_i t_j t_k \tag{3.163}$$

with $\boldsymbol{n}$ and $\boldsymbol{t}$ denoting unit outward normal and tangent vectors, respectively.

A similar expression for boundary stresses may also be developed in three dimensions by interpreting equations (3.158) through (3.160) in three dimensions (Ghosh et al. 1986).

3.5.3 *Nonstationary Thermoelasticity*

The governing equations for nonstationary or transient thermoelasticity problems in homogeneous media can be expressed as

$$T_{,kk} - \frac{1}{\kappa}\dot{T} = -\frac{\bar{Q}}{\kappa} \tag{3.164}$$

and

$$Gu_{i,kk} + (\lambda + G)u_{k,ki} = \gamma T_{,i} \tag{3.165}$$

where $T(\boldsymbol{x}, t)$ and $u_i(\boldsymbol{x}, t)$ are the temperature and displacement fields, respectively, $\bar{Q}(\boldsymbol{x}, t)$ is the heat source density, κ is the thermal diffusivity, and λ and G are the usual Lamé constants of elasticity. Here the coefficient of thermal expansion (α) is assumed to be constant, and

$$\gamma = (3\lambda + 2G)\alpha \tag{3.166}$$

The uncoupled or quasi-coupled thermoelastic equations are valid for cases where the rate of change of volume, $\dot{\varepsilon}_{kk}$, is small such that the term $T\dot{\varepsilon}_{kk}$ can be neglected compared to other terms in the thermal equation (3.164). It is interesting to note that, under such assumptions, the thermal field is independent of the displacement field. However, the displacement and its associated stress fields are dependent on the temperature field. The transient conduction equation is involved here, giving rise to the nonstationary conditions.

The uncoupled nonstationary thermoelasticity problem may be solved by first solving the thermal equation (3.164) and then treating equation (3.165) in the same way as in elastostatics by replacing the body force term with the temperature gradient term $\gamma T_{,i}$. Such an approach, however, involves domain integrals in the BEM formulations for nonstationary thermoelasticity.

Sladek and Sladek (1984, 1989) have also developed a pure boundary formulation for thermoelastic problems. The integral representation of the displacement field may be written as

$$u_k(p, t_F) = \int_0^{t_F} \kappa \int_{\partial B} [T(Q, t)Z_k(p, t_F; Q, t) - q^{(n)}(Q, t)U_k(p, t_F; Q, t)]\, ds_Q dt$$
$$+ \int_{\partial B} [\tau_i(Q, t)U_{ik}(p, Q) - u_i(Q, t)T_{ik}(p, Q)]\, ds_Q + V_k(p, t_F) \tag{3.167}$$

and

$$V_k(p,t_F) = -\int_0^{t_F}\int_B \bar{Q}(q,t)U_k(p,t_F;q,t)\,dA_q\,dt + \int_B U_k(p,t_F;q,t)T(q,0)\,dA_q \tag{3.168}$$

Here the kernels $U_{ik}(p,Q)$, $T_{ik}(p,Q)$, $U_k\,(p,t_F;Q,t)$, and $Z_k(p,t_F;Q,t)$ are (in two-dimensional plane strain problems),

$$U_{ik}(p,Q) = \frac{1}{8\pi G(1-\nu)}[r_{,i}r_{,k} - (3-4\nu)\delta_{ik}\ln r] \tag{3.169}$$

$$T_{ik}(p,Q) = \frac{1-2\nu}{4\pi(1-\nu)r}\Bigg[r_{,k}n_i(Q) - r_{,i}n_k(Q) - \left(\delta_{ik} + \frac{2}{1-2\nu}r_{,i}r_{,k}\right)r_{,j}n_j(Q)\Bigg] \tag{3.170}$$

$$U_k(p,t_F;Q,t) = \frac{mr_{,k}}{2\pi r}(1-e^{-\hat{G}}) \tag{3.171}$$

$$Z_k(p,t_F;Q,t) = \frac{mn_i(Q)}{2\pi}\left[\frac{1}{r^2}(\delta_{ik} - 2r_{,i}r_{,k})(1-e^{-\hat{G}}) + \frac{r_{,i}r_{,k}}{2\kappa t}e^{-\hat{G}}\right] \tag{3.172}$$

where

$$m = \frac{\gamma}{\lambda + 2G} \qquad \text{and} \qquad \hat{G} = \frac{r^2}{4\kappa t} \tag{3.173}$$

and r is the distance between the source point and the field point.

In equations (3.167) and (3.168), $T(q,0)$ and $\bar{Q}(q,t)$ are prescribed functions. Hence, unknowns are placed only on the boundary and the solution at an internal point can be expressed in terms of prescribed quantities and boundary unknowns. The traction τ_i can also be expressed by equations (3.158) and (3.159).

Taking the limit as $p \to P$, the boundary integral equation may be written as

$$\int_{\partial B}\{[u_i(Q,t_F) - u_i(P,t_F)]T_{ik}(P,Q) - \tau_i(Q,t_F)U_{ik}(P,Q)\}\,ds_Q$$

$$= \int_0^{t_F}\kappa\int_{\partial B}[T(Q,t)Z_k(P,t_F;Q,t) - q^{(n)}(Q,t)U_k(P,t_F;Q,t)]\,ds_Q\,dt + V_k(P,t_F) \tag{3.174}$$

As a consequence of the quasi-static approximation, the boundary displacements and tractions enter the boundary equation (3.174) at only the final time. Hence, the solution at any time does not depend on the history of displacements and tractions.

According to equation (3.158), the integral representation of stresses at an internal point can be obtained from that of displacement at an internal point. This requires differentiating the internal equation (3.167) at an internal source point p to determine the displacement gradient. Since the derivatives of the kernel $T_{ik}(P, Q)$ are strongly singular as the field point Q approaches P, special care should be taken. Differentiation may be carried out after integration. The resulting regularized integral representation of stresses becomes

$$\sigma_{ij}(p, t_F) + \gamma T(p, t_F)\,\delta_{ij} = \int_0^{t_F} \kappa \int_{\partial B} [F_{ij}(p, t_F; Q, t) q^{(n)}(Q, t_0)$$

$$- H_{ij}(p, t_F; Q, t) T(Q, t)]\, ds_Q\, dt$$

$$+ \int_{\partial B} \{D_{ijk}(p, Q)[\tau_k(Q, t) + \gamma n_k(Q) T(Q, t)] + T_{ijk}(p, Q) \frac{\partial u_k}{\partial s}(Q, t)\}\, ds_Q$$

$$+ W_{ij}(p, t_F) \qquad (3.175)$$

where

$$W_{ij}(p, t_F) = \int_0^{t_F} \int_B F_{ij}(p, t_F; q, t) \bar{Q}(q, t)\, dA_q\, dt + \int_B F_{ij}(p, t_F; q, t) T(q, t)\, dA_q \qquad (3.176)$$

Here, the kernels $D_{lji}(p, Q)$, $T_{lji}(p, Q)$, $F_{lj}(p, t_F; Q, t)$ and $H_{lj}(p, t_F; Q, t)$ are (for two dimensional applications in plane strain)

$$D_{lji}(p, Q) = \frac{1}{4\pi(1-\nu)r}[(1-2\nu)(\delta_{li} r_{,j} + \delta_{ji} r_{,l} - \delta_{lj} r_{,i}) + 2 r_{,i} r_{,j} r_{,l}] \qquad (3.177)$$

$$T_{lji}(p, Q) = \frac{G}{4\pi(1-\nu)r}\{4\nu\,\delta_{lj} r_{,k}\,\varepsilon_{ki3} + (1-2\nu)[r_{,l}\,\varepsilon_{ji3} + r_{,j}\,\varepsilon_{li3}$$

$$- r_{,k}(\delta_{il}\,\varepsilon_{jk3} + \delta_{ij}\,\varepsilon_{lk3})] - 2 r_{,i} r_{,k}(r_{,l}\,\varepsilon_{jk3} + r_{,j}\,\varepsilon_{lk3})\} \qquad (3.178)$$

$$F_{lj}(p, t_F; Q, t) = \frac{Gm}{\pi}\left\{\left(\frac{\nu}{1-2\nu}\delta_{lj} + r_{,j} r_{,l}\right)\frac{e^{-\hat{G}}}{2\kappa t}\right.$$

$$\left. + \frac{1}{r^2}(\delta_{lj} - 2 r_{,j} r_{,l})(1 - e^{-\hat{G}})\right\} \qquad (3.179)$$

$$H_{lj}(p, t_F; Q, t) = \frac{Gm}{\pi}\Bigg\{ - \left(\frac{\nu}{1-2\nu}\delta_{lj} + r_{,j}r_{,l}\right)\frac{rr_{,i}n_i(Q)}{(2\kappa t)^2}e^{-\hat{G}}$$

$$+\frac{1}{r}[(r_{,j}n_l(Q) + r_{,l}n_j(Q) + (\delta_{lj} - 4r_{,j}r_{,l})r_{,i}n_i(Q)]$$

$$\times\left[\frac{1}{2\kappa t}e^{-\hat{G}} - \frac{2}{r^2}(1 - e^{-\hat{G}})\right]\Bigg\}$$

$$+\frac{\gamma\delta(t_F)}{4\pi(1-\nu)\kappa r}\{[2r_{,j}r_{,l} - (1-2\nu)\delta_{lj}]r_{,i}n_i(Q)$$

$$+ (1-2\nu)[r_{,j}n_l(Q) + r_{,l}n_j(Q)]\} \tag{3.180}$$

where ε_{ijk} is the permutation symbol, δ_{ij} is the kronecker delta and $\delta(t_F)$ is the Dirac delta function; and m and $\hat{G}$ are given in equation (3.173). In the regularized representation of stresses (3.175), the strongest singularity does not exceed that in the integral representation of displacements. The tangential derivatives of displacements are obtained by numerical differentiation of boundary values of displacements.

The stress tensor at a boundary point (P) can now be expressed by following the same procedure as that for stationary thermoelasticity and using the constitutive model (3.158), the traction equation (3.159), and the definition of tangential derivative (3.160). This yields (for plane strain)

$$\sigma_{lj}(P, t_F) = \left[\left(n_l n_j + \frac{\nu}{1-\nu}\tau_l\tau_j\right)n_k + (n_l\tau_j + \tau_l n_j)\tau_k\right]$$

$$\times\left[t_k(P, t_F) + n_k\gamma T(P, t_F)\right]$$

$$+\frac{2G}{1-\nu}\tau_l\tau_j\tau_k\frac{\partial u_k}{\partial s}(P, t_F) - \delta_{lj}\gamma T(P, t_F) \tag{3.181}$$

where the components of the normal and tangent vectors are taken at the source point P. For plane stress problems, ν should be replaced by $\bar{\nu} = \nu/(1+\nu)$.

3.5.3.1 Numerical implementation

For numerical implementation, it is assumed that the temperature and flux distributions are known. The initial temperature distribution $T(q, 0)$ is assumed to be homogeneous and is further assumed (without any loss of generality) to be zero. The internal heat sources $\bar{Q}(q, t)$ are also assumed to be zero. Thus, the domain integrals vanish. Here, we concentrate on the temporal and spatial integrations in the boundary integrals.

Let the time interval $\langle 0, t_F \rangle$ be divided into F time steps or subintervals $\langle t^{1f}; t^{2f} \rangle$ for $f = 1, 2, \ldots, F$. The continuity of these subintervals is satisfied by the conditions $t^{2(f-1)} = t^{1f}$. We shall assume linear time variations of the boundary temperatures and fluxes within each time subinterval, that is,

$$g(Q, t) = \sum_{\alpha=1}^{2} g(Q, t^{\alpha f}) \Psi_\alpha(t) \tag{3.182}$$

where g stands for temperature and flux; furthermore, $t \in \langle t^{1f}, t^{2f} \rangle$ and the interpolation polynomials are given as

$$\Psi_1(t) = \frac{t^{2f} - t}{\Delta t_f} \tag{3.183a}$$

$$\Psi_2(t) = \frac{t - t^{1f}}{\Delta t_f} \tag{3.183b}$$

where Δt_f is the length of the fth step ($\Delta t_f = t^{2f} - t^{1f}$). Substituting equations (3.182) and (3.183) into the boundary equation (3.174), we get,

$$\begin{aligned} &\int_{\partial B} \{[u_i(Q, t_F) - u_i(P, t_f)] T_{ik}(P, Q) - \tau_i(Q, t_f) U_{ik}(P, Q\} \, ds_Q \,. \\ &= \kappa \sum_{f=1}^{F} \sum_{\alpha=1}^{2} \int_{\partial B} \left[T(Q, t^{\alpha f}) \int_{t^{1f}}^{t^{2f}} \Psi_\alpha(t) Z_k(P, t_f; Q, t) \, dt \right. \\ &\left. - q^{(n)}(Q, t^{\alpha f}) \int_{t^{1f}}^{t^{2f}} \Psi_\alpha(t) U_k(P, t_f; Q, t) \, dt \right] ds_Q \end{aligned} \tag{3.184}$$

and the integral representation of stresses becomes

$$\begin{aligned} [\sigma_{lj}(p, t_f) + \gamma \delta_{ij} T(p, t_f)] = &\int_{\partial B} \left\{ [\tau_i(Q, t_f) + \gamma n_i(Q) T(Q, t_f)] D_{lji}(P, Q) \right. \\ &\left. + T_{lji}(P, Q) \frac{\partial u_i}{\partial s}(Q, t_f) \right\} ds_Q \\ &+ \kappa \sum_{f=1}^{F} \sum_{\alpha=1}^{2} \int_{\partial B} \left[q(Q, t^{\alpha f}) \int_{t^{1f}}^{t^{2f}} \Psi_\alpha(t) F_{lj}(p, t_f; Q, t) \right. \\ &\left. - T(Q, t^{\alpha f}) \int_{t^{1f}}^{t^{2f}} \Psi_\alpha(t) H_{lj}(p, t_f; Q, t) \, dt \right] ds_Q \end{aligned} \tag{3.185}$$

The time integrations can be performed analytically and the kernels defined by equations (3.184) and (3.185) may be expressed in terms of elementary and special functions (denoting $U_k^{\alpha f}(r,t_F)=\int_{t^{1f}}^{t^{2f}}\Psi_\alpha(t)U_k(p,t_f;Q,t)\,dt$) as

$$U_k^{1f}(r,t_F)=\frac{mr_{,k}}{2\pi\,\Delta t_f}I_1^f(r,t_f) \tag{3.186a}$$

$$U_k^{2f}(r,t_F)=\frac{mr_{,k}}{2\pi\,\Delta t_f}I_2^f(r,t_F) \tag{3.186b}$$

$$Z_k^{1f}(r,t_F)=\frac{mn_i(Q)}{2\pi\,\Delta t_f}\left(\frac{1}{r}I_1^f(r,t_F)\,\delta_{ik}+r_{,i}r_{,k}I_3^f(r,t_F)\right) \tag{3.186c}$$

$$Z_k^{2f}(r,t_F)=\frac{mn_i(Q)}{2\pi\;\Delta t_f}\left(\frac{1}{r}I_2^f(r,t_F)\,\delta_{ik}+r_{,i}r_{,k}I_4^f(r,t_F)\right) \tag{3.186d}$$

$$\begin{aligned}F_{lj}^{1f}(r,t_F)=\frac{Gm}{\pi\;\Delta t_f}&\left[\frac{1}{1-2\nu}\left(\nu I_3^f(r,t_F)+\frac{1}{r}I_1^f(r,t_F)\right)\delta_{lj}\right.\\&\left.+r_{,l}r_{,j}I_3^f(r,t_F)\right]\end{aligned} \tag{3.186e}$$

$$\begin{aligned}F_{lj}^{2f}(r,t_F)=\frac{Gm}{\pi\;\Delta t_f}&\left[\frac{1}{1-2\nu}\left(\nu I_4^f(r,t_F)+\frac{1}{r}I_2^f(r,t_F)\right)\delta_{lj}\right.\\&\left.+r_{,l}r_{,j}I_4^f(r,t_F)\right]\end{aligned} \tag{3.186f}$$

$$\begin{aligned}H_{lj}^{1f}(r,t_F)=\frac{Gm}{\pi\;\Delta t_f}&\left\{\left[\delta_{lj}\left(\frac{1}{r}I_3^f(r,t_F)-\frac{\nu}{1-2\nu}I_5^f(r,t_F)\right)\right.\right.\\&\left.-r_{,j}r_{,l}\left(I_5^f(r,t_F)+\frac{4}{r}I_3^f(r,t_F)\right)\right]r_{,i}n_i(Q)\\&\left.+\left(r_{,j}n_i(Q)+r_{,l}n_j(Q)\right)\frac{1}{r}I_3^f(r,t_F)\right\}\end{aligned} \tag{3.186g}$$

$$\begin{aligned}H_{lj}^{2f}(r,t_F)=\frac{Gm}{\pi\,\Delta t_f}&\left\{\left[\delta_{lj}\left(\frac{1}{r}I_7^f(r,t_F)+\frac{\nu}{1-2\nu}I_8^f(r,t_F)\right)\right.\right.\\&\left.-r_{,j}r_{,l}\left(I_6^f(r,t_F)-\frac{4}{r}I_7^f(r,t_F)\right)\right]r_{,i}n_i(Q)\\&\left.+\left(r_{,j}n_i(Q)+r_{,l}n_j(Q)\right)\frac{1}{r}I_7^f(r,t_F)\right\}\end{aligned} \tag{3.186h}$$

where $I_a^f(r, t_F)$ for $a = 1,2, \ldots, 8$ are given as follows:

(i) *For* $f < F$:

$$I_1^f(r,t_F) = \frac{1}{2r}\left[\beta_f^2(1-e^{-B}) + \alpha_f^2(1-e^{-A}) - 2\alpha_f\beta_f(1-e^{-B}) + \frac{r^2}{4\kappa}(\beta_f e^{-B} - \alpha_f e^{-A}) + \frac{r^2}{2\kappa}\left(\alpha_f + \frac{r^2}{8\kappa}\right)[Ei(-B) - Ei(-A)]\right] \tag{3.187a}$$

$$I_2^f(r,t_F) = \frac{1}{2r}\left[\beta_f^2(1-e^{-B}) + \alpha_f^2(1-e^{-A}) - 2\alpha_f\beta_f(1-e^{-A}) + \frac{r^2}{4\kappa}(\alpha_f e^{-A} - \beta_f e^{-B}) + \frac{r^2}{2\kappa}\left(\beta_f + \frac{r^2}{8\kappa}\right)[Ei(-A) - Ei(-B)]\right] \tag{3.187b}$$

$$I_3^f(r,t_F) = \frac{1}{4\kappa}\left[\beta_f e^{-B} - \alpha_f e^{-A} + \frac{r^2}{4\kappa}[Ei(-B) - Ei(-A)]\right] - \frac{1}{r^2}[\beta_f^2(1-e^{-B}) + \alpha_f^2(1-e^{-A}) - 2\alpha_f\beta_f(1-e^{-B})] \tag{3.187c}$$

$$I_4^f(r,t_F) = \frac{1}{4\kappa}\left[\alpha_f e^{-A} - \beta_f e^{-B} + \frac{r^2}{4\kappa}[Ei(-A) - Ei(-B)]\right] - \frac{1}{r^2}[\beta_f^2(1-e^{-B}) + \alpha_f^2(1-e^{-A}) - 2\alpha_f\beta_f(1-e^{-A})] \tag{3.187d}$$

$$I_5^f(r,t_F) = \frac{r}{4\kappa^2}[Ei(-A) - Ei(-B)] + \frac{\alpha_f}{\kappa r}(e^{-A} - e^{-B}) \tag{3.187e}$$

$$I_6^f(r,t_F) = \frac{r}{4\kappa^2}[Ei(-A) - Ei(-B)] + \frac{\beta_f}{\kappa r}(e^{-A} - e^{-B}) \tag{3.187f}$$

$$I_7^f(r,t_F) = I_4^f(r,t_F) \tag{3.187g}$$

$$I_8^f(r,t_F) = I_6^f(r,t_F) \tag{3.187h}$$

(ii) *For* $f = F$:

$$I_1^F(r, t_F) = \frac{1}{2r}\left[(\Delta t_F)^2(1 - e^{-E}) + \frac{r^2}{4\kappa}\Delta t_F e^{-E} + \left(\frac{r^2}{4\kappa}\right)^2 Ei(-E)\right] \quad (3.188a)$$

$$I_2^F(r, t_F) = \frac{1}{2r}\left[(\Delta t_F)^2(1 - e^{-E}) - \frac{r^2}{4\kappa}\Delta t_F e^{-E} - \frac{r^2}{2\kappa}\left(\Delta_F + \frac{r^2}{8\kappa}\right)Ei(-E)\right] \quad (3.188b)$$

$$I_3^F(r, t_F) = \frac{1}{4\kappa}\left[\Delta t_F e^{-E} + \frac{r^2}{4\kappa}Ei(-E)\right] - \left(\frac{\Delta t_F}{r}\right)^2(1 - e^{-E}) \quad (3.188c)$$

$$I_4^F(r, t_F) = -\frac{1}{4\kappa}\left[\Delta t_F e^{-E} + \frac{r^2}{4\kappa}Ei(-E)\right] - \left(\frac{\Delta t_F}{r}\right)^2(1 - e^{-E}) \quad (3.188d)$$

$$I_5^F(r, t_F) = -\frac{r}{4\kappa^2}Ei(-E) \quad (3.188e)$$

$$I_6^F(r, t_F) = \frac{1}{\kappa r}\left[2\frac{1-\nu}{1-2\nu}\Delta t_F + \Delta t_F(1 - e^{-E}) - \frac{r^2}{4\kappa}Ei(-E)\right] \quad (3.188f)$$

$$I_7^F(r, t_F) = I_4^F(r, t_F) + \frac{\Delta t_F}{2\kappa} \quad (3.188g)$$

$$I_8^F(r, t_F) = I_6^F(r, t_F) - \frac{1-\nu}{\nu(1-2\nu)}\frac{\Delta t_F}{\kappa r} \quad (3.188h)$$

Since it is very important to know also the asymptotic behavior of the kernels given by equations (3.186) and (3.187) as $r \to 0$, we present the asymptotic expressions for the integrals $I_\alpha^f(r, t_F)$. These are:

(i) *For* $f < F$:

$$I_1^f(r, t_F) = \frac{r}{4\kappa}\left[\Delta t_f + \alpha_f \ln\frac{\alpha_f}{\beta_f} + \frac{r^2}{8\kappa}\left(1 - \frac{\alpha_f}{\beta_f} + \ln\frac{\alpha_f}{\beta_f}\right)\right] + O(r^5) \quad (3.189a)$$

$$I_2^f(r, t_F) = \frac{r}{4\kappa}\left[-\Delta t_f - \beta_f \ln\frac{\alpha_f}{\beta_f} + \frac{r^2}{8\kappa}\left(1 - \frac{\beta_f}{\alpha_f} - \ln\frac{\alpha_f}{\beta_f}\right)\right] + O(r^5) \quad (3.189b)$$

$$I_3^f(r, t_F) = \left(\frac{r}{4\kappa}\right)^2\left(1 - \frac{\alpha_f}{\beta_f} + \ln\frac{\alpha_f}{\beta_f}\right) + O(r^4) \quad (3.189c)$$

$$I_4^f(r, t_F) = \left(\frac{r}{4\kappa}\right)^2 \left(1 - \frac{\beta_f}{\alpha_f} - \ln \frac{\alpha_f}{\beta_f}\right) + O(r^4) \tag{3.189d}$$

$$I_5^f(r, t_F) = \frac{r}{4\kappa^2}\left(\frac{\alpha_f}{\beta_f} - 1 - \ln \frac{\alpha_f}{\beta_f}\right) + O(r^3) \tag{3.189e}$$

$$I_6^f(r, t_F) = \frac{r}{4\kappa^2}\left(1 - \frac{\beta_f}{\alpha_f} - \ln \frac{\alpha_f}{\beta_f}\right) + O(r^3) \tag{3.189f}$$

(ii) *For* $f = F$:

$$I_1^F(r, t_F) = \frac{r}{4\kappa}\left[\Delta t_F + \frac{r^2}{8\kappa}\left(C - \frac{3}{2} + \ln E\right)\right] + O(r^5) \tag{3.190a}$$

$$I_2^F(r, t_F) = -\frac{r}{4\kappa}\left[\Delta t_F(C + \ln E) + \frac{r^2}{8\kappa}\left(C - \frac{5}{2} + \ln E\right)\right] + O(r^5) \tag{3.190b}$$

$$I_3^F(r, t_F) = \left(\frac{r}{4\kappa}\right)^2 \left(C - \frac{1}{2} + \ln E\right) + O(r^4) \tag{3.190c}$$

$$I_4^F(r, t_F) = -\frac{\Delta t_F}{2\kappa} + \left(\frac{r}{4\kappa}\right)^2 \left(\frac{3}{2} - C - \ln E\right) + O(r^4) \tag{3.190d}$$

$$I_5^F(r, t_F) = -\frac{r}{4\kappa^2}(C + \ln E) + O(r^3) \tag{3.190e}$$

$$I_6^F(r, t_F) = \frac{2(1 - 2\nu)}{1 - 2\nu}\frac{\Delta t_F}{\kappa r} - \frac{r}{4\kappa^2}(C - 1 + \ln E) + O(r^3) \tag{3.190f}$$

$$I_7^F(r, t_F) = \left(\frac{r}{4\kappa}\right)^2 \left(\frac{3}{2} - C - \ln E\right) + O(r^4) \tag{3.190g}$$

$$I_8^F(r, t_F) = -\frac{1 - \nu}{\nu}\frac{\Delta t_F}{\kappa r} - \frac{r}{4\kappa^2}(C - 1 + \ln E) + O(r^3) \tag{3.190h}$$

where C is the Euler constant ($C = 0.577\,215\ldots$). Here,

$$\alpha_f = t_F - t_f, \qquad \beta_f = \alpha_f + \Delta t_f \tag{3.191a}$$

and

$$A = \frac{r^2}{4\kappa\alpha_f}, \qquad B = \frac{r^2}{4\kappa\beta_f}, \qquad E = \frac{r^2}{4\kappa\,\Delta t_F} \tag{3.191b}$$

Table 3.12. Interpolation polynomials

Configuration	$Q^1(\rho)$	$Q^2(\rho)$	$P^1(\rho)$	$P^2(\rho)$
$\zeta^b \notin \partial B_q$	$1-\rho/\Delta s$	$\rho/\Delta s$	–	–
$\boldsymbol{\zeta}^b = \boldsymbol{\eta}^{1q}$	$1-\rho/\Delta s$	$\rho/\Delta s$	$-1/\Delta s$	$1/\Delta s$
$\boldsymbol{\zeta}^b = \boldsymbol{\eta}^{2q}$	$\rho/\Delta s$	$1-\rho/\Delta s$	$1/\Delta s$	$-1/\Delta s$

For spatial integration, the boundary contour ∂B is divided into M linear boundary elements ∂B_q ($q = 1, 2, \ldots, M$) with the end points η^{aq} ($a = 1, 2$). Globally, the boundary nodes are also represented as ζ^b ($b = 1, 2, \ldots, m$). The boundary temperature, flux, displacements, and tractions are approximated linearly within each boundary element as

$$f(\boldsymbol{\eta}, t) = \sum_{a=1}^{2} f(\boldsymbol{\eta}^{aq}, t) Q^a(\rho), \qquad Q \in \partial B_q \tag{3.192}$$

where $f(\boldsymbol{\eta}, t)$ stands for the approximated boundary quantity, and $Q^a(\rho)$ are interpolation polynomials in which ρ denotes the local coordinate of the integration point $\boldsymbol{\eta}$ on the element Γ_q. The interpolation polynomials are given in table 3.12 in which Δs denotes the length of the boundary element Γ_q. Thus, the boundary unknowns are localized at a finite number of nodal points on the boundary. Let us consider the BIE at the nodal point $\boldsymbol{\zeta}^b$ and the integral representation of stresses at an internal point $\boldsymbol{y}$. We introduce the following notations:

$$r_i = \eta_i - \zeta_i^b, \qquad r = \sqrt{r_i r_i}, \qquad r_{,i} = \frac{r_i}{r} \tag{3.193a}$$

$$R_i = \eta_i - y_i, \qquad R = \sqrt{R_i R_i}, \qquad R_{,i} = \frac{R_i}{R} \tag{3.193b}$$

and define the integrals

$$U_k^{baq\alpha f}(t_F) = \int_0^{\Delta s} Q^a(\rho) U_k^{\alpha f}(\boldsymbol{r}, t_F)\, d\rho \tag{3.194a}$$

$$Z_k^{baq\alpha f}(t_F) = \int_0^{\Delta s} Q^a(\rho) Z_k^{\alpha f}(\boldsymbol{r}, t_F)\, d\rho \tag{3.194b}$$

$$F_{ij}^{aq\alpha f}(y, t_F) = \int_0^{\Delta s} Q^a(\rho) F_{ij}^{\alpha f}(\boldsymbol{R}, t_F)\, d\rho \tag{3.194c}$$

$$H_{lj}^{aq\alpha f}(y, t_F) = \int_0^{\Delta s} Q^a(\rho) H_{lj}^{\alpha f}(\boldsymbol{R}, t_F)\, d\rho \tag{3.194d}$$

$$D_{lji}^{aq}(y) = \int_0^{\Delta s} Q^a(\rho) D_{lji}(\boldsymbol{R})\, d\rho \tag{3.194e}$$

$$T_{lji}^{1q}(y) = -\, T_{lji}^{2q}(y) = -\frac{1}{\Delta s}\int_0^{\Delta s} T_{lji}(\boldsymbol{R})\, d\rho \tag{3.194f}$$

$$U_{ik}^{baq} = \int_0^{\Delta s} Q^a(\rho) U_{ik}(\boldsymbol{r})\, d\rho \tag{3.194g}$$

$$T_{ik}^{baq} = \int_0^{\Delta s} Q^a(\rho) T_{ik}(\boldsymbol{\eta}, \boldsymbol{\zeta}^b)\, d\rho, \quad \text{if } \boldsymbol{\zeta}^b \notin \Gamma_q \tag{3.194h}$$

$$T_{ik}^{baq} = \int_0^{\Delta s} \rho P^a\, T_{ik}(\boldsymbol{\eta}, \boldsymbol{\zeta})\, d\rho, \quad \text{if } \boldsymbol{\zeta} \in \Gamma_q \tag{3.194i}$$

$$C_{ik}^b = \begin{cases} -\sum_{q=1}^M \sum_{a=1}^2 T_{ik}^{baq}, & \boldsymbol{\zeta}^b \notin \Gamma_q \\ 0, & \boldsymbol{\zeta}^b \in \Gamma_q \end{cases} \tag{3.194j}$$

for $b = 1, 2, \ldots, m$; $a = 1, 2$; $q = 1, 2, \ldots, M$; $\alpha = 1, 2$; and $f = 1, 2, \ldots, F$.

From the asymptotic behavior of the integral kernels, it can be seen that the kernel T_{ik} behaves like ρ^{-1} and the kernels U_{ik} and Z_k^{2F} like $\ln \rho$ on singular elements ($\boldsymbol{\zeta}^b \in \partial B_q$). The integrations over singular elements in U_{ik}^{baq} and T_{ik}^{baq} can be performed analytically. The integral kernel Z_k^{2F} can be decomposed on singular elements into the regular part $dZ_k^{2F} + (mn_k/4\pi\kappa)\ln \rho$ and the logarithmic term $-(mn_k/4\pi\kappa)\ln \rho$. The logarithmic term can be integrated analytically and the rest of the integral $Z_k^{baq2F}(t_F)$ numerically. Since the integrands of the other integrals are bounded even on the singular elements, all the numerical integrations can be carried out by the regular Gaussian quadrature rule.

Taking into account the definitions given by equations (3.194), one obtains the discretized BIE

$$u_i(\boldsymbol{\zeta}^b, t_F) C_{ik}^b + \sum_{q=1}^{M} \sum_{a=1}^{2} [u_i(\boldsymbol{\eta}^{aq}, t_F) T_{ik}^{baq} - t_i(\boldsymbol{\eta}^{aq}, t_F) U_{ik}^{baq}] = X_k^b(t_F) \tag{3.195a}$$

where

$$X_k^b(t_F) = \kappa \sum_{f=1}^{F} \sum_{\alpha=1}^{2} \sum_{q=1}^{M} \sum_{a=1}^{2} [\theta(\boldsymbol{\eta}^{aq}, t^{\alpha f}) Z_k^{baq\alpha f}(t_F) - q(\boldsymbol{\eta}^{aq}, t^{\alpha f}) U_k^{baq\alpha f}(t_F)] \tag{3.195b}$$

Having known the nodal values of displacements and tractions at time t_F, one can evaluate the stresses at this time in any internal source point (p) by

$$\begin{aligned}\sigma_{ij}(p, t_F) = &- \gamma\delta_{ij}\theta(\mathbf{y}, t_F) + \kappa \sum_{f=1}^{F} \sum_{\alpha=1}^{2} \sum_{q=1}^{M} \sum_{a=1}^{2} [q(\boldsymbol{\eta}^{aq}, t^{\alpha f}) F_{lj}^{aq\alpha f}(\mathbf{y}, t_F) \\ &- \theta(\boldsymbol{\eta}^{aq}, t^{\alpha f}) H_{lj}^{aq\alpha f}(\mathbf{y}, t_F)] + \sum_{q=1}^{M} \sum_{a=1}^{2} \{u_i(\boldsymbol{\eta}^{aq}, t_F) T_{lji}^{aq}(\mathbf{y}) \\ &+ [t_i(\boldsymbol{\eta}^{aq}, t_F) + \gamma n_i^q \theta(\boldsymbol{\eta}^{aq}, t_F)] D_{lji}^{aq}(\mathbf{y})\}\end{aligned} \tag{3.196}$$

Finally, the discretized form of the stress tensor at $\boldsymbol{\zeta} \in \partial B_q$ (at a boundary source point P) is given by

$$\begin{aligned}\sigma_{ij}(P, t_F) = \sum_{a=1}^{2} \{&[t_k(\boldsymbol{\eta}^{aq}, t_F) + n_k^q \gamma\theta(\boldsymbol{\eta}^{aq}, t_F)] A_{ljk}^{aq} \\ &+ u_k(\boldsymbol{\eta}^{aq}, t_F) B_{ljk}^{aq} - \delta_{lj} \gamma\theta(\boldsymbol{\eta}^{aq}, t_F) Q^a(\bar{s})\end{aligned} \tag{3.197}$$

where $\bar{s}$ is the local coordinate which defines the position of point $\boldsymbol{\zeta}$ on the boundary element Γ_q and

$$A_{ljk}^{aq} = \left[\left(n_l^q n_j^q + \frac{\nu}{1-\nu} \tau_l^q \tau_j^q n_k^q + (n_l^q \tau_j^q + \tau_l^q n_j^q)\tau_k^q\right]Q^a(\bar{s}) \tag{3.198}$$

$$B_{ljk}^{1q} = -\frac{2G}{1-\nu} \tau_l^q \tau_j^q \tau_k^q \frac{1}{\Delta s}, \qquad B_{ljk}^{2q} = - B_{ljk}^{1q} \tag{3.199}$$

3.5.4 Nonstationary Thermoplasticity

In many thermomechanical problems, such as machining and solidification, the stress level in the body exceeds its yield stress, causing inelastic deformations, and the final state of the body may be significantly influenced by the inelastic fields. As discussed by various researchers (e.g., Richmond and Tien 1971, Tien and Richmond 1982, Heinlein et al. 1984), the stress–strain constitutive model for material behavior must be chosen appropriately for such problems. This model should be simple enough to be tractable and yet include important effects such as rate sensitivity, strain hardening and recovery. Such effects must be correctly modeled for temperatures ranging from ambient room temperature to elevated temperatures near the melting point of the material. Furthermore, temperature dependence of physical properties must also be included in the analysis.

The governing thermal equation remains the same as equation (3.164). Using the additive decomposition of strain rates in equation (3.109) in the context of small strains and deformations, the resultant Navier equation for a nonhomogeneous medium may be expressed (in the absence of body forces) as

$$[\lambda \dot{u}_{k,k}\delta_{ij} + G(\dot{u}_{i,j} + \dot{u}_{j,i})]_{,j} = [(3\lambda + 2G)\dot{\varepsilon}_{kk}^{(T)}]_{,i} + (2G\dot{\varepsilon}_{ij}^{(n)})_{,j} \tag{3.200}$$

The Lamé parameters λ and G are related to Young's modulus (E) and Poisson's ratio (ν) through equation (3.123). The nonelastic strain rates $\varepsilon_{ij}^{(n)}$ are obtained from a suitable plastic or viscoplastic model described by equations (3.112)–(3.114). The components of the traction vector at a point on the boundary are given by

$$\dot{\tau}_i = [\lambda \dot{u}_{k,k}\delta_{ij} + G(\dot{u}_{i,j} + \dot{u}_{j,i})]n_j - (3\lambda + 2G)\dot{\varepsilon}_{kk}^{(T)}n_i - 2G\dot{\varepsilon}_{ij}^{(n)}n_j \tag{3.201}$$

Integral equations for such a problem can be derived following the approach used for the thermoelastic problem in a nonhomogeneous media. The resulting integral equation may be expressed as

$$\begin{aligned}\frac{G(p)}{G_0}\dot{u}_j(p) = &\int_{\partial B}\left(U_{ij}(p,Q)\dot{\tau}_i(Q) - \frac{G(Q)}{G_0}T_{ij}(p,Q)\dot{u}_i(Q)\right)ds_Q \\ &+ \int_B\left[\left(U_{ij,k}(p,q) + U_{kj,i}(p,q)\right)G_{,k}(q)\right. \\ &\left.+ U_{kj,k}(p,q)\lambda_{,i}(q)\right]\dot{u}_i(q)\,dA_q \\ &+ \int_B U_{ij,i}(p,q)(3\lambda(q) + 2G(q))\dot{\varepsilon}_{kk}^{(T)}(q)\,dA_q \\ &+ \int_B 2U_{ij,k}(p,q)G(q)\dot{\varepsilon}_{ik}^{(n)}(q)\,dA_q\end{aligned} \tag{3.202}$$

As in the case of thermoelasticity, the Poisson's ratio ν may be assumed to be constant. For this class of problems, one must handle the domain integrals.

Further details of numerical implementations of nonstationary or transient thermoplasticity problems are discussed later in the context of solidification problems.

References

Abramowitz, M. and Stegun, I. A. (1970). *Handbook of Mathematical Functions*. Dover, New York.

Anand, L. (1982). "Constitutive Equations for the Rate-dependent Deformation of Metals at Elevated Temperatures," *J. Eng. Mater. Technol., ASME*, **104**, 12–17.

Anderson, D. A., Tannehill, J. C., and Pletcher, R. H. (1989). *Computational Fluid Mechanics and Heat Transfer*. Hemisphere Publ. Corp., New York.

Aral, M. M. and Tang, Y. (1989). "A Boundary-only Procedure for Transient Transport Problems With or Without First Order Chemical Reactions," *Appl. Math. Modelling*, **13**, 130–137.

Banerjee, P. K. and Butterfield, R. (1981). *Boundary Element Method in Engineering Science*. McGraw-Hill, London.

Beskos, D. E. (ed.) (1987). *Boundary Element Methods in Mechanics*. North-Holland, Amsterdam.

Brebbia, C. A., Telles, J. C. F., and Wrobel, L. C. (1984). *Boundary Element Techniques—Theory and Applications in Engineering*, Springer-Verlag, Berlin.

Carrier, G. F., Krook, M., and Pearson, C. E. (1966). *Functions of Complex Variable*. McGraw-Hill, New York.

Carslaw, H. S. and Jaeger, J. C. (1986). *Conduction of Heat in Solids*, 2d ed. Clarendon Press, Oxford.

Chan, C. L. and Chandra, A. (1991a). "A Boundary Element Method Analysis of the Thermal Aspects of Metal Cutting Processes," *J. Eng. Ind.*, **113**, 311–319.

Chan, C. L. and Chandra, A. (1991b). "An Algorithm for Handling Corners in Boundary Element Method: Application to Conduction–Convection Equations," *Appl. Math. Modelling*, **15**, 244–255.

Chan, C. L. and Chandra, A. (1991c). "A BEM Approach to Thermal Aspects of Machining Processes and Their Design Sensitivities," *Appl. Math. Modelling*, **15**, 562–575.

Cristescu, M. and Lougignac, G. (1978). "Gaussian Quadrature Formulas for Functions with Singularities in 1/*r* Over Triangles and Quadrangles," *Recent Advances in Boundary Element Methods* (ed. C. A. Brebbia), pp. 375–390. Comp. Mech. Inst., Southampton, UK.

Curran, D. A. S., Cross, M., and Lewis, B. A. (1980). "Solution of Parabolic Partial Differential Equations by the Boundary Element Method Using Discretization in Time," *Appl. Math. Modelling*, **4**, 398–400.

Curran, D. A. S., Lewis, B. A., and Cross, M. (1986). "A Boundary Element Method for the Solutions of the Transient Diffusion Equation in Two Dimensions," *Appl. Math. Modelling*, **10**, 107–113.

Fleuries, J. and Predeleanu, M. (1987). "On the Use of Coupled Fundamental Solutions in BEM for Thermoelastic Problems," *Eng. Anal.*, **4**, 70–74.

Ghosh, N., Rajiyah, H., Ghosh, S., and Mukherjee, S. (1986). "A New Boundary Element Method Formulation for Linear Elasticity," *J. Appl. Mech., ASME*, **53**, 69–76.

Ghosh, S. and Mukherjee, S. (1984). "Boundary Element Method Analysis of Thermoelastic Deformation in Nonhomogeneous Media," *Int. J. Solids Structures*, **20**, 829–843.

Gupta, A., Chan, C. L., and Chandra, A. (1994). "A BEM Formulation for Steady State Conduction–Convection Problems With Variable Velocities," *Num. Heat Transfer*, **25**, 415–432.

Hart, E. W. (1976). "Constitutive Relations for the Non-Elastic Deformation of Metals," *J. Eng. Mater. Technol., ASME*, **98**, 193–202.

Heinlein, M., Mukherjee, S., and Richmond, O. (1984). "A Boundary Element Method Analysis of Temperature Fields and Stresses During Solidification," *Acta Mech.*, **59**, 58–81.

Hrodmadka, T. V. and Lai, C. (1987). *The Complex Variable Boundary Element Method in Engineering Analysis*. Springer-Verlag, Berlin.

Jaswon, M. A. and Symn, G. T. (1977). *Integral Equation Methods in Potential Theory and Elastostatics*. Academic Press, London.

Kwak, B. M. and Choi, J. H. (1987). "Shape Design Sensitivity Analysis Using Boundary Integral Equation for Potential Problems," *Computer Aided Optimal Design: Structural and Mechanical Systems* (ed. C. A. Mota Soares), NATO ASI Series F27. Springer-Verlag, Berlin.

Liggett, J. A. and Liu, P. L.-F. (1983). *The Boundary Integral Equation Method for Porous Media Row*. Allen and Unwin, London.

Lim, J., Chan, C. L., and Chandra, A. (1994). "A BEM analysis for Transient Conduction–Convection Problems," *Int. J. Num. Meth., Heat and Fluid Flow*, **4**, 31–46.

Mukherjee, S. (1982). *Boundary Element Methods in Creep and Fracture*. Elsevier Applied Science, Amsterdam.

Mukherjee, S. and Chandra, A. (1991). "A Boundary Element Formulation for Design Sensitivities in Problems Involving Both Geometric and Material Nonlinearities," *Computers Math. Appl.*, **15**, 245–255.

O'Neill, K. (1983). "Boundary Integral Equation Solution of Moving Boundary Phase Change Problems," *Int. J. Num. Meth. Eng.*, **19**, 1825–1850.

Pima, R. L. G. and Fernandes, J. L. M. (1981). "Some Numerical Integration Formulae Over Triangles and Squares with $1/R$ Singularity," *Appl. Math. Modelling*, **5**, 209–211.

Press, W. H., Flannery, B. P., Tenkolsky, S. A., and Vetterling, W. T. (1986). *Numerical Recipes*. pp. 115–120. Cambridge University Press, New York.

Richmond, O. and Tien, R. H. (1971). "Theory of Thermal Stresses and Air-Gap Formation During the Early Stages of Solidification in a Rectangular Mold," *J. Mech. Phys. Solids*, **19**, 273–284.

Rizzo, F. J. and Shippy, D. J. (1977). "An Advanced Boundary Integral Equation Method for Three-Dimensional Thermoelasticity," *Int. J. Num. Meth. Eng.*, **11**, 1753–1768.

Rolph, W. D. and Bathe, K.-J. (1982). "An Efficient Algorithm for Analysis of Nonlinear Heat Transfer With Phase Changes," *Int. J. Num. Meth. Eng.*, **18**, 119–134.

Skerget, P. and Brebbia, C. A. (1983). "The Solution of Convective Problems in

Laminar Flow," *Proceedings of the Fifth International Conferece of BEM in Engineering*, pp, 251–274. Springer-Verlag, Berlin.

Sladek, J., Sladek, V., and Markechova, I. (1990). "Boundary Element Method Analysis of Stationary Thermoelasticity—Problems in Non-homogeneous Media," *Int. J. Num. Meth. Eng.*, **30**, 505–516.

Sladek, V. and Sladek, J. (1984). "Boundary Integral Equation Method in Two Dimensional Thermoelasticity," *Eng. Anal.*, **1**, 135–148.

Sladek, V. and Sladek, J. (1989). "Computation of Thermal Stresses in Quasi-Static Nonstationary Thermoelasticity using Boundary Elements," *Int. J. Num. Meth. Eng.*, **28**, 1131–1144.

Tanaka, Y., Tonma, T., and Kaji, I. (1986). "On Mixed Element Solutions of Convection–Diffusion Problems in Three Dimensions," *Appl. Math. Modelling*, **10**, 170–175.

Tien, R. H. and Richmond, O. (1982). "Theory of Maximum Tensile Stresses in the Solidifying Shell of a Constrained Rectangular Casting," *J. Appl. Mech., ASME*, **49**, 481–486.

Zabaras, N. and Mukherjee, S. (1987). "An Analysis of Solidification Problems by the Boundary Element Method," *Int. J. Num. Meth. Eng.*, **24**, 1879–1900.

4

Design Sensitivities and Optimization

The first main section of this chapter presents a BEM formulation for the determination of design sensitivity coefficients (DSCs) for "fully nonlinear" (both material and geometric) elasto-viscoplastic problems. As described in chapter 2, problems of interest are those in which elastic strains are small but inelastic strains and rotations can be arbitrarily large. This section is followed by a discussion of numerical implementation and numerical results for design sensitivity coefficients for illustrative examples. The final sections are concerned with optimization problems in elasto-viscoplasticity.

4.1 Design Sensitivity Coefficients (DSCs)

Design sensitivity coefficients are rates of change of response quantities, such as stress or displacement in a loaded body, with respect to design variables. These design variables could be shape or sizing parameters that control the (initial) shape of part or all of the boundary of a body, or they could be boundary conditions, material parameters, and so on. Shape parameters as design variables are of primary concern in this chapter.

DSCs are useful in diverse problems. An example is a design problem where the performance of a modified design can be obtained from that of an initial design using a Taylor series expansion about the initial design. They are useful in solving inverse problems (e.g. Zabaras et al. 1988) and reliability analyses (Ang and Tang 1975). A very important area for the application of DSCs with respect to shape parameters is in optimal shape design. An optimization process starts with a preliminary design and calculation of DSCs for this design. Gradient-based optimization algorithms (mathematical programming methods) (e.g., Vanderplaats 1983) use the preliminary design and its sensitivities to propose a new design. The goal is to optimize an objective function without violating the constraints (typically allowable stresses or displacements) of a problem. This process is carried out in an iterative manner, producing a succession of designs, until an optimal design is obtained. There exists a rich literature on the subject of determination of DSCs for linear problems in mechanics such as elasticity or heat transfer [see, for example, the excellent books by Haug et al. (1986), Haftka and Gürdal (1992), and Sokolowski and Zolesio (1991)] In general, three different approaches have been employed for sensitivity analysis—the finite difference approach (FDA), the adjoint structure approach (ASA), and the direct differentiation approach (DDA). Also, both the finite element method (FEM) and the boundary element method (BEM) have been employed for these analyses by different researchers.

4.1.1 The Finite Difference Approach (FDA)

Conceptually, this is the simplest approach for the determination of sensitivities. Typically, the current design is analyzed and the design function is evaluated. Then, the design parameters are perturbed in succession. For each perturbation, the design functional is reevaluated, and the sensitivity of the design functional with respect to the perturbed parameter is obtained by finite differences. This approach has been tried by many authors [e.g., Choi (1987) for elastic analyses]. The accuracy depends on the magnitude of the perturbation—truncation errors can be significant if the perturbation is too large; round-off errors can prove disastrous if the perturbation is too small.

4.1.2 The Adjoint Structure Approach (ASA)

The adjoint structure approach is an exact approach for the determination of sensitivities and does not involve finite differencing. In this approach, an adjoint system must be prescribed in addition to the physical system. One auxiliary system is defined for each design functional, rather than for each design parameter. The sensitivities of the functional with respect to the entire design vector are calculated directly. Many researchers have used this method (e.g., Dems, 1986, 1987). A discussion of this approach using the FEM is available in the book by Haug et al. (1986). Recently, BEM researchers such as Mota Soares and Leal (1987), Choi and Kwak (1988), Meric (1987), and Aithal and Saigal (1990) have employed the ASA to obtain sensitivities.

4.1.3 The Direct Differentiation Approach (DDA)

The direct differentiation approach also uses analytical methods to yield exact expressions for the sensitivities and avoids the use of finite differences. In this approach, the sensitivities of all the required fields are obtained with respect to design parameters, and these are used to calculate the sensitivities of the design functional via the chain rule. A discussion of this approach, using the FEM, is available in Haug et al. (1986).

4.1.4 Linear Elasticity

Of course, the BEM is of primary concern in this book. A number of papers have been published during the last few years on the subject of calculation of DSCs for linear elasticity problems by the BEM. A representative (but by no means exhaustive) list includes planar (Barone and Yang 1988, Kane and Saigal 1988, Choi and Choi 1990, Saigal 1990, Zhang and Mukherjee 1991a,b, Grabacki 1991, Zhao and Adey 1992), axisymmetric (Saigal et al. 1989, Rice and Mukherjee 1990), and three-dimensional (Barone and Yang 1989, Bonnet 1990, Aithal et al. 1991, Kane et al. 1992) problems.

4.1.5 Nonlinear Problems in Solid Mechanics

Sensitivity analysis of nonlinear problems is greatly complicated by the presence of nonlinearities and also because the original problem, as well as the sensitivity problem, is typically history dependent. Thus, schemes for accurate and efficient integration of sensitivity evolution become indispensable.

The work to date, using the FEM, presents general formulations but numerical results primarily for problems of structural mechanics (trusses, beams, and plates). Arora and his coworkers (e.g., Wu and Arora 1988, Arora and Cardoso 1992, Jao and Arora 1992a,b) and Choi and his coworkers (e.g., Choi and Santos 1987, Santos and Choi 1988) have published papers in this area. For such problems, sizing parameters such as cross-sectional areas of bars or beams and thickness of plates, and shape parameters such as lengths of beams and widths and/or lengths of plates are often chosen as design variables. Both the DDA (Arora and Cardoso 1992, Jao and Arora 1992a,b) and the ASA (Arora and Cardoso 1992, Choi and Santos 1987, Santos and Choi 1988) are employed to formulate sensitivity equations; also, both the total and updated Lagrangian formulations are discussed there. The total Lagrangian formulation is concluded to be more effective (Jao and Arora 1992a,b). Both material and geometric nonlinearities are included in Jao and Arora (1992a,b), but only moderate strain problems are considered in this work due to the use of an endochronic constitutive model. In Choi and Santos (1987) and Santos and Choi (1988), geometrically nonlinear behaviors of structural components are studied; these problems involve typically large displacements and rotations but small strains. Problems of this class are also studied in Wu and Arora (1988), where three different approaches are proposed to calculate limit load sensitivity. Other efforts include those of Dems and Mróz (1989), Haftka et al. (1990), and Tortorelli (1992). Sensitivity of buckling loads for trusses and plates, considering the effect of geometric nonlinearities, is studied in Dems and Mróz (1989) and Haftka et al. (1990). Sensitivity analysis for elastostatic systems with nonlinear constraints using the FEM is investigated in Tortorelli (1992).

As mentioned before, the above FEM research is primarily related to structural mechanics problems. In the continuum mechanics regime, only few research papers have been published to date. Vidal et al. (1991) have used the FEM with consistent tangent operators, together with an implicit scheme, to carry out sensitivity analysis for viscoplastic materials undergoing small deformations. Mukherjee's group (Mukherjee and Chandra 1989, 1991, Zhang et al. 1992a,b, Leu and Mukherjee 1993, 1995) has focused on shape sensitivity analyses for large-deformation problems involving solid continua. These researchers have published a series of papers aimed at accurate determination of DSCs for nonlinear continuum mechanics problems that involve material and/or geometric nonlinearities. Mukherjee and Chandra (1989) have presented a BEM formulation for shape sensitivity analysis for small-strain elasto-viscoplastic problems, and Zhang et al. (1992a) have carried out a numerical implementation for problems of this class. A BEM formulation for fully nonlinear problems with both material and geometric nonlinearities has been proposed by Mukherjee and Chandra (1991), and Zhang et al. (1992b) have published a numerical implementation for these problems. Ongoing work in this area is presented in Leu and Mukherjee (1993, 1995). The first paper discusses an implicit objective integration scheme, in conjunction with the BEM, for fully nonlinear elasto-viscoplastic problems.

4.2 DBEM Sensitivity Formulation

The discussion in this section is limited to plane strain problems of elasto-viscoplasticity. As mentioned before, it is assumed that the elastic strains are small but the inelastic strains and rotations can be arbitrarily large. The sensitivity equations are obtained by the direct differentiation approach (DDA) of the derivative boundary element method (DBEM) equations of section 2.2.8 of this book. The corresponding plane stress sensitivity equations can be obtained by differentiating the DBEM equations in section 2.2.9 (Zhang 1991). They are not repeated here. Unless otherwise indicated, the range of indices in this section is $1, 2$.

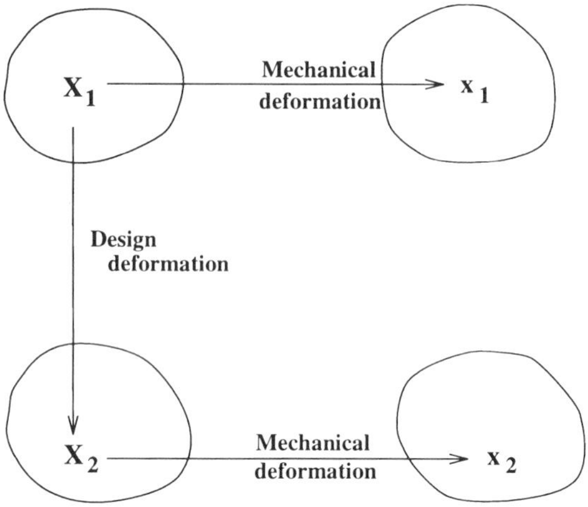

Figure 4.1. Design and mechanical deformations.

4.2.1 Boundary Integral Equations for Sensitivities

Consider a shape design vector $\boldsymbol{b}$ which is defined in the initial configuration $\boldsymbol{X}$ of a body (figure 4.1). It is assumed that the components of the variable $\boldsymbol{b}$ are shape parameters that define the initial (undeformed) shape of the body. Shape perturbations of the undeformed initial configuration $\boldsymbol{X}$ of the body are considered here.

The first step is the differentiation of equation (2.167) with respect to a scalar design variable b, which is a component of the vector $\boldsymbol{b}$. While this approach is mathematically correct (Zhang et al. 1992b), it is more convenient, in the interest of improved numerical accuracy, to first regularize equation (2.167) by the addition subtraction method (see section 2.2.4 and also Leu and Mukherjee 1993, 1995). The regularized form of equation (2.167) becomes

$$\begin{aligned}0=&\int_{\partial B}[U_{ij}(\boldsymbol{b},P,Q)\tau_i^{(s)}(\boldsymbol{b},Q)-W_{ij}(\boldsymbol{b},P,Q)\,\delta_i(\boldsymbol{b},Q)]\,ds(\boldsymbol{b},Q)\\&+2G\int_B U_{ij,k}(\boldsymbol{b},P,q)[d_{ik}^{(n)}(\boldsymbol{b},q)-d_{ik}^{(n)}(\boldsymbol{b},P)]\,da(\boldsymbol{b},q)\\&+2Gd_{ik}^{(n)}(\boldsymbol{b},P)\int_{\partial B}U_{ij}(\boldsymbol{b},P,Q)n_k(\boldsymbol{b},Q)\,ds(\boldsymbol{b},Q)\\&+\int_B U_{ij,k}(\boldsymbol{b},P,q)\,[g_{ki}(\boldsymbol{b},q)-g_{ki}(\boldsymbol{b},P)]\,da(\boldsymbol{b},q)\\&+g_{ki}(\boldsymbol{b},P)\int_{\partial B}U_{ij}(\boldsymbol{b},P,Q)n_k(\boldsymbol{b},Q)\,ds(\boldsymbol{b},Q)\end{aligned}\qquad(4.1)$$

The domain integrals in equation (4.1) are both regular.

Differentiating the above equation (by the chain rule) with respect to b and using a superscribed * to denote the derivative of a variable with respect to $\boldsymbol{b}$, one gets

$$0=\int_{\partial B} [U_{ij}(\boldsymbol{b},P,Q)\overset{*}{\tau}{}_i^{(s)}(\boldsymbol{b},Q) - W_{ij}(\boldsymbol{b},P,Q)\overset{*}{\delta}_i(\boldsymbol{b},Q)]\, ds(\boldsymbol{b},Q)$$

$$+\int_{\partial B} [\overset{*}{U}_{ij}(\boldsymbol{b},P,Q)\tau_i^{(s)}(\boldsymbol{b},Q) - \overset{*}{W}_{ij}(\boldsymbol{b},P,Q)\delta_i(\boldsymbol{b},Q)]\, ds(\boldsymbol{b},Q)$$

$$+\int_{\partial B} [U_{ij}(\boldsymbol{b},P,Q)\tau_i^{(s)}(\boldsymbol{b},Q) - W_{ij}(\boldsymbol{b},P,Q)\,\delta_i(\boldsymbol{b},Q)]\, d\overset{*}{s}(\boldsymbol{b},Q)$$

$$+2G\int_B \overset{*}{V}_{ijk}(\boldsymbol{b},P,q)[d_{ik}^{(n)}(\boldsymbol{b},q) - d_{ik}^{(n)}(\boldsymbol{b},P)]\, da(\boldsymbol{b},q)$$

$$+2G\int_B U_{ij,k}(\boldsymbol{b},P,q)[\overset{*}{d}{}_{ik}^{(n)}(\boldsymbol{b},q) - \overset{*}{d}{}_{ik}^{(n)}(\boldsymbol{b},P)]\, da(\boldsymbol{b},q)$$

$$+2G\int_B U_{ij,k}(\boldsymbol{b},P,q)[d_{ik}^{(n)}(\boldsymbol{b},q) - d_{ik}^{(n)}(\boldsymbol{b},P)]\, d\overset{*}{a}(\boldsymbol{b},q)$$

$$+2G\overset{*}{d}{}_{ik}^{(n)}(\boldsymbol{b},P)\int_{\partial B} U_{ij}(\boldsymbol{b},P,Q)n_k(\boldsymbol{b},Q)\, ds(\boldsymbol{b},Q)$$

$$+2Gd_{ik}^{(n)}(\boldsymbol{b},P)\int_{\partial B} \overset{*}{U}_{ij}(\boldsymbol{b},P,Q)n_k(\boldsymbol{b},Q)\, ds(\boldsymbol{b},Q)$$

$$+2Gd_{ik}^{(n)}(\boldsymbol{b},P)\int_{\partial B} U_{ij}(\boldsymbol{b},P,Q)\overset{*}{n}_k(\boldsymbol{b},Q)\, ds(\boldsymbol{b},Q)$$

$$+2Gd_{ik}^{(n)}(\boldsymbol{b},P)\int_{\partial B} U_{ij}(\boldsymbol{b},P,Q)n_k(\boldsymbol{b},Q)\, d\overset{*}{s}(\boldsymbol{b},Q)$$

$$+\int_B \overset{*}{V}_{ijk}(\boldsymbol{b},P,q)\,[g_{ki}(\boldsymbol{b},q) - g_{ki}(\boldsymbol{b},P)]\, da(\boldsymbol{b},q)$$

$$+\int_B U_{ij,k}(\boldsymbol{b},P,q)\,[\overset{*}{g}_{ki}(\boldsymbol{b},q) - \overset{*}{g}_{ki}(\boldsymbol{b},P)]\, da(\boldsymbol{b},q)$$

$$+\int_B U_{ij,k}(\boldsymbol{b},P,q)\,[g_{ki}(\boldsymbol{b},q) - g_{ki}(\boldsymbol{b},P)]\, d\overset{*}{a}(\boldsymbol{b},q)$$

$$+\overset{*}{g}_{ki}(\boldsymbol{b},P)\int_{\partial B} U_{ij}(\boldsymbol{b},P,Q)n_k(\boldsymbol{b},Q)\, ds(\boldsymbol{b},Q)$$

$$+ g_{ki}(\boldsymbol{b}, P) \int_{\partial B} \overset{*}{U}_{ij}(\boldsymbol{b}, P, Q) n_k(\boldsymbol{b}, Q)\, ds(\boldsymbol{b}, Q)$$

$$+ g_{ki}(\boldsymbol{b}, P) \int_{\partial B} U_{ij}(\boldsymbol{b}, P, Q) \overset{*}{n}_k(\boldsymbol{b}, Q)\, ds(\boldsymbol{b}, Q)$$

$$+ g_{ki}(\boldsymbol{b}, P) \int_{\partial B} U_{ij}(\boldsymbol{b}, P, Q) n_k(\boldsymbol{b}, Q)\, d\overset{*}{s}(\boldsymbol{b}, Q) \qquad (4.2)$$

where $V_{ijk} = U_{ij,k}$. The differentiation with respect to the design variable b is entirely analogous to the concept of the material derivative (usually taken with respect to time) in continuum mechanics. Also, the above approach, using $d\overset{*}{s} = d(ds)/db$ and $d\overset{*}{a} = d(da)/db$, is equivalent to the more commonly used Leibnitz rule. This issue is discussed in section 4.2.10.

The kernels U_{ij} and W_{ij} are given in equations (2.144) and (2.166), respectively. Their differentiated versions (sensitivities) can be obtained from the equations (Zhang et al. 1992a).

$$\overset{*}{U}_{ij}(\boldsymbol{b}, P, Q) = U_{ij,k}(\boldsymbol{b}, P, Q)[\overset{*}{x}_k(Q) - \overset{*}{x}_k(P)] \qquad (4.3)$$

$$\overset{*}{W}_{ij}(\boldsymbol{b}, P, Q) = W_{ij,k}(\boldsymbol{b}, P, Q)[\overset{*}{x}_k(Q) - \overset{*}{x}_k(P)] \qquad (4.4)$$

$$\overset{*}{V}_{ijk}(\boldsymbol{b}, P, Q) = U_{ij,kn}(\boldsymbol{b}, P, Q)[\overset{*}{x}_n(Q) - \overset{*}{x}_n(P)] \qquad (4.5)$$

As discussed before in chapter 2, x_k are the co-ordinates of a material particle at the current configuration (at time t). The quantity $\overset{*}{x}_k$ is a component of the design velocity, which is a consequence of the design deformation (figure 4.1). This issue will be discussed further later in this chapter.

It is important to note that the kernel sensitivities such as $\overset{*}{U}_{ij}$ are known in terms of the design velocities. Also, since $\overset{*}{x}_k(Q) - \overset{*}{x}_k(P) \sim O(r)$, $\overset{*}{U}_{ij}$ and $\overset{*}{W}_{ij}$ are regular and $\overset{*}{V}_{ijk} \sim O(1/r)$. Thus, differentiation with respect to b does not increase the order of the singularity. This implies that all the domain integrals in equation (4.2) are regular!

Discussion of the other sensitivity quantities, such as $\overset{*}{\boldsymbol{\tau}}^{(s)}$, $\overset{*}{\boldsymbol{d}}^{(n)}$, $\overset{*}{\boldsymbol{n}}$, $d\overset{*}{s}$, in equation (4.2) is deferred to the next three subsections (4.2.2, 4.2.3, and 4.2.4)

4.2.2 Boundary Condition Sensitivities

The quantity $\overset{*}{\delta}_i$ in equation (4.2) is the sensitivity of $\delta_i = \partial v_i/\partial s$. This quantity is either prescribed or an unknown of a problem.

The traction sensitivity $\overset{*}{\tau}_i^{(s)}$ is related to physical quantities on the boundary. Referring to section 2.2, the situations for dead and follower loads are as follows.

Dead load. Differentiating equation (2.122) with respect to b, one gets

$$\overset{\circ}{F}{}_i^{(D)}(t) = \overset{*}{\tau}{}_i^{(s)}\, ds + \tau_i^{(s)}\, d\overset{*}{s} \tag{4.6}$$

where the sensitivity of the rate of the dead load, on a surface element, is denoted by a superscribed '∘'.

Follower load. Differentiating equation (2.130) with respect to b, one gets

$$\begin{aligned}\overset{*}{\tau}{}_i^{(s)} &= \overset{\circ}{p}(t)n_i + \dot{p}(t)\overset{*}{n}_i \\ &\quad - \overset{*}{p}(t)n_k h_{ki} - p(t)\overset{*}{n}_k h_{ki} - p(t)n_k \overset{*}{h}_{ki} \\ &\quad + \overset{*}{p}(t)n_i d_{kk} + p(t)\overset{*}{n}_i d_{kk} + p(t)n_i \overset{*}{d}_{kk}\end{aligned} \tag{4.7}$$

Also, the sensitivities of $\boldsymbol{\tau}^{(s)}$ and the Cauchy traction rate $\boldsymbol{\tau}^{(c)}$ can be related by differentiating equation (2.129). The result is

$$\overset{*}{\tau}{}_i^{(s)} = \overset{*}{\tau}{}_i^{(c)} - \overset{*}{n}_m g_{mi} - n_m \overset{*}{g}_{mi} \tag{4.8}$$

where, referring to the definition of $\boldsymbol{g}$ below equation (2.96), one has

$$\overset{*}{g}_{mi} = \overset{*}{\sigma}_{mk}\omega_{ki} + \sigma_{mk}\overset{*}{\omega}_{ki} + \overset{*}{d}_{mk}\sigma_{ki} + d_{mk}\overset{*}{\sigma}_{ki} - \overset{*}{\sigma}_{mi}d_{kk} - \sigma_{mi}\overset{*}{d}_{kk} \tag{4.9}$$

4.2.3 Sensitivities of Inelastic Constitutive Model Equations

Sensitivities of the nonelastic stretching tensor $\boldsymbol{d}^{(n)}$ can be obtained by differentiating a suitable viscoplastic material constitutive model. A general form can be obtained by differentiating equation (2.91) with respect to b. The result is

$$\overset{*}{\boldsymbol{d}}{}^{(n)} = \frac{\partial \boldsymbol{f}}{\partial \boldsymbol{\sigma}} \cdot \overset{*}{\boldsymbol{\sigma}} + \frac{\partial \boldsymbol{f}}{\partial \boldsymbol{q}_p} \cdot \overset{*}{\boldsymbol{q}}_p \tag{4.10}$$

$$\hat{\overset{*}{\boldsymbol{q}}}_p = \frac{\partial \boldsymbol{g}_p}{\partial \boldsymbol{\sigma}} \cdot \overset{*}{\boldsymbol{\sigma}} + \frac{\partial \boldsymbol{g}_p}{\partial \boldsymbol{q}_r} \cdot \overset{*}{\boldsymbol{q}}_r \tag{4.11}$$

where $\hat{\overset{*}{\boldsymbol{q}}}_p = d(\hat{\boldsymbol{q}}_p)/db$ and the tensor products must be taken appropriately. For example (see Zhang et al. 1992a),

$$\left(\frac{\partial \boldsymbol{f}}{\partial \boldsymbol{\sigma}} \cdot \overset{*}{\boldsymbol{\sigma}}\right)_{ij} = \frac{\partial f_{ij}}{\partial \sigma_{mn}}\overset{*}{\sigma}_{mn}$$

4.2.4 *Kinematic and Geometric Sensitivities*

For notational convenience, let the design velocity vector $\overset{*}{\boldsymbol{x}}$ be called $\boldsymbol{v}^{(d)}$. An interesting point to observe here is that while the design derivative commutes with the time derivative, it does not commute with the spatial gradient. In fact, one can show that (Haug et al. 1986), for a function ϕ

$$\frac{d}{db}\left(\frac{\partial \phi}{\partial x_i}\right) = \frac{\partial \overset{*}{\phi}}{\partial x_i} - \frac{\partial \phi}{\partial x_k}\frac{\partial v_k^{(d)}}{\partial x_i} \tag{4.12}$$

Formulas for the sensitivities of various kinematic and geometrical quantities in equation (4.2) are given below.

Design velocity. At a boundary point P, one can use the equations (Zhang et al. 1992b)

$$v_i(P) = \int_A^P \delta_i \, ds + v_i(A) \tag{4.13}$$

$$\overset{*}{v}_i(P) = \int_A^P \overset{*}{\delta}_i \, ds + \int_A^P \delta_i \, d\overset{*}{s} + \overset{*}{v}_i(A) \tag{4.14}$$

where the contour integrations start from some point A with known $\boldsymbol{v}$ and $\overset{*}{\boldsymbol{v}}$. Noting that

$$\overset{*}{v}_i(P) = \frac{d}{dt}(\overset{*}{x}_i(P)) = \frac{dv_i^{(d)}}{dt}$$

time integration of equations (4.13) and (4.14), starting from known initial values of $\boldsymbol{x}(P)$ and $\overset{*}{\boldsymbol{x}}(P)$, yields $\boldsymbol{x}$ and $\overset{*}{\boldsymbol{x}} = \boldsymbol{v}^{(d)}$ at P as functions of time.

An analogous procedure can be adopted for the calculation of the design velocity at an internal point. For this, one needs an integral equation for the velocity of a material point at an internal point p. This is discussed in subsection 4.2.6.

Gradient of design velocity. The first step here is to notice that, from equations (2.18)–(2.20) and taking the reference configuration (at zero time) to be the undeformed one,

$$\dot{\boldsymbol{F}} = (\boldsymbol{d} + \boldsymbol{\omega}) \cdot \boldsymbol{F}, \qquad \boldsymbol{F}(0) = \boldsymbol{I} \tag{4.15}$$

Differentiating the above with respect to b,

$$\overset{\circ}{\boldsymbol{F}} = (\overset{*}{\boldsymbol{d}} + \overset{*}{\boldsymbol{\omega}}) \cdot \boldsymbol{F} + (\boldsymbol{d} + \boldsymbol{\omega}) \cdot \overset{*}{\boldsymbol{F}}, \qquad \overset{*}{\boldsymbol{F}}(0) = 0 \tag{4.16}$$

so that the time histories for $\boldsymbol{F}$ and $\overset{*}{\boldsymbol{F}}$ can be obtained by time integration of $\dot{\boldsymbol{F}}$ and $\overset{\circ}{\boldsymbol{F}}$, respectively.

A useful formula for

$$\frac{d\overset{*}{\boldsymbol{x}}}{d\boldsymbol{x}} = \frac{d\boldsymbol{v}^{(d)}}{d\boldsymbol{x}}$$

can be obtained as follows (Zhang et al. 1992b). With

$$d\boldsymbol{x} = \boldsymbol{F} \cdot d\boldsymbol{X}$$

one gets, by differentiation with respect to b,

$$d\overset{*}{\boldsymbol{x}} = \overset{*}{\boldsymbol{F}} \cdot d\boldsymbol{X} + \boldsymbol{F} \cdot d\overset{*}{\boldsymbol{X}} \qquad \text{and} \qquad \frac{d\overset{*}{\boldsymbol{x}}}{d\boldsymbol{x}} = \overset{*}{\boldsymbol{F}} \cdot \boldsymbol{F}^{-1} + \boldsymbol{F} \cdot \frac{d\overset{*}{\boldsymbol{X}}}{d\boldsymbol{X}} \cdot \boldsymbol{F}^{-1}$$

or, in indicial notation,

$$v_{i,j}^{(d)} = \overset{*}{x}_{i,j} = \overset{*}{F}_{iM} F_{Mj}^{-1} + F_{iM} V_{M,K}^{(d)} F_{Kj}^{-1} \tag{4.17}$$

Equation (4.17) establishes a relationship between $v_{i,j}^{(d)}$ and its initial value $V_{I,J}^{(d)} = \overset{*}{X}_{I,J}$ (where $\boldsymbol{V}^{(d)} = \overset{*}{\boldsymbol{X}}$). Note the similarity of the above equation with equation (2.18) [why is there a second term in equation (4.17)?].

It is very interesting to note that the evolution equation (4.17) for the gradient of the design velocity can also be derived from equation (4.12) by setting $\phi = X_M$. This is left as an exercise for the reader.

The quantities $d\overset{*}{s}$, $d\overset{*}{a}$, $\overset{*}{\boldsymbol{n}}$ and $\overset{*}{\boldsymbol{t}}$. A reinterpretation of the standard equations for $d\overset{*}{S}$ and $d\overset{*}{A}$ [in the initial configuration—see, for example, Haug et al. (1986) and also equation (2.125) in this book], gives

$$d\overset{*}{s} = (v_{k,k}^{(d)} - n_i n_j v_{i,j}^{(d)})\ ds \tag{4.18}$$

$$d\overset{*}{a} = v_{k,k}^{(d)}\, da \tag{4.19}$$

where the first equation must be applied only at a boundary point while the second is typically applied at an internal point. The above equations relate the "material" derivatives of ds and da, with respect to a design variable, to their values at that time. Of course, the evolution equation (4.17) for $v_{i,j}^{(d)}$ and Nanson's formula (2.43), written as

$$\boldsymbol{n} = \frac{\boldsymbol{N} \cdot \boldsymbol{F}^{-1}}{|\boldsymbol{N} \cdot \boldsymbol{F}^{-1}|} = \frac{\boldsymbol{k}}{k} \tag{4.20}$$

must be used to obtain $d\overset{*}{s}$ and $d\overset{*}{a}$ from equations (4.18) and (4.19). An alternative approach for calculation of $\boldsymbol{n}$ is numerical integration of equation (2.124).

Taking the design derivative of the equation $k\boldsymbol{n} = \boldsymbol{k}$, rearranging the terms, and using $\overset{*}{k} = \boldsymbol{k} \cdot \overset{*}{\boldsymbol{k}}/k$, one obtains

$$\overset{*}{\boldsymbol{n}} = \frac{\overset{*}{\boldsymbol{k}}}{k} - \frac{\boldsymbol{n}}{k^2}(\boldsymbol{k} \cdot \overset{*}{\boldsymbol{k}}) \tag{4.21}$$

with

$$\overset{*}{\boldsymbol{k}} = \overset{*}{\boldsymbol{N}} \cdot \boldsymbol{F}^{-1} - k(\boldsymbol{n} \cdot \overset{*}{\boldsymbol{F}} \cdot \boldsymbol{F}^{-1}) \tag{4.22}$$

It is interesting to compare equation (4.21) with the formula given in Hill (1978) for the material derivative of $\boldsymbol{n}$, which is equation (2.124) in this book. It is not difficult to show that if $\overset{*}{\boldsymbol{N}} = 0$, which is the case for other design variables such as sizing and material parameters, equations (4.21) and (4.22) are equivalent to equation (2.124) when the design derivative is replaced by the material derivative. Such, however, is not the case for shape sensitivity problems in which, usually, $\overset{*}{\boldsymbol{N}} \neq 0$.

There still remains the question of efficient evaluation of the initial values of quantities such as $\overset{*}{\boldsymbol{N}}$ and $d\overset{*}{S}$. This discussion is deferred to subsection 4.2.9 where small-strain viscoplasticity problems are considered (see also Zhang et al. 1992a).

4.2.5 Stress Rates and Velocity Gradient Sensitivities on the Boundary

The sensitivity equations for $\dot{\boldsymbol{\sigma}}$ and $\boldsymbol{h}$ on ∂B are obtained by differentiating equations (2.169), (2.170) and (2.90) with respect to b. The resulting equations are

$$\begin{aligned}\overset{\diamond}{\sigma}_{ij} = A_{ijk}\, \overset{*}{\tau}{}_k^{(c)} + B_{ijk}\, \overset{*}{\delta}_k + C_{ijkl}\, \overset{*}{d}{}_{kl}^{(n)} + D_{ij}\, \overset{*}{d}{}_{kk}^{(n)} \\ + \overset{*}{A}_{ijk}\tau_k^{(c)} + \overset{*}{B}_{ijk}\,\delta_k + \overset{*}{C}_{ijkl} d_{kl}^{(n)} + \overset{*}{D}_{ij} d_{kk}^{(n)}\end{aligned} \tag{4.23}$$

$$\overset{*}{h}_{ij} = E_{ijk}\overset{*}{\tau}{}_k^{(c)} + F_{ijk}\overset{*}{\delta}_k + G_{ijkl}\overset{*}{d}{}_{kl}^{(n)} + \overset{*}{E}_{ijk}\tau_k^{(c)} + \overset{*}{F}_{ijk}\,\delta_k + \overset{*}{G}_{ijkl} d_{kl}^{(n)} \tag{4.24}$$

$$\overset{\circ}{\sigma}_{ij} = \overset{\diamond}{\sigma}_{ij} - \overset{*}{\sigma}_{ik}\,\omega_{kj} - \sigma_{ik}\,\overset{*}{\omega}_{kj} + \overset{*}{\omega}_{ik}\,\sigma_{kj} + \omega_{ik}\,\overset{*}{\sigma}_{kj} \tag{4.25}$$

The tensors A_{ijk}, etc., are defined below equations (2.169) and (2.170).

4.2.6 *Sensitivities of Integral Equations at an Internal Point*

A regularized form of the equation for the velocity at an internal point "p" is [see Leu and Mukherjee (1993) and the 3-D BEM equation (2.117)]

$$
\begin{aligned}
v_i(p) = v_i(\tilde{p}) &+ \int_{\partial B} [U_{ij}(\boldsymbol{b},p,Q)\tau_i^{(s)}(\boldsymbol{b},Q) - W_{ij}(\boldsymbol{b},p,Q)\,\delta_i(\boldsymbol{b},Q)]\,ds(\boldsymbol{b},Q) \\
&+ 2G\int_B U_{ij,k}(\boldsymbol{b},p,q)\,[d_{ik}^{(n)}(\boldsymbol{b},q) - d_{ik}^{(n)}(\boldsymbol{b},p)]\,da(\boldsymbol{b},q) \\
&+ 2Gd_{ik}^{(n)}(\boldsymbol{b},p)\int_{\partial B} U_{ij}(\boldsymbol{b},p,Q)n_k(\boldsymbol{b},Q)\,ds(\boldsymbol{b},Q) \\
&+ \int_B U_{ij,k}(\boldsymbol{b},p,q)[g_{ki}(\boldsymbol{b},q) - g_{ki}(\boldsymbol{b},p)]\,da(\boldsymbol{b},q) \\
&+ g_{ki}(\boldsymbol{b},p)\int_{\partial B} U_{ij}(\boldsymbol{b},p,Q)n_k(\boldsymbol{b},Q)\,ds(\boldsymbol{b},Q) \qquad (4.26)
\end{aligned}
$$

In this equation, $\boldsymbol{v}(\tilde{p})$ is the velocity of the boundary point $\tilde{p}$ where a branch cut, extending from point p and pointing in the direction of the positive x_1 axis, intersects the boundary (see Ghosh et al. 1986).

A sensitivity version of the above equation is quite analogous to equation (4.2) for a boundary point, and is not repeated here. For future reference, this equation is called (4.26′). As before (see section 4.2.3), $\overset{*}{v}_i(p)$ can be integrated in time to obtain $v_i^{(d)}(p)$.

Finally, one needs a differentiated version of the velocity gradient equation (2.171) at an internal point. This has the form

$$
\begin{aligned}
\overset{*}{h}_{j\bar{l}}(\boldsymbol{b},p) = &\int_{\partial B} [V_{ij\bar{l}}(\boldsymbol{b},p,Q)\overset{*}{\tau}{}_i^{(s)}(\boldsymbol{b},Q) - Y_{ij\bar{l}}(\boldsymbol{b},p,Q)\overset{*}{\delta}_i(\boldsymbol{b},Q)]\,ds(\boldsymbol{b},Q) \\
&+ \int_{\partial B} [\overset{*}{V}_{ij\bar{l}}(\boldsymbol{b},p,Q)\tau_i^{(s)}(\boldsymbol{b},Q) - \overset{*}{Y}_{ij\bar{l}}(\boldsymbol{b},p,Q)\,\delta_i(\boldsymbol{b},Q)]\,ds(\boldsymbol{b},Q) \\
&+ \int_{\partial B} [V_{ij\bar{l}}(\boldsymbol{b},p,Q)\tau_i^{(s)}(\boldsymbol{b},Q) - Y_{ij\bar{l}}(\boldsymbol{b},p,Q)\delta_i(\boldsymbol{b},Q)]\,d\overset{*}{s}(\boldsymbol{b},Q) \\
&- 2G\overset{*}{d}{}_{ik}^{(n)}(\boldsymbol{b},p)\int_{\partial B} V_{ijk}(\boldsymbol{b},p,Q)n_l(\boldsymbol{b},Q)\,ds(\boldsymbol{b},Q) \\
&- 2Gd_{ik}^{(n)}(\boldsymbol{b},p)\int_{\partial B} \overset{*}{V}_{ijk}(\boldsymbol{b},p,Q)n_l(\boldsymbol{b},Q)\,ds(\boldsymbol{b},Q)
\end{aligned}
$$

$$- 2Gd_{ik}^{(n)}(\boldsymbol{b},p)\int_{\partial B} V_{ijk}(\boldsymbol{b},p,Q)\overset{*}{n}_l(\boldsymbol{b},Q)\,ds(\boldsymbol{b},Q)$$

$$- 2Gd_{ik}^{(n)}(\boldsymbol{b},p)\int_{\partial B} V_{ijk}(\boldsymbol{b},p,Q)n_l(\boldsymbol{b},Q)\,d\overset{*}{s}(\boldsymbol{b},Q)$$

$$- \overset{*}{g}_{mi}(\boldsymbol{b},p)\int_{\partial B} V_{ijm}(\boldsymbol{b},p,Q)n_l(\boldsymbol{b},Q)\,ds(\boldsymbol{b},Q)$$

$$- g_{mi}(\boldsymbol{b},p)\int_{\partial B} \overset{*}{V}_{ijm}(\boldsymbol{b},p,Q)n_l(\boldsymbol{b},Q)\,ds(\boldsymbol{b},Q)$$

$$- g_{mi}(\boldsymbol{b},p)\int_{\partial B} V_{ijm}(\boldsymbol{b},p,Q)\overset{*}{n}_l(\boldsymbol{b},Q)\,ds(\boldsymbol{b},Q)$$

$$- g_{mi}(\boldsymbol{b},p)\int_{\partial B} V_{ijm}(\boldsymbol{b},p,Q)n_l(\boldsymbol{b},Q)\,d\overset{*}{s}(\boldsymbol{b},Q)$$

$$+ \int_B 2G\overset{*}{P}_{ijk\bar{l}}(\boldsymbol{b},p,\mathrm{q})\,[d_{ik}^{(n)}(\boldsymbol{b},q) - d_{ik}^{(n)}(\boldsymbol{b},p)]\,da(\boldsymbol{b},q)$$

$$+ \int_B 2GP_{ijk\bar{l}}(\boldsymbol{b},p,q)[\overset{*}{d}{}_{ik}^{(n)}(\boldsymbol{b},q) - \overset{*}{d}{}_{ik}^{(n)}(\boldsymbol{b},p)]\,da(\boldsymbol{b},q)$$

$$+ \int_B 2GP_{ijk\bar{l}}(\boldsymbol{b},p,q)[d_{ik}^{(n)}(\boldsymbol{b},q) - d_{ik}^{(n)}(\boldsymbol{b},p)]\,d\overset{*}{a}(\boldsymbol{b},q)$$

$$+ \int_B \overset{*}{P}_{ijm\bar{l}}(\boldsymbol{b},p,q)[g_{mi}(\boldsymbol{b},q) - g_{mi}(\boldsymbol{b},p)]\,da(\boldsymbol{b},q)$$

$$+ \int_B P_{ijm\bar{l}}(\boldsymbol{b},p,q)[\overset{*}{g}_{mi}(\boldsymbol{b},q) - \overset{*}{g}_{mi}(\boldsymbol{b},p)]\,da(\boldsymbol{b},q)$$

$$+ \int_B P_{ijm\bar{l}}(\boldsymbol{b},p,q)[g_{mi}(\boldsymbol{b},q) - g_{mi}(\boldsymbol{b},p)]\,d\overset{*}{a}(\boldsymbol{b},q) \qquad (4.27)$$

where $Y_{ij\bar{l}} = W_{ij,\bar{l}}$ and $P_{ijk\bar{l}} = U_{ij,k\bar{l}}$.

Although equation (4.27) is long, it may be evaluated easily. The boundary kernels are regular and the domain integrands are $1/r$ singular. These domain integrals can be evaluated accurately by standard means [e.g., Mukherjee (1982) or Nagarajan and Mukherjee (1993)]. The entire right-hand side of equation (4.27) is known at this stage except for the integrals involving $\overset{*}{g}_{mi}$, which depend on $\overset{*}{h}_{ij}$. Accordingly, iterations are needed over equations (4.2) and (4.27). These iterations are similar to those needed over velocity gradients for BEM analyses of large-strain problems.

4.2.7 *Stress Rate Sensitivities at an Internal Point*

Finally, the sensitivities $\overset{\circ}{\boldsymbol{\sigma}}$ at an internal point are evaluated by differentiating the hypoelastic law (equation 2.89). The result is

$$\overset{\diamond}{\sigma}_{ij} = \lambda \overset{*}{h}_{kk}\,\delta_{ij} + G(\overset{*}{h}_{ij} + \overset{*}{h}_{ji}) - 2G\overset{*}{d}{}^{(n)}_{ij} \tag{4.28}$$

together with equation (4.25) for $\overset{\circ}{\sigma}_{ij}$.

4.2.8 *Sensitivities of Corner and Compatibility Equations*

Equation (2.177) applies if stress is continuous across a corner. In that case, it has proved useful to start from

$$\hat{\sigma}^{+}_{ij} = \hat{\sigma}^{-}_{ij} \tag{4.29}$$

Using the definition $\tau^{(c)}_i = \hat{\sigma}_{ij} n_j$, one can obtain

$$\underline{(\hat{\sigma}_{ij})^{+}(n_j)^{-}} + (\hat{\sigma}_{ij})^{-}(n_j)^{+} = \underline{(\tau^{(c)}_i)^{-}} + (\tau^{(c)}_i)^{+} \tag{4.30}$$

Here, as before, the symbols "+" and "−" denote the two edges of a corner. The above equations should be rewritten in terms of the boundary variables of the corner field point at the two edges. To this end, the boundary algorithm for calculating $\hat{\sigma}_{ij}$ (equation 2.169) and equation (2.129) are needed.

The sensitivity version of equation (4.30) is

$$\underline{(\overset{\diamond}{\sigma}_{ij})^{+}(n_j)^{-} + (\hat{\sigma}_{ij})^{+}(\overset{*}{n}_j)^{-}} + (\overset{\diamond}{\sigma}_{ij})^{-}(n_j)^{+} + (\hat{\sigma}_{ij})^{-}(\overset{*}{n}_j)^{+} = \underline{(\overset{*}{\tau}{}^{(c)}_i)^{-}} + (\overset{*}{\tau}{}^{(c)}_i)^{+} \tag{4.31}$$

Recall that ($\diamond$) denotes the sensitivity of the Jaumann rate of a variable. Again, this should be rewritten in terms of boundary variables. Note that the underlined parts of equations (4.30) and (4.31) are themselves equal.

Another two useful equations, representing continuity of the velocity, are, first,

$$\int_{\partial B} \delta_i(\boldsymbol{b}, Q)\, ds(\boldsymbol{b}, Q) = 0 \tag{4.32}$$

The sensitivity version of this equation is

$$\int_{\partial B} \overset{*}{\delta}_i(\boldsymbol{b}, Q)\, ds(\boldsymbol{b}, Q) + \int_{\partial B} \delta_i(\boldsymbol{b}, Q)\, d\overset{*}{s}(\boldsymbol{b}, Q) = 0 \tag{4.33}$$

These compatibility equations, together with the corner equations and boundary integral equations, constitute the system of equations. Therefore, the system is overdetermined. However, as long as the system is consistent, the solution is unique.

4.2.9 Special Cases—Small-Strain Elasto-viscoplasticity and Linear Elasticity

Elasto-viscoplasticity. When strains and rotations are small (i.e., the problem is materially nonlinear but geometrically linear), the situation simplifies considerably. The changes in the equations, presented above in this section, are discussed briefly below. Details are available in Zhang et al. (1992a).

It is assumed that the spin tensor $\boldsymbol{\omega}$, the tensor $\boldsymbol{g}$, and their sensitivities are very small, and hence are set to zero. The stretch tensor $\boldsymbol{d}$ and its nonelastic counterpart $\boldsymbol{d}^{(n)}$ reduce to the strain rates $\dot{\boldsymbol{\varepsilon}}$ and $\dot{\boldsymbol{\varepsilon}}^{(n)}$, respectively. The tensor $\boldsymbol{F} = \boldsymbol{I}$. The Jaumann rate of a tensor reduces to its material rate (see equation 2.90) and all the different stress measures reduce to the Cauchy stress $\boldsymbol{\sigma}$.

As a consequence of the above, in the basic governing equations (4.1, 4.2, 4.26, and 4.27], the tensors $\boldsymbol{g}$ and $\overset{*}{\boldsymbol{g}}$ are set to zero. The scaled Lagrange traction rate $\boldsymbol{\tau}^{(s)}$ reduces to the ordinary traction rate $\dot{\boldsymbol{\tau}}$ and its sensitivity to $\overset{\circ}{\boldsymbol{\tau}}$.

Kinematic and geometric variables such as $\boldsymbol{v}^{(d)}$, $\boldsymbol{\nabla v}^{(d)}$, ds, da, $\boldsymbol{n}$, $\boldsymbol{t}$, $d\overset{*}{s}$, $d\overset{*}{a}$, $\overset{*}{\boldsymbol{n}}$, and $\overset{*}{\boldsymbol{t}}$ are assumed not to evolve in time but to retain their initial values. In this context, it is useful to note some convenient formulas for the calculation of some of the above quantities from Zhang et al. (1992a). These are given below; in these equations, capital letters denote initial values of the respective quantities.

Let the parametric equations of a curve, which is part or all of the boundary ∂B, be given by

$$X_1 = f_1(\boldsymbol{b}, \eta), \qquad X_2 = f_2(\boldsymbol{b}, \eta) \tag{4.34}$$

where $\eta \in [c, d]$ is a scalar mapping parameter which is independent of $\boldsymbol{b}$. It can be shown (Zhang and Mukherjee 1991a) that

$$\frac{d\overset{*}{S}}{dS} = \frac{\overset{*}{\alpha}}{\alpha} \tag{4.35}$$

$$N_i = \frac{\gamma_{ij}\alpha_j}{\alpha}, \qquad \overset{*}{N}_i = \frac{\gamma_{ij}}{\alpha}\left(\overset{*}{\alpha}_j - \alpha_j \frac{d\overset{*}{S}}{dS}\right) \tag{4.36}$$

where $\boldsymbol{\alpha} = \partial \boldsymbol{X}/\partial \eta$ and α denotes the length of the vector $\boldsymbol{\alpha}$. Also, $\gamma_{11} = \gamma_{22} = 0$, $\gamma_{12} = -\gamma_{21} = 1$. Since $T_1 = -N_2$ and $T_2 = -N_1$, the formulas for the tangent vector and its sensitivity have exactly the same forms as equation (4.36) with the Kronecker delta δ_{ij} replacing γ_{ij}.

Linear elasticity. Formulations for classical linear elasticity are obtained from the equations of small-strain elasto-viscoplasticity by setting the nonelastic strain $\boldsymbol{\varepsilon}^{(n)}$ to zero. Also, the equations are written in terms of the physical variables rather than their rates. Sensitivity analysis of linear elastic problems is fairly standard. The reader is referred to papers such as Zhang and Mukherjee (1991a) or Barone and Yang (1988) for discussions of these problems by the DDA of the BEM.

4.2.10 Leibnitz Rule, Calculation of Geometric Sensitivities, and Related Issues

Leibnitz rule and equation (4.2). The differentiation process used to obtain equation (4.2) from equation (4.1) (here called the "$d\overset{*}{s}$" rule) is somewhat nonstandard. It is illuminating, in this context, to show the equivalence of this approach to the standard Leibnitz formula for differentiating under the integral sign. Considering (in either two or three dimensions)

$$I = \int_B f(\boldsymbol{x}, b)\, dv$$

the Leibnitz rule gives an expression for $\overset{*}{I} = \partial I/\partial b$ as [see Fung (1977) for the corresponding formula for the time derivative of I]

$$\overset{*}{I} = \int_B \frac{\partial f}{\partial b}(\boldsymbol{x},\ b)\ dv + \int_{\partial B} f(\boldsymbol{x},\ b) v_i^{(d)} n_i\, ds$$

Using the divergence theorem,

$$\begin{aligned}\overset{*}{I} &= \int_B \frac{\partial f}{\partial b}(\boldsymbol{x}, b)\, dv + \int_B (f(\boldsymbol{x}, b) v_i^{(d)})_{,i}\, dv \\ &= \int_B \left[\frac{\partial f}{\partial b}(\boldsymbol{x}, b)\, dv + v_i^{(d)} f_{,i}(\boldsymbol{x}, b) + f(\boldsymbol{x}, b) v_{i,i}^{(d)}\right] dv\end{aligned}$$

$$\overset{*}{I} = \int_B \overset{*}{f}(\boldsymbol{x}, b)\, dv + \int_B f(\boldsymbol{x}, b)\, d\overset{*}{v} \tag{4.37}$$

Equation (4.37) is the "$d\overset{*}{s}$" rule. Equation (4.19) (relating dv to $d\overset{*}{v}$) and the definition of the material derivative of f [see equation (4.38) below] have been used in the above.

The design velocity at P and at p. While the quantity $\overset{*}{\boldsymbol{X}}(P)$, at a boundary point, is relatively easy to compute from the design motion of the boundary, such is not the case for $\overset{*}{\boldsymbol{X}}(p)$ at an internal point p. The internal points must move in a manner compatible with the boundary motion, but this motion is not unique and additional assumptions are necessary to tie the motion of internal points in B to the motion of the boundary ∂B. Also, dependent variables such as stress or displacement can depend on b in an explicit as well as in an implicit manner through $X_i(p)$. Thus, for example, the stress field inside a hollow elastic disk (with circular concentric cutout), subjected to external pressure, is a function of its position as well as the disk radii "a" and "b". Here, let the radius b be a design variable. A consequence of this fact is that for a dependent variable $\chi(\boldsymbol{X}(p), b)$, one gets a material derivative

$$\overset{*}{\chi}(X_i(p), b) = \left.\frac{\partial \chi}{\partial b}\right|_{X_i(p)} + \frac{\partial \chi}{\partial X_i(p)} \overset{*}{X}_i(p) \tag{4.38}$$

Obviously, an assumption regarding the design velocity at an internal point will affect the value of $\overset{*}{\chi}$.

$d\overset{*}{S}/dS$ on an ellipse. Let an ellipse be parameterized by the equations

$$X_1 = a \cos \eta, \qquad X_2 = b \sin \eta \tag{4.39}$$

Using equation (4.35), with "b" the design variable,

$$\left|\frac{\partial \boldsymbol{X}}{\partial \eta}\right| = \alpha = \sqrt{a^2 \sin^2 \eta + b^2 \cos^2 \eta}$$

$$\overset{*}{\alpha} = \frac{b \cos^2 \eta}{\sqrt{a^2 \sin^2 \eta + b^2 \cos^2 \eta}}$$

The result is

$$\frac{d\overset{*}{S}}{dS} = \frac{b \cos^2 \eta}{a^2 \sin^2 \eta + b^2 \cos^2 \eta} \tag{4.40}$$

Equation (4.18) (at time $t = 0$), which is more general than equation (4.35), can also be used to derive the above equation (4.40).

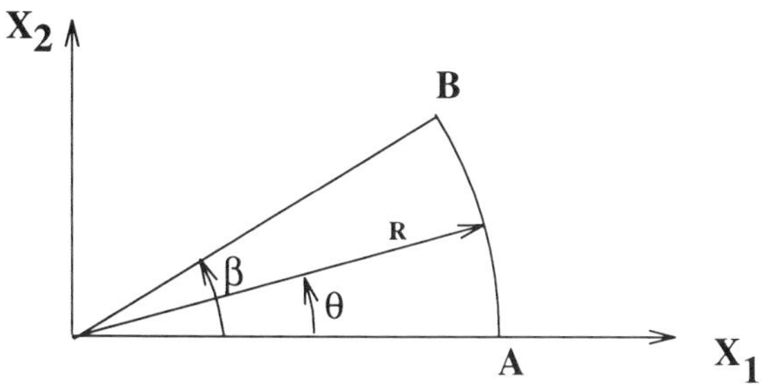

Figure 4.2. A wedge.

$d\overset{*}{S}/dS$ on a wedge. Consider the wedge with circular boundary shown in figure 4.2. Here, on AB, $X_1 = R\cos\theta$, $X_2 = R\sin\theta$. Let β be the design variable and assume a linear variation $\theta = \eta\beta$. Now, it is easy to show that

$$\left|\frac{\partial X}{\partial \eta}\right| = \alpha = \beta R, \qquad \overset{*}{\alpha} = R$$

and

$$\frac{d\overset{*}{S}}{dS} = \frac{1}{\beta} \tag{4.41}$$

where, again, equation (4.35) is used.

The above result is independent of the linearity assumption, that is, it remains the same with $\theta = g(\eta)\beta$, where $g(\eta)$ is a sufficiently smooth function of η.

In this example, use of equation (4.35) makes the above calculations quite simple. The same result, of course, can also be obtained from equation (4.18) (at time $t = 0$). Here, one must consider a generic radius ρ such that

$$X_1 = \rho\cos\theta, \qquad X_2 = \rho\sin\theta$$

with $\theta = g(\eta)\beta$, $0 \leq \eta_1 \leq 1$, and $g(0) = 0$, $g(1) = 1$,

$$\overset{*}{X}_1 = -\rho\sin\theta\, g(\eta) = -X_2 g(\eta)$$
$$\overset{*}{X}_2 = \rho\cos\theta\, g(\eta) = X_1 g(\eta)$$

It is very important to remember that $g(\eta)$ is an implicit function of X_1 and X_2, since $g(\eta) = \theta(X_1, X_2)/\beta$.

Using $\theta_{,1} = -X_2/\rho^2$ and $\theta_{,2} = X_1/\rho^2$, one gets

$$\overset{*}{X}_{1,1} = \frac{\sin^2\theta}{\beta}, \qquad \overset{*}{X}_{2,2} = \frac{\cos^2\theta}{\beta}$$

$$\overset{*}{X}_{1,2} = -g(\eta) - \frac{\sin\theta\ \cos\theta}{\beta}$$

and

$$\overset{*}{X}_{2,1} = g(\eta) - \frac{\sin\theta\ \cos\theta}{\beta}$$

Finally, using the above, and $N_1 = \cos\theta$, $N_2 = \sin\theta$ in equation (4.18), leads to the result given in equation (4.41).

$d\overset{*}{S}/dS$ and $d\overset{*}{A}/dA$ on a quarter disc. Consider a quarter of a disk as shown in figure 4.3.

Let "a" be the design variable and ρ a generic radius. This time, while

$$\overset{*}{\rho}(a)=1, \qquad \overset{*}{\rho}(b)=0$$

$\overset{*}{\rho}$ is not unique inside the disk and on the lines AB and CD. For example, one can make different assumptions regarding the variation of $\overset{*}{\rho}$ as a function of ρ, such as

$$\text{Linear:} \qquad \overset{*}{\rho} = \frac{b-\rho}{b-a} \tag{4.42}$$

$$\text{Quadratic:} \qquad \overset{*}{\rho} = \left(\frac{b-\rho}{b-a}\right)^2 \tag{4.43}$$

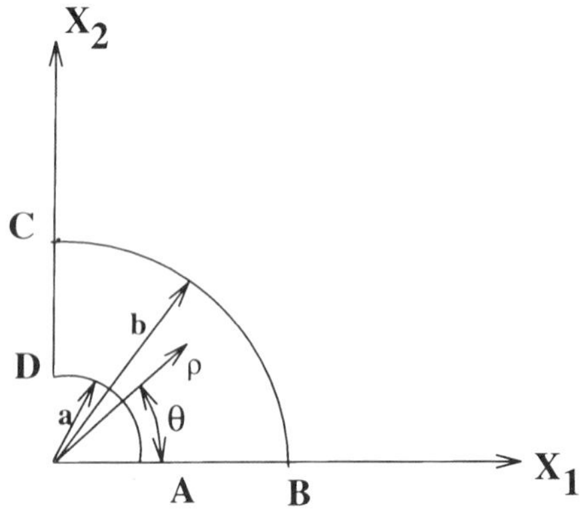

Figure 4.3. Quarter of a disk.

The value of $d\overset{*}{A}/dA$ at a generic point inside the disk depends on the assumption about $\overset{*}{\rho}$ above. For the linear assumption,

$$\overset{*}{X}_1 = \overset{*}{\rho} \cos\theta = \frac{b\cos\theta \; - X_1}{b-a}$$

$$\overset{*}{X}_2 = \overset{*}{\rho} \sin\theta = \frac{b\sin\theta \; - X_2}{b-a}$$

Using the formula $d\overset{*}{A}/dA = \overset{*}{X}_{K,K}$ (see equation 4.19) and remembering that θ is a function of X_1 and X_2, one gets the result

$$\frac{d\overset{*}{A}}{dA} = -\frac{2}{b-a} + \left(\frac{b}{b-a}\right)\frac{1}{\rho} \tag{4.44}$$

With the quadratic assumption,

$$\frac{d\overset{*}{A}}{dA} = \frac{(b-\rho)(b-3\rho)}{(b-a)^2\rho} \tag{4.45}$$

In a similar manner, one can compute $d\overset{*}{S}/dS$ on the line AB in figure 4.3. This time,

Linear: $\quad \overset{*}{X}_1 = \dfrac{b-X_1}{b-a}, \qquad X_2 = \overset{*}{X}_2 = 0$

Quadratic: $\quad \overset{*}{X}_1 = \left(\dfrac{b-X_1}{b-a}\right)^2, \qquad X_2 = \overset{*}{X}_2 = 0$

so that, from equation (4.18) applied at $t = 0$,

Linear: $$\frac{d\overset{*}{S}}{dS} = \frac{d\overset{*}{X}_1}{dX_1} = -\frac{1}{b-a} \tag{4.46}$$

Quadratic: $$\frac{d\overset{*}{S}}{dS} = \frac{d\overset{*}{X}_1}{dX_1} = \frac{2(X_1-b)}{b-a} \tag{4.47}$$

It is observed from equations (4.44) and (4.45) that $d\overset{*}{A}/dA$ at an internal point typically depends on the assumption regarding the distribution of the design velocity. This is consistent with equation (4.38). The situation with $d\overset{*}{S}/dS$ depends on the nature of the boundary parameterization. For example, equation (4.39), with "b" as the design variable, represents a unique mapping between points on the original and the perturbed ellipses, and therefore yields a unique answer for $d\overset{*}{S}/dS$. The situation is quite different for the line AB in figure 4.3, as can be seen from equations (4.46) and (4.47).

Sensitivity of an integral. The final topic in this section is the calculation of the sensitivity of a simple integral by the "$d\overset{*}{S}$" rule.

Let

$$I_1 = \int_a^b \rho \, d\rho = \frac{b^2 - a^2}{2}$$

and

$$I_2 = \int_a^r \rho \, d\rho = \frac{r^2 - a^2}{2}$$

It is obvious that with "a" as the design variable,

$$\overset{*}{I}_1 = -a \qquad \text{and} \qquad \overset{*}{I}_2 = r\overset{*}{r} - a \tag{4.48}$$

Using the "$d\overset{*}{S}$" rule,

$$\overset{*}{I}_1 = \int_a^b \overset{*}{\rho} \, d\rho + \int_a^b \rho \, d\overset{*}{\rho} \tag{4.49}$$

With either the linear or the quadratic assumption for $\overset{*}{\rho}$ (equation 4.42) or 4.43), one gets the same answer from equation (4.49). This is

$$\overset{*}{I}_1 = -a$$

which is expected since I_1 is only a function of the end points and therefore independent of the assumption on the design velocity $\overset{*}{\boldsymbol{\rho}}$.

For the second integral I_2, however, with

$$\overset{*}{I}_2 = \int_a^r \overset{*}{\rho} \, d\rho + \int_a^r \rho \, d\overset{*}{\rho} \tag{4.50}$$

the linear assumption for $\overset{*}{\rho}$ (equation 4.42) yields

Linear: $$\overset{*}{I}_2 = r\left(\frac{b - r}{b - a}\right) - a \tag{4.51}$$

but with the quadratic assumption (equation 4.43)

$$\text{Quadratic:} \qquad \overset{*}{I}_2 = r\left(\frac{b-r}{b-a}\right)^2 - a \tag{4.52}$$

which are different except at $r = a$ and at $r = b$.

The last two results are consistent with equation (4.48) for $\overset{*}{I}_2$ with

$$\text{Linear:} \qquad \overset{*}{r} = \frac{b-r}{b-a} \tag{4.53}$$

and

$$\text{Quadratic:} \qquad \overset{*}{r} = \left(\frac{b-r}{b-a}\right)^2 \tag{4.54}$$

This time, the value of $\overset{*}{I}_2$ is a function of the assumption on the design velocity $\overset{*}{\rho}$.

4.3 Numerical Implementation

Numerical implementation involves two key steps—discretization of equations and solution strategy. The approach here is quite analogous to that discussed earlier for the solution of the mechanics problem by the BEM (section 2.4.2).

4.3.1 Discretization of Equations

The plane strain sensitivity equations (4.2), (4.26′), and (4.27) are discretized in a manner that is quite analogous to their standard BEM counterparts. Of course, for plane stress problems the corresponding equations from Zhang (1991) must be used. The approach used to obtain the numerical results, presented later in this chapter, is briefly outlined below.

The boundary of the body ∂B is subdivided into piecewise quadratic, conforming boundary elements. The sensitivities $\overset{*}{\boldsymbol{\tau}}^{(s)}$ and $\overset{*}{\boldsymbol{\delta}}$, and the scalar $d\overset{*}{s}/ds$ are assumed to be piecewise quadratic on the boundary elements. The domain of the body is divided into Q4 internal cells. The tensors $\overset{*}{\boldsymbol{d}}^{(n)}$ and $\overset{*}{\boldsymbol{g}}$ and scalar $d\overset{*}{a}/da$ are interpolated on these internal cells.

As has been mentioned before, singular integrals must be evaluated with great care for these problems. Logarithmically singular integrands are integrated with log-weighted Gaussian integration formulas on the boundary elements. Two methods have been employed for the numerical evaluation of $1/r$ singular domain integrals. These approaches are discussed in Mukherjee (1982, pp. 91–92), and Nagarajan and Mukherjee (1993). The number of Gauss points used for both regular and log-singular boundary integrals is 10. For regular domain integrals, as well as $1/r$ singular domain integrals which are transformed to regular form, the number of Gauss points used is 4×4. It is very important to mention here that use of the regularized equations (4.2), (4.26′), and (4.27) (rather than their more strongly singular forms) is absolutely crucial for the success of this numerical scheme.

Sensitivity equations at corners, and sensitivity versions of compatibility equations, are assembled together with the DBEM equations. The final discretized forms of the DBEM equations, for the mechanics and sensitivity problems, have the forms

$$[A]\{\delta\} + [B]\{\tau^{(s)}\} = \{C_1\} \tag{4.55}$$

and

$$[A]\{\overset{*}{\delta}\} + [B]\{\overset{*}{\tau}{}^{(s)}\} = \{C_2\} \tag{4.56}$$

A very important fact that should be noted here is that the coefficient matrices $[A]$ and $[B]$, in the two equations above, are identical. This results in considerable savings in computer time for sensitivity analysis by this approach.

4.3.2 *Solution Strategy*

The solution strategy for the complete problem (mechanics as well as sensitivity analysis) is outlined below. Reference is made here to the integral equations in section 4.2. Of course, their discretized counterparts must be used for the numerical calculations.

The algorithm (for the sensitivity problem) for going from time t to $t + \Delta t$ is illustrated in figure 4.4. The complete solution strategy is as follows.

1. Solve the elasticity problem. Usually one starts the problem with small values of boundary conditions.
2. Obtain the tensors $d_{ik}^{(n)}$ and $\overset{*}{d}{}_{ik}^{(n)}$ from the constitutive equations and the derivatives of the constitutive equations, given in equations (2.91) and (4.10), respectively.
3. Solve the DBEM equations for the mechanics problem.
   - 3.1. If $t = 0$, assume $h_{ij} = 0$. Otherwise use the values from the previous time step.
   - 3.2. Solve equation (4.1) with updated values of $\tau_i^{(s)}$ and δ_i on the boundary.

3.3. Calculate h_{ij} for both boundary and internal nodes using equations (2.170) and (2.171), respectively.

4. Solve the DBEM equations for the sensitivity problem.
   4.1. If $t = 0$, assume $\overset{*}{h}_{ij} = 0$. Otherwise use the values from the previous time step.
   4.2. Solve equation (4.2) with updated values of $\overset{*}{\tau}_i^{(s)}$ and $\overset{*}{\delta}_i$ on the boundary.
   4.3. Calculate $\overset{*}{h}_{ij}$ for both boundary and internal nodes using equations (4.24) and (4.27), respectively.
   4.4. Check convergence of $\overset{*}{h}_{ij}$. If it is not satisfied, use the calculated value of $\overset{*}{h}_{ij}$ as the new updated value and go to step 4.2.
5. Calculate boundary values of $\hat{\sigma}_{ij}$ and $\dot{\sigma}_{ij}$, as well as $\overset{\diamond}{\sigma}_{ij}$ and $\overset{\circ}{\sigma}_{ij}$ from equations (2.169), (2.90), (4.23), and (4.25), respectively.
6. Calculate internal values of $\hat{\sigma}_{ij}$ and $\dot{\sigma}_{ij}$, as well as $\overset{\diamond}{\sigma}_{ij}$ and $\overset{\circ}{\sigma}_{ij}$ from equations (2.89), (2.90), (4.28), and (4.25), respectively.
7. Obtain v_i and $\overset{*}{v}_i$ on the boundary ∂B by integrating δ_i and $\overset{*}{\delta}_i$ along the boundary using equations (4.13) and (4.14). The internal values of v_i and $\overset{*}{v}_i$ can be obtained using equations (4.26) and (4.26′).
8. Use equations (4.15) and (4.16) to calculate $\dot{\boldsymbol{F}}$ and $\overset{\circ}{\boldsymbol{F}}$.
9. Integrate the rate quantities $\dot{\sigma}_{ij}$, $\overset{\circ}{\sigma}_{ij}$, v_i, $\overset{*}{v}_i$, $\dot{\boldsymbol{F}}$ and $\overset{\circ}{\boldsymbol{F}}$ in time to obtain σ_{ij}, $\overset{*}{\sigma}_{ij}$, u_i, $\overset{*}{u}_i$, $\boldsymbol{F}$, and $\overset{*}{\boldsymbol{F}}$ at $t + \Delta t$. These quantities are obtained both on the boundary and in the domain.
10. Determine whether the geometry of the body needs to be updated. If yes, go to the next step; otherwise, go to step 2 until the simulation ends.
11. Update the geometry of the body.
    11.1. Use equation (4.17) to obtain $v_{i,j}^{(d)}$ first.
    11.2. Determine $d\overset{*}{s}$ and $d\overset{*}{a}$ on the new geometry from equations (4.18) and (4.19).
    11.3. Update x_i and $\overset{*}{x}_i$ adding u_i and $\overset{*}{u}_i$, respectively, to their values at the previous step.
    11.4. Update n_i and $\overset{*}{n}_i$ using equations (4.20) and (4.21).
    11.5. Update all the kernels and derivatives of the kernels that appear in equations (4.1), (2.171), (4.2), and (4.27).

For classical elastoplastic material models, $\overset{*}{d}{}_{ij}^{(n)}$ depends on $\overset{\diamond}{\sigma}_{ij}$ as well as on the stress components and their sensitivities. This requires iterations over $\overset{*}{d}{}_{ij}^{(n)}$ within each time step. The sensitivity problem, however, still has approximately the same level of complexity as the original elastoplastic problem. For large-strain problems, iterations over $\overset{*}{d}{}_{ij}^{(n)}$ may be carried out within the iteration scheme for $\overset{*}{h}_{ij}$.

Thus, large-strain sensitivity problems of elastoplasticity and elasto-

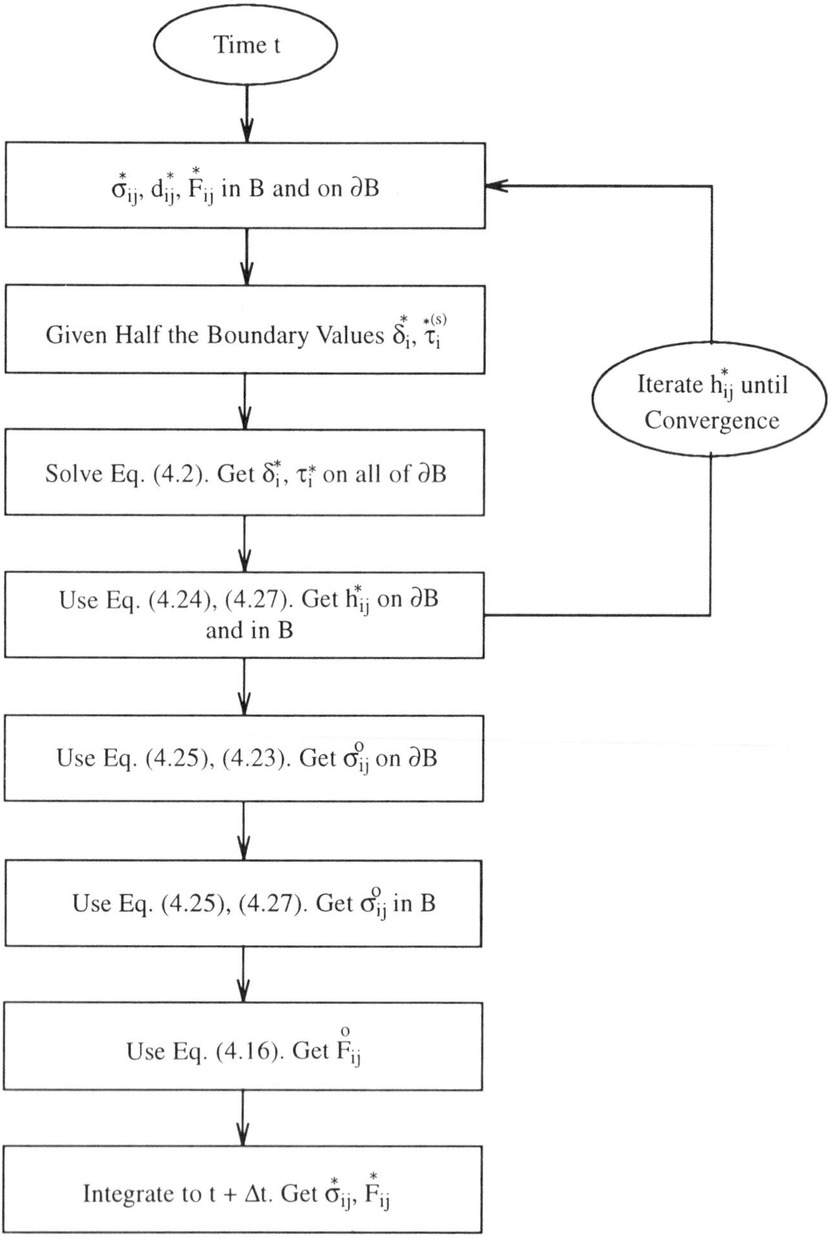

Figure 4.4. Solution strategy for large-deformation sensitivity problems.

viscoplasticity are expected to require less than twice the computational effort needed for the regular BEM analysis including both geometric and material nonlinearities when sensitivity with respect to one design variable is needed. In a typical design environment, however, sensitivities with respect to a large number of shape design variables are desired. It is interesting to note here that the determination of sensitivities with respect to additional design variables does not require solutions of new matrix systems. The coefficient matrices remain the same for all cases, only the right-hand sides change. Hence, for the slight increase in costs due to additional evaluations of the right-hand side, it is possible to simultaneously track the sensitivities with respect to several design variables.

4.4 Numerical Results for Sample Problems

A computer program for numerically calculating sensitivities for general two-dimensional (plane strain or plane stress) problems has been developed by Mukherjee's group at Cornell. All the numerical results presented in this section have been obtained from this 2D program. The viscoplastic constitutive model that has been used for these calculations is that from Anand (1982). The equations for this model are given in section 2.4.1. The sensitivity equations for this model are easy to derive by following the approach of equations (4.10) and (4.11). These are not detailed here. The material parameters for the constitutive model, for Fe-0.05 carbon steel at 1173 K, are given in section 2.4.1.

4.4.1 One-Dimensional Problems

The problem. The example considered here is a rectangular plate of initial length L, of unit width. The problem can be plane strain or plane stress. The plate is deformed to a current length l by a prescribed displacement history $\bar{u}$.

Figure 4.5 shows two such plates of initial lengths L_1 and L_2. A generic material point X_1, in this simple one-dimensional situation, moves to x_1 and X_2 moves to x_2, where

$$x_1 = \frac{l_1}{L_1} X_1 \tag{4.57}$$

and

$$x_2 = \frac{l_2}{L_2} X_2 \tag{4.58}$$

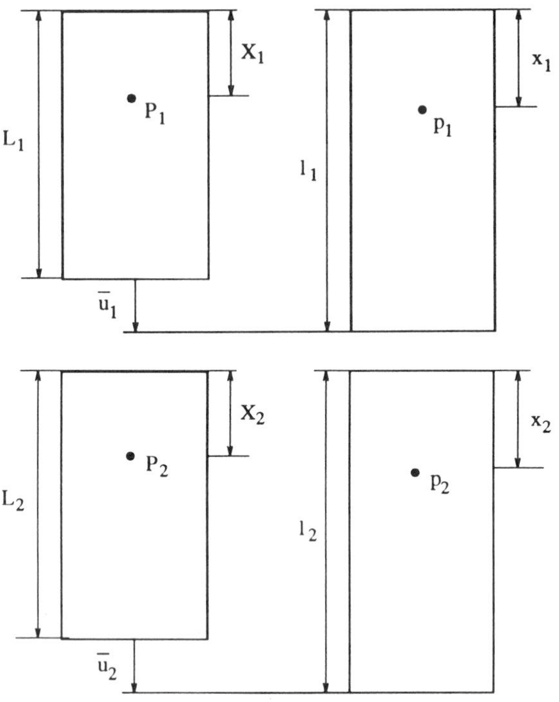

Figure 4.5. Illustrative one-dimensional problem.

The design deformation $X_1 \rightarrow X_2$ is assumed to be of the form

$$X_2 = X_1 \frac{L_2}{L_1} \tag{4.59}$$

which is a linear assumption. As discussed before, this assumption affects the values of the sensitivities inside the body (i.e., for $0 < X_1 < L_1$) but not on its boundary $X_1 = 0$ and $X_1 = L_1$.

As a consistency check, one can write

$$\overset{*}{x} = \frac{dx}{dX_D} \cdot \frac{dX_D}{dL} \tag{4.60}$$

where L is the design variable and the design deformation $X_D = X_2 - X_1$. It is easy to show that

$$\frac{dx}{dX_D} = \lim_{\Delta X \to 0} \frac{x_2 - x_1}{X_2 - X_1} = 1 + \overset{*}{\bar{u}} \tag{4.61}$$

$$\frac{dX_D}{dL} = \lim_{\Delta L \to 0} \frac{X_2 - X_1}{L_2 - L_1} = \frac{X}{L} \tag{4.62}$$

so that, from equation (4.60)

$$\overset{*}{x} = \frac{X}{L}(1 + \overset{*}{\bar{u}}) \tag{4.63}$$

Now,

$$\overset{*}{u} = \lim_{\Delta L \to 0} \frac{u_2 - u_1}{L_2 - L_1} = \frac{X}{L}\overset{*}{\bar{u}}$$

so that

$$\overset{*}{x} = \overset{*}{X} + \overset{*}{u}$$

as expected. Also,

$$\overset{*}{l} = 1 + \overset{*}{\bar{u}}$$

Consistency checks for other formulas for 1D strain. A one-dimensional strain problem is a plane strain problem with only a single component of the displacement at every point in the body. Let this component be the displacement in the x_1 direction. The only nonzero components of $\boldsymbol{F}$ and $\boldsymbol{d}$, therefore, are

$$F_{11} = \frac{l}{L} \tag{4.64}$$

and

$$d_{11} = \frac{v}{l} \tag{4.65}$$

Now,

$$\dot{F}_{11} = \frac{v}{L} \tag{4.66}$$

$$\overset{*}{F}_{11} = \frac{\overset{*}{l}}{L} - \frac{l}{L^2} \tag{4.67}$$

and

$$\overset{*}{d}_{11} = \frac{\overset{*}{v}}{l} - \frac{v}{l^2}\overset{*}{l} \tag{4.68}$$

In the above, v is the velocity component in the x_1 direction. Also, $\omega = 0$. From equations (4.57) and (4.63),

$$\overset{*}{x}_1 = \overset{*}{x} = \frac{x}{l}(1 + \overset{*}{\bar{u}}) \tag{4.69}$$

Various formulas presented before can now easily be checked for this simple one-dimensional case. Equation (4.15) for $\dot{\boldsymbol{F}}$ is obviously valid. The right- and left-hand sides of equation (4.16) become $\overset{*}{v}/L - v/L^2$.

The formula (4.17) for $v_{i,j}^{(d)}$, with its initial value $d\overset{*}{X}/dX = 1/L$, gives

$$\frac{d\overset{*}{x}}{dx} = v_{1,1}^{(d)} = \frac{1 + \overset{*}{\bar{u}}}{l}$$

which is consistent with equation (4.69).

From equations (4.18) and (4.19) ($d\overset{*}{s}/ds$ is evaluated on edges parallel to the x_1 axis),

$$\frac{d\overset{*}{s}}{ds} = \frac{1 + \overset{*}{\bar{u}}}{l}, \qquad \frac{d\overset{*}{a}}{da} = \frac{1 + \overset{*}{\bar{u}}}{l}$$

which are correct since, in this case, s can be identified with x, that is (see equation 4.69),

$$\overset{*}{s} = \frac{s}{l}(1 + \overset{*}{\bar{u}})$$

1D stress and 1D strain. The problems described here are a one-dimensional strain problem in which the only nonzero displacement is u_1 and a one-dimensional stress problem with the only nonzero stress σ_{11}. The plane strain version of the two-dimensional DBEM computer program is employed to solve the one-dimensional strain problem, while the plane stress version is used to solve the one-dimensional stress problem. In both cases, a constant velocity $v_0 = 2 \times 10^{-3}$ m/s is applied in the x_1 direction with a very small value of initial strain in this direction. The design variable is the initial length L in the x_1 direction. The initial dimensions of the plate are 2 m × 2 m. The constant time step Δt for explicit time integration is 0.01 s. These tests are chosen to validate the methodology and the algorithm of the DBEM approach, especially the updating of the geometry as well as quantities such as $\overset{*}{x}$, $d\overset{*}{s}/ds$, and $d\overset{*}{a}/da$ during a simulation. These quantities are updated every second in these simulations.

Direct solution for the one-dimensional strain problem. A direct solution is obtained by time integration of the one-dimensional equations. For the one-dimensional strain problem being considered here,

$$v_{1,1} = d_{11} = \frac{v_0}{l}, \qquad \text{rest of } v_{i,j} = 0, \omega_{ij} = 0$$

From the hypoelastic law (2.89) (and equation 2.90),

$$\dot{\sigma}_{11} = \hat{\sigma}_{11} = \frac{2(1-\nu)}{1-2\nu} G \frac{v_0}{l} - 2Gd_{11}^{(n)} \tag{4.70}$$

$$\dot{\sigma}_{22} = \hat{\sigma}_{22} = \frac{2\nu}{1-2\nu} G \frac{v_0}{l} - 2Gd_{22}^{(n)}, \qquad \dot{\sigma}_{33} = \dot{\sigma}_{22} \tag{4.71}$$

The rest of the stress rates vanish.

For the corresponding sensitivity equations,

$$\overset{*}{\bar{u}} = 0, \qquad \overset{*}{l} = 1, \qquad \frac{d\overset{*}{s}}{ds} = \frac{1}{l}, \qquad \frac{d\overset{*}{a}}{da} = \frac{1}{l}$$

and

$$\overset{\circ}{\sigma}_{11} = -\frac{2(1-\nu)}{1-2\nu} G \frac{v_0}{l^2} - 2G\overset{*}{d}{}_{11}^{(n)} \tag{4.72}$$

$$\overset{\circ}{\sigma}_{22} = -\frac{2\nu}{1-2\nu} G \frac{v_0}{l^2} - 2G\overset{*}{d}{}_{22}^{(n)} \tag{4.73}$$

$$\overset{\circ}{\sigma}_{33} = \overset{\circ}{\sigma}_{22} \tag{4.74}$$

The quantities $\boldsymbol{d}^{(n)}$ and $\overset{*}{\boldsymbol{d}}{}^{(n)}$ are obtained from the constitutive equations (2.197)–(2.200) together with the derivatives of these equations with respect to the design variable L. These sensitivity equations, derived from the above constitutive model, are given in Zhang (1991). Similar equations can easily be derived for the one-dimensional stress problem.

Numerical results. The physical situation, as well as the results, are shown in figures 4.6 and 4.7. The boundary conditions, in terms of the rate quantities and their sensitivities, also appear in these figures. The DBEM model has nonstandard boundary variables, and these must be prescribed such that a unique solution is obtained (Zhang et al. 1992a). Information can be lost

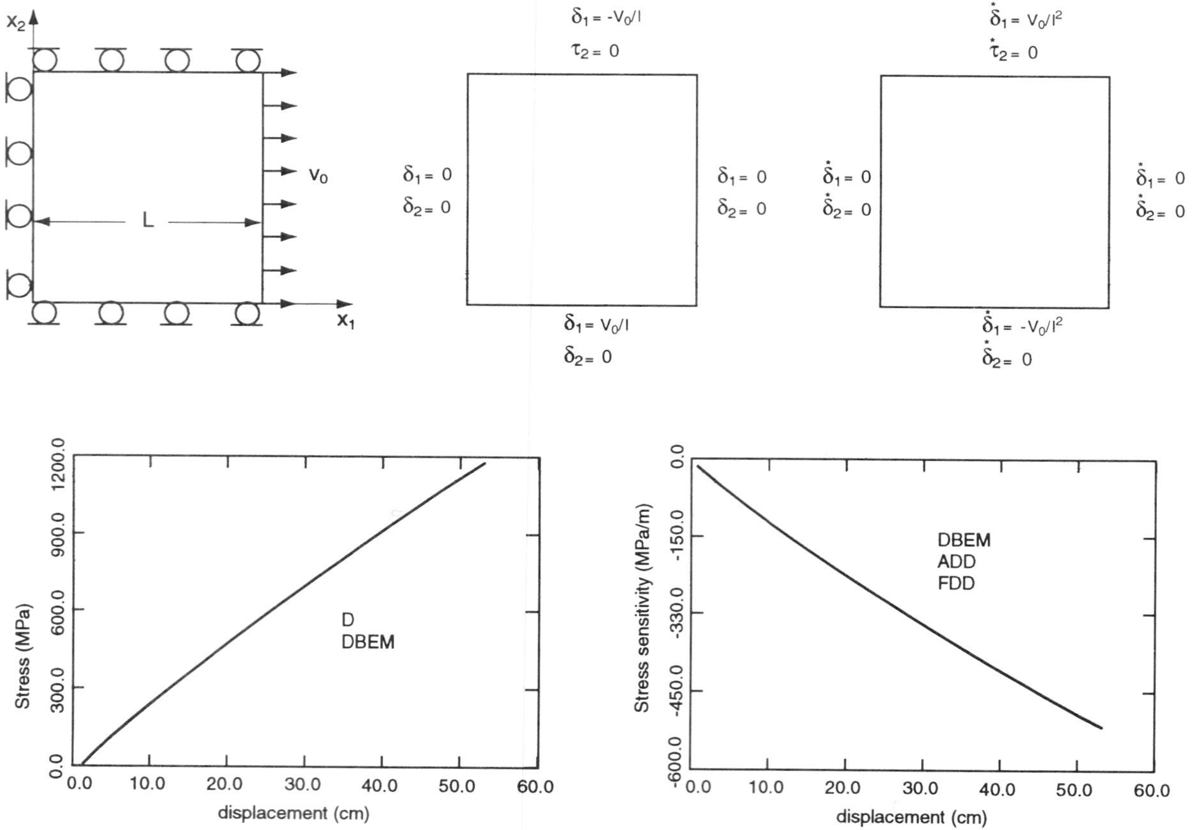

Figure 4.6. Results for 1D strain. $v_0 = 2 \times 10$ (m/s). FDD, finite difference of direct solutions; D, direct; DBEM, derivative BEM; ADD, analytical differentiation of direct solutions (from Zhang et al. 1992b).

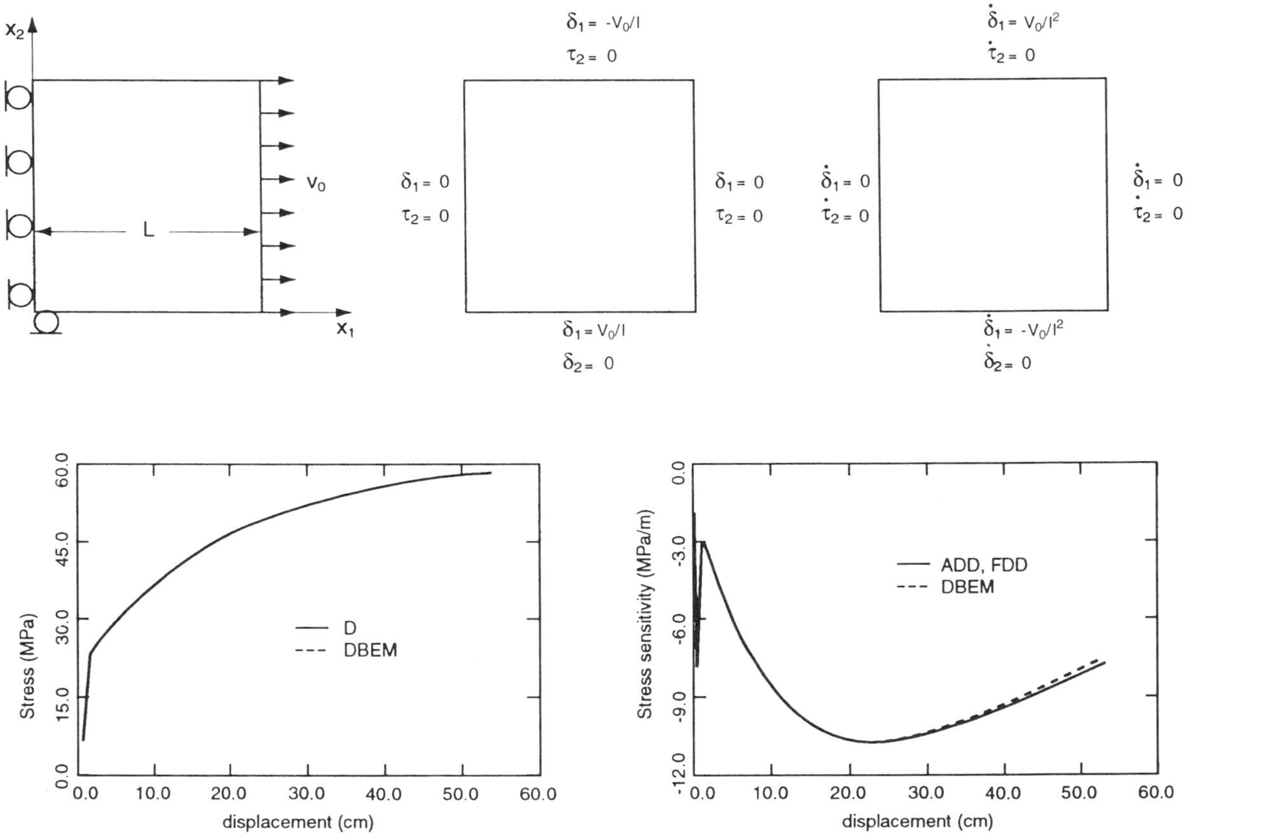

Figure 4.7. Results for 1D stress. $v_0 = 2 \times 10$ (m/s). FDD, finite difference of direct solutions; D, direct; DBEM, derivative BEM; ADD, analytical differentiation of direct solutions (from Zhang et al. 1992b).

if a zero δ_i is prescribed on part of a boundary instead of a constant v_i. This can be remedied by using constraint equations like equation (2.168). Such constraint equations are recommended in general. Here, for a simple geometrical situation, δ_s, the tangential component of $\boldsymbol{\delta}$, is prescribed on certain boundaries of the square.

Another concern is rigid body motion. This DBEM formulation gives unique results which are unaffected by rigid body translations, as long as $\boldsymbol{\Delta} = \partial \boldsymbol{u}/\partial s$, $\boldsymbol{\tau}$ and $\boldsymbol{\sigma}$ are sought. Of course, calculation of displacements requires specification of the displacement of some point in the structure. An imposed rigid body rotation, however, changes the values of δ_n, the normal component of $\boldsymbol{\delta}$, on ∂B, without changing the stresses in the body. Thus, rigid body rotations must be eliminated by the boundary conditions in order to get a unique solution from the DBEM equations.

In summary, a DBEM model, plus a prescribed displacement at one point in the body, must be consistent with a corresponding usual BEM model in terms of $\boldsymbol{v}$ and $\boldsymbol{\tau}^{(s)}$ on ∂B.

Figure 4.6 shows the stress–displacement plot in the x_1 direction (σ_{11} as a function of u_1) and the corresponding sensitivity plot ($\overset{*}{\sigma}_{11}$ versus u_1) for the one-dimensional strain problem. Here, "D" refers to the direct solution, "FDD" to the finite difference of direct solutions, and "ADD" to the analytical differentiation of a direct solution. The direct solution is a time-marching solution obtained by integrating the one-dimensional equations given in the previous section. The FDD solution is obtained with $\Delta L = 0.001$ m. The DBEM numerical results are obtained employing four quadratic boundary elements (one for each edge) and four internal Q4 elements with one internal point at the center of the plate.

The corresponding situation for one-dimensional stress problems is depicted in figure 4.7. The DBEM numerical solutions on the four plots are seen to agree almost perfectly with the direct solutions, which can be regarded as semianalytical solutions of the problems. From a practical point of view, the one-dimensional strain problem is somewhat unrealistic in that it is overconstrained and requires very large stresses for the deformation shown in figure 4.6. However, this example serves as a good check for the computer program.

The sensitivity plot for 1D stress, in figure 4.8, is considerably more complicated than its 1D strain counterpart, especially in the elasto-viscoplastic transition region. This is seen more clearly from figure 4.9.

To better understand figure 4.8, stress–displacement plots for two values of initial length ($L = 2$ m and $L = 2.5$ m) are shown in figure 4.9. It is seen that in the elastic region $\overset{\circ}{\sigma}_{11} = -Ev_0/L^2$ ($\overset{\circ}{\sigma}_{11}$ is the slope of the $\overset{*}{\sigma}_{11}$ curve in figure 4.8) and in the fully developed plastic region $\overset{\circ}{\sigma}_{11} \approx -E_T v_0/L^2$ in terms of the tangent modulus E_T, which is, of course, less than E. The

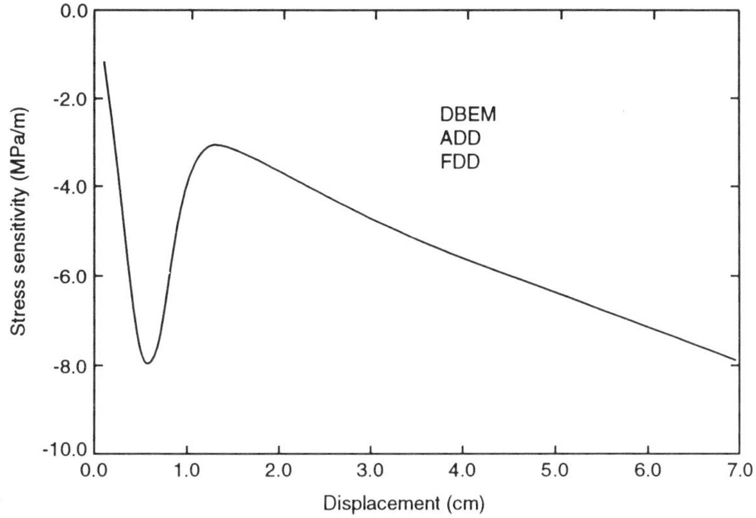

Figure 4.8. The early region of the 1D stress sensitivity plot from figure 4.7 (from Zhang et al. 1992a).

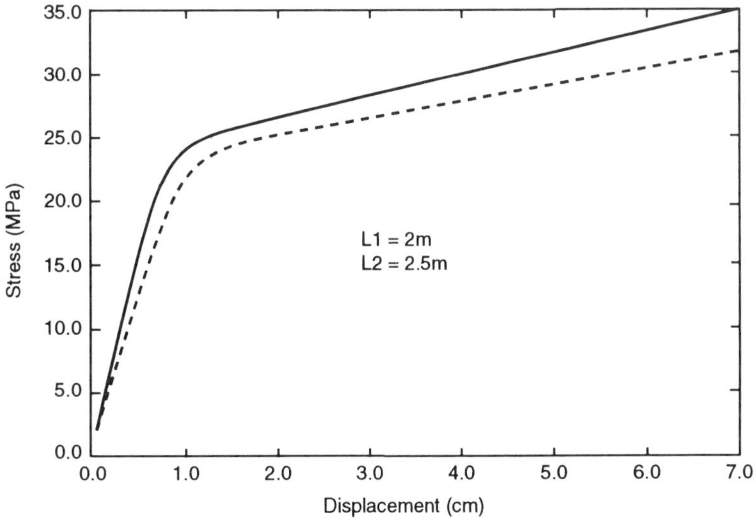

Figure 4.9. Stress for the 1D stress problem, for two different values of the initial length of the square. Solutions from the direct method (from Zhang et al. 1992a).

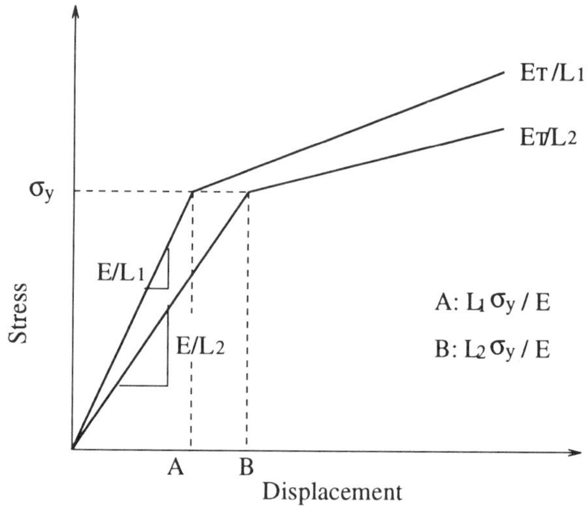

Figure 4.10. Schematic diagram for ideal elastoplasticity, for two specimens of initial lengths L_1 and L_2, respectively.

transition displacement, however, is also a function of L ($u_y = L\sigma_Y/E$ for an idealized elastoplastic model with a yield stress σ_Y). This causes shifting of the curve for $L = 2.5$ m, relative to that for $L = 2$ m. Consequently, the two curves come closer together in the transition region, leading to a positive value of $\mathring{\sigma}_{11}$ in this region. Figure 4.10 shows a schematic figure to illustrate this effect for ideal elastoplasticity.

4.4.2 A 2D Problem—Simple Shearing Motion

A simple shearing motion, at constant applied shearing rate, as shown in figure 4.11, is defined by

$$x_1(t) = X_1 + (\dot{\gamma}tX_2) = X_1 + \gamma X_2$$

$$x_2(t) = X_2$$

where, as before, x_i and X_i ($i = 1, 2$) are the deformed and undeformed Cartesian coordinates, respectively; $\dot{\gamma}$ and γ are the shearing rate and the amount of shear, respectively; and t is the elapsed time. Note that the undeformed configuration is a square with unit sides.

The shearing strain rate $\dot{\gamma}$ is taken as a design variable in this example. This could be classified as a design variable of the boundary condition type.

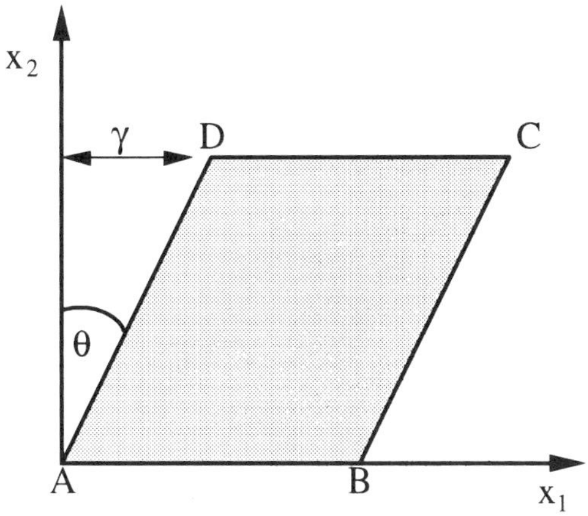

Figure 4.11. The simple shearing motion problem.

Therefore, here a superscribed star means:

$$(\overset{*}{\ }) = \frac{\partial(\ )}{\partial \dot{\gamma}}$$

The boundary conditions for the mechanics problem and sensitivity problem, referred to figure 4.11, are as follows.

(i) Boundary conditions for the mechanics problem

on AB and CD, $\delta_1 = 0, \quad \delta_2 = 0$
on BC, $\delta_1 = \dot{\gamma}\cos\theta, \quad \delta_2 = 0$
on DA, $\delta_1 = -\dot{\gamma}\cos\theta, \quad \delta_2 = 0$

(ii) Boundary conditions for the sensitivity problem

on AB and CD, $\overset{*}{\delta}_1 = 0, \quad \overset{*}{\delta}_2 = 0$
on BC, $\overset{*}{\delta}_1 = \cos\theta - \dot{\gamma}\overset{*}{\theta}\sin\theta, \quad \overset{*}{\delta}_2 = 0$
on DA, $\overset{*}{\delta}_1 = -\cos\theta + \dot{\gamma}\overset{*}{\theta}\sin\theta, \quad \overset{*}{\delta}_2 = 0$

In the above,

$$\theta = \tan^{-1}(\dot{\gamma}t), \qquad \overset{*}{\theta} = \frac{t}{1 + (\dot{\gamma}t)^2}$$

The geometric sensitivities for this problem are usually nonlinear functions of certain quantities and their sensitivities. For example, on the boundary BC (see figure 4.11)

$$\overset{*}{\boldsymbol{n}} = -\overset{*}{\theta}\sin\theta\boldsymbol{e}_1 - \overset{*}{\theta}\cos\theta\boldsymbol{e}_2, \quad \frac{d\overset{*}{s}}{ds} = t\sin\theta\cos\theta$$

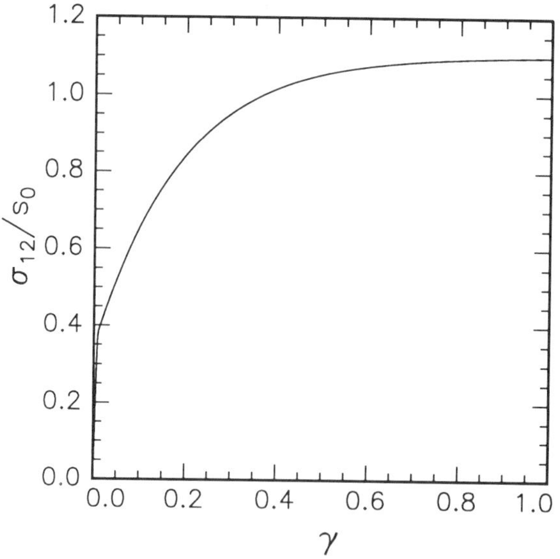

Figure 4.12. Normalized shearing stress response for the simple shear problem. The DBEM and direct solutions coincide (from Leu and Mukherjee 1993).

where $\boldsymbol{e}_1$ and $\boldsymbol{e}_2$ are the unit vectors along the x_1 and x_2 axes, respectively. Those on other boundaries can be obtained by symmetry.

The area sensitivities over the domain can be shown to be

$$\frac{d\overset{*}{a}}{da} = 0$$

Assume that the prescribed $\dot{\gamma}$ is equal to 0.01/s, and the simulation time is 100 seconds. Here, 2 boundary elements for each side, and 16 Q4 internal cells are employed in the BEM discretization. The numerical results are depicted in figures 4.12, 4.13, and 4.14. In these figures, the shearing stress and its sensitivity have been normalized with respect to $s_0 = 48.11$ MPa, which is the initial value of the internal variable in the Anand (1982) model. The mechanics and sensitivity problems are solved by the DBEM with 1000 time increments. A direct solution for this problem is easy to obtain. Time integration for the direct solution of the mechanics problem is carried out by an explicit Euler forward method with 10 000 time increments.

The direct sensitivity solutions are calculated by two methods:

(a) Integrating the sensitivity version of the equations for the mechanics problem, with 10 000 time increments.

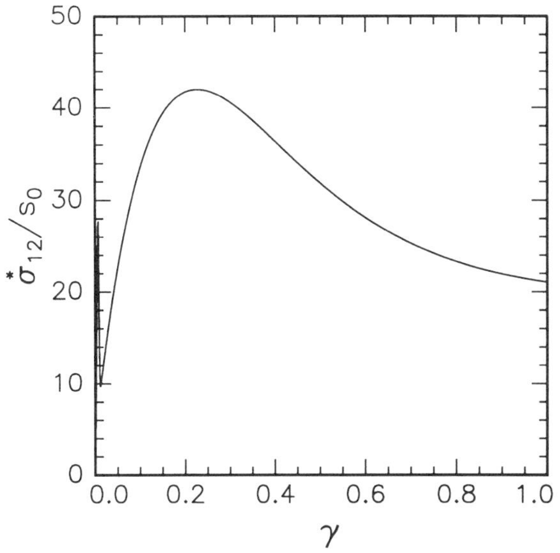

Figure 4.13. Normalized shearing stress sensitivity for the simple shear problem. The DBEM and direct solutions coincide (from Leu and Mukherjee 1993).

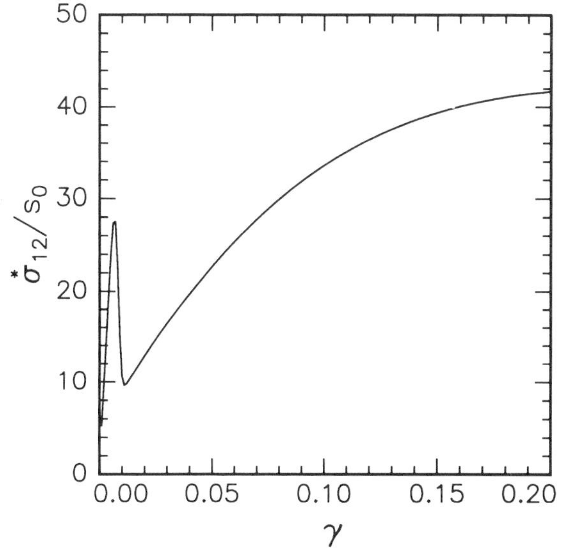

Figure 4.14. A kink occurs in the transition zone. This figure shows the early part of figure 4.13 in more detail (from Leu and Mukherjee 1993).

(b) The finite difference results of two neighboring mechanics problems solved by the direct method. The base problem has a design variable of $\dot{\gamma} = 0.01/\mathrm{s}$, and the perturbed one of $\dot{\gamma} = 0.010\,01/\mathrm{s}$.

The numerical results from the two approaches (a) and (b) coincide within plotting accuracy. A solution with 1000 increments is also excellent, except for a small difference at the peak value in figure 4.13. Note that the geometric sensitivities are updated numerically by the formulas given in section 4.2.4. Those values, together with the geometry, are updated every 4 seconds. This verifies the correctness of these updating formulas. Another set of results, obtained using the exact formulas for the geometric sensitivities, has also been compared with the above BEM results. They match perfectly. However, in general, the exact formulas are not available and in this case the geometric sensitivities can only be obtained numerically. It is also found that, for such a uniform deformation problem, no iterations are necessary for solving the system of equations. This implies that the DBEM is very accurate.

There is a kink at the beginning of figure 4.13. This is shown more clearly in Figure 4.14. This phenomenon is analogous to that described in section 4.4.1 and shown in figure 4.8. It is caused here also by the elasto-viscoplastic transition.

4.4.3 Axisymmetric Problems

The numerical example presented in this section is that of a long cylinder in plane strain, as shown in figure 4.15 with initial inner and outer radii $a_0 = 1$ m and $b_0 = 1.2$ m, respectively. Two sets of boundary conditions are

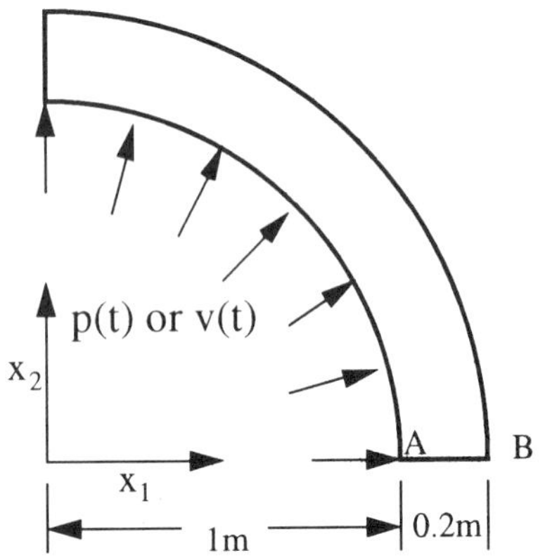

Figure 4.15. The cylinder problems—prescribed pressure or velocity on the inner surface.

being considered here. The first is that of prescribed pressure rate on the inner surface, and the second is that of prescribed velocity on the inner surface. For both cases, the original inner radius a_0 is taken as a design variable. The direct solutions for both cases are given in Leu and Mukherjee (1995), where an approach analogous to that presented by Rajiyah and Mukherjee (1987) is used. The direct method uses a cylindrical coordinate system and the fact that the problem is axisymmetric. The sensitivity solution is obtained both by analytical differentiation of the direct solution with respect to the inner radius "a", as well as by the finite differencing of two neighboring direct solutions. These two results coincide within plotting accuracy. Direct solutions are called reference solutions in all the figures in this section.

Due to symmetry, only a quarter of the cylinder is modeled. The boundary mesh consists of three equal quadratic elements along the radial direction and seven equal elements along the circumferential direction. By connecting all the boundary nodes, there are 84 Q4 internal cells.

(a) Prescribed pressure. Suppose that the prescribed pressure rate is 2 MPa/s and the simulation time is 8 seconds. The geometric sensitivities at zero time, for this problem, are discussed in section 4.2.10 (the initial values of a and b must be used in these equations). Also, the linear assumption (equation 4.42) is used. The time increment is 0.1 seconds. The geometry, geometric sensitivities, kernels, kernel sensitivities, and boundary conditions are updated every second.

Figures 4.16 and 4.17 show the results for the mechanics and sensitivity problems, where u_r is the radial displacement at point A of the inner surface (see figure 4.15). The DBEM solution is obtained without any iterations for the velocity and the velocity gradients. These are approximated by their values at the previous time step. The CPU time on an IBM 3090 supercomputer is 82.98 seconds. Figure 4.18 shows the evolution of the positions and design velocities along line AB (see figure 4.15). It is seen that at the end of the simulation the design velocities are not smooth. Also, the stress sensitivity has a similar trend. To overcome this problem, a smoothing technique is adopted—assuming linear distribution of the design velocities along the radial direction (as in the original distribution, but with values at both ends obtained from the BEM solutions). Based on this smoothing assumption, the tangential stress sensitivities along line AB (see figure 4.15), at the end of the simulation, are reported in table 4.1, together with the direct solution.

It is important to mention here that obtaining accurate solutions for shape sensitivities, at the end of a large-deformation simulation process, is not an easy task [see, for example, page 71 of Jao and Arora (1992b), where both

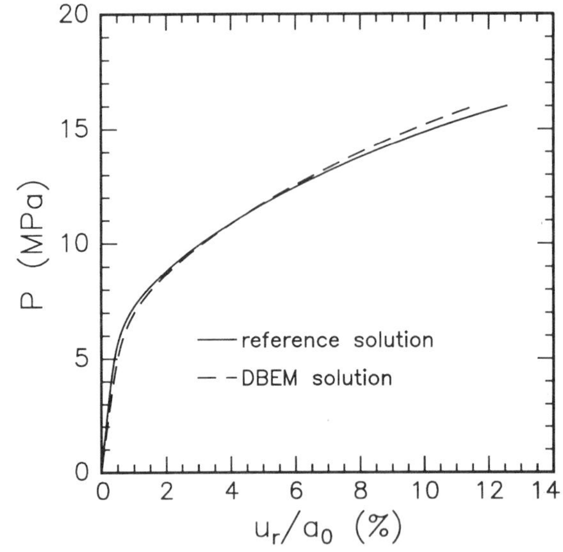

Figure 4.16. Inner pressure as a function of normalized displacement at point A for the pressure-prescribed cylinder problem (from Leu and Mukherjee 1993).

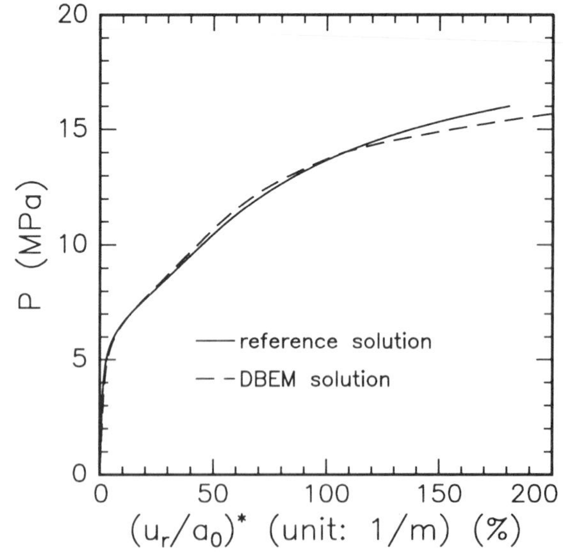

Figure 4.17. Inner pressure as a function of sensitivity of normalized displacement at point A for the pressure-prescribed cylinder problem (from Leu and Mukherjee 1993).

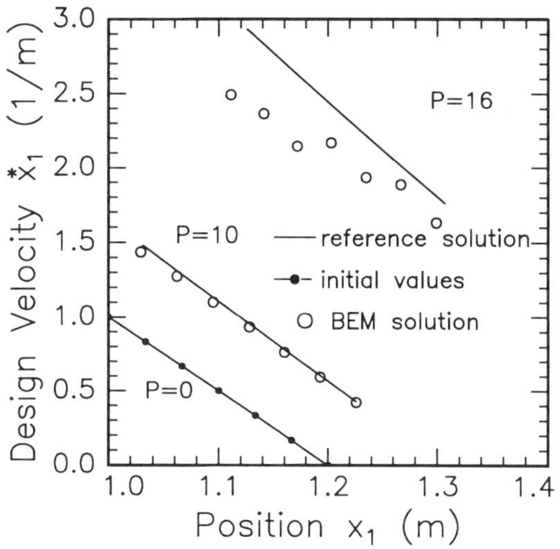

Figure 4.18. Evolutions of positions and design velocities along line *AB*.

displacement and stress sensitivities are presented at moderate strains]. In general, stress sensitivities are more difficult to obtain accurately than those of displacements. Considering the above, the results in table 4.1 at around 11% strain (tangential) are considered to be quite encouraging. Research on improving the accuracy of sensitivities is currently in progress.

(b) Prescribed velocity. Suppose that the prescribed velocity on the inner surface is $v_{in} = 0.01$ m/s and the simulation time is 10 seconds. The geometric sensitivities and boundary conditions are the same as those for the prescribed

Table 4.1. Tangential stress sensitivities along line *AB* (see figure 4.15) at the end of the simulation (Unit = MPa/m)

Node	DBEM solution	Direct solution
1 (at *A*)	716.15	898.85
2	708.33	906.85
3	985.23	913.13
4	921.73	917.79
5	1139.42	920.94
6	1034.34	922.70
7 (at *B*)	999.33	923.21

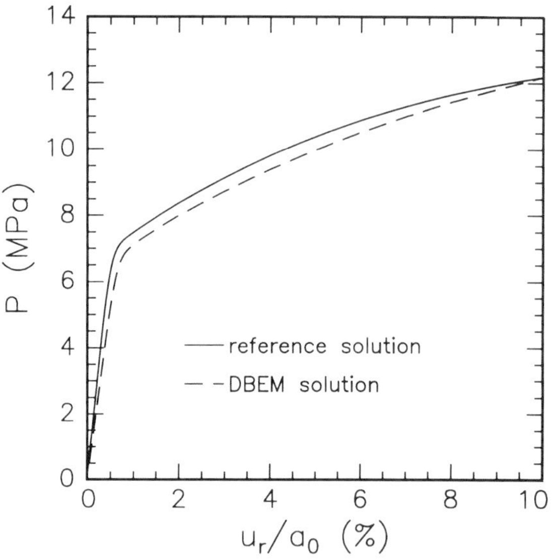

Figure 4.19. Inner pressure as a function of normalized displacement at point A for the velocity-prescribed cylinder problem.

pressure problem except for the boundary conditions on the inner surface, which are

$$\delta_1 = \frac{x_2}{a^2} v_{in}, \qquad \delta_2 = -\frac{x_1}{a^2} v_{in}$$

and

$$\overset{*}{\delta}_1 = -\frac{x_2}{a^3} v_{in}, \qquad \overset{*}{\delta}_2 = \frac{x_1}{a^3} v_{in}$$

where a is the current inner radius.

Numerical results for the mechanics and sensitivity problems are shown, respectively, in figures 4.19 and 4.20. The DBEM results are seen to be in good agreement with the reference (direct) solutions. The CPU time, on an IBM 3090 supercomputer, is 122.23 seconds.

The linear relationship between u_r/a_0 and $(u_r/a_0)^*$, in figure 4.20, can be proved as follows. By definition,

$$\left(\frac{u_r(t)}{a_0}\right)^* = \lim_{\Delta a_0 \to 0} \frac{\left(\dfrac{u_r'(t)}{a_0 + \Delta a_0}\right) - \left(\dfrac{u_r(t)}{a_0}\right)}{\Delta a_0}$$

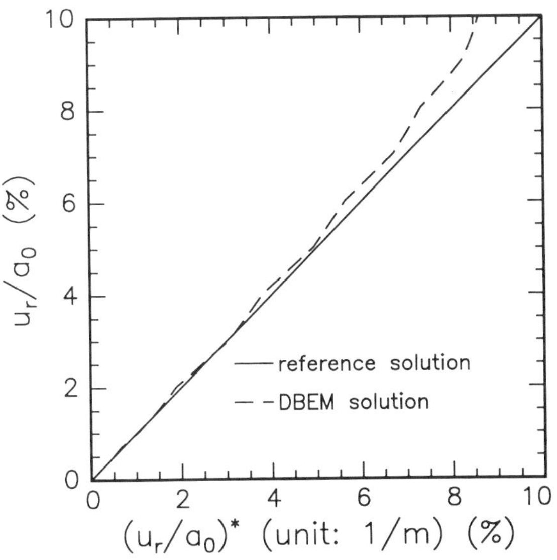

Figure 4.20. Normalized displacement at point A as a function of its negative sensitivity for the velocity-prescribed cylinder problem.

where

$$u_r(t) = v_{in}t, \qquad u_r'(t) = v_{in}t$$

are the radial displacements on the inner surface corresponding to the original and perturbed designs, respectively, at time t. Using the equations for $u_r'(t)$ and $u_r(t)$, one obtains

$$\left(\frac{u_r}{a_0}\right)^* = -\left(\frac{u_r}{a_0}\right)\frac{1}{a_0}$$

4.5 Design Optimization

An optimal design for a problem is generally achieved by an optimizer through an iterative process. Typically, a gradient-based optimization algorithm (mathematical programming approach) starts with a preliminary design, and its DSCs. The goal is to optimize an objective function (e.g., minimize weight or make stresses as uniform as possible) without violating the constraints (typically allowable stresses or displacements) for the design. The iterative process produces a succession of designs until it converges to the optimal one. A typical algorithm for this process is shown in figure 4.21.

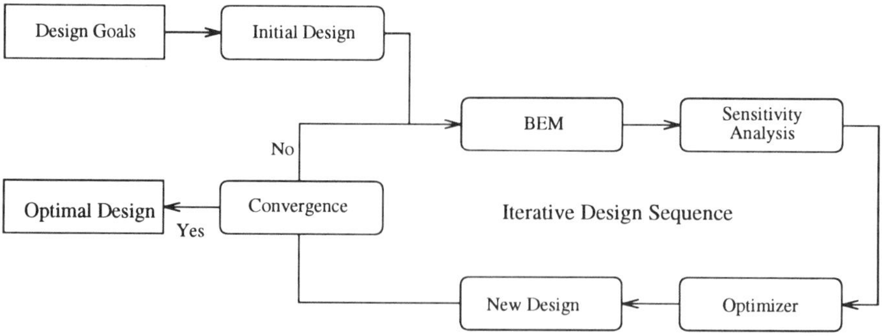

Figure 4.21. Optimal design methodology.

It should be pointed out here that while objective functions and constraints in these problems are often nonlinear functions of their arguments, the physical problem can be linear (e.g., linear elasticity) or nonlinear (e.g., elasto-viscoplasticity, large strains, and rotations). There is an extensive literature on the subject of optimization of linear problems in mechanics such as those governed by Laplace's equation and linear elasticity. Examples are books by Banichuk (1990) and a more recent book by Haftka and Gürdal (1992). Various mathematical approaches such as the calculus of variations, sequential linear programming, and sequential quadratic programming have been employed in the optimizer of figure 4.21 to solve these nonlinear programming problems, that is, extremizing a nonlinear functional (or functionals) without violating linear and nonlinear constraints. Examples are Vanderplaats (1983), Arora and Tseng (1987), and Schittkowski (1986).

Shape optimal design in linear elasticity is typically associated with the optimal shape of an object, for example, the shape of a notch or a cutout in a plate to minimize stress concentration. Optimal shape design of an object, however, is but a small part of a broad class of design problems. One can consider, for example, a general class of situations where a product or a process is optimized with respect to geometry, environment, or material:

	Product	Process
Geometry	×	×
Environment	×	×
Material	×	×

Here, a process could mean a manufacturing process that can be optimized with respect to geometry, environment (i.e., mechanical and/or thermal boundary conditions, body forces), or material. Thus, for example, one can think of optimal die shapes for extrusion, optimal preform or intermediate die shapes for forging, or optimal design of beads for sheet metal forming.

Objective functions to be minimized for such problems could be, for example, the total energy required for the process, or the maximum principal residual stress in some region of the product.

The above discussions assume separate designs for a product and a process. One can also, in the future, consider problems of concurrent optimal design of a process as well as the product!

Research on optimal design of problems with physical or geometrical nonlinearities is at its early stages. Based on the gradient approach to optimization, the first step in the design of solid forming processes is the calculation of sensitivities of continuum mechanics problems with large strains and large rotations. This issue has been discussed, at length, earlier in this chapter. To date, numerical results, using the algorithm shown in figure 4.21, have not yet been obtained for problems of optimal shape design for elasto-viscoplastic problems with large strains and large rotations. Recently, however, numerical calculations have been carried out for optimal shape design for elasto-viscoplastic problems with small strains and rotations (Wei et al. 1994). This problem is discussed at length in section 4.6 of this chapter. Work is currently in progress at Cornell University on optimal design of large-deformation problems using DSCs (i.e., using the algorithm shown in figure 4.21).

It should be mentioned here that some recent attempts have been made in the area of optimal design of metal forming processes. These approaches do not use the algorithm shown in figure 4.21. Instead, they use other concepts such as backward tracing or minimum plastic work. A short summary of this work is given below.

Kobayashi, Oh, and Altan (1989) have investigated issues of preform design of metal forming processes by using a backward tracing algorithm with a rigid-plastic material model. Han, Grandhi, and Srinivasan (1993) have published a paper on optimum design of intermediate forging dies that is also based on a backward tracing algorithm with a rigid-plastic material model. Richmond and Devenpeck (1962) had introduced the idea of ideal forming based on the concept of minimum plastic work. More recently, using the concept of the minimum plastic work path of Hill (1986), Chung and Richmond (1992a,b) have extended the ideal forming theory and have attempted to bridge the gap between analysis and optimal design of forming processes.

It is not the purpose in this book to discuss optimization algorithms at length. However, the subroutine N0ONF (available from the IMSL library) has been used to obtain the numerical results presented in section 4.6 and serves as an example of a typical optimization algorithm. This subroutine is based on the subroutine NLPQL, a FORTRAN code developed by Schittkowski (1986). A brief description of this algorithm follows.

A typical nonlinear optimization problem is stated as follows:

$$\min \phi(\boldsymbol{b}), \qquad \boldsymbol{b} \in R^n$$

subject to

$$\begin{aligned} &h_j(\boldsymbol{b}) = 0, && \text{for } j = 1, \ldots, m \\ &g_k(\boldsymbol{b}) \geq 0, && \text{for } k = 1, 2, \ldots, p \\ &b_{il} \leq b_i \leq b_{iu}, && \text{for each component } b_i \text{ of } \boldsymbol{b} \end{aligned}$$

where $\phi(\boldsymbol{b})$ is the objective function, $\boldsymbol{b}$ is the design variable vector with n components. b_{il} and b_{iu} are lower and upper bounds for each component of $\boldsymbol{b}$, and $h_j(\boldsymbol{b})$ and $g_k(\boldsymbol{b})$ are equality and inequality constraints, respectively.

The sequential quadratic programming algorithm NLPQL uses a quadratic approximation of the Lagrangian and linearization of the constraints to define a sequence of subproblems. This requires the evaluation of a positive definite approximation of the Hessian of ϕ.

Let $\boldsymbol{d}_k$ be the solution of a subproblem at the kth iterative step. A line search is used to find a new design, $\boldsymbol{b}_{k+1}$, defined as

$$\boldsymbol{b}_{k+1} = \boldsymbol{b}_k + \lambda_k \boldsymbol{d}_k, \qquad 0 < \lambda_k < 1 \quad (\text{no sum on } k)$$

such that the augmented Lagrangian function has a lower function value at the new design. Here, λ_k is the line search or step length parameter.

The iterative process stops when the Kuhn–Tucker optimality conditions are satisfied within an acceptable tolerance. Schittkowski (1986) shows that, under some mild assumptions, the algorithm converges globally; that is, starting from an arbitrary initial point, at least one accumulation point of the iterates will satisfy the Kuhn–Tucker optimality conditions.

Coupling of the optimizer with the mechanics and sensitivity calculations (figure 4.21) is straightforward. One must input the functions ϕ, $\partial\phi/\partial b_i$, h_j, g_k, $\partial h_j/\partial b_i$, and $\partial g_k/\partial b_i$ at each iteration. Typically, ϕ, h_j, and g_k depend on quantities such as stress or displacement and are, therefore, implicit as well as explicit functions of $\boldsymbol{b}$. The gradients of these functions, with respect to b_i, are obtained from the sensitivities, such as $d\boldsymbol{\sigma}/d\boldsymbol{b}$, by the chain rule of differentiation. It is important to note that, for elasto-viscoplastic problems, the above functions and their gradients are evaluated at a preset time T from the start of the simulation. This optimizer performs very well for the problems considered in this work (see section 4.6), provided that the sensitivities are obtained with sufficient accuracy.

4.6 Optimization of Plates with Cutouts

This section is concerned with plates of arbitrary shape, in plane stress, with a traction free cutout inside the plate. The plate material is either linear elastic or elasto-viscoplastic. Only small-strain small-rotation problems are considered here (see Wei et al. 1994).

4.6.1 Parametrization of Cutout Boundary

With the global X_1 and X_2 axes centered at the center of a cutout, a variety of smooth curves (e.g., a circle, an ellipse, or a rectangle with rounded corners) can be represented by the equations (Sadegh 1988–89, Lekhnitski 1968)

$$X_1 = a(\cos\theta + \varepsilon\cos 3\theta) \tag{4.75}$$

$$X_2 = a(\beta\sin\theta - \varepsilon\sin 3\theta) \tag{4.76}$$

where a controls the size, ε the shape, and β the aspect ratio of the smooth cutout.

For example, with $\beta = 1$, a slight variation of the above equations is

$$X_1 = \frac{a}{1+\varepsilon}(\cos\theta + \varepsilon\cos 3\theta) \tag{4.77}$$

$$X_2 = \frac{a}{1+\varepsilon}(\sin\theta - \varepsilon\sin 3\theta) \tag{4.78}$$

where the points $(a, 0)$ (with $\theta = 0$) and $(0, a)$ (with $\theta = \pi/2$) are points on the curve. Considering a fixed and ε variable, $\varepsilon = -0.15$, for example, gives a square with rounded corners and $\varepsilon = 0$, of course, gives a circle.

4.6.2 Objective Functions and Constraints

The optimization problem is set up as

$$\min \phi(b_i) \tag{4.79}$$

subject to the constraints

$$f_i(b_j) \geq 0 \tag{4.80}$$

Various choices are possible for the objective function ϕ. Two of these are

$$\phi_1 = \int_{\partial B_0} \tau_i u_i \, dS \tag{4.81}$$

$$\phi_2 = \frac{1}{L} \int_{\partial B_c} (\sigma_{tt}(S) - \bar{\sigma}_{tt})^2 \, dS \tag{4.82}$$

where ∂B_0 is the outer boundary, ∂B_c is the cutout boundary, σ_{tt} is the tangential stress on the cutout boundary, $\bar{\sigma}_{tt}$ is the mean value of σ_{tt}, and L is the total length of the cutout boundary.

Equations (4.79) and (4.81) express the requirement of minimizing the external work done on the body (on a traction-free cutout, $\tau = 0$), and equations (4.79) and (4.82) express the requirement of minimizing the variance of the tangential stress on the cutout, thereby requiring the tangential stress on the cutout to be as uniform as possible. *It is very important to note that, for time-dependent elasto-viscoplastic problems, an objective function is defined here as the value of ϕ at a fixed time T from the start of the deformation process.* The constraints used in these problems are related to bounds on the shape design variables.

4.6.3 *Elastic Shape Optimization*

These numerical examples have been solved in order to check the performance of the computer program against known elastic solutions.

Minimize external work. In this case, the objective function is ϕ_1 from equation (4.81), so that

$$\overset{*}{\phi}_1 = \int_{\partial B_0} \overset{*}{\tau}_i u_i \, dS + \int_{\partial B_0} \tau_i \overset{*}{u}_i \, dS + \int_{\partial B_0} \tau_i u_i \, d\overset{*}{S} \tag{4.83}$$

The plate is square, and the cutout shape is defined by equations (4.75) and (4.76) with $\beta = 1$, a fixed, and ε the only design variable. Thus, the points $[a(1+\varepsilon), 0]$ and $[0, a(1+\varepsilon)]$ lie on the cutout. A quarter of the plate is modeled because of symmetry (figure 4.22 and table 4.2). Uniform biaxial loading $\tau_1^\infty = \tau_2^\infty = 1$ MPa is applied to the plate. The constraints used in this problem are

$$-0.15 \leq \varepsilon \leq 0.10 \tag{4.84}$$

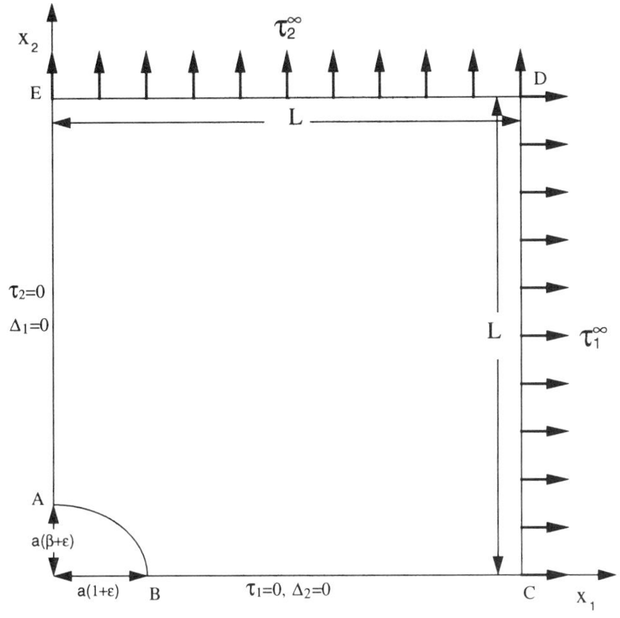

Figure 4.22. Optimal shape design of a cutout in a plate (from Wei et al. 1994).

The expression for $d\overset{*}{S}/dS$ on cutout boundary AB can be obtained easily from equations (4.77) and (4.78), together with equation (4.35). The expressions for the normal $\boldsymbol{N}$ and its sensitivity $\overset{*}{\boldsymbol{N}}$ on AB are obtained from equation (4.36). Using a linear approximation for the design velocities on the lines EA and BC, one gets

$$\text{on } BC: \quad \overset{*}{X}_1 = \frac{a(L - X_1)}{L - a(1+\varepsilon)}, \qquad \overset{*}{X}_2 = 0, \qquad \frac{d\overset{*}{S}}{dS} = -\frac{a}{L - a(1+\varepsilon)} \tag{4.85}$$

$$\text{on } AE: \quad \overset{*}{X}_1 = 0, \qquad \overset{*}{X}_2 = \frac{a(L - X_2)}{L - a(1+\varepsilon)}, \qquad \frac{d\overset{*}{S}}{dS} = -\frac{a}{L - a(1+\varepsilon)} \tag{4.86}$$

Design velocities and $d\overset{*}{S}/dS$ are zero on the lines CD and DE in figure 4.22.

The shape of the cutout, at different iterations, is shown in figure 4.23. The converged solution, after seven iterations, is $\varepsilon = -0.015\,15$. The well-known optimal analytical solution for a cutout in an infinite elastic plate (Banichuk 1990) is a circle with $\varepsilon = 0$.

Table 4.2. Geometrical and loading parameters for various optimization problems for cutouts in plates

Problem	Parameters
Elastic problem (a)	$L = 30$ m
	$a = 2/(1 + \varepsilon_0) = 2.353$ m
	$\beta = 1$
	$\varepsilon \equiv$ design variable
	$\tau_1^\infty = \tau_2^\infty = 1$ MPa
Elastic problem (b)	$L = 5$ m
	$a = 1$ m
	$\varepsilon = 0$
	$\beta \equiv$ design variable
	$\tau_1^\infty = 4$ MPa
	$\tau_2^\infty = 3$ MPa
Elasto-viscoplastic problem	$L = 5$ m
	$a = 1$ m
	$\varepsilon = 0$
	$\beta \equiv =$ design variable
	$\tau_1^\infty = S(t) = 8 + 4t$ MPa (t in seconds)
	$\tau_2^\infty = 0.75S(t)$, $T = 4$ s
	$\Delta t = 0.2$ s for $0 \leq t \leq 2$ s; 0.05 s for $2 \leq t \leq 4$ s

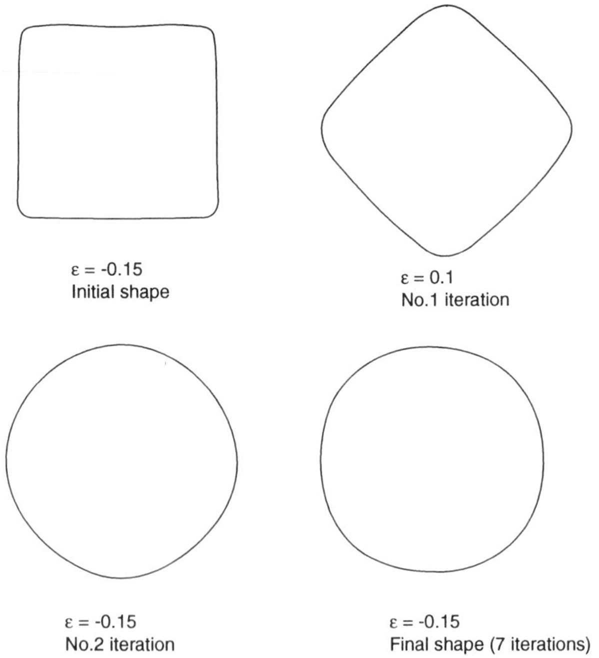

Figure 4.23. Shapes of cutout at different iterations for the elastic optimization problem (a) (from Wei et al. 1994).

Minimize variance of tangential stress around cutout boundary. In this case, the objective function is ϕ_2 from equation (4.82) so that

$$\overset{*}{\phi}_2 = -\frac{\overset{*}{L}}{L}\phi_2 + \frac{1}{L}\int_A^B 2(\sigma_{tt} - \bar{\sigma}_{tt})[\overset{*}{\sigma}_{tt} - \overset{*}{\bar{\sigma}}_{tt}]\, dS$$
$$+\frac{1}{L}\int_A^B (\sigma_{tt} - \bar{\sigma}_{tt})^2\, d\overset{*}{S} \tag{4.87}$$

where

$$\overset{*}{\bar{\sigma}}_{tt} = -\frac{\overset{*}{L}}{L}\bar{\sigma}_{tt} + \frac{1}{L}\int_A^B \overset{*}{\sigma}_{tt}\, dS + \frac{1}{L}\int_A^B \sigma_{tt}\, d\overset{*}{S}$$

$$L = \int_A^B dS \qquad \text{and} \qquad \overset{*}{L} = \int_A^B d\overset{*}{S}$$

The plate is square, as before. This time, an elliptical cutout boundary is modeled as

$$X_1 = a\cos\theta, \qquad X_2 = b\sin\theta \tag{4.88}$$

with a fixed and $b = a\beta$ the design variable. The constraint used in this problem is

$$0.3 \leq \beta \leq 1.0 \tag{4.89}$$

Referring to figure 4.22 and table 4.2, this time $a = 1.0$, $\varepsilon = 0$, $L = 5$ m, $\tau_1^\infty = S = 4$ MPa, and $\tau_2^\infty = (3/4)\ S = 3$ MPa.

Once again, the expression for $d\overset{*}{S}/dS$ over the cutout boundary AB can be obtained easily. A linear approximation is used for the design velocity on the line EA. The design velocity is zero on the rest of the boundary $BCDE$. The actual formulas are given later.

The results for successive iterations are given in table 4.3. The convergent solution is $\beta = 0.756$. It is well known (Banichuk 1990) that the optimal analytical solution for a cutout in an infinite elastic plate, in this case, is an ellipse with $\beta = 0.75$.

4.6.4 Elasto-viscoplastic Shape Optimization

Constitutive model. The viscoplastic constitutive model due to Anand (1982) is used in these calculations. The constitutive equations and material parameters used here are given in section 2.4.1.

Table 4.3. Values of β and ϕ_2 at different iterations for the elastic shape optimization problem (b)[a]

Number of iterations	$\beta = b/a$	ϕ_2 (MPa^2)
1	1.000	2.702 00
2	0.300	15.289 00
3	0.814	0.179 60
4	0.738	0.017 20
5	0.756	0.000 98

[a]CPU time = 29.78seconds on an IBM 3090 supercomputer.

Numerical issues. Demands on numerical accuracy in these problems are quite stringent, especially for the calculation of the rates of sensitivities of stresses. It has proved very useful to check numerical results for benchmark problems against semianalytical solutions.

Such a benchmark problem, that of a circular plate with a concentric circular cutout, subjected to spatially uniform external pressure that increases in time at a constant rate, is described in detail by Zhang et al. (1992a). A semianalytical solution for this problem is also given in that paper. As discussed by Zhang et al., semianalytical solutions for sensitivities have been obtained both by analytical differentiation (ADD) and finite differencing (FDD) of direct solutions of the corresponding mechanics problem. These solutions agree within plotting accuracy [see figure 9 of Zhang et al. (1992a)]. Since time integration can be fairly expensive, it appears best, at first, to apply a large pressure suddenly on the boundary of a disk and compare the BEM rates against semianalytical ones. An example of such an exercise is given below. The problem chosen here is that of an annular circular disk, of inside and outside radii $a = 1.0$ m and $b = 1.5$ m, respectively, with a suddenly applied external pressure of 12 MPa. The inner radius is the design variable. The mesh and material model used are exactly the same as those for the corresponding problem by Zhang et al. (1992a). The results are shown in table 4.4. It should be emphasized that, while the problem is axisymmetric, the BEM program is a general two-dimensional one.

In table 4.4, an "analytical" solution is an exact solution for an elastic annular disk while a "semianalytical" solution is obtained by numerical time integration of a system of differential equations governing the elasto-viscoplastic deformation of an axisymmetric hollow disk. These equations, in cylindrical coordinates, are given as equations 30 to 33 in Zhang et al. (1992a). It should be noted here that the "semianalytical" sensitivity equations are obtained by analytical differentiation of the mechanics equations. These "semianalytical" solutions can be considered, for practical

Table 4.4. Comparison of tangential stress σ_{tt}, $\overset{*}{\sigma}_{tt}$, $\dot{\sigma}_{tt}$, and $\overset{\circ}{\sigma}_{tt}$ around a cutout in a circular disk: BEM and analytical or semianalytical solutions (see Zhang et al. 1992a)

Location around circular cutout, θ (degrees)	Elastic				Elasto-viscoplastic			
	σ (MPa)		$\overset{*}{\sigma}_{tt}$ (MPa/m)		$\dot{\sigma}_{tt}$ (MPa/s)		$\overset{\circ}{\sigma}_{tt}$ (Mpa/(m-s)	
	BEM	Analytical	BEM	Analytical	BEM	Semianalytical	BEM	Semianalytical
0	−43.21	−43.2	−69.28	−69.12	207.77	205.01	2081.1	2072.7
9	−43.19	−43.2	−69.03	−69.12	204.39	205.01	2097.3	2072.7
18	−43.20	−43.2	−69.26	−69.12	204.97	205.01	2051.1	2072.7
27	−43.20	−43.2	−69.02	−69.12	202.14	205.01	2070.9	2072.7
36	−43.19	−43.2	−69.25	−69.12	203.15	205.01	2028.2	2072.7
45	−43.20	−43.2	−69.02	−69.12	201.38	205.01	2063.3	2072.7
54	−43.19	−43.2	−69.25	−69.12	203.16	205.01	2028.6	2072.7
63	−43.20	−43.2	−69.02	−69.12	202.15	205.01	2072.0	2072.7
72	−43.20	−43.2	−69.26	−69.12	204.99	205.01	2052.4	2072.7
81	−43.19	−43.2	−69.04	−69.12	204.41	205.01	2099.4	2072.7
90	−43.21	−43.2	−69.28	−69.12	207.80	205.01	2083.3	2072.7

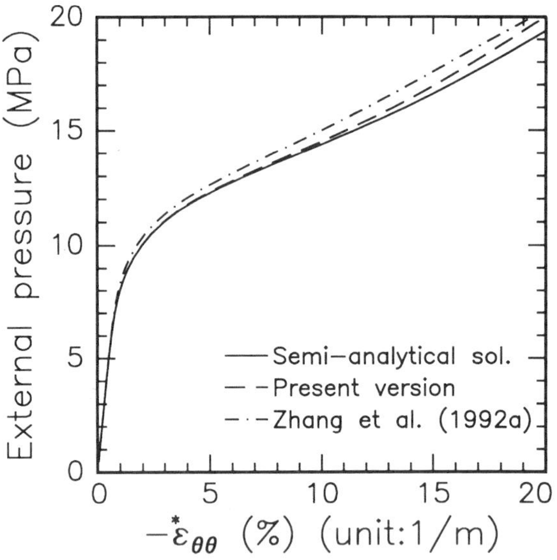

Figure 4.24. External pressure as a function of $\overset{*}{\varepsilon}_{\theta\theta}(A)$ for a hollow disk. The simulation ends at $T = 4$ s when $\varepsilon_{\theta\theta} = -4.8$ percent [see also Zhang et al. (1992a), figure 9] (from Wei et al. 1994).

purposes, to be exact. Of course, they are used to benchmark the numerical DBEM solutions for special illustrative problems, and are not available for general two-dimensional problems. Finally, "BEM" in table 4.4 refers to numerical solutions from the present, general, 2D DBEM approach.

The sensitivity history for a problem of pressurization of a disk, in plane stress, has been presented in Zhang et al. (1992a, figure 9). In this example, once again, the disk has inner and outer radii 1 m and 1.5 m, respectively, and the inner radius is the design variable. This time, however, the disk is subjected to an external pressure rate of 5 MPa/s. It is seen that the numerically calculated sensitivity history in Zhang et al. (1992a) is not very accurate.

There are two differences between the results presented in this section and those in Zhang et al. The first is the use of the correct formula (equation 4.44) for $d\overset{*}{A}/dA$ for this problem. The second is the use of the regularized equations that are presented in section 4.2. Numerical results for the sensitivity history are shown in figure 4.24. Here, the "semianalytical" solution can be considered exact (see above) while "Zhang et al. (1992a)" refers to the use of the $O(1/r)$ singular DBEM equations in that paper (see also equation 2.167) together with the correction given in equation (4.44) above. Finally, the "present version" uses the regularized DBEM equations given in this book. This version is the most accurate. The importance of regularization is evident from figure 4.24.

Constraints, geometry, and loading history. The problem considered here is the elasto-viscoplastic version of the second elasticity problem above (minimize variance of tangential stress around cutout boundary). The objective function is now ϕ_2 (equation 4.82) evaluated at some fixed time T into the deformation history, and the elliptical cutout is defined by equation (4.88) with a fixed and $b = a\beta$ the design variable. The constraint equation used here is

$$0.5 \leq \beta \leq 1.0 \tag{4.90}$$

Referring to figure 4.22 and table 4.2, a fixed remote loading history, for the period $0 \leq t \leq T$, is applied to the plate. The question being asked here is: For what values of β (i.e., shape of the cutout) is the tangential stress around the cutout, at time T, as uniform as possible? Except for the case $\tau_1^\infty = \tau_2^\infty$ (see figure 4.22), for which the optimal cutout shape is a circle (with $\beta = 1$), the elasto-viscoplastic solution is expected to differ from the elastic solution. This is because, for $\tau_1^\infty \neq \tau_2^\infty$, the rate of stress relaxation around the elliptical cutout will not be uniform. Also, the optimal value of β is a function of the loading history and the final time T. These issues are discussed further later in this section.

Geometric sensitivities and mesh. The choice of mesh, especially the internal cells, is crucial for the solution of this class of problems. The best approach is to use adaptive meshing during the iterative optimization process. Work along these lines is currently in progress.

In this work, however, a fixed mesh is chosen, based on numerical experimentation of problems with known elastic and elasto-viscoplastic solutions. As mentioned earlier, the benchmark elasto-viscoplastic problem of a plate with a circular cutout and uniform remote loading has proved to be invaluable for this propose. The mesh chosen, both boundary elements and internal cells, is shown in figure 4.25. In this figure, all boundary nodes are also nodes for internal cells, except on lines CD and DE where mid-boundary nodes are not connected to internal cells.

With b as the design variable, the sensitivities of geometrical quantities, used in these calculations, are as follows (see figure 4.26):

(a) Boundary

On BC, CD, and DE,

$$\overset{*}{X}_1 = \overset{*}{X}_2 = 0, \qquad \frac{d\overset{*}{S}}{dS} = 0, \qquad \frac{d\overset{*}{A}}{dA} = 0 \tag{4.91}$$

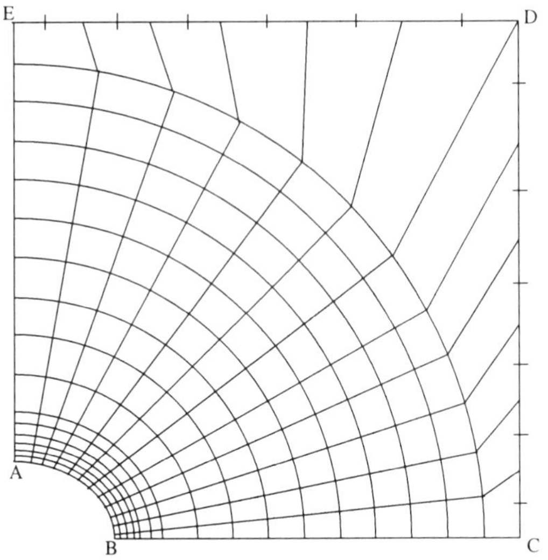

Figure 4.25. The boundary elements and internal cells used for the elasto-viscoplastic problem (from Wei et al. 1994).

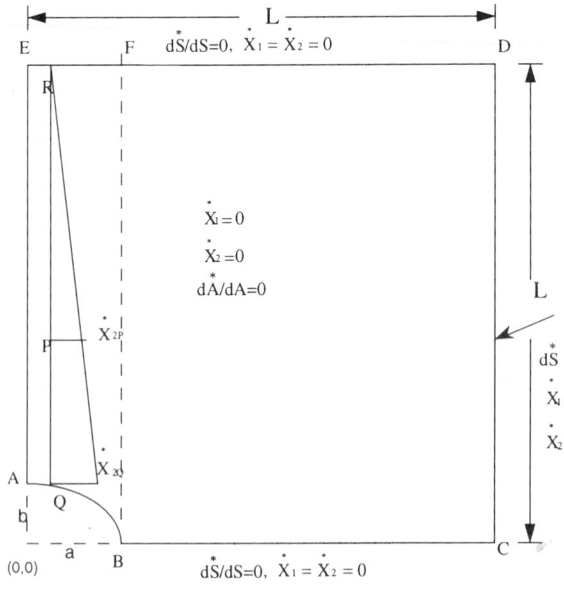

Figure 4.26. Sensitivities of geometrical quantities (from Wei et al. 1994).

On EA,

$$\overset{*}{X}_1 = 0, \qquad \overset{*}{X}_2 = \frac{L - X_2}{L - b} \quad \text{(linear assumption)}$$

$$\frac{d\overset{*}{S}}{dS} = -\frac{1}{L - b}, \qquad \frac{d\overset{*}{A}}{dA} = -\frac{1}{L - b} \tag{4.92}$$

On AB,

$$X_1 = a \cos \theta, \qquad X_2 = b \sin \theta$$

$$\overset{*}{X}_1 = 0, \qquad \overset{*}{X}_2 = \sin \theta = \frac{X_2}{b}$$

$$\frac{d\overset{*}{S}}{dS} = \frac{b^3 X_1^2}{a^4 X_2^2 + b^4 X_1^2}, \qquad \frac{d\overset{*}{A}}{dA} = \frac{1}{b} - \frac{X_2^2}{b^3} \tag{4.93}$$

(b) Internal points
Inside the rectangle $BCDF$,

$$\overset{*}{X}_1 = \overset{*}{X}_2 = 0, \qquad \frac{d\overset{*}{A}}{dA} = 0 \tag{4.94}$$

Inside the region $BFEA$, using a linear assumption for the velocity $\overset{*}{X}_2$ (see figure 4.26)

$$\overset{*}{X}_{2P} = \left(\frac{L - X_{2P}}{L - X_{2Q}} \right) \overset{*}{X}_{2Q} \tag{4.95}$$

and

$$\frac{d\overset{*}{A}}{dA} = -\frac{\sqrt{1 - X_{1P}/a^2}}{L - b\sqrt{1 - X_{1P}^2/a^2}} \tag{4.96}$$

where the quantities X_{2P}, and so on, are defined in figure 4.26. On the line FB,

$$\frac{d\overset{*}{A}}{dA} = 0 \tag{4.97}$$

It is interesting to note that while the value of $d\overset{*}{A}/dA$ inside the region $BFEA$ matches the values on its boundary lines EA and FB, such is not the case on the lines AB and EF. This is expected since a separate kinematic assumption for the design velocities must be made inside the region $BFEA$. In the calculations reported next, the values of $d\overset{*}{A}/dA$ on the lines AB and EF are those from equation (4.96).

Numerical results. Consider first figure 4.22 (see, also, table 4.2), with $\varepsilon = 0$, $b = a/\beta$, $\tau_1^\infty = S$, and $\tau_2^\infty = 0.75S$. Elastic stress concentrations for this problem, at points A and B, can easily be shown to be (Timoshenko and Goodier 1970)

$$\hat{\sigma}_{11}(A) = \frac{\sigma_{11}(A)}{S} = 2\frac{b}{a} + 0.25, \qquad \hat{\sigma}_{22}(B) = \frac{\sigma_{22}(B)}{S} = 1.5\frac{a}{b} - 0.25 \tag{4.98}$$

As is well known (Banichuk 1990), if $b/a = \tau_2^\infty/\tau_1^\infty$ (here 0.75), the load ratio

$$\frac{\sigma_{tt} \text{ on ellipse}}{S} = 1.75$$

so that the tangential stress σ_{tt} is uniform around the ellipse. This is the optimal solution for the elastic problem, as discussed before.

For the elasto-viscoplastic problem, however, the case $\tau_2^\infty/\tau_1^\infty = b/a$ does not lead to uniform relaxation of stress around the ellipse. Perhaps a useful way to see this is to define "apparent" stress concentrations at points A and B in figure 4.22. Again, for the case $\tau_1^\infty = S$ and $\tau_2^\infty = 0.75S$, one gets, for the elastic case,

$$\frac{\sigma_{11}(A)}{\sigma_{11}(\infty)} = 2\frac{b}{a} + 0.25, \qquad \frac{\sigma_{22}(B)}{\sigma_{22}(\infty)} = 2\frac{a}{b} - 0.333 \tag{4.99}$$

Thus, for example, for $b/a = 0.75$, the aforementioned numbers are 1.75 and 2.333, so one might reasonably expect the tangential stress to relax faster at B than at A in the elasto-viscoplastic case. It seems reasonable to expect, therefore, that the optimal value of b for the elliptical cutout (with a fixed) should be a number other than 0.75 for the elasto-viscoplastic case. This value of b must be such that, starting with a nonuniform distribution of σ_{tt} around the ellipse, the tangential stress becomes uniform at a fixed (prechosen) time T into the elasto-viscoplastic deformation process. Of course, this optimal value of b would depend on the choice of T and the loading history.

With the above-mentioned preamble, the central result of this section is presented in table 4.5. Here, for each value of b, $T = 4$ seconds. For the case $\beta = 0.75$, for example, the strain $\varepsilon_{11}(A, T)$ (see figure 4.22) equals 1.125

Table 4.5. Values of β and $\phi_2(T)$ at different iterations for the elasto-viscoplastic shape optimization problem[a]

Number of iterations	$\beta = b/a$	ϕ_2 (MPa^2)
1	1.000	31.566 00
2	0.500	14.133 00
3	0.724	0.410 76
4	0.697	0.031 75
5	0.690	0.010 52

[a]CPU time = 1.278 hours on an IBM 3090 supercomputer.

percent. Table 4.5 shows successive values of β, starting with the circle $\beta = 1$, for the shape optimization problem for the elasto-viscoplastic case. The iterations are seen to converge, at the fifth iteration, to the value $\beta = 0.69$, with a corresponding very low value of ϕ_2. As discussed before in this paper, the optimization algorithm used here converges when the Kuhn–Tucker optimality conditions are satisfied within an acceptable tolerance (Schittkowski 1986).

It is useful to comment further on some of the details of this problem. Table 4.6 shows the stress concentrations at A and at B, for different values of β, for the elastic, as well as the elasto-viscoplastic, solutions at time T. Table 4.7 shows the correlation of the "apparent" stress concentration at A and at B to the stress relaxation at these points in the elasto-viscoplastic case. It is seen that the point with the larger "apparent" stress concentration experiences larger relaxation of stress. The results for $\beta = 0.75$ (the elastic optimal solution) are included in table 4.7. In this case, A and B have the

Table 4.6. Stress concentrations at A and B (figure 4.22) for different values of β[a]

	Elastic				Elasto-viscoplastic	
	$\sigma_{11}(A)/S(0)$		$\sigma_{22}(B)/S(0)$		$\hat{\sigma}_{11}(A,T)$	$\hat{\sigma}_{22}(B,T)$
β	Analytical	BEM	Analytical	BEM	BEM	BEM
1.000	2.250	2.380	1.250	1.224	1.619	0.943
0.500	1.250	1.206	2.750	2.874	1.097	1.627
0.724	1.698	1.714	1.822	1.885	1.324	1.250
0.697	1.644	1.653	1.902	1.972	1.298	1.284
0.690	1.630	1.636	1.924	1.998	1.290	1.294

[a] $\hat{\sigma}_{11}(A,T) = \sigma_{11}(a,T)/(S(T); \hat{\sigma}_{22}(B,T) = \sigma_{22}(B,T)/S(T)$. Numerical results are obtained from the mesh in figure 4.25.

Table 4.7. Correlation of stress relaxation at points A and B (figure 4.22) with the "apparent" elastic stress concentrations at these points

	"Apparent" stress concentration elastic : analytical		Drop in stress concentration elasto-viscoplastic : BEM	
β	$\sigma_{11}(A)/\sigma_{11}(\infty)$	$\sigma_{22}(B)/\sigma_{22}(\infty)$	$\hat{\sigma}_{11}(A,T)/\hat{\sigma}_{11}(A,0)$	$\hat{\sigma}_{22}(B,T)/\hat{\sigma}_{22}(B,0)$
1.000	2.250	1.667	0.680	0.770
0.500	1.250	3.667	0.910	0.566
0.724	1.698	2.430	0.772	0.663
0.697	1.644	2.536	0.785	0.651
0.690	1.630	2.571	0.789	0.648
0.750	1.750	2.333	1.761	0.675

[a]Numerical results are obtained from the mesh in figure 4.25.

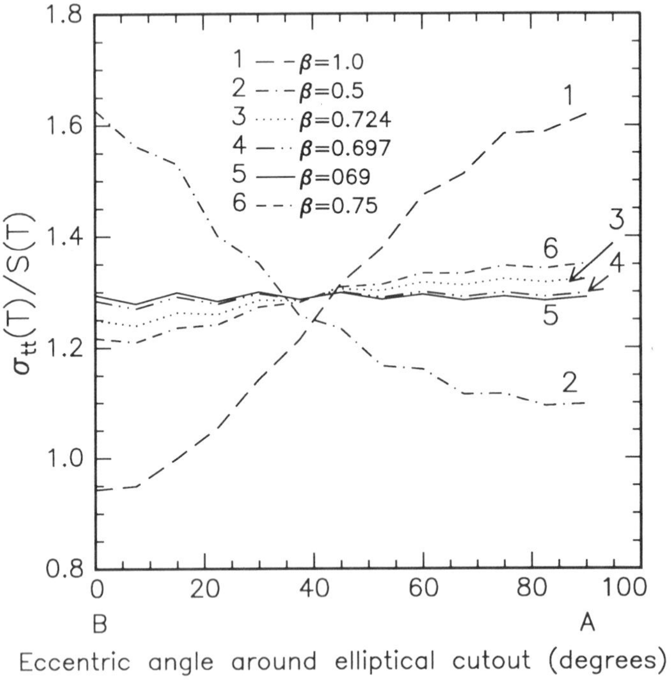

Figure 4.27. Tangential stress concentration around elliptical cutout, at final time T, for different values of β (from Wei et al. 1994).

Table 4.8. Stress sensitivities at A and B (figure 4.22) for different values of β[a]

	Elastic				Elasto-viscoplastic	
	$\overset{*}{\sigma}_{11}(A)/S(0)$		$\overset{*}{\sigma}_{22}(B)/S(0)$		$\overset{*}{\sigma}_{11}(A,T)/S(T)$	$\overset{*}{\sigma}_{22}(B,T)/S(T)$
β	Analytical	BEM	Analytical	BEM	BEM	BEM
1.000	2.0	2.540	−1.500	−1.828	1.039	−1.114
0.500	2.0	2.247	−6.000	−6.158	0.892	−2.485
0.724	2.0	2.321	−2.862	−3.181	0.901	−1.318
0.697	2.0	2.305	−3.088	−3.400	0.894	−1.398
0.690	2.0	2.301	−3.151	−3.467	0.893	−1.423
0.750	2.0	2.339	−2.667	−2.983	0.909	−1.250

[a]Numerical results are obtained from the mesh in figure 4.25. Units = 1/m.

same initial value of stress concentration, but point B, with the higher "apparent" stress concentration, experiences a larger relaxation of stress.

Figure 4.27 shows the tangential stress concentration around the ellipse, at time T, for various values of β. It is seen that the distribution for $\beta = 0.75$ is not uniform, while that for $\beta = 0.69$ is uniform at time T.

Finally, something should be said about the stress sensitivities, accurate calculations of which are essential for the success of the optimization process. Differentiating equation (4.98) with respect to the design variable b, one gets, for the elastic problem,

$$\frac{\overset{*}{\sigma}_{11}(A)}{S} = \frac{2}{a}, \qquad \frac{\overset{*}{\sigma}_{22}(B)}{S} = -1.5\frac{a}{b^2} \tag{4.100}$$

Table 4.8 shows the numerical and analytical values of these quantities, for the elastic problem, for various values of β, as well as the numerical values of these quantities at time T. The numerical values are calculated with the mesh shown in figure 4.25. A finer mesh would improve the accuracy of the results. This has been demonstrated in Zhang and Mukherjee (1991a, figure 6b).

References

Aithal, R. and Saigal, S. (1990). "Adjoint Structure Approach for Shape Sensitivity Analysis Using BEM," *ASCE J. Eng. Mech.*, **116**, 2663–2680.

Aithal, R., Saigal, S., and Mukherjee, S. (1991). "Three-dimensional Boundary Element Implicit Differentiation Formulation for Design Sensitivity Analysis," *Math. Computer Modelling*, **15**, 1–10.

Anand, L. (1982). "Constitutive Equations for the Rate-dependent Deformation of Metals at Elevated Temperatures," *ASME, J. Eng. Mater. Technol.*, **102**, 12–17.

Ang, A. H.-S. and Tang, W. H. (1975). *Probability Concepts in Engineering Planning and Design*. Wiley, New York.

Arora, J. S. and Cardoso, J. B. (1992). "Variational Principle for Shape Sensitivity Analysis," *AIAA J.*, **30**, 538–547.

Arora, J. S. and Tseng, C. H. (1987). *User Manual for IDESIGN: Version 3.5*. Optimal Design Laboratory, College of Engineering, The University of Iowa, Iowa City.

Banichuk, N. V. (1990). *Introduction to Optimization of Structures*, Springer-Verlag, New York.

Barone, M. R. and Yang, R.-J. (1988). "Boundary Integral Equations for Recovery of Design Sensitivities in Shape Optimization," *AIAA J.*, **26**, 589–594.

Barone, M. R. and Yang, R.-J. (1989). "A Boundary Element Approach for Recovery of Shape Sensitivities in Three-dimensional Elastic Solids," *Computer Meth App. Mech. Eng.*, **17**, 69–82.

Bonnet, M. (1990). "Shape Differentiation of Regularized BIE: Application to 3D Crack Analysis by the Virtual Crack Extension Approach," *Boundary Elements in Mechanical and Electrical Engineering* (ed. C. A. Brebbia and A. Chaudouet). Computational Mechanics Publications and Springer-Verlag, Berlin.

Choi, J. H. and Choi, K. K. (1990). "Direct Differentiation Method for Shape Design Sensitivity Analysis Using Boundary Integral Formulation," *Computers Structures*, **34**, 499–508.

Choi, J. H. and Kwak, B. M. (1988). "Boundary Integral Equation Method for Shape Optimization of Elastic Structures," *Int. J. Num. Meth. Eng.*, **26**, 1579–1595.

Choi, K. K. (1987). "Shape Design Sensitivity Analysis and Optimal Design of Structural Systems," *Computer Aided Optimal Design: Structural and Mechanical Systems* (ed. C. A. Mota Soares), vol. 2, pp. 54–108. NATO ASI Series, Lisbon.

Choi, K. K. and Santos, J. L. T. (1987). "Design Sensitivity Analysis of Nonlinear Structural System; Part I: Theory," *Int. J. Num. Meth. Eng.*, **24**, 2039–2055.

Chung, K. and Richmond, O. (1992a). "Ideal Forming—I, Homogeneous Deformation with Minimum Plastic Work," *Int. J. Mech. Sci.*, **34**, 575–591.

Chung, K. and Richmond, O. (1992b). "Ideal Forming—II, Sheet Forming with Optimum Deformation," *Int. J. Mech. Sci.*, **34**, 617–633.

Dems, K. (1986). "Sensitivity Analysis in Thermal Problems—I: Variation of Material Parameters within a Fixed Domain," *J. Thermal Stresses*, **9**, 303–324.

Dems, K. (1987). "Sensitivity Analysis in Thermal Problems—II: Structural Shape Variation," *J. Thermal Stresses*, **10**, 1–16.

Dems, K. and Mróz, Z. (1989). "Sensitivity of Buckling Load and Vibration Frequency with Respect to Shape of Stiffened and Unstiffened Plates," *Mech. Structures Machines*, **17**, 431–457.

Fung, Y. C. (1977). *A First Course in Continuum Mechanics*, 2d ed. Prentice Hall, Englewood Cliffs, N.J.

Ghosh, N., Rajiyah, H., Ghosh, S., and Mukherjee, S. (1986). "A New Boundary

Element Method Formulation for Linear Elasticity, *ASME, J. Appl. Mech.*, **53**, 69–76.
Grabacki, J. (1991). "Boundary Integral Equations in Sensitivity Analysis," *Appl. Math. Modelling*, **15**, 170–181.
Haftka, R. T. and Gürdal, Z. (1992). *Elements of Structural Optimization*, 3d ed. Kluwer Academic Publishers, Dordrecht, The Netherlands.
Haftka, R. T., Cohen, G. A., and Mróz, Z. (1990). "Derivatives of Buckling Loads and Vibration Frequencies with Respect to Stiffness and Initial Strain Parameters," *ASME, J. Appl. Mech.*, **57**, 18–24.
Han, C. S., Grandhi, R. V., and Srinivasan, R. (1993). "Optimum Design of Forging Die Shape Using Nonlinear Finite Element Analysis," *AIAA J.*, **31**, 774–781.
Haug, E. J., Choi, K. K., and Komkov, V. (1986). *Design Sensitivity Analysis of Structural Systems*. Academic Press, New York.
Hill, R. (1978) "Aspects of Invariance in Solid Mechanics," *Advances in Applied Mechanics*, **18** (ed. C.-S. Yih). Academic Press, New York.
Hill, R. (1986). "Extremal Paths of Plastic Work and Deformation," *J. Mech. Phys. Solids*, **34**, 511–523.
Jao, S. Y. and Arora, J. S. (1992a). "Design Sensitivity Analysis of Nonlinear Structures Using Endochronic Constitutive Model. Part 1: General Theory," *Comput. Mech.*, **10**, 39–57.
Jao, S. Y. and Arora, J. S. (1992b). "Design Sensitivity Analysis of Nonlinear Structures Using Endochronic Constitutive Model. Part 2: Discretization and Applications," *Comput. Mech.*, **10**, 59–72.
Kane, J. H. and Saigal, S. (1988). "Design Sensitivity Analysis of Solids Using BEM," *ASCE J. Eng. Mech.*, **114**, 1703–1722.
Kane, J. H., Zhao, G., Wang, H., and Guru Prasad, K. (1992). "Boundary Formulations for Three-dimensional Continuum Structural Shape Sensitivity Analysis," *ASME, J. Appl. Mech.*, **59**, 827–834.
Kobayashi, S., Oh, S. I., and Altan, T. (1989). *Metal Forming and the Finite Element Method*. Oxford University Press, New York.
Lekhnitski, S. G. (1968). *Anisotropic Plates*, 2d ed. (transl. by S. W. Tsai and T. Cheron). Gordon and Breach, New York.
Leu, L.-J. and Mukherjee, S. (1993). "Sensitivity Analysis and Shape Optimization in Nonlinear Solid Mechanics," *Eng. Anal. Boundary Elem.*, **12**, 251–260.
Leu, L.-J. and Mukherjee, S. (1995). "Sensitivity Analysis in Nonlinear Solid Mechanics by the Boundary Element Method with an Implicit Scheme," *J. Eng. Anal. Design*, **2**, 33–55.
Meric, R. A. (1987). "Material and Load Optimization by the Adjoint Variable Method," *ASME, J. Heat Transfer*, **109**, 782–784.
Mota Soares, C. A. and Leal, R. P. (1987). "Mixed Elements in Sensitivity Analysis of Structures," *Computer Aided Optimal Design: Structural and Mechanical Systems* (ed. C. A. Mota Soares), vol. 1, pp. 442–436. NATO ASI Series, Lisbon.
Mukherjee, S. (1982). *Boundary Element Methods in Creep and Fracture*. Elsevier Applied Science Publishers, Barking, Essex, UK.
Mukherjee, S. and Chandra, A. (1989). "A Boundary Element Formulation for Design Sensitivities in Materially Nonlinear Problems," *Acta Mech.*, **78**, 243–253.

Mukherjee, S. and Chandra, A. (1991). "A Boundary Element Formulation for Design Sensitivities in Problems Involving Both Geometric and Material Nonlinearities," *Math. Computer Modelling*, **15**, 245–255.

Nagarajan, A. and Mukherjee, S. (1993). "A Mapping Method for Numerical Evaluation of Two-dimensional Integrals with $1/r$ Singularity," *Comput. Mech.*, **12**, 19–26.

Rajiyah, H. and Mukherjee, S. (1987). "Boundary Element Analysis of Inelastic Axisymmetric Problems with Large Strains and Rotations," *Int. J. Sol. Struct.*, **23**, 1679–1698.

Rice, J. R. and Mukherjee, S. (1990). "Design Sensitivity Coefficients for Axisymmetric Elasticity Problems by Boundary Element Methods," *Eng. Anal. Boundary Elem.*, **7**, 13–20.

Richmond, O. and Devenpeck, M. L. (1962). "A Die Profile for Maximum Efficiency in Strip Drawing," *Proc. Fourth U.S. Nat. Cong. Appl. Mech.*, 1053–1057.

Sadegh, A. M. (1988–89). "On the Green's Functions and Boundary Integral Formulation of Elastic Planes with Cutouts," *Mech. Structures Machines*, **16**, 293–311.

Saigal, S. (1990). "Reanalysis for Structural Modifications in Boundary-element Response and Design Sensitivity Analysis," *AIAA J.*, **28**, 323–328.

Saigal, S., Borggaard, J. T., and Kane, J. H. (1989). "Boundary Element Implicit Differentiation Equations for Design Sensitivities of Axisymmetric Structures," *Int. J. Solids Structures*, **25**, 527–538.

Santos, J. L. T. and Choi, K. K. (1988). "Sizing Design Sensitivity Analysis of Nonlinear Structural Systems; Part II: Numerical Method," *Int. J. Num. Meth. Eng.*, **26**, 2097–2114.

Schittkowski, K. (1986). "NLPQL: a FORTRAN Subroutine Solving Constrained Nonlinear Programming Problems," *Ann. Oper. Res.*, **5**, 485–500.

Sokolowski, J. and Zolesio, J.-P. (1991). *Introduction to Shape Optimization—Shape Sensitivity Analysis*, Springer-Verlag, New York.

Timoshenko, S. P. and Goodier, J. N. (1970). *Theory of Elasticity*, 3d ed. McGraw-Hill, New York.

Tortorelli, D. A. (1992). "Sensitivity Analysis for Nonlinear Constrained Elastostatic Systems," *Int. J. Num. Meth. Eng.*, **33**, 1643–1660.

Vanderplaats, G. N. (1983). *Numerical Optimization Techniques for Engineering Design*. McGraw-Hill, New York.

Vidal, C. A., Lee, H.-S., and Haber, R. B. (1991). "The Consistent Tangent Operator for Design Sensitivity of History-dependent Response," *Comput. Syst. Eng.*, **2**, 509–523.

Wei, X., Leu, L.-J., Chandra, A., and Mukherjee, S. (1994). "Shape Optimization in Elasticity and Elasto-viscoplasticity by the Boundary Element Method," *Int. J. Solids Structures*, **31**, 533–550.

Wu, C. C. and Arora, J. S. (1988). "Design Sensitivity Analysis for Non-linear Buckling Load," *Comput. Mech.*, **3**, 129–140.

Zabaras, N., Mukherjee, S., and Richmond, O. (1988). "An Analysis of Inverse Heat Transfer Problems with Phase Changes Using an Integral Method," *ASME, J. Heat Transfer*, **110**, 554–561.

Zhang, Q. (1991). "Shape Design Sensitivity Analysis by the Boundary Element Method," Ph.D. thesis, Cornell University, Ithaca, N.Y.

Zhang, Q. and Mukherjee, S. (1991a). "Design Sensitivity Coefficients for Linear

Elastic Bodies with Zones and Corners by the Derivative Boundary Element Method," *Int. J. Solids Structures*, **27**, 983–998.

Zhang, Q. and Mukherjee, S. (1991b). "Second Order Design Sensitivity Analysis for Linear Elastic Problems by the Derivative Boundary Element Method," *Computer Meth. Appl. Mech. Eng.*, **86**, 321–335.

Zhang, Q., Mukherjee, S., and Chandra, A. (1992a). "Design Sensitivity Coefficients for Elasto-viscoplastic Problems by Boundary Element Methods," *Int. J. Num. Meth. Eng.*, **34**, 947–966.

Zhang, Q., Mukherjee, S., and Chandra, A. (1992b). "Shape Design Sensitivity Analysis for Geometrically and Materially Nonlinear Problems by the Boundary Element Method," *Int. J. Solids Structures*, **29**, 2503–2525.

Zhao, Z. and Adey, R. A. (1992). "A Numerical Study on the Elements of Shape Optimum Design," *Eng. Anal. Boundary Elem.*, **9**, 339–349.

5

Planar Forming Processes

Forming is one of the most common manufacturing processes. This chapter focuses on the applications of the boundary element method (BEM) to analyses of various planar forming processes, for example, plane strain extrusion, slab rolling, or profile rolling.

The boundary element formulations for elastoplastic and elasto-viscoplastic problems involving large strains and rotations are considered first. The elastic strains are assumed to remain small. The rotations and nonelastic strains, however, are allowed to be large. Bulk compressibility effects, which are often ignored for finite-strain problems of metallic solids, are explicitly included here.

An updated Lagrangian approach is adopted for the BEM analyses of forming processes. The frictional conditions at the tool–workpiece interface are introduced through a smeared interface element. For several forming problems, such as rolling, the frictional stresses change direction at the neutral region, and the location of this region is not known a priori. The effectiveness of the BEM in analyzing this class of planar forming problems involving both material and geometric nonlinearities with complicated interface conditions is discussed. Several practical applications are considered and issues relating to design sensitivities as well as optimization of these processes are discussed.

5.1 Introduction

Metal forming processes such as extrusion, rolling, and sheet metal forming generally subject the workpiece to finite strains and displacements. The components of elastic strain in these examples are generally limited to about 10^{-3} since the elastic moduli of metals are typically about three orders of magnitude larger than the yield stress. Thus, the nonelastic strain components, which can be of the order of unity, greatly dominate the elastic strains.

A considerable body of literature (e.g., McMeeking and Rice 1975, Lee et al. 1977, Onate and Zienkiewicz 1983, Chandra and Mukherjee 1984a,b, Dawson 1984, Kobayashi et al. 1989) exists where the finite element method has been used to analyze metal forming problems using rigid-plastic, elasto-plastic, flow-type, and elasto-viscoplastic material models. While the finite element method has been very successful in several engineering applications, it is quite sensitive to aspect ratios of individual elements. This requires remeshing where severe deformation is involved. Moreover, the secondary variables obtained through numerical differentiation in a finite element technique are inherently less accurate than the primary variables. In a typical displacement formulation, the secondary variables are stresses and, in a problem involving both material and geometric nonlinearities, the stresses at the present time essentially drive the problem through future time steps. If the deformation is volume conserving (as is typically the case for large plastic deformations), a FEM formulation (unless special care is taken) tends to lock and the discrete model appears to be much stiffer than the actual continua it is supposed to represent. These are the major limitations of typical FEM approaches that stand in the way of robust and cost-effective, yet accurate, FEM analyses of metal forming processes. The cost-effective FEM analyses (e.g., Onate and Zienkiewicz 1983, Dawson 1984, Kobayashi et al. 1989) typically treat the metal forming process as viscous flow. Thus, very critical quantities like residual stresses are difficult to recover from such analyses. Also, the viscous flow approaches are more suitable for steady-state processes. In many forming applications (e.g., forging), the transient phenomena are important. Even when the process is essentially steady state (e.g., extrusion), the transients may have significant influence on several characteristics (e.g., microstructural and damage evolutions, defect nucleations, residual stresses) of the final product.

Rolling is a very common metal working process in industry and has very interesting features. For more than half a century, both analytical and experimental investigations have been carried out on rolling. The slab method, based on a simplified equilibrium of force, was first suggested by von Kármán (1925). Orowan (1943) and Hill (1959), among others, derived approximate solutions of the equilibrium equations using various assumptions. In later years, slip-line [Alexander (1972), among others] and upper bound [Avitzur (1980) among others] methods were introduced to rolling analysis. The slab, slip-line, and upper bound methods have been widely used in theoretical analyses of metal forming in general and have gained popularity in industrial design of rolling mills. However, owing to the complexities of deformation involved in metal forming processes, particularly in rolling, various degrees of simplifications and idealizations have become necessary. Accordingly, the methods cited above have provided useful but limited information on metal deformation in rolling.

For rolling problems, the boundary conditions are typically not well known at the tool–workpiece interfaces. A unique feature of deformation in rolling is the existence of a neutral point (or region) along the roll–workpiece interface where the tangential relative velocity between the deforming material and the roll becomes zero. The frictional stresses change direction at the neutral point (or region). The location of this point (or region) is not known a priori, however, and this adds considerable complications to modeling of such processes. In order to determine the effects quantitatively, the relationship between the geometrical change of the workpiece and the friction condition at the interface must be established. This aspect has been investigated by various researchers. Avitzur (1969) and Lee and Altan (1972) used upper bound solutions, and Chen and Kobayashi (1978) developed a variational formulation for finite element analysis using a rigid-plastic material model.

In recent years, forming processes have also been used increasingly in manufacturing profiled workpieces. Some of the well-known advantages of profile forming over metal cutting are a shorter production period, reduced raw material requirements, and a more favorable fibrous and strain-hardened structure of the processed material. Among the different types of forming processes used to manufacture tooth-profiled workpieces, rolling has attained the broadest range of applications in the manufacturing industry. One of the most important applications of profile rolling is in gear manufacturing and cold rolling of involute profiles. It is a very effective and efficient way of manufacturing gears (Lange and Kurz 1984). However, the profile rolling process, conceived and developed in the factory, has not been investigated adequately. As a result, even today, the design of the profile rolling process is mostly empirical.

A treatment of the applications of the BEM to analysis and design sensitivity studies of this class of planar metal forming problems involving complicated interface conditions, in addition to material and geometric nonlinearities, is the main purpose of this chapter. Issues relating to design optimizations of such processes are then discussed in light of the insights gained from these analyses and design sensitivity studies.

5.2 Interface Conditions in Planar Forming Problems

A review of the planar BEM formulations for elastoplastic and elasto-viscoplastic problems involving large strains and rotations is presented in chapter 2 of this book. This section is devoted to a discussion of common interface conditions in forming processes and their implementation within a BEM formulation.

5.2.1 *General Equations*

The starting point here is the equation for the traction at a point on the boundary of a body $(i, j = 1, 2)$

$$\tau_i = \sigma_{ij} n_j = p(t) n_i + q(t) t_i \tag{5.1}$$

where $p(t)$ is the normal and $q(t)$ is the tangential component of the traction vector τ. This analysis is an extension of the follower load calculation that was discussed in section 2.2.2 of chapter 2.

Differentiating equation (5.1) with respect to time, using equation (2.124) for $\dot{\boldsymbol{n}}$ and an analogous equation for $\dot{\boldsymbol{t}}$ (Hill 1978),

$$\dot{t}_i = -(t_i t_k - \delta_{ik}) v_{k,j} t_j \tag{5.2}$$

one gets the equation

$$\begin{aligned} \dot{\sigma}_{ij} n_j = \dot{p}(t) n_i + \dot{q}(t) t_i - p(t) n_k v_{k,i} + q(t) v_{i,k} t_k \\ - q(t) t_i (n_m v_{m,k} n_k + t_m v_{m,k} t_k) + \sigma_{ij} n_m v_{m,j} \end{aligned} \tag{5.3}$$

Next, one uses the definition of the Jaumann rate of the Cauchy stress (equation 2.90) and the equation relating $\tau^{(s)}$ and $\tau^{(c)}$ (equation 2.98) to obtain

$$\tau_i^{(s)} = \dot{p}(t) n_i + \dot{q}(t) t_i - p(t) n_k v_{k,i} + q(t) v_{i,k} t_k + p(t) d_{kk} n_i \tag{5.4}$$

or, in direct form,

$$\boldsymbol{\tau}^{(s)} = \dot{p}(t)\boldsymbol{n} + \dot{q}(t)\boldsymbol{t} - p(t)\boldsymbol{n} \cdot \boldsymbol{\nabla v} + q(t)\frac{\partial \boldsymbol{v}}{\partial s} + p(t)\,\mathrm{tr}(\boldsymbol{d})\boldsymbol{n} \tag{5.5}$$

which, of course, reduces to equation (2.130) if $q(t) = 0$.

At this stage, it is useful to write equation (5.5) above, and related equations, in local Cartesian coordinates with base vectors e_α and e_β where e_α coincides with the outward unit normal and e_β coincides with the anticlockwise unit tangent vector at a point P on the boundary ∂B (figure 5.1).

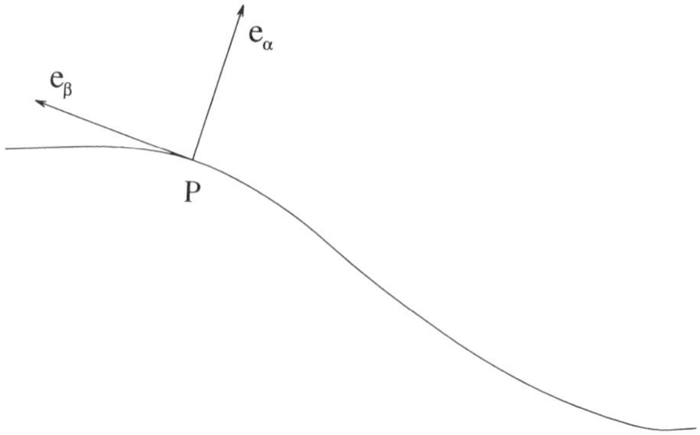

Figure 5.1. Local coordinate system for planar forming problem.

In these coordinates (no summation over repeated indices is assumed in the following for Greek indices)

$$p(t) = \tau_\alpha = \sigma_{\alpha\alpha}, \qquad q(t) = \tau_\beta = \sigma_{\alpha\beta} \tag{5.6}$$

and

$$\tau_\alpha^{(c)} = \hat{\sigma}_{\alpha\alpha}, \qquad \tau_\beta^{(c)} = \hat{\sigma}_{\alpha\beta} \tag{5.7}$$

Now, equation (5.4) in local coordinates becomes

$$\tau_\alpha^{(s)} = \dot{\sigma}_{\alpha\alpha} + \sigma_{\alpha\beta} h_{\beta\beta} + \sigma_{\alpha\beta} h_{\alpha\beta} \tag{5.8}$$

$$\tau_\beta^{(s)} = \dot{\sigma}_{\alpha\beta} - \sigma_{\alpha\beta} h_{\beta\beta} + \sigma_{\alpha\beta} h_{\beta\beta} \tag{5.9}$$

which are particularly simple.

It is also useful to transform equation (2.98) into these local coordinates. One gets

$$\tau_\alpha^{(s)} = \tau_\alpha^{(c)} - \sigma_{\alpha\beta} h_{\beta\alpha} + \sigma_{\alpha\alpha} h_{\beta\beta} \tag{5.10}$$

$$\tau_\beta^{(s)} = \tau_\beta^{(c)} - \sigma_{\alpha\alpha}\omega_{\alpha\beta} - \sigma_{\beta\beta} d_{\alpha\beta} + \sigma_{\alpha\beta} d_{\beta\beta} \tag{5.11}$$

Finally, eliminating $\tau_\alpha^{(s)}$ between equations (5.8) and (5.10) and $\tau_\beta^{(s)}$ between equations (5.9) and (5.11), leads to the simple equations for the Cauchy traction rates:

$$\tau_\alpha^{(c)} = \dot{\sigma}_{\alpha\alpha} + 2\sigma_{\alpha\beta} d_{\alpha\beta} \tag{5.12}$$

$$\tau_\beta^{(c)} = \dot{\sigma}_{\alpha\beta} + d_{\alpha\beta}(\sigma_{\beta\beta} - \sigma_{\alpha\alpha}) \tag{5.13}$$

Various physical situations can now be easily examined. In general, one is concerned with velocity components ν_α and ν_β and traction rate components $\tau_\alpha^{(s)}$ and $\tau_\beta^{(s)}$. In a plane strain BEM formulation [see equation (2.143)], two BEM equations exist at P. Two other equations relating the above four physical quantities are needed. A discussion of specific situations follows.

5.2.2 *Follower Load*

This situation arises when $q(t) = 0$ in equation (5.1). Physically, one gets this case, for example, for sheet forming under applied pressure when $p(t)$ is prescribed (Chandra and Mukherjee 1985) or extrusion in the absence of friction when $p(t) = \sigma_{\alpha\alpha}$ is unknown (Chandra and Mukherjee 1987). In the latter case, with a stationary die, $\nu_\alpha = 0$ as is discussed later.

With $\tau_\alpha = \sigma_{\alpha\alpha} = p$, $\tau_\beta = \sigma_{\alpha\beta} = q = 0$, equations (5.8) and (5.9) become

$$\tau_\alpha^{(s)} = \dot{p}(t) + p(t) h_{\beta\beta} \tag{5.14}$$

$$\tau_\beta^{(s)} = -p(t) h_{\alpha\beta} \tag{5.15}$$

5.2.3 *Sheet Forming*

One typically has to model either sticking or slipping conditions. Assuming Coulomb friction (other friction laws can also be considered), under sticking conditions, ν_α and ν_β are prescribed (equal to that of the punch). Under slipping conditions, ν_α is prescribed.

$$\tau_\beta = \pm\mu\tau_\alpha \qquad \text{or} \qquad \sigma_{\alpha\beta} = \pm\mu\sigma_{\alpha\alpha}$$

where μ is the coefficient of friction. The sign above, of course, is determined by the direction of $\tilde{\nu}_\beta$, the tangential velocity of the sheet relative to that of the punch.

Now, equations (5.8) and (5.9) become

$$\tau_\alpha^{(s)} = \dot{\sigma}_{\alpha\alpha} + \sigma_{\alpha\alpha} h_{\beta\beta} - h_{\alpha\beta}\left(\frac{\tilde{\nu}_\beta}{|\tilde{\nu}_\beta|}\right)\mu\sigma_{\alpha\alpha} \tag{5.16}$$

$$\tau_\beta^{(s)} = -\frac{\tilde{\nu}_\beta}{|\tilde{\nu}_\beta|}\mu\dot{\sigma}_{\alpha\alpha} - \sigma_{\alpha\alpha} h_{\beta\beta} - h_{\beta\beta}\left(\frac{\tilde{\nu}_\beta}{|\tilde{\nu}_\beta|}\right)\mu\sigma_{\alpha\alpha} \tag{5.17}$$

One can eliminate $\dot{\sigma}_{\alpha\alpha}$ between equations (5.16) and (5.17) to get a constraint equation between $\tau_\alpha^{(s)}$ and $\tau_\beta^{(s)}$.

Sticking conditions prevail as long as $|\tau_\beta| < \mu|\tau_\alpha|$. Otherwise one gets slip.

It is important to mention here that the stick–slip situation discussed in this section is generic in the sense that these equations can apply to other metal forming processes such as extrusion, rolling, and forging provided a Coulomb friction law is adopted. Equations (5.8) and (5.9), however, have not been used before to model metal forming by the BEM. Instead, equations (5.10) and (5.11) have been used, and now assumptions regarding the Cauchy traction rate $\tau_\beta^{(c)}$ (or some other assumptions) become necessary. These issues are discussed further in the following subsections.

5.2.4 *Extrusion*

It is assumed here that the die is stationary and slipping conditions always prevail. Now

$$\nu_\alpha = 0 \tag{5.18}$$

If Coulomb friction is assumed to prevail, one can use equations (5.16) and (5.17) above with $\widetilde{\nu}_\beta = \nu_\beta$ since the die is stationary.

It is also common to assume a friction law, in rate form for the Cauchy traction $\tau_\beta^{(c)}$ (Chandra and Mukherjee 1987). In this case, it is convenient to employ equation (5.11) for $\tau_\beta^{(s)}$ in terms of $\tau_\beta^{(c)}$. Possibilities for friction laws are

(a) $$\tau_\beta^{(c)} = \hat{\sigma}_{\alpha\beta} = c_1 \nu_\beta \tag{5.19}$$

(b) $$\tau_\beta^{(c)} = \hat{\sigma}_{\alpha\beta} = c_2 h_{\beta\alpha} \tag{5.20}$$

The second case assumes that the Cauchy traction rate is proportional to the velocity gradient $\nu_{\beta,\alpha}$.

5.2.5 *Slab Rolling*

Physically, at the start of the roll bite, the workpiece is stationary. Hence, slipping conditions prevail with

$$\tau_\beta = \frac{\nu_\beta^{(R)}}{|\nu_\beta^{(R)}|} \mu \tau_\alpha \tag{5.21}$$

where $\nu_\beta^{(R)}$ is the tangential velocity of the roll relative to workpiece. As the rolling process continues, the workpiece accelerates until it locally reaches the velocity of the roll. This situation is called the neutral point or region.

At the neutral point, $\tau_\beta = 0$. Beyond this point, the workpiece tends to move faster than the roll and the direction of τ_β reverses since the roll imparts a braking action on the workpiece.

Chen and Kobayashi (1978) and Li and Kobayashi (1982) [see also Chandra (1989)] have proposed the following equations for the rolling process:

$$\widetilde{\nu}_\alpha = 0 \tag{5.22}$$

$$\tau_\beta = -\frac{\widetilde{\nu}_\beta}{|\widetilde{\nu}_\beta|}\mu\tau_\alpha\left[\frac{2}{\pi}\tan^{-1}\left(\frac{|\widetilde{\nu}_\beta|}{a}\right)\right] \tag{5.23}$$

where $\widetilde{\boldsymbol{v}}$ is the velocity of the workpiece relative to that of the roll at P and a is a constant which is several orders of magnitude smaller than $\nu_\beta^{(R)}$. Thus, this equation approximates equation (5.21) at the start of the roll bite and also models the neutral points when $\widetilde{\nu}_\beta = 0$.

One can obtain traction rates in a rate formulation by differentiating equation (5.23) with respect to time. This is not convenient, since acceleration would appear in the resulting equations. One option is to discard equation (5.23) and employ, instead, the stick–slip equations discussed above for sheet forming. Alternatively, one can numerically estimate $\tau_\beta^{(c)}$ from equation (5.7) and use this estimate in equation (5.11) for $\tau_\beta^{(s)}$. Details are given in Chandra (1989a,b).

5.3 Numerical Implementation for Planar Cases

Most of the earlier procedures regarding numerical implementation of the small-deformation BEM equations (Banerjee and Butterfield 1981, Mukherjee 1982) also apply here to the case of large deformation. A discretized version of equation (2.118) for plane stress or plane strain can now be obtained as

$$\begin{aligned} C_{ij}(P_M)v_i(P_M) &= \sum_{N_S}\int_{\Delta S_N}[U_{ij}(P_M,Q)\tau_i^{(s)}(Q) - T_{ij}(P_M,Q)v_i(Q)]\,dS(Q) \\ &+ \sum_{n_i}\int_{\Delta A_n}[\lambda U_{ij,i}(P_M,q)\,d_{kk}^{(n)}(q) \\ &+ 2GU_{ij,k}(P_M,q)\,d_{ik}^{(n)}(q)]\,dV(q) \\ &+ \sum_{n_i}\int_{\Delta A_n}U_{ij,m}(P_M,q)g_{mi}(q)\,dV(q) \end{aligned} \tag{5.24}$$

where the boundary of the body in the reference configuration ∂B^0 is divided into N_S boundary segments and the interior into n_i internal cells, and $v_i(P_M)$ are the components of velocities at a point P which coincides with node M. Suitable shape functions must now be chosen for the variation of tractions and velocities on the surface element ΔS_N and the variation of the nonelastic strain rates and velocities over an internal cell ΔA_n. Following essentially the same procedure as that for small-strain problems, one obtains the algebraic system

$$[A]\{v\} + [B]\{\tau^{(s)}\} = \{b\} \tag{5.25}$$

where the coefficient matrices $[A]$ and $[B]$ contain integrals of the kernels and shape functions. The vector $\{b\}$ contains the contributions of various quantities from the three domain integrals. For a well-posed problem, half of the set of variables $\{v\}$ and $\{\tau^{(s)}\}$ are known. The unknown half of the set is then obtained in terms of the known half. Equations for the velocity field (2.117) and the velocity gradients (2.135) at an internal point can also be discretized in a similar fashion. It should be noted that, once boundary velocities and traction rates are known, velocities as well as velocity gradients at internal points can be obtained through algebraic evaluations without requiring any further matrix solutions.

5.3.1 *Objective Stress Rates for Problems Involving Large Shear Strains*

Many forming processes such as slab rolling and profile rolling may involve very large shear strains. For such problems involving induced material anisotropy, it is well known (e.g., Dienes 1979, Nagtegaal and De Jong 1982, Lee et al. 1983, Atluri 1984, Johnson and Bamman 1984, Chandra and Mukherjee 1986a) that integration of material rates of stresses obtained from their Jaumann derivatives results in spurious oscillations of shear stresses for monotonically increasing shear strains. This stems from the use of the spin tensor ω in the definition of Jaumann rate (equation 2.85).

In order to avoid any spurious results for forming problems involving large shear strains, the stress integration strategy proposed by Chandra and Mukherjee is used here.

This scheme may be illustrated by viewing $\hat{\boldsymbol{\sigma}}(=\phi(\boldsymbol{d}))$ in equation (2.89) as an unspecified objective rate (instead of Jaumann rate) of stress. We can proceed from $\phi(\boldsymbol{d})$ to obtain the components of Cauchy stress in a desired global spatially fixed basis by using the equation

$$\sigma(t) = R(t)\sigma(t-\Delta t)R^{\mathrm{T}}(t) + R(t)\left[\int_{t-\Delta t}^{t} R^{\mathrm{T}}(\tau)\phi(d)R(\tau)d\tau\right]R^{\mathrm{T}}(t) \tag{5.26}$$

where $\boldsymbol{F} = \partial \boldsymbol{x}/\partial \boldsymbol{X} = \boldsymbol{R} \cdot \boldsymbol{U}$; $\sigma(t - \Delta t)$ is the value of the Cauchy stress at time $t - \Delta t$; and $\boldsymbol{R}^{\mathrm{T}}$ denotes the transpose of $\boldsymbol{R}$. The symbols in the above equation are matrices corresponding to the appropriate tensors.

The rationale behind this proposal is to observe that the expression $R^{\mathrm{T}} \phi(d) R$ delivers the components of $\phi(\boldsymbol{d})$ in a local basis which is rotating with respect to the fixed global basis with R the measure of this rotation. Integration is then carried out with respect to an observer in the rotating basis. Finally, premultiplication by R and postmultiplication by R^{T}, at any time, delivers the Cauchy stress components in the desired global basis.

5.3.1.1 Relationship with the Dienes rate (Dienes 1979)

Defining the Cauchy stress $\bar{\sigma}$ in the rotating basis (Johnson and Bamman 1984) and using equation (5.26),

$$\bar{\sigma} = R^{\mathrm{T}} \sigma R = \sigma(t - \Delta t) + \int_{t-\Delta t}^{t} R^{\mathrm{T}} \phi(d) R \, d\tau \tag{5.27}$$

Proceeding in a manner similar to Dienes (1979), one differentiates equation (5.27) with respect to time. Comparing these expressions and taking note of equation (2.89) results in the equation

$$\phi(\boldsymbol{d}) = \hat{\boldsymbol{\sigma}} = \dot{\boldsymbol{\sigma}} - \boldsymbol{\Omega}\boldsymbol{\sigma} + \boldsymbol{\sigma}\boldsymbol{\Omega} \tag{5.28}$$

so that $\hat{\boldsymbol{\sigma}}$, for this hypoelastic model, is the Dienes rate of the Cauchy stress, with $\boldsymbol{\Omega} = \dot{\boldsymbol{R}}\boldsymbol{R}^{\mathrm{T}}$.

Thus, equation (5.26) can be regarded as an integral form of equation (2.89) and helps to clarify the physical interpretation of the Dienes rate. Goddard and Miller (1966) also presents an equation similar to equation (5.26) for the inverse of Jaumann derivative. This rate is also called the Green–Naghdi rate by Johnson and Bamman (1984).

5.3.1.2 Relationship with Rolph and Bathe's model (Willam 1984)

Starting from the relationship

$$2\boldsymbol{d} = \dot{\boldsymbol{F}} \cdot \boldsymbol{F}^{-1} + \boldsymbol{F}^{-\mathrm{T}} \cdot \dot{\boldsymbol{F}}^{\mathrm{T}} \tag{5.29}$$

and using polar decomposition of $\boldsymbol{F}$,

$$2\boldsymbol{d} = \boldsymbol{R} \cdot (\dot{\boldsymbol{U}} \cdot \boldsymbol{U}^{-1} + \boldsymbol{U}^{-1} \cdot \dot{\boldsymbol{U}}) \cdot \boldsymbol{R}^{\mathrm{T}} \tag{5.30}$$

This time a special situation is considered in which the directions of principal stretches remain fixed in the body during deformation. Thus, one may decompose $U = Q\Lambda(t)Q^{\mathrm{T}}$, where $\Lambda(t)$ is diagonal and Q is orthogonal but independent of time. In this case, one can show that

$$R^{\mathrm{T}}DR = \frac{D}{Dt}(\ln U) \tag{5.31}$$

where $\ln U = Q \ln \Lambda Q^{\mathrm{T}}$, so that, from equation (2.89),

$$R^{\mathrm{T}}\phi(D)R = \frac{D}{Dt}[\lambda \operatorname{tr}(\ln U)I + 2G \ln U] \tag{5.32}$$

From equation (5.27), the left-hand side of the above equation (5.32) equals $\dot{\bar{\sigma}}$. Hence,

$$\sigma = \lambda(\operatorname{tr} E)I + 2GE \tag{5.33}$$

where $E = R \ln UR^{\mathrm{T}}$ is the logarithmic or Hencky strain. The above equation is the model of Rolph and Bathe (Willam 1984). Thus, it can be shown that for an isotropic linear hypoelastic material, for deformations in which the directions of principal stretches remain fixed in the body, equations (2.89), (5.26) and (5.33) are all equivalent. Equations (2.89) and (5.26), of course, are equivalent under more general conditions as shown before.

5.3.1.3 Elastoplasticity with finite rotations

Elastoplastic problems in materials exhibiting induced anisotropy typically involve tensors such as the back stress α. A typical evolution equation for small-strain small rotation elastoplasticity might be of the form (Lee et al. 1983)

$$\dot{\alpha} = g[\dot{\bar{\varepsilon}}^{(p)}]\, \dot{\varepsilon}^{(p)} \tag{5.34}$$

where $\dot{\varepsilon}^{(p)}$ is the plastic strain rate and $\dot{\varepsilon}^{(p)}$ is a suitable invariant of $\dot{\varepsilon}^{(p)}$. Large-strain generalizations of equation (5.12) usually involve replacing $\dot{\alpha}$ with a suitable objective rate of α and $\dot{\varepsilon}^{(p)}$ with $\boldsymbol{d}^{(p)}$, the plastic part of the rate of deformation tensor. Thus, the evolution of such tensors must be considered in addition to that of the stress. Also, in general, the function $\phi(\boldsymbol{d})$ in the hypoelastic law (2.89) might involve the stress as well.

It is proposed that for such problems, a modified form of equation (2.89) with elastic rotations $\boldsymbol{R}^{(e)}$ be adopted. Thus, for a small time step Δt, one may write

$$\bar{T}_t = \bar{T}_{t-\Delta t} + [R^{(e)T}HR^{(e)}]_{t-\Delta t}\Delta t \tag{5.35}$$

$$T_t = [R^{(e)}\bar{T}R^{(e)T}]_t \tag{5.36}$$

where the tensor $\boldsymbol{T}$ can be the Cauchy stress, the back stress or some other suitable tensor internal variable and $\boldsymbol{H}$ is a function such that for small-strain small-rotation problems $\dot{\boldsymbol{T}} = \boldsymbol{H}$.

The function $\boldsymbol{H}$ must be suitably interpreted by replacing $\dot{\varepsilon}^{(e)}$ by $d^{(e)}$, and so on. Also, Δt is a small time increment. The above equations are, strictly speaking, correct in the limit $\Delta t \to 0$. Operationally, of course, one must use small, finite time increments Δt. Also, these equations must be used in a march forward time integration procedure.

The need for the use of elastic rotations arises from the nature of elastic and plastic deformations. As with elastic strains, elastic rotations with respect to a virgin configuration are remembered by the solid. It is common to assume that for plasticity analysis it is not necessary to use variables involving the virgin configuration of the material prior to plastic flow (Lee et al. 1983).

Various researchers have been looking into constitutive descriptions for plastic and elastic rotations. This research is still at an early stage. Thus, for the present work, it is assumed that $\boldsymbol{R} = \boldsymbol{R}^e$. Further, it may be shown (Atluri 1984, Chandra and Mukherjee 1986a) that

$$\boldsymbol{\Omega} = \boldsymbol{\omega} - \tfrac{1}{2}\boldsymbol{R} \cdot [\dot{\boldsymbol{U}} \cdot \boldsymbol{U}^{-1} - \boldsymbol{U}^{-1} \cdot \dot{\boldsymbol{U}}] \cdot \boldsymbol{R}^{\mathrm{T}} \tag{5.37}$$

so that in the updated Lagrangian frame, with $\boldsymbol{R} = \boldsymbol{U} = \boldsymbol{I}$, we get $\boldsymbol{\Omega} = \boldsymbol{\omega}$. Thus, the BEM equations developed in chapter 2 still remains valid, in an updated Lagrangian frame of reference, with $\hat{\boldsymbol{\sigma}}(= \phi(\boldsymbol{d}))$ in equation (2.89) viewed as the Dienes rate. Integrations of stresses and proper tensors used in the state variable representation of material behavior are now carried out using equations (5.35) and (5.36).

5.3.1.4 Solution strategy

A solution strategy for these planar forming problems may now be developed. This can be best explained by referring to figure 5.2.

The presence of velocity gradients in the boundary traction rates and in some of the domain integrals requires iterations within each time step. As mentioned before, the unknown velocity gradient components occur in a domain integral for elastoplastic problems, even for the case of small strain. Here, this situation occurs whether elastoplastic or elasto-viscoplastic constitutive models are used. The solution strategy for these problems can be described as follows:

1. The elasticity problem is first solved to obtain initial displacements. For this step, equation (5.25) is solved with $\{b\} = 0$ for the unknown components of displacements and tractions in terms of the prescribed ones.

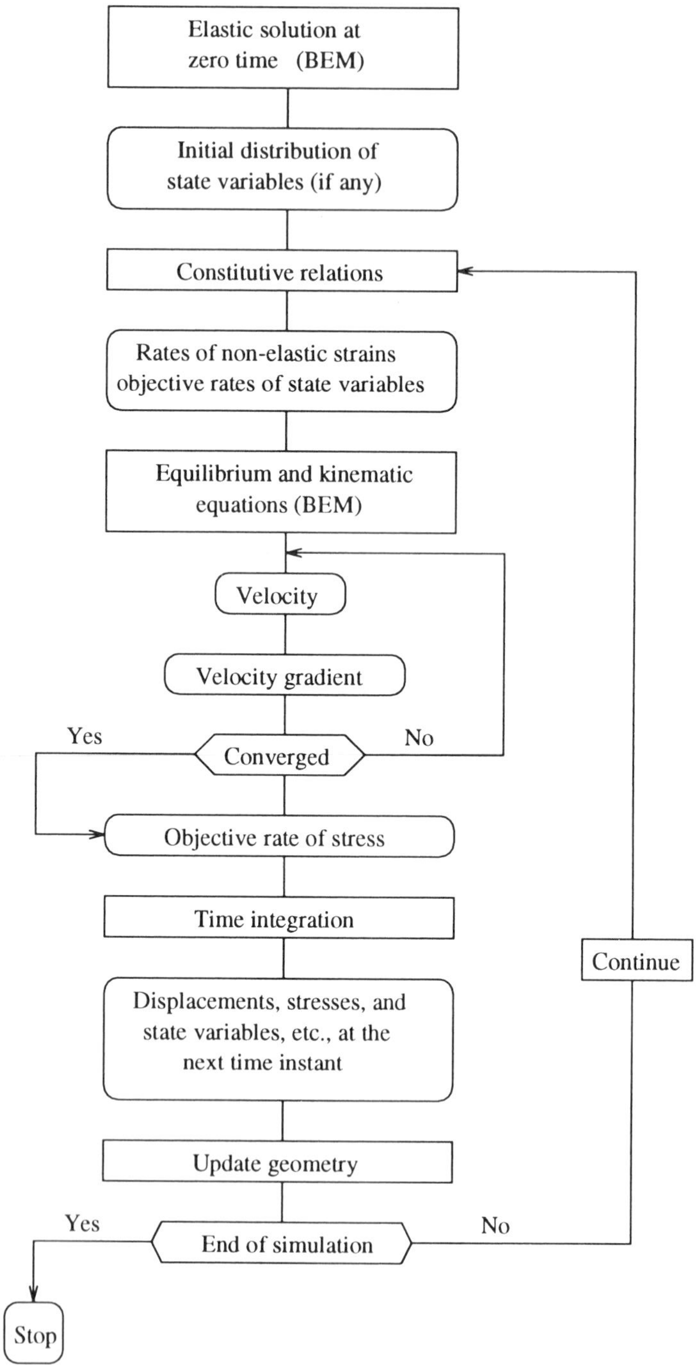

Figure 5.2. Flow chart for the solution of large-strain elasto-viscoplastic problems.

2. The initial values of the displacement gradients are obtained from a truncated version of equation (2.135) with $d_{ik}^{(n)} = 0$, $v_{k,n} = 0$, and $\boldsymbol{v}$ and $\boldsymbol{\tau}^{(s)}$ replaced by $\boldsymbol{u}$ and $\boldsymbol{\tau}$ in the rest of the equation. The initial stresses are determined from strains through Hooke's law. The displacement gradients are used to form the deformation gradient matrix $\boldsymbol{F}$. This matrix is decomposed according to the polar decomposition theorem $\boldsymbol{F} = \boldsymbol{RU}$, and $\boldsymbol{R}$ is obtained.
3. The tensor $d_{ik}^{(n)}$ at zero time is obtained from the constitutive equations.
4. The first approximation to $v_j(P)$ at $t = 0$ is calculated from the boundary equation for velocity with $v_{k,l}$ set to zero. It must be noted that the velocity gradients occur in two of the domain integrals, as well as in the expression for $\dot{\tau}_i$.
5. The first approximation to $v_{j,T}(p)$ at $t = 0$ is obtained from the velocity gradient equation with all the velocity gradients in the integrals on the right-hand side set to zero. The $v_{j,T}(p)$ values are extrapolated to get $v_{j,T}(P)$.
6. The velocity gradients are inserted into the boundary equation and the full equation is solved to determine a second approximation to $v_j(P)$. This $v_j(P)$ and the first approximation of velocity gradients are then used to obtain a second approximation for $v_{j,T}(p)$.
7. Step (6) is repeated until the required convergence is achieved and $v_j(P)$ is determined at zero time.
8. The iterations are completed with $v_j(p)$, $v_{j,l}(p)$, and $\hat{\sigma}_{ji}$, and $\hat{\varepsilon}_{ij}^{(a)}$ are calculated in the fixed global basis at zero time. The material derivative of $\boldsymbol{F}$ is determined from the equation $\dot{\boldsymbol{F}} = \boldsymbol{LF}$, where $L_{ij} = v_{i,j}$.
9. Time integration is performed next. An explicit Euler-type scheme with proper time step controls (Kumar et al. 1980) is used to find the relevant quantities, including $\boldsymbol{F}$, at time Δt. $\boldsymbol{F}$ is decomposed into $\boldsymbol{RU}$ at time Δt, and $\boldsymbol{R}$ is obtained at Δt.
10. The objective rates of the Cauchy stress and the anelastic strain are integrated in time. Here,

$$[\bar{\sigma}]_{t=\Delta t} = [\bar{\sigma}]_{t=0} + ([R]^{\mathrm{T}}_{t=\Delta t}[\hat{\sigma}]_{t=0}[R]_{t=\Delta t})\,\Delta t \tag{5.38}$$

where $[\bar{\sigma}]$ represents the stress components in a frame that has rotated by $[R]$.

The Cauchy stress components in the laboratory frame may then be obtained as

$$[\sigma]_{t=\Delta t} = ([R][\bar{\sigma}][R]^{\mathrm{T}})_{t=\Delta t} \tag{5.39}$$

and similarly for $[\varepsilon^{(a)}]$.

Thus, the relevant quantities (displacements, displacement gradients, stresses, anelastic strains, etc.) are found at $t = \Delta t$. The computation then marches forward in time by suitably updating geometry and kernels. For cases involving relatively small shear strains, time marching can also be carried out with the material derivatives of the tensor (obtained from objective, e.g., Jaumann, derivatives).

In many forming processes, the critical stresses are the stresses on the boundary. The most convenient way to find stress rate and velocity gradients at a point P on the boundary ∂B (where it is locally smooth) of a planar body is to use the following equations:

$$\hat{\sigma}_{ij} = \lambda(v_{k,k} - d_{kk}^{(n)})\,\delta_{ij} + G(v_{i,j} + v_{j,i}) - 2Gd_{ij}^{(n)} \tag{5.40}$$

$$\tau_i^{(c)} = n_j \hat{\sigma}_{ji} \tag{5.41}$$

$$\frac{\partial v_i}{\partial s} = v_{i,j} t_j \tag{5.42}$$

Once the velocity field on the boundary is known, the tangential derivative $\partial v_i / \partial s$ at P is obtained by numerically differentiating to velocity field along the boundary. Then, the set of equations (5.40)–(5.42) involves seven unknowns ($\hat{\sigma}_{11}$, $\hat{\sigma}_{12}$, $\hat{\sigma}_{22}$, $v_{1,1}$, $v_{1,2}$, $v_{2,1}$, $v_{2,2}$) and seven equations. The boundary stress rates $\hat{\sigma}_{ij}$ and velocity gradients $v_{i,j}$ at any point P on the boundary can be obtained by solving this 7×7 system at that point. The material rate of the Cauchy stress can then be obtained from its Jaumann rate (identifying $\hat{\sigma}_{ij}$ as the Jaumann rate) by using equation (2.90). The stretch $\boldsymbol{d}$ and spin $\boldsymbol{\omega}$ can also easily be obtained. Alternatively, $\hat{\sigma}_{ij}$ may be identified with the Dienes rate (Chandra and Mukherjee 1986a,b) and integrated through equations (5.35) and (5.36). In that case, the rotation rate $\dot{\boldsymbol{R}}$ is distinct from the spin $\boldsymbol{\omega}$. However, both of them can easily be calculated.

5.4 Applications to Forming Problems

Applications of the above-cited BEM formulations to various planar forming problems are presented in this section. Interface modeling processes are discussed and numerical results obtained by BEM are presented.

For planar problems, the BEM program uses straight boundary elements and polygonal internal cells. The velocity and traction rate are taken to be piecewise linear over the boundary elements, while the nonelastic deformation rate and the velocity gradients are assumed to be piecewise constant over the internal cells. The values of the boundary variables are assigned at nodes that lie at the intersections of boundary segments. Possible discontinuities

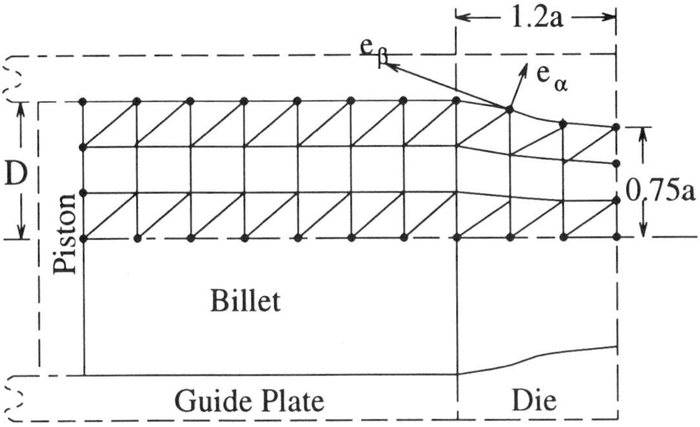

Figure 5.3. Geometry of the plane strain extrusion problem.

in tractions are taken care of by placing a "zero length" element between nodes and assigning different values of traction at each of those nodes. All integrations of kernels are carried out analytically.

5.4.1 Plane Strain Extrusion

Extrusion is a commonly used forming process. The shearing strains in an extrusion process are much smaller than unity. Accordingly, Jaumann rates utilizing the spin tensor can be used for objective rates of tensor quantities.

Interface conditions at the die–workpiece boundary are considered here for problems of plane strain extrusion. The assumptions are best explained in terms of a local coordinate system (α, β, γ) shown in figure 5.3. The origin of this coordinate system is positioned on the die–workpiece interface. The α axis is tangential to the die surface and the α axis is the outward normal to the die surface.

A consequence of the assumption of plane strain is that

$$v_\gamma = 0, \sigma_{\alpha\gamma} = \sigma_{\gamma\alpha} = \sigma_{\beta\gamma} = \sigma_{\gamma\beta} = 0 \tag{5.43}$$

It is further assumed that the contact or lubricant layer adjacent to the die surface cannot provide any resistance to tensile or compressive deformation. Thus, the scheme is equivalent to having an interface or bond element with zero stiffness in the direction tangential to the contact surface. Therefore,

$$\sigma_{\beta\beta} = 0 \tag{5.44}$$

so that only pressures and shear loadings get transferred across the interface and the stretch or compression of either the die or the workpiece is not transferred to the other. Another assumption made here is that the material is incompressible, so that

$$d_{\alpha\alpha}+d_{\beta\beta}=0 \qquad (\text{since } d_{\gamma\gamma}=0) \tag{5.45}$$

Using the above assumptions, zero normal velocity ($v_\alpha = 0$), and the fact that the normal has components (1,0,0) in the local coordinate system, the load correction equation in the local coordinate system becomes

$$\tau_\beta^{(s)} = \hat{\sigma}_{\alpha\beta} - \sigma_{\alpha\alpha}\omega_{\alpha\beta} + \sigma_{\alpha\beta}d_{\beta\beta} \tag{5.46}$$

where a superposed hat (∧) indicates the Jaumann rate. The other traction component is assumed to vanish, that is,

$$\tau_\alpha^{(c)} = 0 \tag{5.47}$$

The final assumptions relate to friction,

$$\hat{\sigma}_{\alpha\beta} = \frac{G_s}{h}\mu v_\alpha \tag{5.48}$$

(where g_s and h are the shear modulus and the height of the interface element, respectively, and μ is the coefficient of friction), and

$$\omega_{\alpha\beta} = -\kappa v_\beta \tag{5.49}$$

where κ is the local curvature.

The rate of the traction component, $\tau_\beta^{(s)}$, has now been obtained in terms of the tangential velocity v_β, its gradient in the tangential direction, $v_{\beta,\beta}$, and the nonzero stress components, $\sigma_{\alpha\beta}$ and $\sigma_{\alpha\alpha}$.

5.4.1.1 Numerical results for plane strain extrusion

Numerical results obtained from the BEM analyses of plane strain extrusion problems are presented in this section. These problems involve both geometric and material nonlinearities. Moreover, handling of the boundary conditions at the die interface requires special care for the extrusion problem. In the present BEM analysis, this can be incorporated quite easily through "load correction" (equations 5.46–5.49). The material used is commercially pure aluminum at 24°C. The material parameters for Hart's model along with the experimental basis can be found in Alexopoulos (1981). The parameter $G_s/h = 1.268 \times 10^5$ MPa/m.

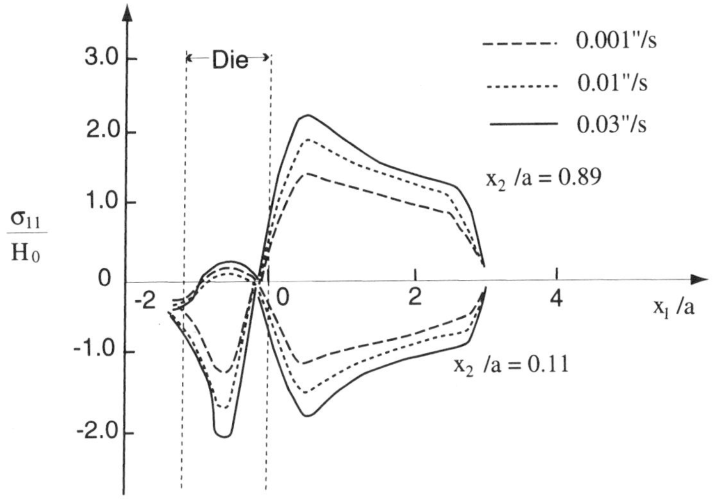

Figure 5.4. Steady-state distribution of σ_{11} at different piston velocities for two values of $X_2(\mu = 0)$.

The geometry of planar extrusion considered here is shown in figure 5.3, and numerical results are presented in figures 5.4–5.7. Figure 5.4 shows the steady-state distributions of the longitudinal stress σ_{11} for three different velocities in the absence of friction. In particular, the centerline of the workpiece is chosen to be the x_1 axis and the stress distributions are shown for material points in the deformed configuration, which initially had the same relative ordinate ($x_2/a = 0.11$ and 0.89) in the billet. Maximum residual tensile stress in an extruded workpiece is of crucial importance in design, since this is the primary potential source for crack initiation and growth. It is seen from figure 5.4 that the rate dependence of this quantity is quite significant. As the piston velocity is tripled from 0.254 mm/s to 0.762 mm/s, the BEM analysis predicts a change of 15.5 percent in the maximum longitudinal tensile stress in the workpiece. This compares well with the 17 percent change predicted by the FEM analysis (Chandra and Mukherjee 1984b). The faster the billet is forced through the die, the less time there is for stresses to relax at material points in the workpiece as they move through the die. Consequently, the maximum longitudinal tensile stress upon exit from the die increases substantially with the speed of extrusion. Predictions of crucial effects like rate dependence of residual stresses is only possible through a detailed analysis using a realistic elasto-viscoplastic model for material behavior.

Another important feature of elasto-viscoplastic analysis of plane strain extrusion is that, following a peak value, the magnitude of σ_{11} decreases as

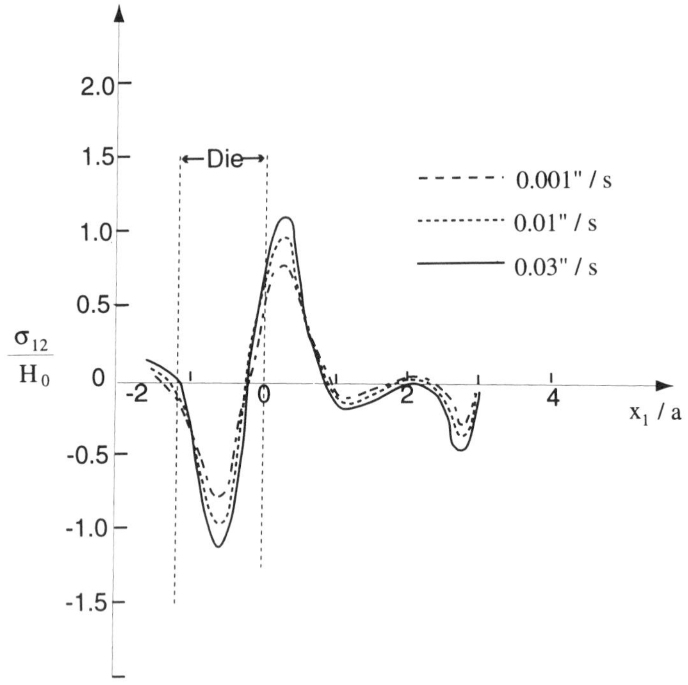

Figure 5.5. Steady-state distribution of σ_{12} as functions of piston velocity ($\mu = 0$).

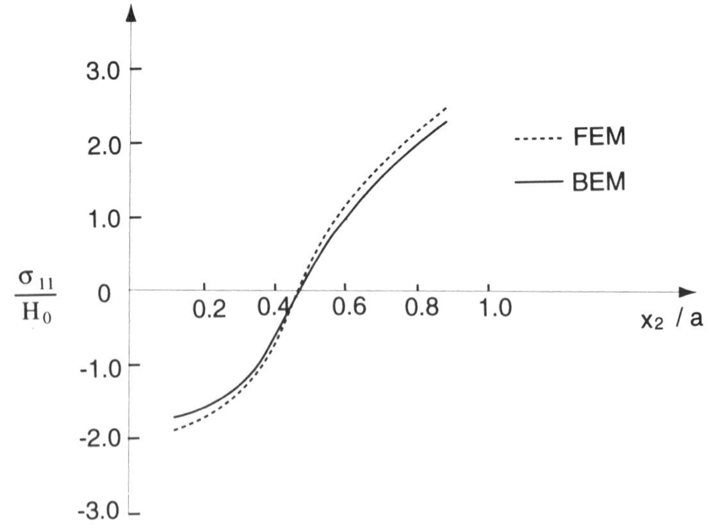

Figure 5.6. Comparisons of residual stress distributions from BEM and FEM (piston velocity = 0.762 mm/s, $\mu = 0$).

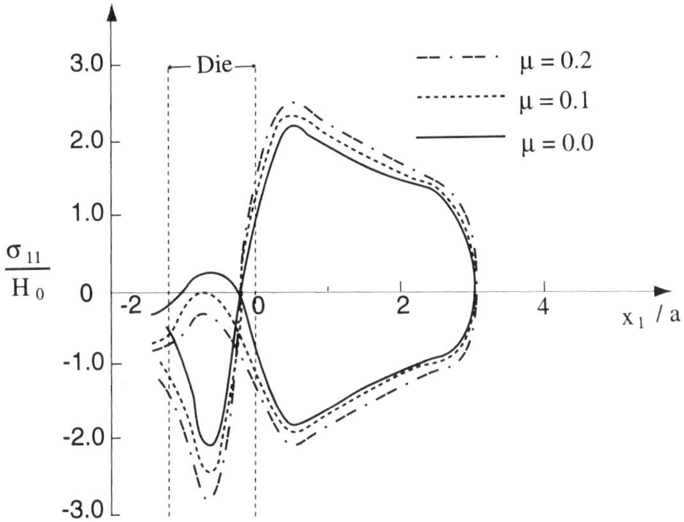

Figure 5.7. Steady-state distributions of σ_{11} as functions of friction coefficient for two values of X_2 (piston velocity = 0.762 mm/s).

a function of x_1 in most of the billet that has passed through the die. This is a result of stress relaxation in the workpiece after it is deformed, and the BEM predictions for such relaxations compare very well with those predicted by FEM analysis (Chandra and Mukherjee 1984b).

The results for the steady-state distributions of shearing stress σ_{12} are shown in figure 5.5. It is seen that there is a marked variation of shearing stress inside and in the neighborhood of the die. Residual longitudinal stress distributions over a cross-section ($x_1 = 0.4a$ from the die exit) from the two methods are shown in figure 5.6. It should be noted that these residual stress distributions must be self-equilibrating and, for the residual stresses obtained from the current BEM analysis, the error in satisfying equilibrium is less than 5 percent.

The effect of friction on the results is depicted in figure 5.7. Three different values of friction are chosen for a piston velocity of 0.762 mm/s. The presence of friction increases the longitudinal stress peaks significantly. Of course, the friction model with constant coefficient of friction that is used here is crude at best, and much work remains to be done on modeling friction in metal forming processes. The present analysis does show, however, the effects of friction. The BEM predictions match quite well with FEM predictions of Chandra and Mukherjee (1984c).

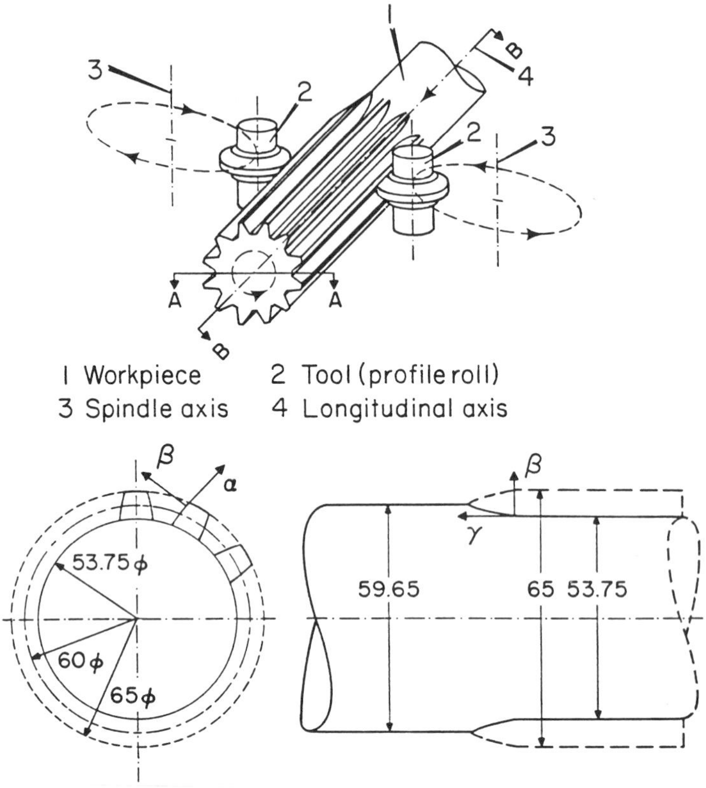

Figure 5.8. Schematic diagram of the profile rolling process.

5.4.2 Profile Rolling of Gears

The problem of profile rolling is considered next. This forming process involves rolling contact. Moreover, the shear strains can be high. Accordingly, care needs to be taken to avoid any spurious stress oscillations.

Boundary conditions at the tool–workpiece interface are considered next. The working principle of the cold profile rolling process is shown in figure 5.8. Both rolling tools have the profile of the tooth space to be formed on the workpiece. These two profiled rolls rotate like planets about their orbital axes and can rotate either in the same direction (down-profile rolling) or in opposite directions (up-profile rolling) with respect to the feed of the workpiece. Due to this arrangement, the tools come into brief contact with the workpiece. The contact, occurring as a sudden stroke for a short period, is made once per tool revolution about the orbital axis. The work material is pushed radially from the tooth base toward the tooth top. During this forming operation, the workpiece does not rotate. As the tool moves away from it, the workpiece is turned by a tooth pitch and can also be advanced simultaneously in the longitudinal direction. Hence, the tool rolls and slides on the workpiece, and it is this combination of rolling and sliding that makes interface modeling quite difficult for profile rolling.

A considerable body of literature (Campos et al. 1982, Martins and Oden 1983, Pires and Oden 1983, Torstenfelt 1983, Bathe and Chaudhary 1985) exists where sliding contact is handled in the FEM formulation by modifying the functional used in the variational principle. This typically introduces a Lagrange multiplier, or a penalty function, and gives rise to a mixed formulation. Zolti (1983) introduces an orthotropic gap element. It has been shown by Chandra and Mukherjee (1984b,c) that a gap or interface element can also be simulated within the original finite element through load correction. This smearing of the interface element keeps the geometric modeling simple (no need for extra interface elements) and retains the advantages of displacement formulations in FEM (positive definite stiffness matrix, ease of specifying boundary conditions, etc.).

Mukherjee and Chandra (1987) also applied the load correction technique to BEM. Since both displacements and tractions appear as unknowns, a BEM formulation is essentially a mixed one and is easily applicable in a contact situation. Here, the strategy of Mukherjee and Chandra is extended to include rolling contact situations as well.

The particular rolling contact model used here is the one proposed by Bhargava et al. (1985) and is based on the theory of Merwin and Johnson (1963). It was observed by Bhargava et al. that, for indentation loading (without rolling), the normal displacements under the center of contact are substantially larger than predicted by the Hertzian theory. This larger displacement must be accompanied by an increase in the contact width in order to maintain continuity of the contacting surfaces. Also, the Merwin and Johnson analysis makes no attempt to satisfy the equilibrium requirements while rolling. The stress components acting at any depth below the surface are uncoupled from those acting at any other depth. Thus, any region that remains elastic in the Merwin and Johnson analysis is devoid of residual stresses, and equilibrium for a plane surface free from traction is attained at the end of every contact sequence by completely relaxing elastically the radial and shear residual stresses. Bhargava et al. make the analysis rigorous with respect to equilibrium and continuity requirements. The normal displacement differences of points at the center and edges of the two surfaces are required to be the same. This three-point continuity requirement increases the contact width. Accordingly, the Hertzian pressure distribution is modified. The total applied load, however, remains unaltered:

$$P = \int_{-a'}^{a'} p_0' \left[1 - \left(\frac{x^2}{a'^2} \right) \right]^{0.5} dx = \frac{\pi p_0' a'}{2} = \frac{\pi p_0 a}{2} \tag{5.50}$$

Here, P is the total applied load; p_0 and $2a$ are the Hertzian peak pressure and contact width, respectively; p'_0 is the modified pressure peak; and $2a'$ is the modified contact width.

The interface conditions for profile rolling of gears can be best explained in terms of a local coordinate system (α, β, γ). The origin of this coordinate system is positioned at the tool–workpiece interface. Figure 5.8b shows a section of the workpiece cut perpendicular to the direction of the feed. The β axis is tangential to the interface at this section, and the α axis is outward normal to the workpiece surface at this section. The γ axis is perpendicular to this section and is along the direction of the axial feed of the workpiece. Figure 5.8c shows another section of the workpiece parallel to the direction of the feed. As seen in figure 5.8c, the γ axis is also tangential to the planetary motion of the tool and represents the direction of rolling.

Accordingly, the rolling contact occurs in the α–γ plane. Here, plane strain conditions are assumed for rolling contact. The rolling contact is also assumed to be frictionless. Bhargava et al. (1985) translate the modified pressure distribution along the rolling direction. In profile rolling, however, the displacement boundary conditions are specified. As shown in figure 5.8c, the depth of bite can also vary within a single forming stroke. Hence, the rolling contact situation is modeled by translating the appropriate displacement distribution in the α–γ plane along the γ direction. Unlike the model of Bhargava et al., the pressure distribution and total load are not known a priori in this case but are calculated for each increment of translation. Once the complete pressure distribution is obtained from the displacement distribution, the rolling contact model of Bhargava et al. is used to obtain the corresponding displacements, and the procedure is repeated until convergence. Upon convergence, the pressure distribution is integrated in space over the entire bite and a resultant load is obtained. The rate of this loading is stored and is later used as $\tau_\alpha^{(c)}$ for analysis of deformation in the α–β plane.

Most of the sliding at the tool–workpiece interface, as well as the deformation of the workpiece, is expected to occur in the α–β plane. It is also the deformation pattern in the α–β plane that determines the shape of the gear teeth. For the analysis in the α–β plane, it is assumed that $v_\gamma = 0$. Thus, a plane strain situation arises.

Hence, the model of profile rolling presented here is a combination of the two plane strain models. The rolling contact between the tool and the workpiece occurs in the α–γ plane, whereas the sliding of the material that directly influences the deformation pattern and the shape of the gear teeth takes place in the α–β plane. Thus, the following BEM analysis and the results pertain to the α–β plane and assume plane strain conditions in that plane. However, the rolling contact model (in the α–γ plane) determines the applied traction at the interface and thus influences the BEM analysis in the α–β plane.

A consequence of the plane strain assumption in the α–β plane is that $v_\gamma = 0$ and $\sigma_{\alpha\gamma} = \sigma_{\gamma\alpha} = \sigma_{\beta\gamma} = \sigma_{\gamma\beta} = 0$. It is further assumed that the simulated interface element has zero stiffness in the direction tangential to the contact surface; therefore, only pressures and shear loadings get transferred across the interface. Hence, $\sigma_{\beta\beta} = 0$ at the interface.

Using the above assumptions, zero normal velocity at the interface ($v_\alpha = 0$), and the fact that the normal has components (1,0,0) in the local coordinate system, the load correction equation in the local coordinate system at the interface can be written as

$$\tau_\beta^{(s)} = \tau_\beta^{(c)} - \sigma_{\alpha\alpha}\omega_{\alpha\beta} - \sigma_{\alpha\beta}d_{\alpha\alpha} - \sigma_{\alpha\beta}(d_{\alpha\alpha} + d_{\beta\beta}) \tag{5.51}$$

and

$$\tau_\alpha^{(s)} = \tau_\alpha^{(c)} + \sigma_{\alpha\beta}\omega_{\alpha\beta} - \sigma_{\alpha\beta}d_{\alpha\beta} - \sigma_{\alpha\alpha}d_{\alpha\alpha} - \sigma_{\alpha\alpha}(d_{\alpha\alpha} + d_{\beta\beta}) \tag{5.52}$$

The local rotation rate on the interface is defined as

$$\omega_{\alpha\beta} = -\kappa v_\beta \tag{5.53}$$

where κ is the local curvature. It should be noted, however, that equations (5.51)–(5.53) are valid only in an updated Lagrangian frame.

Here, $\tau_\alpha^{(c)}$ is obtained from the rolling contact model in the α–γ plane. As mentioned earlier, the rolling contact in the α–γ plane is assumed to be frictionless. Once the time history of the pressure distribution is obtained from the rolling contact model, the time derivative of the pressure distribution and the resulting rate of loading are obtained. Next, the work-equivalent traction rates are obtained at nodal points in the α–β plane and are transformed to the appropriate coordinate system to get $\tau_\alpha^{(c)}$.

Much research effort is currently focused upon the development of appropriate sliding mechanisms. Considering the development of our algorithm, we should use a friction model that is physically realistic and easily extendable as more information becomes available. Coulomb's law of friction with μ_s as the static coefficient of friction and μ_d as the dynamic (or kinetic) coefficient of friction fulfills these criteria.

For the particular problem considered here, it is assumed that sliding occurs in the α–β plane only. If τ_β represents the developed tangential tractions along the interface, we assume that there is no relative motion between two adjacent particles on the tool and the workpiece as long as $|t_\alpha| \leq \mu_s \tau_\alpha$ (τ_α = compressive normal traction). The maximum traction of static friction is the smallest force necessary to start motion. During motion, the magnitude of the tangential traction resisted by friction is $\mu_d \tau_\alpha$ (with $\mu_d \leq \mu_s$). The motion continues as long as the frictional traction developed greater than or equal to $\mu_d \tau_\alpha$, that can actually be resisted. Once the developed tangential traction drops below the dynamic friction $\mu_d \tau_\alpha$, the relative motion between the tool–workpiece interfaces ceases until such time that, again, the developed tangential traction exceeds the frictional capacity.

Since the material is incompressible and the normal at the tool–workpiece interface has components $(1, 0, 0)$, the developed tangential traction rate can be obtained as

$$\tau_\beta^{(c)} = \hat{\sigma}_{\alpha\beta} \tag{5.54}$$

The developed tangential traction (τ_β) is compared to the frictional capacity to determine sticking or sliding situations.

5.4.2.1 Numerical results for profile rolling

The rolling contact problem is solved first and iterated until convergence is obtained. The normal traction rate $\tau_\alpha^{(c)}$ is obtained for later use in the plane strain BEM analysis. In the BEM analysis, the velocity gradient in the previous time step is used as a first guess at each time step and iterated until convergence.

The geometry of the plane strain forming problem is shown in figure 5.9.

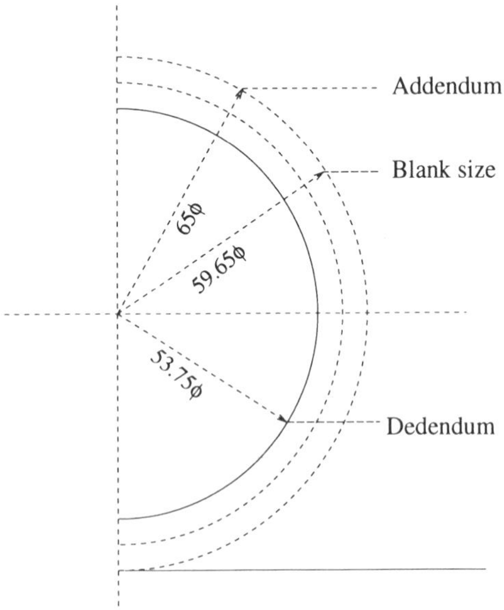

Figure 5.9. Workpiece geometry for profile rolling.

Initial blank diameter is 59.65 mm. The pitch diameter is 60 mm. Addendum and dedendum diameters are 65.0 mm and 53.75 mm, respectively. The teeth have a pressure angle of 20 degrees. The material used is AISI 1045 heat-treatable steel. As shown by Eggert and Dawson (1987) and Mukherjee (1982), the material data (Bardes 1978, Lange and Kurz 1984) are fitted to Hart's state variable model. The state variable model parameters for AISI 1045 steel are found to be:

$$E = 1.07 \times 10^5 \text{ MPa } (3 \times 10^7 \text{ psi}), \qquad \nu = 0.3$$
$$\lambda = 0.15, \qquad M = 7.8, \qquad m = 5.0$$
$$\underset{\sim}{M} = 1.2 \times 10^6 \text{ MPa } (1.74 \times 10^8 \text{ psi})$$
$$\overset{(*)}{d} = 1.4 \times 10^{-41}\,\text{s}^{-1} \quad \text{at} \quad \sigma_S = 68.95 \text{ MPa } (10^4 \text{ psi})$$
$$d_0 = 3.15\,\text{s}^{-1} \quad \text{at} \quad \sigma_0 = 68.95 \text{ MPa } (10^4 \text{ psi})$$
$$\beta = 0.910 \times 10^3 \text{ MPa } (0.132 \times 10^6 \text{ psi}), \qquad \delta = 1.20$$

The initial values of the state variables are:

$$H(\boldsymbol{x}, 0) = 600 \text{ MPa } (87 \times 10^3 \text{ psi})$$

$$\varepsilon_{ij}^{(a)}(\boldsymbol{x}, 0) = 0.0$$

As the tool comes into contact with the workpiece at the tooth spaces, the material is pushed in toward the dedendum. As a result, material movement is induced and the material between two spaces is pushed out to the addendum, thus forming the gear teeth.

Figure 5.9 shows a symmetric half of the workpiece geometry. As the tool forms a particular tooth space, the two adjoint tooth flanks are partially formed. A tooth flank becomes fully developed when both of its adjoint tooth spaces are formed completely. As shown in figure 5.8, two formed tools situated diametrically opposite each other are used. Three consecutive tooth spaces are formed in the symmetric half of the blank to study mutual influences of neighboring tooth spaces. This also develops the fully formed tooth flanks. Thus, the analysis for the formation of three consecutive tooth spaces is considered sufficient to understand the basic mechanics of profile rolling of gears.

Figure 5.10 shows the radial forces as a function of tool position for two different tool penetration velocities. As the tool penetrates into the workpiece, the radial forces rise rapidly. For down-profile rolling, the peak occurs about 5 to 10 degrees before the tool reaches its lowest position (as shown in figure 5.10a). As observed in figure 5.10b for up-profile rolling, the peak radial force occurs just after the tool attains its lowest position. It can be distinctly recognized that in down-profile rolling, the forming process

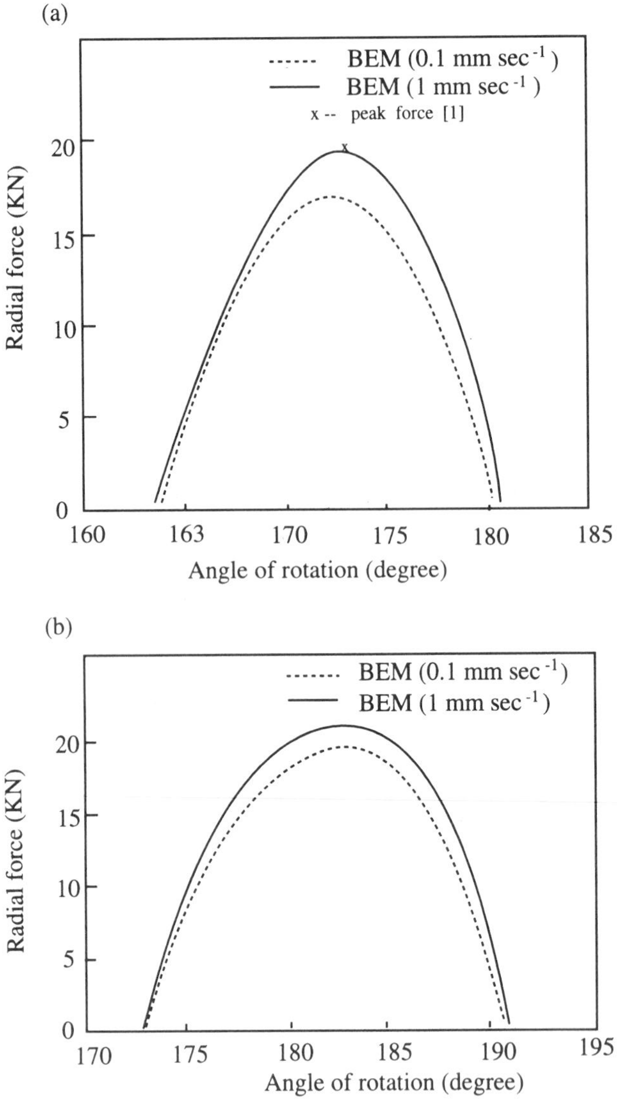

Figure 5.10. Radial forming forces in profile rolling ($\mu_s = \mu_d = 0$): (a) down-profile rolling; (b) up-profile rolling.

is completed shortly after the tool passes through its lowest point. On the other hand, for up-profile rolling, the maximum force occurs at this point. The resistance to rolling is greater in up-profile rolling. This is manifested in the higher maximum radial force for up-profile rolling. In down-profile rolling, the contact zone between the tool and the workpiece is larger compared to the corresponding contact zone in up-profile rolling. After

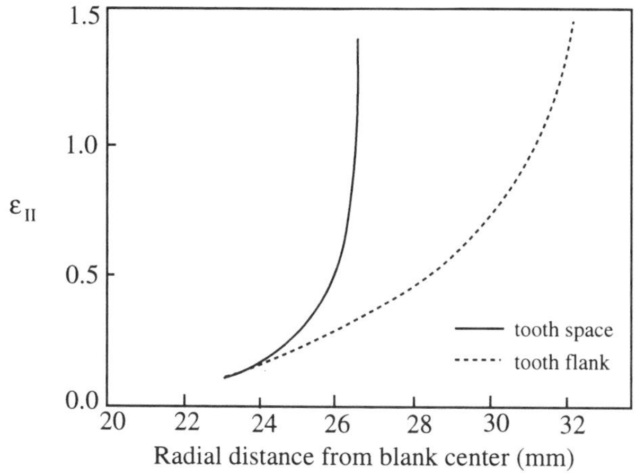

Figure 5.11. Radial variation of equivalent strain in profile rolling ($\mu_s = \mu_d = 0$).

passing through the maxima, the radial forces decrease rapidly for both down- and up-profile rolling. This drastic fall can be ascribed to the rapid reduction in tool bite. The peak force obtained from the BEM analysis for down-profile rolling compares well with the peak force obtained by Lange and Kurz (1984). The BEM analysis predicts the peak force to be 20 kN, while Lange and Kurz observed the peak force to be 21.78 kN. This difference is quite small, considering the fact that the material parameters used here only approximate the behavior of the material used by Lange and Kurz. For both down- and up-profile rolling, increasing the total penetration velocity by an order of magnitude increases the peak radial force by about 12 percent.

As the tool penetrates the workpiece, it induces extreme deformations near the periphery. Figure 5.11 shows the variations of equivalent strain ε_{II} (second invariant of Almansi strain) over radial lines over a tooth space, as well as a tooth flank. In both cases, the strain is highly concentrated near the periphery. Along a radial line over a tooth space, ε_{II} drops from a peak value of 1.42 to 0.1 as we move just 4 mm inside. Along a radial line over a tooth flank, ε_{II} also drops from 1.44 to less than 0.1 over a radial distance of 9.5 mm. The magnitude and distribution of ε_{II} are found to change very little as the tool velocity is varied.

Figure 5.12(a) shows the variations of σ_{II} (second invariant of Cauchy stress) along radial lines over a tooth space and a tooth flank. In both situations, σ_{II} rises sharply as we approach the periphery. As the radial distance decreases from 26.875 mm (periphery of tooth space) to 21 mm, σ_{II}/H drops from a peak value of 3.8 to about 0.9. Along a radial line over the tooth flank, σ_{II}/H drops from 3.9 at the tip of the tooth to 1.75 at the root of the tooth. It is evident that severe deformation near the periphery

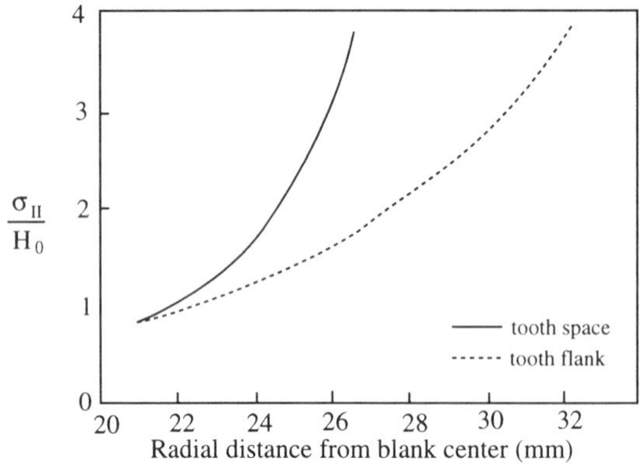

(a)

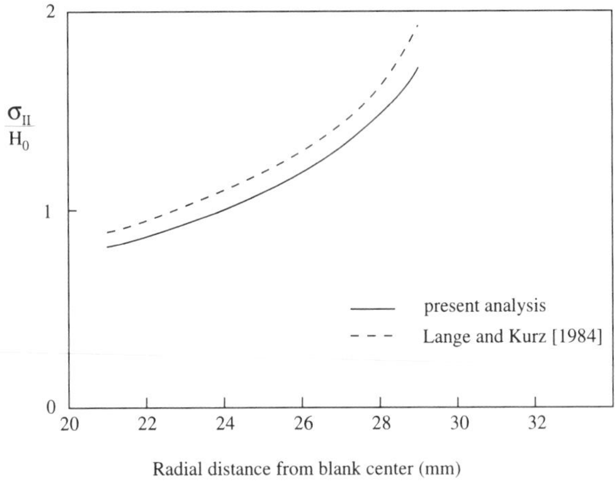

(b)

Figure 5.12. Radial variation of equivalent stress in profile rolling ($\mu_s = \mu_d = 0$). (a) Variations along tooth space and tooth flank. (b) Comparisons with Lange and Kurz (1982).

results in very high strain hardening of the material at the periphery. This is desirable since the strain hardening hardens the gear teeth. This also improves the endurance limit of the formed gear teeth.

Lange and Kurz (1984) formed the tooth spaces up to a total penetration depth of 1.6 mm. At this point, the tooth space is only partially formed. Figure 5.12(b) shows the comparison between the σ_{II} distributions obtained by Lange and Kurz and those obtained from the BEM analysis at a depth of penetration of 1.6 mm. Maximum discrepancy between the two curves is less than 12 percent.

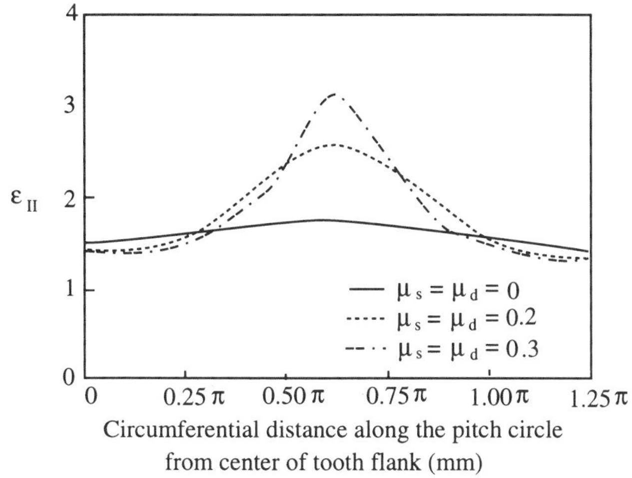

Figure 5.13. Circumferential variation of equivalent strain in profile rolling at different friction coefficients.

The effect of friction on the circumferential distribution of ε_{II} is shown in figure 5.13. The origin represents the center of a tooth flank, and a circumferential distance of 1.25π represents the center of the next tooth space. For complete sticking at the tool–workpiece interface ($\mu = \infty$), there is no flow of the material in the circumferential direction. At tooth spaces, the material is pushed inward and, as a consequence, the material in between is pushed up. However, at high friction, the material cannot flow easily in the circumferential direction. Hence, extreme shear deformation occurs at the boundary between a tooth space and tooth flank and can induce cracking. Proper lubrication of the interface alleviates this problem. Lubricating the interface reduces the coefficient of friction and smooths the circumferential distribution of ε_{II}. As seen in figure 5.13, ε_{II} is higher at the boundary between a tooth space and tooth flank than at the centers of the space and flank regions. At $\mu = 0.0$, the distribution is almost uniform, with ε_{II} equal to 1.76 at the boundary between the two regions. At $\mu = 0.2$, however, the distribution gets sharper, with ε_{II} equal to 2.58 at the region boundary. At $\mu = 0.3$, the peak value of ε_{II} rises to about 3.125. Thus, at higher frictions, the shearing action gets localized rapidly.

5.4.3 Plane Strain Slab Rolling

A unique feature of slab rolling processes (shown schematically in figure 5.14) is the existence of a neutral point (or region) along the tool–workpiece interface where the tangential relative velocity between the deforming

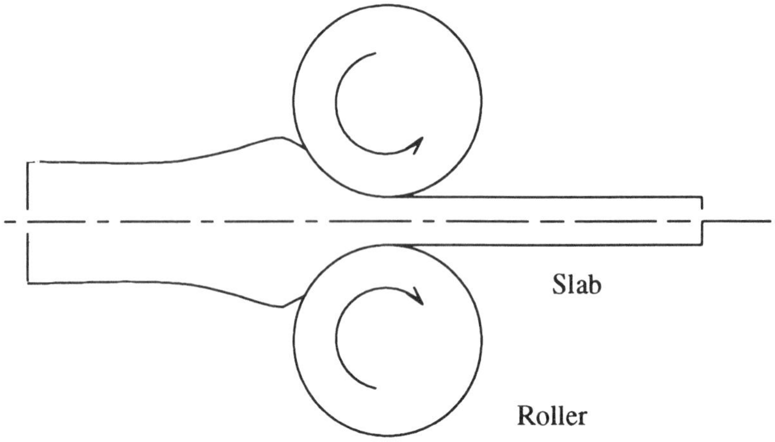

Figure 5.14. Schematic diagram of plane strain rolling.

material and the roll becomes zero. The frictional stresses usually change direction at this neutral point (or region). The location of this neutral point or zone, however, is not known a priori. Accordingly, care must be taken to model this phenomenon appropriately.

In addition to plane strain assumptions, the sliding contact at the roll–workpiece interface can be represented as

$$\sigma_{\alpha\beta} = -\mu_s \sigma_{\alpha\alpha}\left[\frac{2}{\pi}\tan^{-1}\left(\frac{|V_r|}{a}\right)\right]\frac{V_r}{|V_r|} \tag{5.55}$$

(Chen and Kobayashi 1978, Li and Kobayashi 1982), where V_r is the relative velocity between the roll and the workpiece in the tangential direction (α-direction), a is a constant several orders of magnitude smaller than the roll velocity, and μ_s is the static coefficient of friction. At the beginning of the roll bite, the velocity of the workpiece is zero. Hence, V_r is large (equal to the tangential velocity of the roll surface). Thus, $\sigma_{\alpha\beta} = -\mu_s \sigma_{\alpha\alpha}$, which represents the case of static friction. As the process of rolling continues, the workpiece is driven by the rolls and the relative velocity between the roll and the workpiece tends to decrease. Finally, a point (or region) is reached where the workpiece has reached the velocity of the roll. This point (or region), where the relative velocity at the roll–workpiece interface is zero, is called the neutral point (or region). At the neutral point, there is no shear transfer between the roll and the workpiece. Thus, $\sigma_{\alpha\beta}$ at this point drops to zero. Beyond this point, the workpiece tends to move faster than the roll, and the sense of $\sigma_{\alpha\beta}$ at the interface reverses. Accordingly, the developed tangential traction also reverses direction at this point.

While the friction models of Chen and Kobayashi (1978) and Li and Kobayashi (1982) represent the phenomenon of a neutral point quite well, they relate tangential traction rather than the tangential traction rate to the relative velocity. This is consistent with Coulomb's law of friction. The BEM formulation, however, is in rate form. In the case of plane strain rolling, we can also show that

$$\tau_{\beta}^{(c)} = \hat{\sigma}_{\alpha\beta} \tag{5.56}$$

since the material is incompressible and the normal at the tool–workpiece interface has components (1,0,0). Thus, knowing $\sigma_{\alpha\beta}$ and the size of the time step, we can estimate $\tau_{\beta}^{(c)}$. This estimated value of $\tau_{\beta}^{(c)}$ is incorporated into the load correction equations and the assembled BEM equation is solved for the unknown velocities and traction rates (Chandra and Mukherjee 1987). The estimate of $\tau_{\beta}^{(c)}$ is updated, and the scheme marches forward in time.

5.4.3.1 Numerical results for slab rolling

Time integration for the rolling problem is carried out by an explicit Euler-type method and an implicit ABM method with proper controls (Chandra 1986). The solution strategy, in essence, consists of marching forward in real time with suitable updating of the configuration of the body. The presence of velocity gradients in the boundary traction rates and in the domain integral for geometric nonlinearity requires iterations within each time step. Body force rate is assumed to be zero. However, this assumption can be waived without any difficulty. In the BEM analysis, the velocity gradient in the previous time step is used as a first guess at each time step and is iterated until convergence (Chandra and Mukherjee 1984a, 1985, 1987, Mukherjee and Chandra 1987). At the very first time step, the velocity of the workpiece is assumed to be zero, and the friction model (Chen and Kobayashi 1978, Li and Kobayashi 1982) is used to obtain the value of the developed tangential traction (τ_{β}). Later, at any time step, the velocities of the contacting nodes of the workpiece in the previous time step are used to estimate V_r and a new estimate of τ_{β} is obtained. Knowing the value of τ_{β} at the beginning of the time step, the size of the time step, and the desired value of τ_{β} at the end of the time step, $\tau_{\beta}^{(c)}$ during the time step is estimated and used in the BEM analysis.

The geometry of the plane strain rolling problem considered here is shown in figure 5.14. The initial thickness of the workpiece is 10 mm. The horizontal contact length of the roll bite is 40 mm. Final thickness of the workpiece is 7.5 mm. The material used is commercially pure aluminum at a temperature of 24°C. The details of the parameters for Hart's model for this material and their significance are given by Alexopoulos (1981), Chandra and Mukherjee (1984b), and Chandra (1986). The material parameters are:

$$
\begin{aligned}
E &= 48.276 \times 10^3 \text{ MPa } (7 \times 10^6 \text{ psi}), \ \nu = 0.298 \\
\lambda &= 0.15, \ M = 7.8, \ m = 5 \\
\mathcal{M} &= 298.62 \times 10^3 \text{ MPa } (43.3 \times 10^6 \text{ psi}) \\
\overset{*}{d}_{ST} &= 5.05 \times 10^{-15} \text{ s}^{-1} \text{ at } \sigma_s \\
&= 68.965 \text{ MPa } (10^4 \text{ psi}) \text{ and } T_B = 298 \text{ K} \\
d_0 &= 0.315 \times 10^3 \text{ s}^{-1} \text{ at } \sigma_0 = 68.965 \text{ MPa } (10^4 \text{ psi}) \\
\beta &= 103.45 \text{ MPa } (1.5 \times 10^4 \text{ psi}), \ \delta = 2.8
\end{aligned}
$$

Initial values of state variables are

$$
\begin{aligned}
\overset{*}{\sigma}(\boldsymbol{x}, 0) &= 41.38 \text{ MPa } (6 \times 10^3 \text{ psi}) \\
\varepsilon_{ij}^{(a)}(\boldsymbol{x}, 0) &= 0.0
\end{aligned}
$$

The process of rolling is non-steady-state at the beginning of a bite. It becomes steady-state as a material is rolled through the gap between the two rolls. The rollers have a constant angular velocity and compress the material as it is driven through the roll bite. As the workpiece comes into contact with the rollers, an interfacial friction force develops and the workpiece is drawn into the roll bite. Free-surface conditions are assumed on each side of the bite.

Figure 5.15 shows the variations in Almansi strains along the centerline of the workpiece as it passes through the roll bite. The thickness reduction

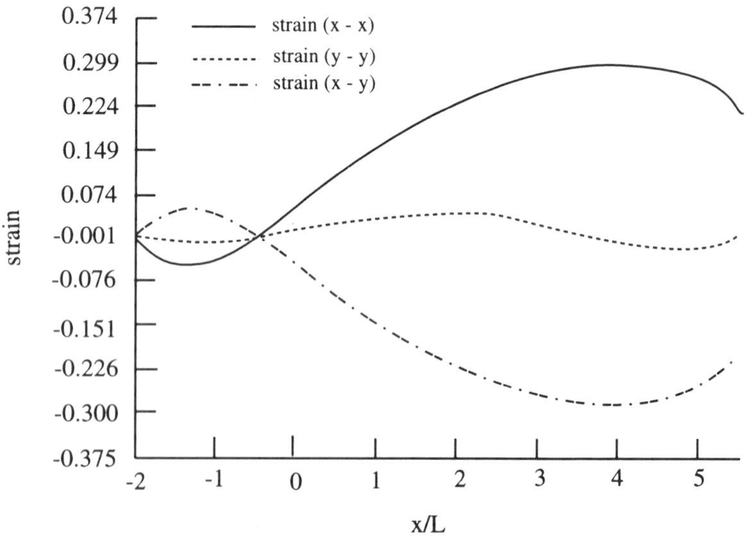

Figure 5.15. Variations of strains along the centerline.

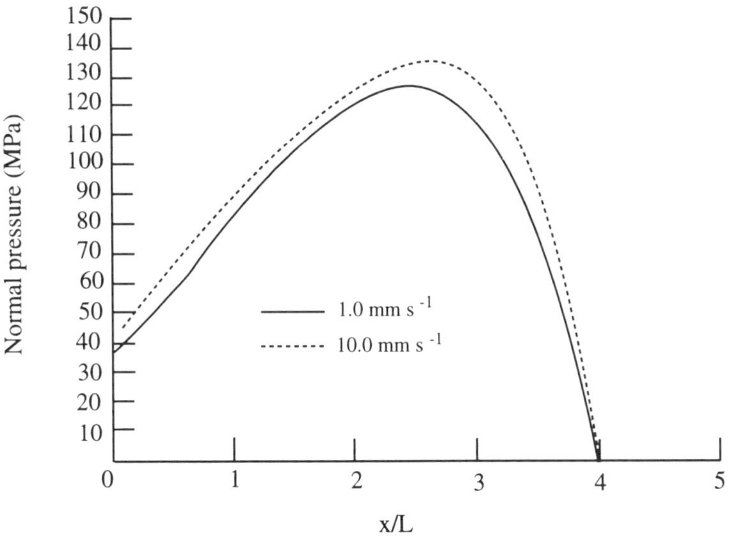

Figure 5.16. Distributions of normal pressures different roll velocities of 1.0 and 10.0 mm/s ($\mu = 0.10$).

here is 25 percent. It is interesting to note the small amount of material buildup just before the entrance into the roll bite. Due to this phenomenon, ε_{xx} is initially negative before entrance into the roll bite. As the material is gradually drawn between the rollers and the slab thickness is reduced, ε_{xx} becomes tensile. It reaches a peak value of 0.29 and drops slightly as the material passes by the rollers and the thickness increases due to springback. The strain variations shown in figure 5.15 agree qualitatively quite well with the strain fields obtained by Dawson (1987) from a viscous flow-type analysis. To avoid numerical instabilities, Dawson constrained the material velocity beyond the exit from the roll bite to be horizontal. In the present analysis, however, traction-free conditions are assumed. This allows proper springback of the material.

The normal pressure distributions between the roll and the workpiece at different roll velocities are shown in figure 5.16. For 25 percent reduction, the distribution is of the friction hill type. As the roll velocity increases from 1 to 10 mm/s, the peak normal pressure changes from 128 to 137 MPa. Also, the position of the peak shifts slightly toward the exit of the roll bite from $x/L = 2.5$ to $x/L = 2.67$. The normal pressure distributions compare quite well qualitatively to those obtained by Li and Kobayashi (1982) and Mori et al. (1982) from a rigid-plastic finite element analysis.

Figure 5.17 shows the shear stress distributions at the roll–workpiece interface at two different roll velocities. As the workpiece enters the roll bite,

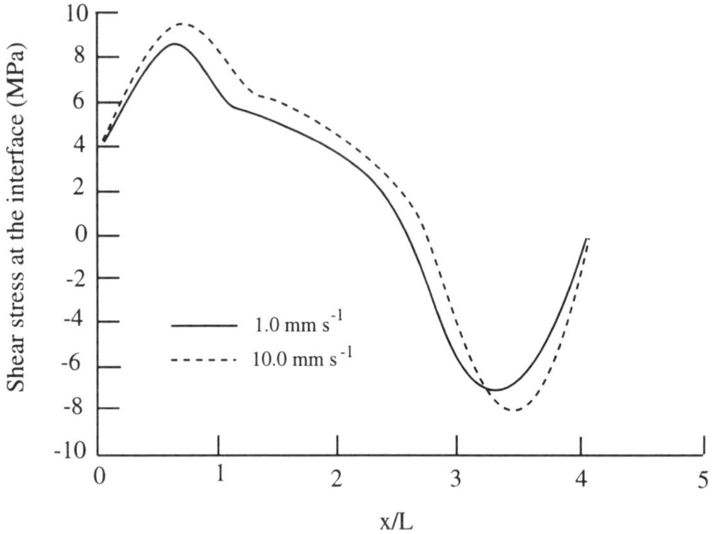

Figure 5.17. Distributions of shear stress at the roll-workpiece interface ($\mu = 0.10$).

the relative velocity between the workpiece and the roll is large. Accordingly, the shear stress is large. Once the workpiece has entered the bite, the normal pressure increases and the relative velocity decreases. Initially, the increase in normal pressure dominates and the shear stress goes up to a peak value of 6.7 MPa for a roll velocity of 1.0 mm/s (for $\mu = 0.1$). Then, the decrease in relative velocity causes the shear stress to drop, and at $x/L = 2.58$ the shear stress changes direction. At this point, the relative velocity between the roll and the workpiece is negligible and this point (or region) is called the neutral point (or region). The shear stress reaches a value of -6.66 MPa before rising to zero at the exit from the roll bite. At a higher roll velocity of 10 mm/s, the neutral point shifts toward the exit and occurs at $x/L = 2.71$. The stress peaks also increase to +9.6 and -7.67 MPa. An increase in interface friction also moves the neutral point toward the exit. As the friction coefficient goes from 0.10 to 0.15, as shown in figure 5.18, the position of the neutral point moves from $x/L = 2.7$ to $x/L = 3.0$. For a roll velocity of 10 mm/s, the first peak rises from +9.6 to +13.33 MPa. However, the other peak changes from -7.67 MPa to only -8.33 MPa. The qualitative nature (or shape) of the shear stress distribution along the roll bite is not affected much by variations in roll velocity or interface friction.

Figure 5.19 shows the residual longitudinal stress distribution across the cross-section of the workpiece at the exit from the roll bite. The residual σ_{xx} is tensile at the top and bottom faces and compressive at the center. As the roll

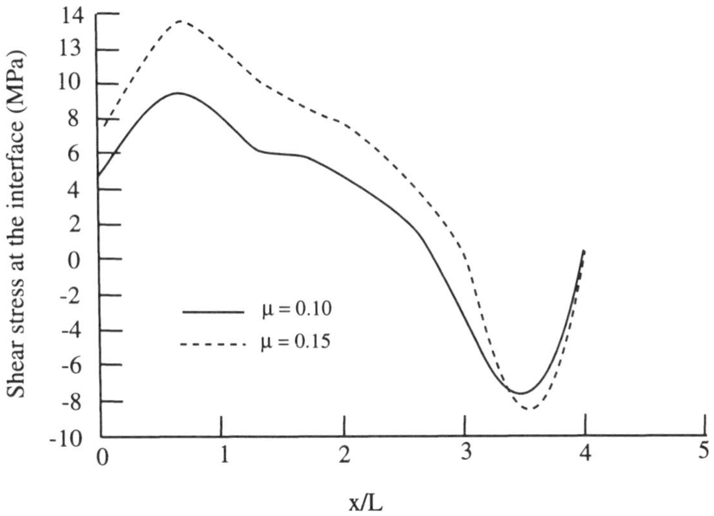

Figure 5.18. Variations of shear stress at the interface with the frictional condition (roll velocity = 10 mm/s).

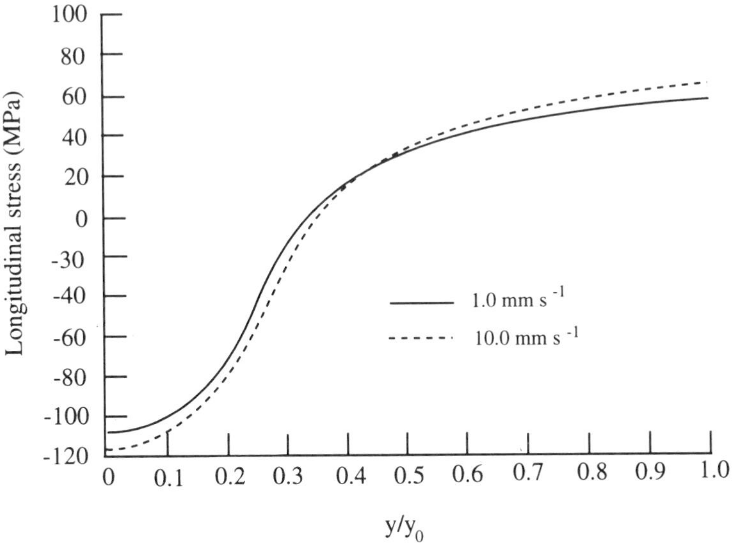

Figure 5.19. Residual stress distributions over a cross-section in plane-strain rollup as functions of roll velocity.

velocity increases from 1.0 to 10 mm/s, the residual stress peaks change from +55.0 to +61.6 MPa at the top and bottom faces and from −106.7 to −118.32 MPa at the center. Also, the residual stresses should be self-equilibrating. The distributions in figure 5.19 satisfy equilibrium with errors of less than 10 percent.

5.4.4 Plane Strain Sheet Forming

Sheet forming is another widely used technique for fabricating light-weight structures with high aspect ratios. Such structural parts are used extensively in automotive as well as aerospace industries.

In this section, a simple plane strain bulging process is considered to illustrate the applications of the BEM formulation in analyzing sheet forming problems. In most sheet forming processes the sheet thickness is of the order of 1 mm, while its in-plane dimensions are of the order of 100 mm or more. Accordingly, sheet forming processes have been investigated by various researchers (e.g., Wang and Wenner 1982, Chung and Richmond 1989c) in the context of shell theory. Such analyses are efficient and can reveal many useful insights. However, any shell theory application requires an assumption for the strain distribution across the thickness of the sheet (e.g., Love–Kirchhoff hypothesis) and the analysis is obliged to satisfy such a priori assumptions. Hence, a deforming shell representation of a sheet forming process cannot capture the actual strain and residual stress distributions across the thickness of the sheet. Such deformation and strain fields, however, are important for determining the springback of the formed part in its unloaded state.

To obtain an accurate springback prediction, one must also discretize the sheet along its thickness. This calls for a full three-dimensional analysis for a general sheet forming process. For special cases, a plane strain or an axisymmetric representation is also feasible. As a first step, such a plane strain problem is investigated here. It may be further simplified by restricting our attention to a bulging problem. For a bulging problem, the scaled Lagrange traction rates in the boundary equation (2.118) may be specified in terms of the current normal pressure and its rate through equations (5.14)–(5.15). More realistic sheet forming processes involving a punch and a die with boundary conditions ranging from complete sticking to relatively free sliding at the punch–sheet interface can be modeled by using equations (5.16) and (5.17) as illustrated in section 5.2.3. The boundary conditions at the die–sheet interface may be represented by completely fixing the boundary nodes (in both directions) at the die–sheet interface for a punch stretching problem. If the blank is allowed to be drawn in, it can also be represented by imposing an appropriate back stretching traction (τ_β) on the boundary nodes at the die–sheet interface.

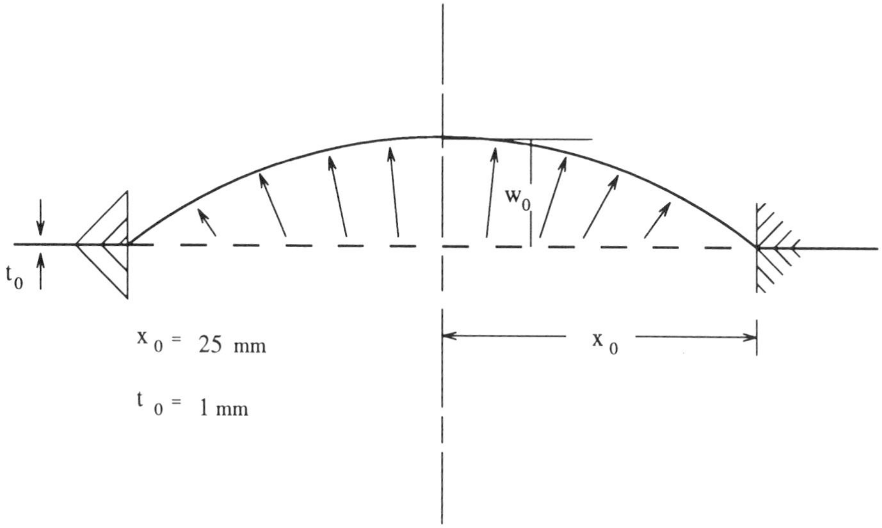

Figure 5.20. Schematic of plane strain bulge test.

5.4.4.1 Numerical results for plane strain sheet forming

As shown in figure 5.20, a sheet of thickness 1 mm and length 50 mm is clamped completely at both ends. The sheet is then deformed by applying normal pressure (modeled as follower traction) which is increased at a rate of 0.5 MPa/s (72.5 psi/s). The boundary conditions for this plane strain bulging problem is specified through equations (5.14) and (5.15) by assigning $\dot{p}(t) = 0.5$ MPa/s, and its cumulated value at any instant of time is represented by $p(t)$.

The longitudinal strain distributions along the sheet at different stages of deformation are shown in figure 5.21. For the case of plane strain bulging, the strain distribution remains almost uniform throughout the sheet. At $w_0/x_0 = 0.7$, ε_{11} is 0.325. Hence, unlike axisymmetric bulging (Wang and Wenner 1982), the material will not tend to thin at the center. Thus, higher dome heights may be obtained from the same blank size under plane strain conditions. The solid line denotes the results obtained from the BEM analysis and the dotted line shows the results obtained from a finite element analysis (Chandra and Mukherjee 1985) of the same problem. As observed in figure 5.21, the BEM and the FEM predictions are quite close.

Figure 5.22 shows the increase in longitudinal strain at the center ($x = 0$) with bulge height. It is observed that the slope of the ε_{11} vs. w_0/x_0 curve increases as bulge height is increased. Thus, as the deformation proceeds, the strain at the center increases at an accelerating rate and the structure

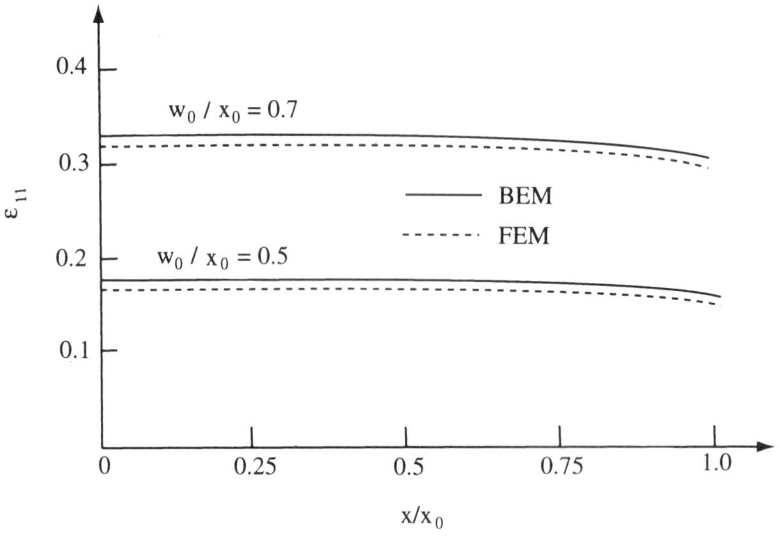

Figure 5.21. Longitudinal strain distribution at different bulge heights.

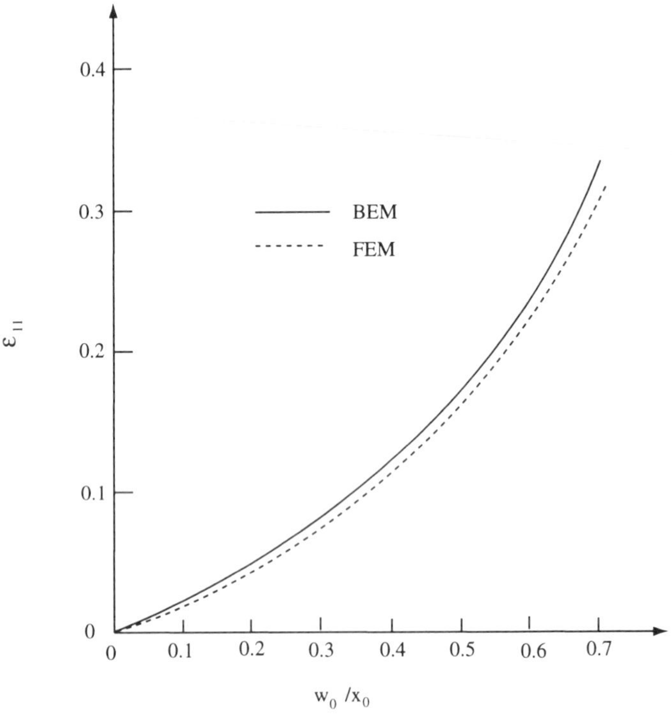

Figure 5.22. Variation of central longitudinal strain with bulge height.

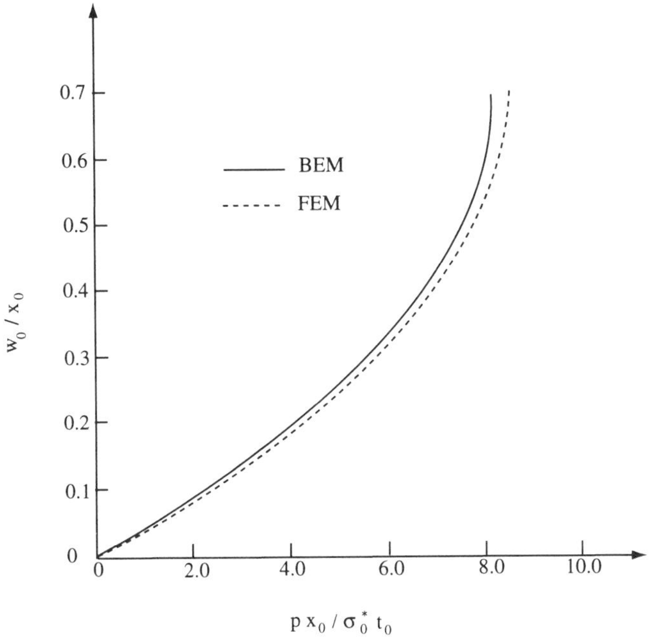

Figure 5.23. Variation of bulge with pressure.

becomes more and more prone to failure. The solid line representing the BEM analysis and the dotted line representing the FEM analysis agree quite well.

The rise in bulge height with increase in pressure is shown in figure 5.23. It is seen that for normalized pressure in excess of 7.0, the bulge height is quite sensitive to small changes in pressure. This sets the stage for potential instability. Once again, it is observed that the BEM and FEM results are in good agreement.

The secondary variables such as strains and stresses are obtained pointwise in the BEM analysis through analytic differentiation of the relevant primary variables. These quantities (especially $\varepsilon^{(a)}$, σ, and H) play a very important role in this class of problems and their accuracy is critical to the stability of the time integration scheme. The higher obtainable accuracy for these secondary variables permitted bigger time steps in the BEM analysis than those allowable in the FEM analysis. This resulted in improved computing times for the BEM analysis. To obtain a normalized bulge height of $w_0/x_0 = 0.7$, the BEM program (with 28 boundary nodes and 10 internal cells) needed 205 seconds of CPU time on an IBM 370/168 whereas the FEM program (with 105 nodes and 40 elements) required 484 seconds.

5.5 Concurrent Preform and Process Design for Formed Products

It is evident from the previous discussions in chapter 2 and in this chapter that over the last decade significant advancements have been made in analytical, computational (both by BEM and FEM), and experimental abilities to understand the fundamental mechanics of various sheet and bulk forming processes. Improvement in product and process designs, however, hinges on the ability to make rational decisions, based on the synthesis of various pieces of information obtained from analyses and experiments, to arrive at an integrated design that is effective and efficient.

An analytical or experimental investigation of a forming process studies the object of interest in great detail under a given configuration with specified geometric, material, and process parameters. Assuming that one has the ability to predict the performance for a specific configuration, the question then arises of how to modify the configuration to improve its performance. It is then the designer's job to extract the important insights from these studies and to arrive at the best or optimal configuration for the intended purpose.

This brings us to the all-important question, "What if?," which arises again and again in design. In fact, the design process is not complete until all such pertinent questions have been asked and satisfactorily answered. To a very large extent, the quality of the final design depends on how extensively and how effectively these "What if?"s have been asked and answered.

In conventional practice, the design of formed products can be classified in two broad categories: (i) product design and (ii) manufacturing design. The product design is typically concerned with the functional attributes of the final formed configuration in a given environment or service condition. Such a design is best carried out using optimization procedures in which, typically, one or more objective functions are extremized without violating certain constraints. In recent years, various researchers have used nonlinear programming techniques to seek the optimum shape and material properties for the intended purpose. A crucial ingredient for the success of such optimization techniques is the ability for accurate determination of the design sensitivity coefficients (DSCs). These DSCs are rates of quantities, such as displacements, stresses, or temperatures with respect to design variables. These design variables may be parameters that control the shape of part or all of the body, material properties, or other suitable quantities. A typical use of sensitivities in shape optimization procedures starts from a nominal design. Design sensitivities are then obtained, and improved designs are sought for specified objective functions. It typically takes a number of steps to arrive at the final optimized design, with each step requiring analysis of the problem, as well as determination of the relevant sensitivities. A review of the BEM formulations for the determination of DSCs in fully nonlinear elasto-viscoplastic problems is presented in chapter 3.

Manufacturing design, on the other hand, is concerned with the process of realization of the desired final configuration from the available raw material. In this phase, the designer attempts to optimize the manufacturing process with respect to geometry, material, and environment (i.e., mechanical and/or thermal boundary conditions, along with their spatial and temporal variations, body forces). Thus, for example, one can think of optimization of microstructure development and residual stress distributions with respect to die shape for extrusion, preform configuration for forging, or draw bead configuration in sheet forming. Other examples might include optimum cooling profile for the solidification of a casting of desired shape, optimum process conditions in welding to obtain the desired microstructure, or optimum tool profile and cutting conditions in a machining operation. The state of a formed product (as well as many other manufactured products) depends to a large extent on its processing history. This complicates the manufacturing design even further. While one is interested in the end product of a manufacturing process, the designer now has to optimize a sequence of states during the forming operation. Such an activity requires concurrent optimization of the preform geometry and the time history of the processing conditions. Accordingly, simultaneous optimization of the preform shape and the forming process parameters is the focus of discussion in this section.

Over the past few years, Kobayashi et al. (1989) have investigated the issues of preform design for metal forming processes using a rigid-plastic material model. Preform design essentially refers to the design of an initial shape of the workpiece that, when it has undergone an associated forming process, forms the required product shape with desired properties without any formation of defects and without excessive waste of materials. Kobayashi et al. (1989), using a backward tracing algorithm in conjunction with a rigid-plastic finite element analysis, have investigated preform designs for several forming processes such as closed-die forging, shell nosing, rolling, and sheet forming. This is a novel approach. However, the backward tracing algorithm requires several re-analyses in one time step and can become computationally very expensive. More importantly, design of an optimal preform shape requires simultaneous determination of optimum process conditions. The work of Kobayashi et al., however, addresses a narrower problem and has been concerned with the determination of the best preform shape under a prespecified set of process conditions.

Richmond and Devenpeck (1962) introduced the concept of ideal forming based on the concept of minimum plastic work path and investigated die profiles for maximum efficiency in strip drawing processes. Using the generalization of the concept of minimum plastic work path by Hill (1986),

Chung and Richmond (1989a,b,c) have extended the ideal forming theory and have attempted to bridge the gap between analysis and design optimization of forming processes. Based on a generalized deformation theory of plasticity, they have proposed an ideal forming theory in order to obtain direct information regarding optimum process parameters for forming. They prescribe that individual material elements deform along minimum plastic work paths. Then, assuming that formability of local material elements is optimum along minimum plastic work paths, the ideal global process is defined as the one having such local deformations optimally distributed in the final shape. For example, the ideal forming conditions may be obtained by requiring the most uniform strain distributions in final products without having boundary shear tractions. As part of the solution, their technique also provides the history of evolutions of intermediate shapes and process conditions. It is interesting to note that the ideal forming theory of Chung and Richmond prescribes the deformations of individual material elements along minimum plastic work paths. Accordingly, care must be taken to ensure global compatibility. Since the strain history is already prescribed by the minimization of plastic work, appropriate rotations are chosen to make the total displacement gradients compatible (Chung and Richmond 1990).

It is evident from the literature that the issues regarding designs of forming processes have been investigated in three major veins. The work of Kobayashi et al. (1989) focuses on preform design for given process conditions. The work of Chung and Richmond (1989a,b,c) seeks the paths of minimum plastic work to improve process efficiency for a forming operation. The design sensitivity studies (described in chapter 3) attempt to understand the consequences of varying different design parameters. While these investigations provide excellent insights into the forming processes, each one of them considers only specific aspects of the process.

Accordingly, the development of a unified approach to obtain an integrated procedure for designing formed products and forming processes is the primary focus of this section. Such an approach requires that issues regarding the quality of the formed product as well as the efficiency of the forming process be addressed simultaneously.

5.5.1 The Concept of Reverse Forming

At this point, we depart from the conventional practice of starting from the initial state of the raw material and attempting to devise a forming process to transform this initial state into the desired state of the final product. We note that once the product design is completed, the desired state of the final product is well characterized. For our manufacturing design, we start

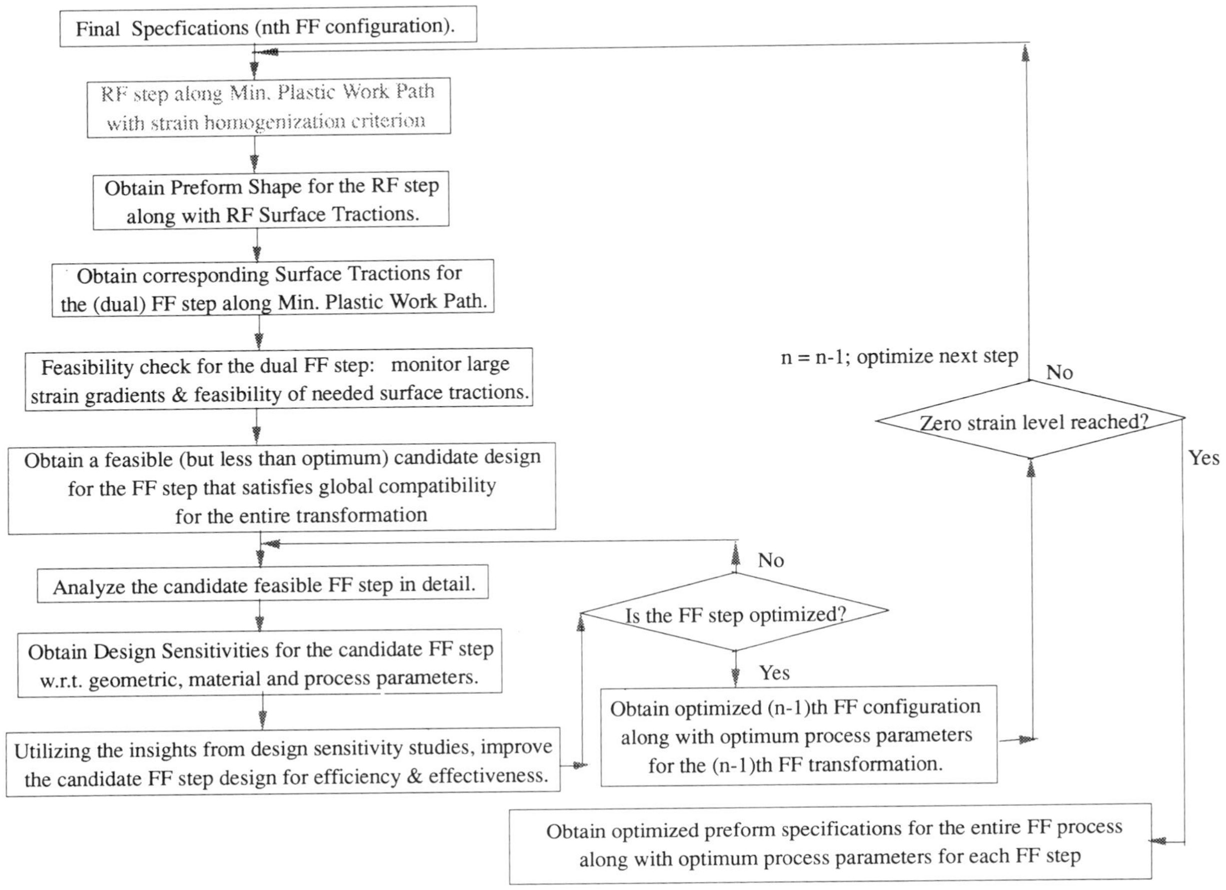

Figure 5.24. Schematic for the proposed design strategy.

from this final configuration. The preform specifications and the process parameters are the end results of our proposed design strategy.

As shown in figure 5.24, we first investigate a hypothetical process of "reverse forming" (RF). In the context of this discussion, reverse forming (RF) is essentially a backward tracing algorithm. For the reverse forming (RF) process, the initial configuration is given by the final product specifications in terms of the geometric shape, the desired material properties (hardness, yield strength, toughness, etc.), and the desired mechanical attributes (residual stress distribution, accumulated plastic strain distribution, etc.). We then attempt to transform, through a hypothetical RF process, this initial configuration into a preform configuration. For the proposed study, the preform (for the entire process) is defined to be the configuration at which the plastic strain distribution is zero at every spatial location. A desired plastic strain distribution is prescribed for the formed product, which is the initial state of the RF process. If the entire forward forming (FF) process consists of n steps, then the jth step of the RF process corresponds to the $(n-j+1)$th step of the FF process. For both forward and reverse forming processes, the jth step defines the transformation that converts the jth configuration into the $(j+1)$th configuration.

It is important to note here that requiring material elements to deform along paths of minimum plastic work is equivalent to using the generalized deformation theory of plasticity (Chung and Richmond 1989a,b,c). Accordingly, any deformation that is constrained to occur along a minimum plastic work path is reversible.

5.5.2 *Integrated Design Algorithm*

The proposed strategy for simultaneous design of the preform and the forming process parameters is outlined in figure 5.24. The specific steps of the integrated design algorithm are given below.

5.5.2.1 Step 1: reverse forming along minimum plastic work path

For a formed product, typically the geometry of the final part is well defined. The desired strength of the part determines the amount of plastic straining (along with associated strain hardening) needed from the virgin material. Knowing the intended service conditions, desired residual stress distributions for the final part may also be obtained. Here, we start from the desired characteristics of the final product.

Starting from the initial state of the RF process (which is the final or nth configuration of the forward forming process), we attempt to reduce the strain level in the body by a predetermined amount. This amount may be spatially uniform or any desired admissible distribution of strain. The material properties are assumed to be isotropic, and the generalized deformation theory of plasticity is used to represent the material model for the RF process. This simplifies the search for the minimum plastic work paths.

Using the generalized deformation theory (Chung and Richmond, 1989a), deformation paths for material elements are assumed. The plastic work then becomes dependent on the displacements. Among the infinite number of possible ways of assuming deformation paths, the one requiring minimum plastic work is chosen to reach the specified strain levels for individual material elements. This will allow us to obtain a new RF configuration for the specified plastic strain decrement. A detailed discussion of the concept of minimum plastic work path and the associated generalized deformation theory may be found in the work of Hill (1986) and Chung and Richmond (1989a,b,c).

Since the concept of minimum plastic work path is applied locally to individual material elements, the rotation tensors are adjusted appropriately to make the total displacement gradients compatible and ensure global compatibility for the entire RF step. A homogenization constraint limiting strain gradients in the neighborhood of any material element is also used to avoid any undesired strain localization. It should be noted here that the strain levels typically decrease during reverse forming. Accordingly, a stable strain-hardening material will tend to soften during a reverse forming process. Since the second RF configuration [$(n-1)$th FF configuration] at the end of the first RF step is obtained by transformation along the minimum plastic work path along with appropriate adjustments of rotations to satisfy compatibility requirements, the second RF configuration is expected to be the optimum in terms of efficiency of the first RF step (or the nth FF step).

5.5.2.2 Step 2: feasibility check for the forward forming step

Once the desired configuration at the end of the RF step is obtained, we focus on the corresponding FF step. At this point, two major issues regarding the feasibility of the FF step need to be addressed.

(i) In the RF process, the concept of minimum plastic work path is utilized in a local sense for individual material elements. As discussed in section 5.5.2.1, global compatibility requirements are satisfied through appropriate choices of rotations that make the total displacement gradients compatible. It has also been observed (Chung and Richmond 1989a,b,c) that a tendency exists for strain localization when individual material elements deform along minimum plastic work paths without any other constraints. In section 5.5.2.1, the use of a homogenization constraint was proposed to avoid localization of strain. In addition, the strain distribution is very closely monitored to identify large strain gradients and to adopt measures for avoiding them.

(ii) Using the BEM formulation presented in chapter 2, the surface traction vectors for the RF step can easily be determined. It is noted here that the deformation in the RF step has occurred along paths of minimum plastic work. Accordingly, the RF step is reversible by reversing the directions of all external loadings. By reversing the directions of the surface traction vectors in the RF step, a set of surface traction conditions may be obtained for the corresponding FF step along the minimum plastic work paths. For real-life forming operations, however, only specific types of boundary conditions can be applied physically. There also exist practical ranges for the realizable magnitudes of these boundary conditions. Accordingly, the surface traction conditions for the corresponding FF step along minimum plastic work paths may not be physically realizable due to hardware and other limitations. Thus, the physical realization of the corresponding FF step cannot be guaranteed.

The work paths traced by individual elements are also checked for any strain localization. At this stage, the directions of surface tractions obtained from the ideal RF step are reversed to determine the surface tractions needed for the corresponding ideal FF step. It is important to note here that deformation along minimum plastic work path is equivalent to using generalized deformation theory of plasticity. This makes the ideal RF step as well as the ideal FF step reversible. The feasibility of these surface tractions will also be examined at this stage. Thus, at the end of step 2, all the difficulties associated with realizing the corresponding ideal FF step along the minimum plastic work path will be identified.

5.5.2.3 Step 3: analysis of a feasible forward forming step

Starting from the end configuration of the RF step (also called the preform for the corresponding FF step), a forward forming operation is now carried out. An elasto-viscoplastic material model with internal variables (e.g. Hart 1976, Anand 1982) can be used to analyze the forward forming step. The surface traction vectors obtained in step 2 are now utilized as boundary conditions for the FF step.

We start with a nominal design of the FF step that is feasible. A strain homogenization criterion requiring smooth variations in strain and limiting strain gradients will be imposed to avoid strain localization. Such constraints may be introduced through a penalty function or a Lagrange multiplier technique. If the surface traction vectors obtained in step 2 cannot be easily realized, we start from a perturbed feasible set of surface tractions. The global compatibility requirements are enforced for the entire duration of the FF step. A detailed description of the stress and deformation fields during the FF step are obtained. The work paths followed in this nominal FF step are traced in detail and are compared to the minimum plastic work path. Since a different rate-dependent state variable material model is used, and the global compatibility is strictly enforced during the FF step, any possible deviation of the end configuration of this nominal FF step from the desired end configuration of the corresponding ideal FF step (along minimum plastic work paths) should be monitored carefully.

5.5.2.4 Step 4: design sensitivities of the forward forming step

At this stage, an ideal FF step (representing an unconstrained optimum) along minimum plastic work paths has been obtained. Realizing that this ideal FF step may not be feasible in practice, we have also obtained and analyzed in detail a feasible candidate design for the FF step. The task of the designer now is to obtain an optimum that satisfies all the imposed constraints. In order to approach this task in a rational way, the designer needs to gather a thorough understanding of the effects of perturbing various design parameters on the state of the FF step.

To facilitate the insight gathering mission of the designer, at this stage, we will investigate the design sensitivities of the feasible candidate FF step with respect to various geometric, material, and process parameters. In this context, the design sensitivities are the derivatives of quantities like stresses, inelastic strains, and so on, with respect to appropriate geometric, material, or process variables.

Details of the design sensitivity calculations are presented in chapter 4. Knowing the velocities and their tangential derivatives along with the appropriate Lagrange rates of tractions from the analysis in step 3, we can directly evaluate the sensitivities of traction rates and tangential derivatives of velocities on the boundary of the domain using equations (4.2)–(4.5). The sensitivities of the Lagrange rates of traction may be written as

$$\begin{aligned}\overset{*}{\tau}{}_i^{(s)} = {} & \mathring{p}(t)n_i + \dot{p}(t)\overset{*}{n}_i + \mathring{q}(t)t_i + \dot{q}(t)\overset{*}{t}_i \\ & - \overset{*}{\dot{p}}(t)n_k h_{ki} - p(t)\overset{*}{n}_k h_{ki} - p(t)n_k \overset{*}{\dot{h}}_{ki} \\ & + \overset{*}{\dot{q}}(t)h_{ik}t_k + q(t)\overset{*}{\dot{h}}_{ik}t_k + q(t)h_{ik}\overset{*}{\dot{t}}_k \\ & + \overset{*}{\dot{p}}(t)d_{kk}n_i + p(t)\overset{*}{\dot{d}}_{kk}n_i + p(t)d_{kk}\overset{*}{n}_i \end{aligned} \tag{5.57}$$

where the sensitivity $\overset{*}{n}_i$ of the normal vector is given by equations (4.21–4.22) and $\overset{*}{t}_i$ can be obtained in analogous fashion. Also, a superscribed $^\circ$ above $p(t)$ and $q(t)$ denotes the sensitivity of $\dot{p}(t)$ and $\dot{q}(t)$, respectively. Finally, the current normal $\boldsymbol{n}$ is related to the normal $\boldsymbol{N}$ in the initial configuration by Nanson's formula

$$\boldsymbol{n} = \frac{\boldsymbol{N} \cdot \boldsymbol{F}^{-1}}{|\boldsymbol{N} \cdot \boldsymbol{F}^{-1}|} \tag{5.58}$$

For forming applications, equation (5.57) is specialized to an appropriate local coordinate system. Using sensitivity versions of the boundary conditions discussed in section 5.2 and setting the sensitivities to zero where the actual quantities are specified, the boundary conditions may be determined for calculating sensitivities of forming processes with respect to geometric, material, or process parameters. Sensitivities of nonelastic strain rates and stress rates on the boundary may also be obtained from equations (4.10), (4.11) and (4.23), (4.25), respectively. The internal equations (4.26)–(4.28) may be used to obtain the sensitivities of velocities, velocity gradients, and stress rates at an internal point in the domain of the workpiece.

Design sensitivities of the end configuration of the FF step are investigated in detail with respect to shape variations in the corresponding preform for the FF step, variations in the temporal and spatial distributions of the process parameters (surface traction vectors, friction at die–workpiece interfaces, etc.), and variations in material parameters. These sensitivity investigations will allow us to identify ways of obtaining desired final configurations while satisfying the compatibility and feasibility requirements for the FF step.

In this phase, one can also investigate design sensitivities of several parameters such as residual stress distributions in the end product, nonelastic strains, and so on, with respect to appropriate geometric, material, and process variables. Particular attention is given in this phase to sensitivities with respect to the preform shape parameters.

The shape of the preform for the FF step, which is one of the primary design variables, is defined on the boundary. The same is true for several other design parameters (e.g., surface tractions, friction conditions) of crucial interest. A BEM sensitivity formulation, by its ability to handle boundary variations in this class of elasto-viscoplastic problems involving large strains and rotations, provides an effective and efficient tool for such purposes. In our work, we will be very interested in the sensitivities of boundary stresses, which the BEM algorithm can deliver very efficiently as well.

5.5.2.5 Step 5: optimization of the forward forming step

The goal of this step will be to optimize the FF step under consideration. This implies concurrent optimization of the process parameters, as well as the geometric and material specifications of the preform for the FF step under consideration.

If the product design is assumed to be completed and only the manufacturing design is being considered, the effectiveness of the entire forward forming process will be ensured by imposition of a set of constraints relating to the quality of the final product. These constraints will require that the end configuration of the entire forming process satisfy certain geometric and material specifications derived from the product design and prescribed a priori. Such specifications may also include desired residual stress distributions.

In our approach, product specifications will determine the nth FF configuration [end state of the $(n-1)$th FF step or the initial state for the first RF step]. The $(n-1)$th configuration of the beginning configuration for the $(n-1)$th FF transformation will then be optimized along with the process parameters for the nth FF step.

The final or $(n-1)$th FF step needs to produce the predetermined end configuration. This FF step must also satisfy global compatibility requirements throughout the entire step, and the traction and displacement boundary conditions needed for the FF step must be realizable in practice. The strain distribution is also required to be continuous and a predetermined level of effective strain must be attained in the FF step. These requirements constitute the constraints for the optimization problem.

As discussed before, the efficiency of the FF step is measured by the amount of work needed to perform that step. Accordingly, the objective for the optimization procedure will be to expend a minimum amount of work in the FF step. Thus, we will attempt to minimize the deviations from the path of minimum plastic work in the FF step. This constitutes the goal of the optimization procedure.

The geometric and material parameters for the preform and the process parameters (displacement and traction boundary conditions) are the principal design variables for the forward forming step.

As discussed in step 1, the RF step, with a predetermined admissible strain distribution and deformations along the minimum plastic work paths of individual material elements, represents the most efficient transformation. The FF step obtained by reversing the RF transformation is essentially an optimum design, but it may not be realizable in practice. Thus, the corresponding FF step exists in the dual space.

In step 3, we have also obtained a candidate feasible design for the FF step. The actual optimum is now bounded and will lie between these two designs of the FF step, one lying in the feasible space and the other lying in the dual space.

The optimization procedure will start from the feasible design of the FF step. Utilizing the stress and deformation analysis performed in step 3 and the design sensitivities obtained in step 4, we will seek to optimize the FF step. As a first attempt in designing forming processes, optimization with regard to efficiency will be carried out first. The effectiveness of the FF step is determined by the end configuration. For the process design, satisfaction of the specifications for the end configuration will be guaranteed through constraints.

Existing optimization codes such as ADS (Vanderplaats 1983) may be used to carry out optimization of representative forward forming steps. At this point, we will seek to optimize the FF step with respect to both process parameters and preform specifications. The concepts of minimum plastic work paths, utilized in the RF process, give us a goal in terms of efficiency. Utilizing the insights gained from the sensitivity studies, we will now attempt to gradually modify the nominal feasible design to reach the goal for efficiency and seek an optimum constrained by feasibility and compatibility requirements. The sensitivity calculations will also allow us to make intelligent trade-offs when the efficiency goal (representing an unconstrained optimum) set by the RF step cannot be made through a feasible process. Thus, at the end of step 5, we expect to obtain a Pareto-optimal design of the FF step in terms of both efficiency and effectiveness.

As discussed earlier, the final or $(n-1)$th FF step will be optimized first. The optimized preform for the $(n-1)$th FF step will constitute the target end configuration to be achieved by the $(n-2)$th FF step and the procedure will continue until the preform for the entire process along with the optimized process parameter history has been obtained.

Integrated design of the forward forming process. Sections 5.5.2.1–5.5.2.5 describe the procedure for optimizing a particular step of the forward forming process. Starting from the desired product specifications (nth configuration of the FF process), the technique given above will be used to simultaneously optimize the final or $(n-1)$th transformation, as well as the $(n-1)$th configuration. Using the five steps outlined in sections 5.5.2.1–5.5.2.5, the $(n-2)$th transformation along with its beginning $[(n-2)\text{th}]$ configuration will now be optimized in the same manner. As shown in figure 5.24, the design procedure will continue until a desirable preform for the entire process (at the zero strain level) has been obtained along with its associated processing history.

Due to the nature of the design strategy, the design procedure can be terminated at any stage as may be desired. For a general forming process, the preform shape obtained through this procedure may not necessarily match the most economic billet shape. At that point, the designer may consider variations in the preform shape to make it more economic. The effects of such variations may also be studied using the sensitivity investigations. Iterations of the above steps may then be used to generate an optimized process plan considering the cost of producing preform shapes.

Long-term goal—concurrent product and process design. An optimized manufacturing design of formed products is the current goal of the proposed project. For this purpose, it has been assumed that the product design has already been completed.

5.5.2.6 Issues relating to concurrent product and process design

It is also important to note here that an optimized product design and an optimized manufacturing design do not guarantee the realization of an optimum overall design. Sequential optimization of the product design and the manufacturing design tends to produce less than optimum designs, as conflicts are resolved in an ad-hoc fashion and numerous iterations are needed to eventually arrive at a design that is effective and efficient. A superior procedure will consider the product design and the manufacturing design issues concurrently. For example, the shape of a part being formed may be simultaneously optimized with respect to both product design and manufacturing design goals. Thus, a design procedure consists of a product design phase and a manufacturing design phase, but one also needs the ability to consider both of these aspects simultaneously. Accordingly, the long-term goal of the proposed collaborative project is to investigate the issues regarding concurrent design of formed products along with the forming processes for realizing them.

This may be achieved by varying and extending the design strategy outlined in sections 5.5.2.1–5.5.2.5. Instead of introducing the specifications for the end configuration through constraints in section 5.5.2.5, the product specifications may be introduced through an additional objective function. The product design issues will determine the effectiveness objective for the forming process.

In designing each of the FF steps, the designer will then attempt to obtain an end configuration for the FF step that is as close as possible to the prespecified end configuration for that step. Accordingly, minimizing the difference between the desired and the obtained end configurations, in addition to the efficiency considerations discussed before, will be the dual objectives of the optimization procedure. Costs of satisfying the effectiveness objective will be weighed against potential benefits of product quality and performance. The entire optimization problem may then be treated as one involving multiple goals. Optimization based on both of these objectives relating to product performance issues and manufacturing issues will make the formed product at the end of the FF step both effective and efficient.

References

Alexander, J. M. (1972). "On the Theory of Rolling," *Proc. Royal Soc. (London), Series A*, **326**, 535–563.

Alexopoulos, P. S. (1981). "An Experimental Investigation of Transient Deformation Based on a State Variable Approach," Ph.D. thesis, Department of Materials Science and Engineering, Cornell University, Ithaca, N.Y.

Anand, L. (1982). "Constitutive Equations for the Rate-dependent Deformation of Metals at Elevated Temperatures," *J. Eng. Mater. Technol., ASME*, **104**, 12–17.

Atluri, S. N. (1984). "On Constitutive Relations at Finite Strain: Hypoelasticity and Elasto-plasticity with Isotropic or Kinematic Hardening," *Computer Meth. Appl. Mech. Eng.*, **43**, 137–171.

Avitzur, B. (1969). "Bulge in Hollow Disk Forging," Tech. Rept. AFML-TR-70-19.

Avitzur, B. (1980). *Metal Forming: The Applications of Limit Analysis*. Marcel Dekker, New York.

Banerjee, P. K. and Butterfield, R. (1981). *Boundary Element Methods in Engineering Science*. McGraw-Hill, London.

Bardes, B. P. (ed.) (1978). *Metals Handbook*, vol. 1. ASM, Metals Park, Ohio.

Bathe, K. J. and Chaudhary, A. (1985). "A Solution Method for Planar and Axisymmetric Contact Problems," *Int. J. Num. Meth. Eng.*, **21**, 65–88.

Bhargava, V., Hahn, G. T., and Rubin, C. A. (1985). "An Elastic-Plastic Finite Element Model of Rolling Contact: (1) Analysis of Single Contacts; (2) Analysis of Repeated Contacts," *J. Appl. Mech, ASME*, **52**, 67–82.

Campos, L.T., Oden, J.T., and Kikuchi, N. (1982). "A Numerical Analysis of a Class of Contact Problems with Friction in Elastostatics," *Computer Meth. Appl. Mech. Eng.*, **34**, 821–845.

Chandra, A. (1986). "A Generalized Finite Element Analysis of Sheet Metal Forming with an Elastic-Viscoplastic Material Model," *J. Eng. Ind., ASME*, **108**, 9–15.

Chandra, A. (1989a). "Profile Rolling of Gears: A Boundary Element Analysis," *J. Eng. Ind., ASME*, **111**, 48–55.

Chandra, A. (1989b). "Simulation of Rolling Processes by the Boundary Element Method," *Comput. Mech.*, **4**, 443–451.

Chandra, A. (1992). "A Boundary Element Formulation for Design Sensitivities in Thermoplastic Problems Involving Nonhomogeneous Media," *Eng. Anal.*, **10**, 49–57.

Chandra, A. and Mukherjee, S. (1984a). "Boundary Element Formulations for Large Strain–Large Deformation Problems of Viscoplasticity," *Int. J. Solids Structures*, **20**, 41–53.

Chandra, A. and Mukherjee, S. (1984b). "A Finite Element Analysis of Metal Forming Problems with an Elastic-Viscoplastic Material Model," *Int. J. Num. Meth. Eng.*, **20**, 1613–1628.

Chandra, A. and Mukherjee, S. (1984c). "A Finite Element Analysis of Metal Forming Processes with Thermomechanical Coupling," *Int. J. Mech. Sci.*, **26**, 661–676.

Chandra, A. and Mukherjee, S. (1985). "A Boundary Element Formulation for Sheet Metal Forming," *Appl. Math. Modeling*, **9**, 175–182.

Chandra, A. and Mukherjee, S. (1986a). "An Analysis of Large Strain Viscoplasticity Problems Including the Effects of Induced Material Anisotropy," *J. Appl. Mech., ASME*, **53**, 77–82.

Chandra, A. and Mukherjee, S. (1986b). "A Boundary Element Formulation for Large Strain Problems of Compressible Plasticity," *Int. J. Eng. Anal.*, **3**, 71–78.

Chandra, A. and Mukherjee, S. (1987). "A Boundary Element Analysis of Metal Extrusion Processes," *J. Appl. Mech., ASME*, **54**, 335–340.
Chen, C. C. and Kobayashi, S. (1978). "Rigid Plastic Finite Element Analysis of Ring Compression," *ASME, Appl. Num. Meth. Forming Processes, AMD*, **28**, 163–174.
Chung, K. and Richmond, O. (1989a). "A Generalized Deformation Theory of Plasticity Based on Minimum Work Paths," Division Report 52-89-05, ALCOA, Pittsburgh, Pa.
Chung, K. and Richmond, O. (1989b). "Ideal Forming, Part I: Homogeneous Deformation with Minimum Plastic Work," Division Report 52-89-01, ALCOA, Pittsburgh, Pa.
Chung, K. and Richmond, O. (1989c). "Ideal Forming, Part II: Sheet Forming with Most Uniform Deformation," Division Report 52-89-02, ALCOA, Pittsburgh, Pa.
Chung, K. and Richmond, O. (1990). "Mechanics of Ideal Forming Theory," Division Report 52-90-12, ALCOA, Pittsburgh, Pa.
Dawson, P. R. (1984). "A Model for the Hot or Warm Forming of Metals with Special Use of Deformation Mechanism Maps," *Int. J. Mech. Sci.*, **26**, 227–244.
Dawson, P. R. (1987). "On Modeling Mechanical Property Changes During Flat Rolling of Aluminum," *Int. J. Solids Structures*, **23**, 947–968.
Dawson, P. R. and Thompson, E. G. (1977). "Steady State Thermomechanical Finite Element Analysis of Elasto-Viscoplastic Metal Forming Processes," *Numerical Methods for Manufacturing Processes*, pp. 167–182. ASME, New York.
Dienes, J. K. (1979). "On the Analysis of Rotation and Stress Rate in Deforming Bodies," *Acta Mech.*, **32**, 217–232.
Eggert, G. M. and Dawson, P. R. (1987). "On the Use of Internal Variable Constitutive Equations in Transient Forming Processes," *Int. J. Mech. Sci.*, **29**, 95–113.
Goddard, J. D. and Miller, C. (1966). "An Inverse of the Jaumann Derivative and Some Applications in the Rheology of Viscoelastic Fluids," *Rheol. Acta*, **5**, 177–184.
Hart, E. W. (1976). "Constitutive Relations for the Non-Elastic Deformation of Metals," *J. Eng. Mater. Technol., ASME*, **98**, 193–202.
Hill, R. (1959). "Some Basic Principles in the Mechanics of Solids Without Natural Time," *J. Mech. Phys. Solids*, **7**, 209–225.
Hill, R. (1978). "Aspects of Invariance in Solid Mechanics," *Advances in Applied Mechanics, 18* (ed. Chia-Shun Yih). Academic Press, New York.
Hill, R. (1986). "External Paths of Plastic Work and Deformation," *J. Mech. Phys. Solids*, **6**, 1–8.
Johnson, G. C. and Bamman, D. J. (1984). "A Discussion of Stress Rates in Finite Deformation Problems," *Int. J. Solids Structures*, **20**, 725–737.
Kármán, T. von (1925). "On the Theory of Rolling," *Z. Angew. Math. Mech.*, **5**, 130–141.
Kobayashi, S., Oh, S.-I., and Altan, T. (1989). *Metal Forming and the Finite Element Method*. Oxford University Press, New York.
Kumar, V., Morjaria, M., and Mukherjee, S. (1980). "Numerical Integration of Some Stiff Constitutive Models of Inelastic Deformation," *J. Eng. Mater. Technol., ASME*, **102**, 92–96.
Lange, K. and Kurz, N. (1984). "Theoretical and Experimental Investigations of the

'Groß' Cold Shape-Rolling Process," Institute of Metal Forming, University of Stuttgart, Germany.

Lee, C. H. and Altan, T. (1972). "Influence of Flow Stress and Friction Upon Metal Flow in Upset Forging of Rings and Cylinders," *J. Eng. Ind.*, **94**, 1149–1155.

Lee, E. H., Mallet, R. L., and Wertheimer, T. B. (1983). "Stress Analysis for Anisotropic Hardening in Finite Deformation Plasticity," *J. Appl. Mech., ASME*, **50**, 554–560.

Lee, E. H., Mallet, R. L., and Yang, W. H. (1977). "Stress and Deformation Analysis of Metal Extrusion Processes," *Computer Meth. Appl. Mech. Eng.*, **10**, 339–353.

Li, G. J. and Kobayashi, S. (1982). "Rigid-Plastic Finite Element Analysis of Plane Strain Rolling," *J. Eng. Ind., ASME*, **104**, 55–64.

Martins, J. A. C. and Oden, J. T. (1983). "A Numerical Analysis of a Class of Problems in Elastodynamics with Friction," *Computer Meth. Appl. Mech. Eng.*, **40**, 327–360.

McMeeking, R. M. and Rice, J. R. (1975). "Finite Element Formulation for Problems of Large Elastic-Plastic Deformation," *Int. J. Solids Structures*, **11**, 601–616.

Merwin, J. E. and Johnson, K. L. (1963). "An Analysis of Plastic Deformation in Rolling Contact," *Proc. Inst. Mech. Eng.*, **177**, 667–673.

Mori, K., Osakada, K., and Oda, T. (1982). "Simulation of Plane-Strain Rolling by the Rigid-Plastic Finite Element Method," *Int. J. Mech. Sci.*, **24**, 519–527.

Mukherjee, S. (1982). *Boundary Element Methods in Creep and Fracture*. Elsevier Applied Science, Barking, Essex, U.K.

Mukherjee, S. and Chandra A. (1987). "Nonlinear Solid Mechanics," *Boundary Element Methods in Mechanics* (ed. D. E. Beskos), pp. 285–331. North-Holland, Amsterdam.

Nagtegaal, J. C. and De Jong, J. E. (1982). "Some Aspects of Non-Isotropic Workhardening in Finite Strain Plasticity," *Proceedings of the Workshop on Plasticity of Metals at Finite Strain* (ed. E. H. Lee and R. L. Mallet), pp. 65–102. Division of Applied Mechanics, Stanford University, Stanford, Calif., and Department of Mechanical Engineering, Aeronautical Engineering and Mechanics, R.P.I., Troy, N.Y.

Onate, E. and Zienkiewicz, O. C. (1983). "A Viscous Shell Formulation for the Analysis of Thin Sheet Metal Forming," *Int. J. Mech. Sci.*, **25**, 305–335.

Orowan, E. (1943). "The Calculation of Roll Pressure in Hot and Cold Flat Rolling," *Proc. Inst. Mech. Eng.*, **150**, 140–167.

Pires, E. B. and Oden, J. T. (1983). "Analysis of Contact Problems with Friction Under Oscillating Loads," *Computer Meth. Appl. Mech. Eng.*, **39**, 337–362.

Rajiyah, H. and Mukherjee, S. (1987). "Boundary Element Analysis of Inelastic Axisymmetric Problems with Large Strains and Rotations," *Int. J. Solids Structures*, **23**, 1679–1698.

Richmond, O. and Devenpeck, M. L. (1962). "A Die Profile for Maximum Efficiency in Strip Drawing," *Proc. Fourth U.S. Natl. Congr. Appl. Mech.*, pp. 1053–1057.

Torstenfelt, B. (1983). "Contact Problems with Friction in General Purpose Finite Element Computer Programs," *Computers Structures*, **16**, 487–493.

Vanderplaats, G. N. (1983). *Numerical Optimization Techniques for Engineering Design*. McGraw-Hill, New York.

Wang, N. M. and Wenner, M. L. (1982). "Effects of Strain-Hardening Representation in Sheet Metal Forming Calculations of 2036-T4 Aluminum," *Formability of Metallic Materials–2000 AD*, ASTM, STP-753, pp. 84–104.

William, K. J. (Ed.), (1984). *Constitutive Equations: Macro and Computational Aspects*. ASME, New York.

Zolti, E. (1983). "A Finite Element Procedure to Time Dependent Contact Analysis," *Computers Structures*, **17**, 555–561.

6

Axisymmetric Forming Processes

As discussed in chapter 5, interface conditions at the tool–workpiece interfaces play crucial roles in various forming operations. Axisymmetric upsetting and ring compression tests provide useful means for analytically investigating the interfacial friction conditions and validating them against experimental observations. A unique feature of deformation in ring compression, as well as in rolling and forging, is the existence of a neutral point (or region) along the tool–workpiece interface where the tangential relative velocity between the deforming material and the die becomes zero. The frictional stresses usually change direction across a neutral point (or region). The location of this point (or region), however, is not known a priori and this causes considerable difficulty in modeling such processes. This chapter discusses the applicability of BEM to this class of problems. Numerical results are presented for axisymmetric ring compression and extension problems. Salient features of these processes are then discussed in view of the numerical results.

6.1 Introduction

The BEM is a general-purpose method for numerical solution of boundary and initial-value problems involving linear as well as nonlinear operators and complicated explicit or implicit boundary conditions. It is very tolerant of aspect ratio degradation, can yield secondary variables as accurate as the primary ones, and avoids any "locking" for incompressible deformation fields. Due to these inherent features, the BEM has been used widely in recent years to analyze stress and deformation fields in problems involving both material and geometric nonlinearities. Chapter 2 of this book presents the fundamentals of a BEM formulation for fully nonlinear axisymmetric problems. Chapter 5 presents the applications to planar forming problems, and this chapter is devoted to axisymmetric forming problems. Here, we are particularly interested in problems with neutral points (or regions) across which the tangential traction reverses its direction on the tool–workpiece interface.

6.2 Interface Conditions for Axisymmetric Forming Problems

Following the development in section 5.2 for planar forming problems, the starting-point here is the equations for traction components at a point on the boundary of a body. Referring to figure 2.5 and equations (2.160) and (5.4), the Lagrange rates of traction components may be expressed (no summation over Greek indices) as

$$\begin{aligned}\tau_\rho^{(s)} &= \dot{p}(t)n_\rho + \dot{q}(t)t_\rho - p(t)[n_\rho v_{\rho,\rho} + n_\zeta v_{\zeta,\rho}] \\ &\quad + q(t)[v_{\rho,\rho}t_\rho + v_{\rho,\zeta}t_\zeta] + p(t)n_\rho[d_{\rho\rho} + d_{\theta\theta} + d_{\zeta\zeta}]\end{aligned} \tag{6.1a}$$

$$\begin{aligned}\tau_\zeta^{(s)} &= \dot{p}(t)n_\zeta + \dot{q}(t)t_\zeta - p(t)[n_\rho v_{\rho,\zeta} + n_\zeta v_{\zeta,\zeta}] \\ &\quad + q(t)[v_{\zeta,\rho}t_\rho + v_{\zeta,\zeta}t_\zeta] + p(t)n_\zeta[d_{\rho\rho} + d_{\theta\theta} + d_{\zeta\zeta}]\end{aligned} \tag{6.1b}$$

where $p(t)$ is the normal and $q(t)$ is the tangential component of the traction vector τ, and a field point is represented by coordinates (ρ, θ, ζ) as shown in figure 6.1.

The equation (6.1) may now be written in local Cartesian coordinates with base vectors e_α and e_β, where e_α coincides with the outward normal and e_β coincides with the anticlockwise unit tangent vector at a point (figure 6.1). The γ direction is identical to the global θ direction.

In this coordinate system,

$$p(t) = \tau_\alpha = \sigma_{\alpha\alpha}, \qquad q(t) = \tau_\beta = \sigma_{\alpha\beta} \tag{6.2}$$

and

$$\tau_\alpha^{(c)} = \hat{\sigma}_{\alpha\alpha}, \qquad \tau_\beta^{(c)} = \hat{\sigma}_{\alpha\beta} \tag{6.3}$$

where $\boldsymbol{\tau}^{(c)}$ represents the Cauchy rate of the traction and $\hat{\boldsymbol{\sigma}}$ denotes an objective rate of the Cauchy stress. Depending on the particular problem at hand and the level of shear strain expected, $\hat{\boldsymbol{\sigma}}$ may be viewed as the Jaumann rate of the Dienes rate (also called the Green–Naghdi rate). Issues regarding the choice of objective stress rates are discussed in Chapter 5 for planar forming problems. These issues are very similar for axisymmetric problems.

Source Coords (R, O, Z)
Field Coords (ρ, θ, ζ)
$r^2 = \rho^2 + R^2 + \zeta^2 + Z^2 - 2R\rho\cos\theta - 2Z\zeta$

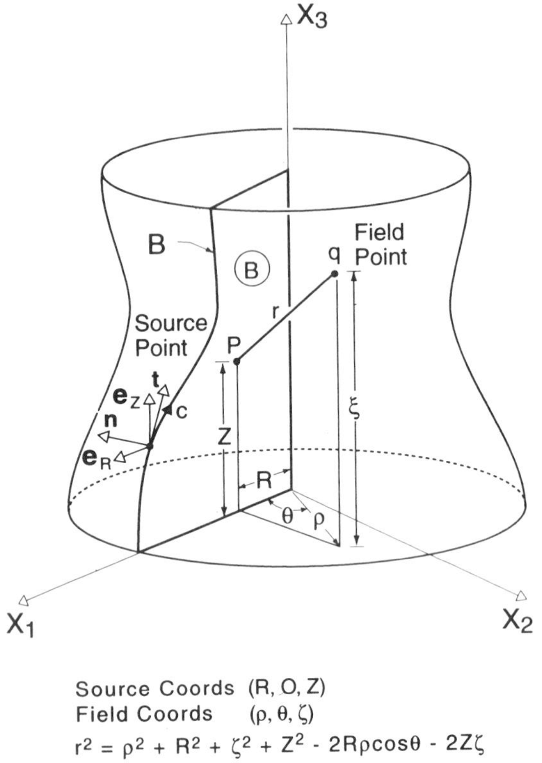

Figure 6.1. Schematic of an axisymmetric problem.

Various physical situations can now easily be examined. In general, one is concerned with velocity components v_α and v_β and traction rate components $\tau_\alpha^{(s)}$ and $\tau_\beta^{(s)}$. In an axisymmetric BEM formulation (equations (2.151)–(2.160), two BEM equations exist at P. Two other equations relating the above four physical quantities are needed. A discussion of specific situations follows.

6.2.1 Axisymmetric Ring Compression

It is well known (e.g., Chen and Kobayashi 1978, Carter and Lee 1986, Chandra and Srivastava 1991) that in axisymmetric upsetting and ring compression problems, the inner diameter of the ring increases and expands outward with the outer diameter for low values of friction coefficients at the tool–workpiece interface. As the friction coefficient is increased beyond a certain threshold value (which depends on material characteristics and the aspect ratio of the ring), the inner diameter of the ring deforms inward in a direction opposite to that of the outer diameter of the ring. This suggests the existence of a neutral plane (or region) across which the tangential traction at the interface reverses its direction, and such a phenomenon must be represented in the interface model for an axisymmetric ring compression problem.

Chen and Kobayashi (1978) have proposed the following equations for the axisymmetric ring compression tests (figure 6.2):

$$\widetilde{v}_\alpha = 0 \tag{6.4}$$

$$\tau_\beta = -\frac{\widetilde{v}_\beta}{|\widetilde{v}_\beta|}\mu\tau_\alpha\left[\frac{2}{\pi}\tan^{-1}\left(\frac{|\widetilde{v}_\beta|}{a}\right)\right] \tag{6.5}$$

where, $\widetilde{\boldsymbol{v}}$ is the velocity of the workpiece relative to that of the die at P and a is a constant which is several orders of magnitude smaller than the tangential velocity of the die relative to the workpiece. Thus, equation (6.5) models the change in direction of the frictional traction across the neutral plane (where $\widetilde{v}_\beta = 0$), while equation (6.4) ensures that the workpiece can not penetrate the die.

The traction rates in a rate formulation may be obtained by differentiating equation (6.5) with respect to time. However, this is not convenient, since acceleration would appear in the resulting differentiated versions of equation (6.5). In the past, Chandra and Srivastava (1991) have numerically estimated $\tau_\beta^{(c)}$ from equation (6.3). $\tau_\beta^{(s)}$ may now be expressed as

$$\tau_\beta^{(s)} = \tau_\beta^{(c)} - \sigma_{\alpha\alpha}\omega_{\alpha\beta} - \sigma_{\beta\beta}d_{\alpha\beta} + \sigma_{\alpha\beta}d_{\beta\beta} \tag{6.6}$$

and can be estimated with known $\tau_\beta^{(c)}$ from equation (6.6).

6.2.2 *Axisymmetric Extrusion*

For extrusion problems, the shear strain is usually much smaller than unity. Accordingly, the objective stress rate $\hat{\boldsymbol{\sigma}}$ may be safely interpreted as the Jaumann rate. It is also assumed here that the die is stationary and rigid and slipping conditions always prevail.

Referring to the local coordinate system in figure 6.3,

$$v_\alpha = 0 \tag{6.7}$$

If Coulomb friction is assumed to prevail, one can use

$$\tau_\alpha^{(s)} = \dot{\sigma}_{\alpha\alpha} + \sigma_{\alpha\alpha}h_{\beta\beta} - h_{\alpha\beta}\left(\frac{\widetilde{v}_\beta}{|\widetilde{v}_\beta|}\right)\mu\sigma_{\alpha\alpha} \tag{6.8}$$

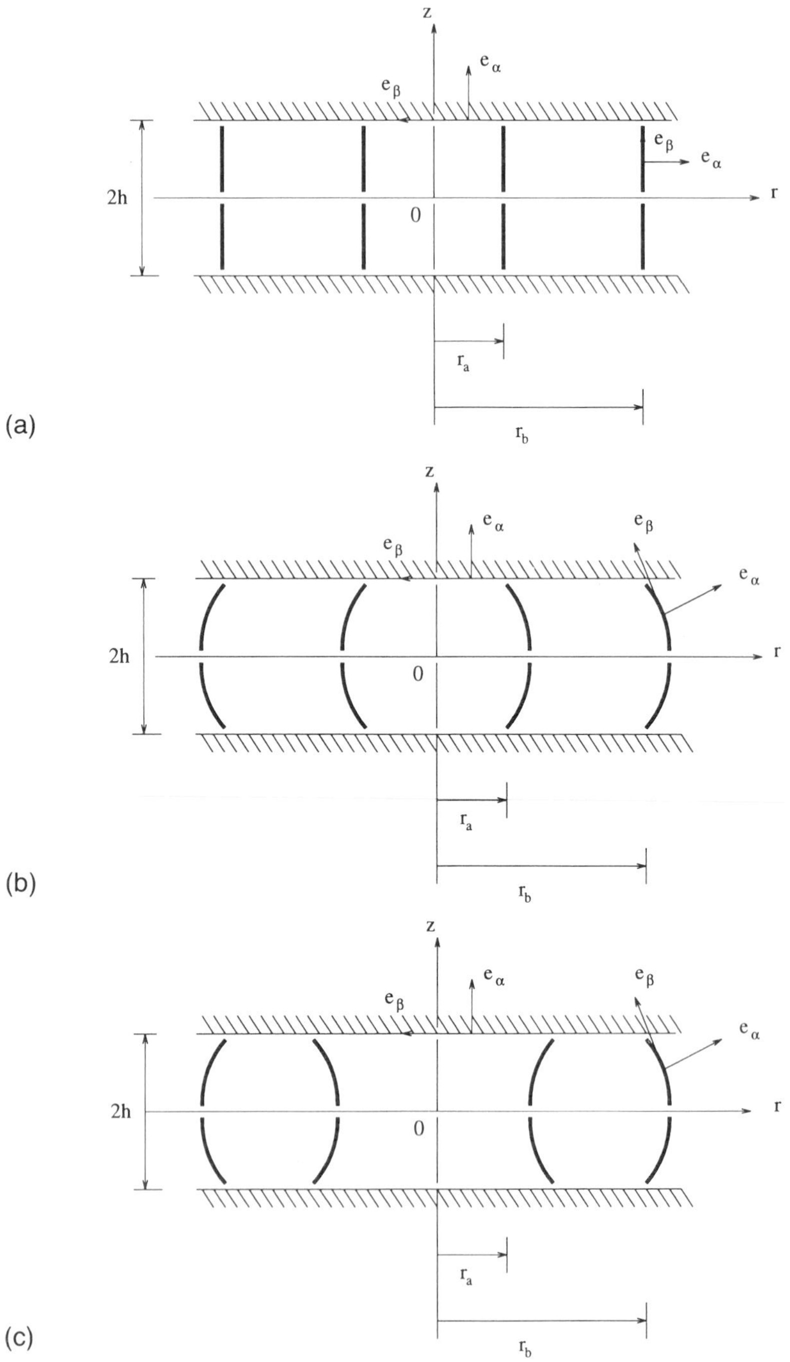

Figure 6.2. Geometry of the axisymmetric upsetting problem: (a) initial configuration; (b) Schematic of deformed configuration (low friction); (c) Schematic of deformed configuration (high friction).

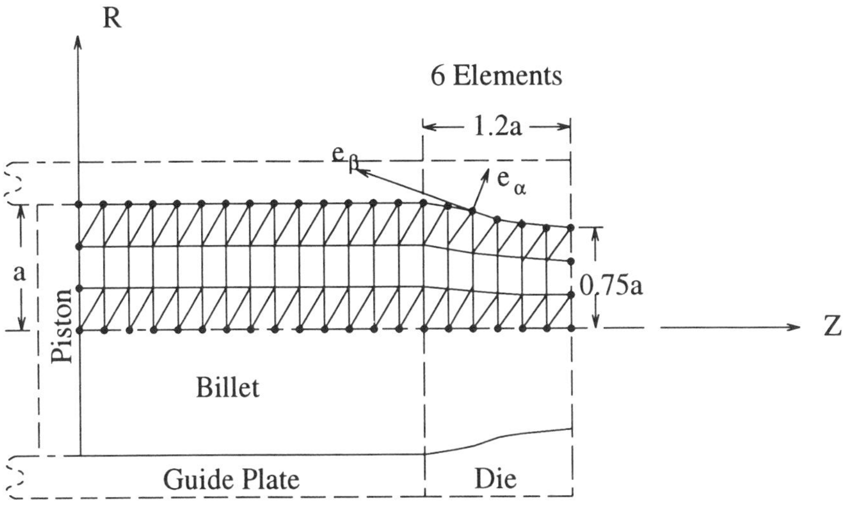

Figure 6.3. Geometry of the axisymmetric extrusion problem.

and

$$\tau_\beta^{(s)} = -\frac{\tilde{v}_\beta}{|\tilde{v}_\beta|}\mu\dot{\sigma}_{\alpha\alpha} - \sigma_{\alpha\alpha}h_{\alpha\beta} - h_{\beta\beta}\left(\frac{\tilde{v}_\beta}{|\tilde{v}_\beta|}\right)\mu\sigma_{\alpha\alpha} \tag{6.9}$$

with $\tilde{v}_\beta = v_\beta$ since the die is rigid and stationary.

As discussed in section 5.2.4 for planar extrusion problems, a friction law may also be assumed in the rate form, for the Cauchy traction rate $\tau_\beta^{(c)}$ (Chandra and Srivastava, 1991). In this case, it is convenient to express $\tau_\beta^{(s)}$ in terms of $\tau_\beta^{(c)}$. Possible friction laws may be expressed as

$$\tau_\beta^{(c)} = \hat{\sigma}_{\alpha\beta} = c_1 v_\beta \tag{6.10a}$$

or

$$\tau_\beta^{(c)} = \hat{\sigma}_{\alpha\beta} = c_2 h_{\beta\alpha} \tag{6.10b}$$

6.3 Numerical Implementation for Axisymmetric Cases

For numerical implementation, the first step, as usual, is to divide the boundary ∂B of an R–Z section of an axisymmetric body into N_s boundary segments and the interior into n_i internal cells. Denoting by $v_i(P_M)$ the components of the velocity at a point P which correspond with node M, a discretized version of the boundary equation [equation (2.150) with $p \to P$; no sum over ρ, θ, ζ; and $j = 1$ and 3] may be written as

$$
\begin{aligned}
C_{ij} v_i(P_M) = {} & \sum_{N_s} \int_{\Delta c} [U_{\rho j} \tau_\rho^{(s)} + U_{\zeta j} \tau_\zeta^{(s)} - T_{\rho j} v_\rho - T_{\zeta j} v_\zeta] \rho \, dc \\
& + \sum_{n_i} \int_{\Delta A_n} \Bigg[\lambda \left(U_{\rho j,\rho} + U_{\zeta j,\zeta} + \frac{U_{\rho j}}{\rho} \right) d^{(n)} + 2G \Bigg(U_{\rho j,\rho} d_{\rho\rho}^{(n)} \\
& + U_{\rho j,\zeta} d_{\rho\zeta}^{(n)} + U_{\zeta j,\zeta} d_{\zeta\rho}^{(n)} + U_{\zeta j,\zeta} d_{\zeta\zeta}^{(n)} + \frac{U_{\rho j} d_{\theta\theta}^{(n)}}{\rho} \Bigg) \Bigg] \rho \, d\rho \, d\zeta \\
& + \sum_{n_i} \int_{\Delta A_n} \Big(U_{\rho j,\rho} [\sigma_{\rho\rho} d_{\rho\rho} + \sigma_{\rho\zeta} (d_{\rho\zeta} - \omega_{\rho\zeta}) - \sigma_{\rho\rho} d] \\
& + U_{\rho j,\zeta} [\sigma_{\rho\rho} d_{\rho\zeta} + \sigma_{\rho\zeta} d_{\zeta\zeta} - \sigma_{\zeta\zeta} \omega_{\rho\zeta} - \sigma_{\rho\zeta} d] \\
& + U_{\zeta j,\rho} [\sigma_{\rho\rho} \omega_{\rho\zeta} + \sigma_{\rho\zeta} d_{\rho\rho} + \sigma_{\zeta\zeta} d_{\rho\zeta} - \sigma_{\zeta\rho} d] \\
& + U_{\zeta j,\zeta} [\sigma_{\rho\zeta} (d_{\rho\zeta} + \omega_{\rho\zeta}) + \sigma_{\zeta\zeta} d_{\zeta\zeta} - \sigma_{\zeta\zeta} d] + \frac{U_{\rho j}}{\rho} [\sigma_{\theta\theta} d_{\theta\theta} - \sigma_{\theta\theta} d] \Big)
\end{aligned}
\tag{6.11}
$$

A similar discretized equation may also be written for the velocity gradients $v_{j,\bar{l}}(p)$ by differentiating equation (2.150) at an internal source p and discretizing. Differentiating equation (2.150), the left-hand side becomes $v_{j,\bar{l}}(p)$. On the right-hand side, the boundary integral becomes

$$
\int_{\partial B^o} [U_{\rho j,\bar{l}} \tau_\rho^{(L)} + U_{\zeta j,\bar{l}} \tau_\zeta^{(L)} - T_{\rho j,\bar{l}} v_\rho - T_{\zeta j,\bar{l}} v_\zeta] \rho \, dc
$$

In the above expression, the derivatives have been moved under the integral sign, since p is an internal point, Q is a boundary point, and the above integral is regular. This, however, is not the case for the domain integrals in equation (2.150), which are in general $1/r$ singular. The derivatives $\partial I / \partial X_{\bar{l}}$, where I is either of the domain integrals on the right-hand side of equation (2.150), may be evaluated by the technique of Bui (1978). This approach gives rise to free terms. The appropriate free terms from the various derivatives of the displacement kernels, for the cases $p \notin$ on the axis of symmetry and $p \in$ on the axis of symmetry are given in tables 2.4 and 2.5, respectively. It is interesting to note the existence of free terms for some first derivatives of $U_{\rho 1}$ and $U_{\rho 3}$ if $p \in$ on the axis of symmetry.

Thus, for example, the explicit form of the equation for $v_{1,\bar{1}}(p_k)$ may be written as (at the kth internal source point, no sum over ρ or ζ, $p_k \notin x_3$ axis).

$$
\begin{aligned}
v_{1,\bar{1}}(p_k) = & \sum_{N_s} \int_{\Delta c} [U_{\rho 1,\bar{1}} \tau_\rho^{(L)} + U_{\zeta\ 1,\bar{1}} \tau_\zeta^{(L)} - T_{\rho 1,\bar{1}} v_\rho - T_{\zeta 1,\bar{1}} v_\zeta] \rho\, dc \\
& + \sum_{n_i} \int_{\Delta A_n - \Delta A \eta(p_k)} \left[\lambda \left(U_{\rho 1,\rho\bar{1}} + U_{\zeta 1,\zeta\bar{1}} + \frac{U_{\rho 1,\bar{1}}}{\rho} \right) d^{(n)} \right. \\
& + 2G \left(U_{\rho 1,\rho\bar{1}} d_{\rho\rho}^{(n)} + U_{\rho 1,\zeta\bar{1}} d_{\rho\zeta}^{(n)} + U_{\zeta 1,\rho\bar{1}} d_{\zeta\rho}^{(n)} + U_{\zeta 1,\zeta\bar{1}} d_{\zeta\zeta}^{(n)} \right. \\
& \left. \left. + \frac{U_{\rho 1,\bar{1}} d_{\theta\theta}^{(n)}}{\rho} \right) \right] \rho\, d\rho\, d\zeta + \lambda \left(\frac{5 - 8\nu}{16(1-\nu)G} - \frac{1}{16(1-\nu)G} \right) d^{(n)}(p_k) \\
& + \frac{5 - 8\nu}{8(1-\nu)} d_{\rho\rho}^{(n)}(p_k) - \frac{1}{8(1-\nu)} d_{\zeta\zeta}^{(n)}(p_k) \\
& + \sum_{n_i} \int_{\Delta A_n - \Delta A_\eta(p_k)} \{ U_{\rho 1,\rho\bar{1}} [\sigma_{\rho\rho} d_{\rho\rho} + \sigma_{\rho\zeta}(d_{\rho\zeta} - \omega_{\rho\zeta}) - \sigma_{\rho\rho} d] \\
& + U_{\rho 1,\zeta\bar{1}} [\sigma_{\rho\rho} d_{\rho\zeta} + \sigma_{\rho\zeta} d_{\zeta\zeta} - \sigma_{\zeta\zeta} \omega_{\rho\zeta} - \sigma_{\rho\zeta} d] \\
& + U_{\zeta 1,\rho\bar{1}} [\sigma_{\rho\rho} \omega_{\rho\zeta} + \sigma_{\rho\zeta} d_{\rho\rho} + \sigma_{\zeta\zeta} d_{\rho\zeta} - \sigma_{\zeta\rho} d] \\
& + U_{\zeta 1,\zeta\bar{1}} [\sigma_{\rho\zeta}(d_{\rho\zeta} + \omega_{\rho\zeta}) + \sigma_{\zeta\zeta} d_{\zeta\zeta} - \sigma_{\zeta\zeta} d] \\
& + \frac{U_{\rho 1,\bar{1}}}{\rho} [\sigma_{\theta\theta} d_{\theta\theta} - \sigma_{\theta\theta} d] \} \rho\, d\rho\, d\zeta \\
& + \frac{5 - 8\nu}{16G(1-\nu)} [\sigma_{\rho\rho}(p_k) d_{\rho\rho}(p_k) + \sigma_{\rho\zeta}(p_k)(d_{\rho\zeta}(p_k) \\
& - \omega_{\rho\zeta}(p_k)) - \sigma_{\rho\rho}(p_k) d(p_k)] \\
& - \frac{1}{16G(1-\nu)} [\sigma_{\rho\zeta}(p_k)(d_{\rho\zeta}(p_k) + \omega_{\rho\zeta}(p_k)) \\
& + \sigma_{\zeta\zeta}(p_k) d_{\zeta\zeta}(p_k) - \sigma_{\zeta\zeta}(p_k) d(p_k)]
\end{aligned} \tag{6.12}
$$

Similarly, other components of the velocity gradient (for $p_k \notin x_3$ axis and $p_k \in x_3$ axis) may be evaluated. In equation (6.12), $\Delta A_\eta(p_k)$ is a circle of small radius η, centered at p_k, in the plane of the generator of the axisymmetric solid. The Cauchy principal values of these integrals, with $\eta \to 0$, must now be evaluated accurately. This, in general, is a formidable task since the integrands are $1/r^2$ singular. Rajiyah and Mukherjee (1987) have developed a novel analytical-numerical technique for accurately determining these integrals, and a similar approach is pursued here. A detailed discussion of the analytical-numerical scheme for singular integration of $U_{ij,kl}$ over internal cells may be found in Rajiyah and Mukherjee (1987).

Suitable shape functions must now be chosen for the variation of velocity and traction rates along boundary elements and for the variation of $d_{ij}^{(n)}$ and $v_{i,j}$ over internal cells. It should be noted here that $\tau_i^{(s)}$ denotes Lagrangian traction rate. Thus, the present formulation involves velocity gradients in the domain integrals, as well as in the boundary integrals of equations (6.11) and (6.12). As is discussed in the section on interface modeling, the Cauchy traction rate $\tau_i^{(c)}$ in the tangential direction at the die–workpiece interface is obtained from a friction model. In the present work, the domain shape functions for $v_{i,j}$ are used to approximate those quantities at the boundary nodes, instead of using separate shape functions for $v_{i,j}$ on the boundary.

Let us now consider the time step from t to $t + \Delta t$. As shown in figure 6.4, σ_{ij} is known explicitly at time t. For viscoplastic material models, $d_{ij}^{(n)}$ is also known explicitly. As a first guess, d_{ij} and ω_{ij} (on the boundary and in the interior) from the previous time step are used. $\tau_i^{(c)}$ from the previous time step is also used to evaluate $\tau_i^{(L)}$. Equation (6.11) is solved first. After the matrix solution, an updated value for $\tau_i^{(c)}$, and hence for $\tau_i^{(s)}$, is obtained. New values of d_{ij} and ω_{ij} (at time t) on the boundary are obtained using the boundary algorithm given in equations (2.163)–(2.165) for velocity gradients and stress rates. Velocity gradients and hence d_{ij} and ω_{ij} (at time t) at internal points are evaluated from equation (6.12). These new values are incorporated in the discretized boundary equation (6.11) and iterated until convergence is achieved for $v_j(p)$. It should be noted here that the coefficient matrices for each time step may be stored after the LDU decomposition. Thus, iterations over the boundary equation (6.11) require repeated right-hand-side substitutions only. For time-independent plasticity, iterations over $d_{ij}^{(n)}$ will be required. Such iterations may also be carried out within the above strategy without added complications (e.g., Banerjee and Butterfield 1981). For problems involving large shear strains (particularly if shear strain is greater than 1), an integration scheme similar to that described in step 10 of section 5.3.4.1 may also be adopted for axisymmetric problems.

The boundary stress rates, at any time, are best obtained from a boundary algorithm rather than attempting to take the limit of the equation for $v_{j,T}$ as $p \rightarrow P$. This approach, requiring tangential differentiation of the velocity components at a boundary point, was first suggested by Rizzo and Shippy (1968) and was later extended to materially and geometrically nonlinear problems by Mukherjee and coworkers (1982, 1987).

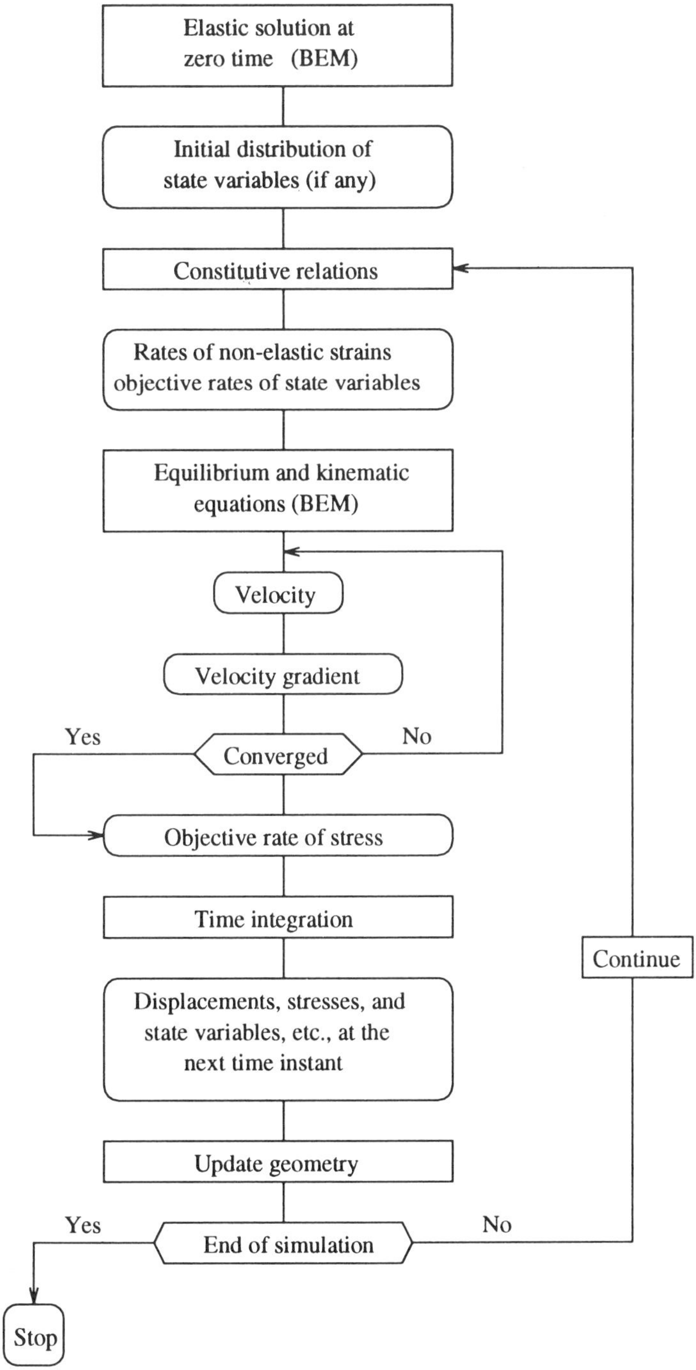

Figure 6.4. Flow chart for the solution of large-strain elasto-viscoplastic problems.

The normal and tangential components of the traction rate vector are first calculated at some point P on ∂B (P is assumed to lie at a point on ∂B where it is locally smooth). Now, in the local coordinate system n–c (where c is tangential to ∂B at P), $\hat{\sigma}_{nn}$ and $\hat{\sigma}_{nc}$ (in equation (6.3)) provide the normal and shearing components of Jaumann rates of Cauchy stress at P. For ring compression and axisymmetric extrusion, the n–c system may be identified with the appropriate α–β system at the interface. The normal and tangential components v_c and v_n are calculated next, and their tangential derivatives ($\partial v_c/\partial c$ and $\partial v_n/\partial c$) are obtained at P by numerical differentiation along the boundary element. The constitutive equations may then be written as

$$\frac{\partial v_c}{\partial c} = d_{cc} = \frac{1}{E}[\hat{\sigma}_{cc} - \nu(\hat{\sigma}_{nn} + \hat{\sigma}_{\theta\theta})] + d_{cc}^{(n)} \tag{6.13}$$

and

$$\frac{v_R}{R} = d_{\theta\theta} = \frac{1}{E}[\hat{\sigma}_{\theta\theta} - \nu(\hat{\sigma}_{nn} + \hat{\sigma}_{cc})] + d_{\theta\theta}^{(n)} \tag{6.14}$$

The spin is given as

$$\omega_{cn} = \omega_{RZ} = \frac{\hat{\sigma}_{nc}}{2G} + d_{nc}^{(n)} - \frac{\partial v_n}{\partial c} \tag{6.15}$$

The nonelastic deformation rates $d_{RR}^{(n)}$, $d_{\theta\theta}^{(n)}$, $d_{ZZ}^{(n)}$, and $d_{RZ}^{(n)}$ are known at P from the Cauchy stresses through an appropriate constitutive model. In the above, ω_{cn} is the spin at P, which is invariant with respect to coordinate transformation. Now, $d_{cc}^{(n)}$ and $d_{nc}^{(n)}$ are obtained as

$$d_{cc}^{(n)} = d_{RR}^{(n)} C_R^2 + d_{ZZ}^{(n)} C_Z^2 + 2 d_{RZ}^{(n)} C_R C_Z \tag{6.16}$$

and

$$d_{nc}^{(n)} = (d_{ZZ}^{(n)} - d_{RR}^{(n)}) C_R C_Z + d_{RZ}^{(n)} (C_R^2 - C_Z^2) \tag{6.17}$$

where

$$C_R = \mathbf{C} \cdot \mathbf{e}_R \qquad \text{and} \qquad C_Z = \mathbf{C} \cdot \mathbf{e}_Z \tag{6.18}$$

The quantities $\hat{\sigma}_{cc}$, $\hat{\sigma}_{\theta\theta}$, and ω_{RZ} are determined from equations (6.13)–(6.15). $\hat{\sigma}_{RR}$, $\hat{\sigma}_{ZZ}$, and $\hat{\sigma}_{RZ}$ may then be evaluated from stress transformations:

$$\hat{\sigma}_{RR} = \hat{\sigma}_{nn} C_Z^2 + \hat{\sigma}_{cc} C_R^2 - 2\hat{\sigma}_{nc} C_R C_Z \tag{6.19}$$

$$\hat{\sigma}_{ZZ} = \hat{\sigma}_{cc} - \hat{\sigma}_{RR} + \hat{\sigma}_{nn} \tag{6.20}$$

$$\hat{\sigma}_{RZ} = (\hat{\sigma}_{cc} - \hat{\sigma}_{nn}) C_R C_Z + \hat{\sigma}_{nc}(C_R^2 - C_Z^2) \tag{6.21}$$

6.4 Applications to Axisymmetric Forming

The axisymmetric BEM program used in the present study utilizes straight boundary elements and polygonal internal cells. The velocity and traction rates are taken to be piecewise quadratic over the boundary elements, while the nonelastic deformation rates and the velocity gradients are assumed to be piecewise linear over the internal cells. Possible discontinuities in traction rates are taken care of by placing "zero length" elements between nodes and assigning different values of traction rates at each of these nodes. Details on integrations of kernels are presented in chapter 2.

The particular material model used here is that due to Anand (1982). The particular material parameters used here are for Fe–0.05 carbon steel in a temperature range of 1173–1573 K and a strain rate range of 1.4×10^{-4} to 2.3×10^{-2}/s. Once again, for the sake of brevity, discussions on the material model are avoided here. Details of the material model along with the particular values of material parameters may be found in references by Anand (1982).

The FEM calculations are carried out using the algorithm of Bathe and Chaudhary (1985) incorporated in the finite element analysis code ADIMA. A similar strategy for FEM analysis has also been used by Carter and Lee (1986). An elasto-plastic material model (with isotropic work hardening) depicting the elasto-viscoplastic response of Fe–0.05 carbon steel in a temperature range of 1173–1573 K and at a strain rate of 10^{-3}/s is used in the FEM calculations. For the BEM analysis, the parameters for Anand's model were chosen to fit the response of the elasto-plastic model undergoing a tension test at a strain rate of 10^{-3}/s.

6.4.1 *Axisymmetric Upsetting and Ring Compression*

The ring geometry is chosen with a ratio of height to inner diameter to outer diameter ($h : d_i : d_0$) of 2 : 3 : 6. The diameter-to-height ratios ($d : h$) for the solid cylindrical specimens are 1.0, 2.0, and 3.0 (Chandra and Srivastava 1991).

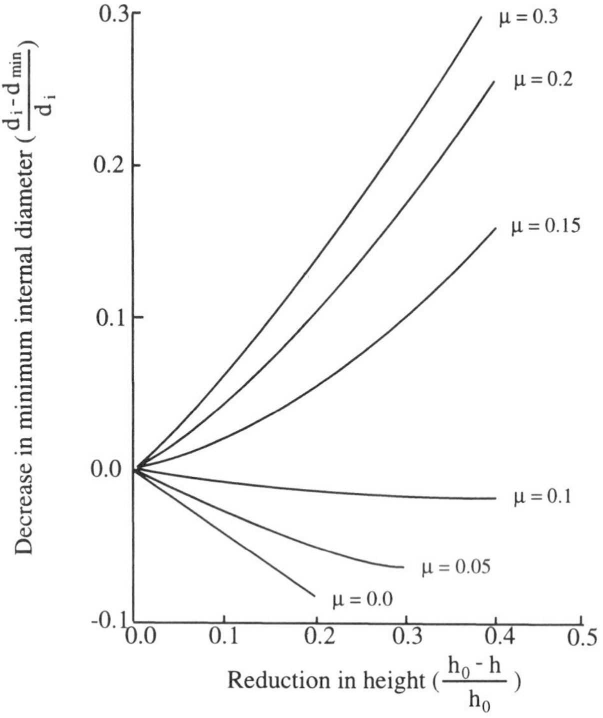

Figure 6.5. Ring compression test calibration curves for different friction coefficients.

Figure 6.5 shows the changes in minimum internal diameter as functions of reduction in height during ring compression tests with different friction conditions at the interface. During a single ring compression test, the coefficient of friction is assumed to be constant spatially and temporally. When there is no friction at the interface and the workpiece can slide freely, the inner diameter, as well as the outer diameter, expands without bulging as the ring is compressed. For $\mu = 0.05$, both the inner periphery and the outer periphery bulge outward from the axis. For $\mu = 0.1$, however, the expansion of the inner diameter increases by only 2 percent. With further increases in the coefficient of friction, the deformed shape of the ring changes drastically. At $\mu = 0.15$, the inner periphery bulges inward while the outer periphery bulges outward, indicating the existence of a "neutral plane" in between. At a 40 percent reduction in height, the internal diameter reduces by 16 percent. The inward bulging of the inner surface becomes more pronounced with further increases in the friction coefficient. A similar trend is also observed by Chen and Kobayashi (1978) and Carter and Lee (1986).

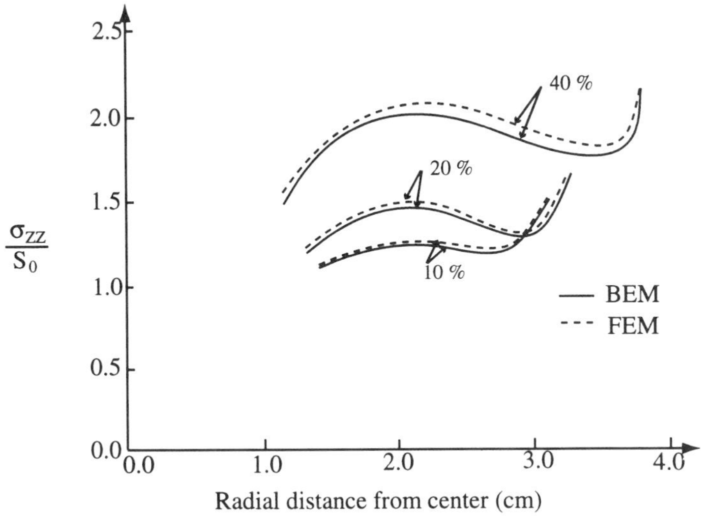

Figure 6.6. Pressure distributions along the interface during ring compression test at different reductions in height ($\mu = 0.2$).

Figure 6.6 shows the variation in pressure distributions at the interface at different stages of deformation. At a 10 percent reduction in height, the pressure distribution is almost uniform, with a slight increase at the outer periphery. With further deformation, the die pressure at a distance of about one-third of $(d_0 - d_i)$ from the inner periphery also increases, and for a 40 percent reduction in height shows a peak value of 1.98 for σ_{ZZ}/S_0 in this region. The pressure distributions at various stages of deformation obtained from the FEM analysis match quite well with those obtained from the BEM analysis. The maximum discrepancy in σ_{ZZ}/S_0 is less than 7 percent.

Figure 6.7 shows the load–displacement curves for upsetting of solid cylinders with different aspect ratios (d/h) at two different values of the friction coefficient. It is observed that the friction condition at the interface plays an important role in the upsetting process. For $d/h = 3$, the load requirement for a 40 percent reduction in height increases by 11.6 percent as μ is increased from 0.15 to 0.2. For $d/h = 2$, the increase in load requirement for a 40 percent reduction in height, as μ is increased from 0.15 to 0.2, is about 12.4 percent. As aspect ratio is changed from $d/h = 2$ to $d/h = 3$, the load requirement for a 40 percent reduction in height increases by 18.8 percent for $\mu = 0.2$ and by 19.7 percent for $\mu = 0.15$. The load–deflection curves also compare well qualitatively to the experimental curves for 6061-T6 aluminum reported by Carter and Lee (1986).

Figure 6.8 shows the pressure distribution at the interface for a 40 percent reduction in height during upsetting of cylinders with $\mu = 0.2$ for two

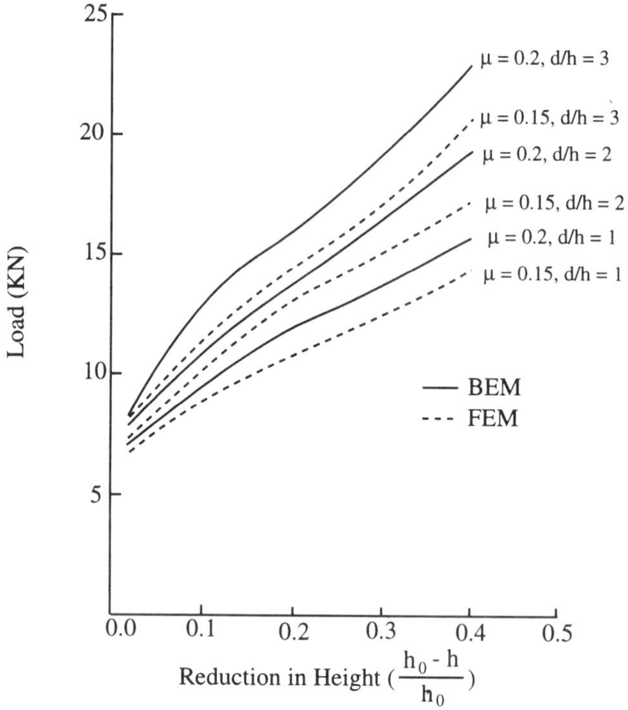

Figure 6.7. Load–displacement curves for upsetting of solid cylinders.

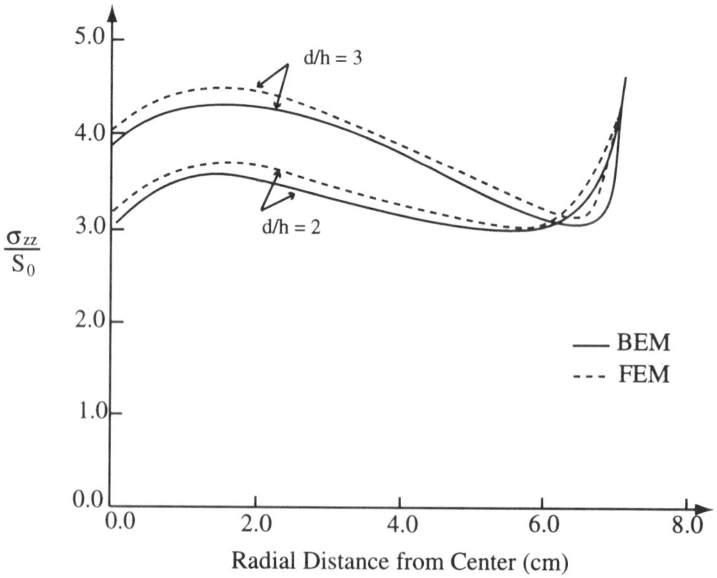

Figure 6.8. Pressure distributions along the interface during upsetting of solid cylinders ($\mu = 0.2$) at 40 percent reduction in height.

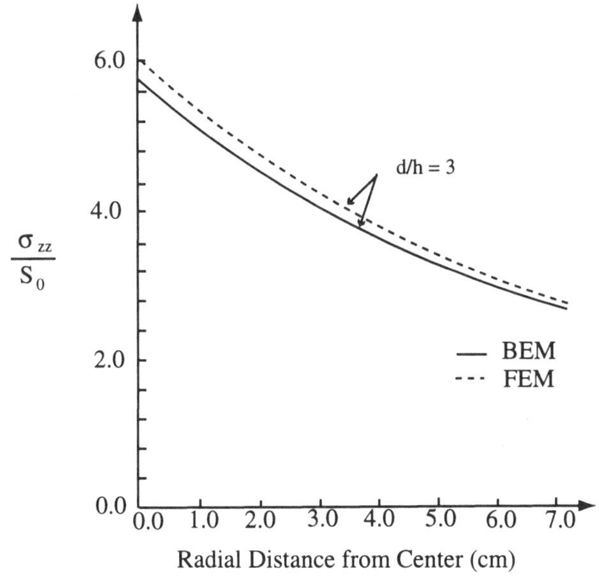

Figure 6.9. Pressure distributions along the interface during upsetting of solid cylinders (with barreling eliminated) at 40 percent reduction in height ($\mu = 0.2$).

different aspect ratios. As d/h is increased from 2 to 3, the pressure profile shifts upward and the peak moves from $\sigma_{ZZ}/S_0 = 3.56$ to 4.33. This is consistent with experimental observations. As d/h increases, the normal stress also increases due to a higher friction contribution. The high values of σ_{ZZ}/S_0 observed at the outside diameter provide a possible cause for sticking at the outside periphery. In general, the pressure profiles predicted by the BEM analysis match quite well with those obtained from the FEM analysis.

Figure 6.9 represents the pressure distribution along the interface for the same situation as in figure 6.5 ($d/h = 3$) except that the outer edge is constrained to remain straight. Hence, no barreling is permitted. In this case, the pressure distribution looks very similar to that obtained from friction hill analysis. When barreling was permitted, a relatively rigid cone of material developed under the die. Such a zone is not observed in this case. The pressure distribution obtained from FEM analysis also matches well with that obtained from the BEM analysis.

6.4.2 Axisymmetric Extrusion

The problem of axisymmetric extrusion is considered next (Chandra and Saigal 1991). Referring to figure 6.3, it is assumed that sliding occurs in the α–β plane only. A friction model of the type

$$\hat{\sigma}_{\alpha\beta} = \frac{G_s}{h}\mu v_\beta \tag{6.22}$$

is considered in the present analysis. Here, G_s and h are the shear modulus and height of the interface element, respectively, and μ represents the coefficient of friction. It is also assumed that $\sigma_{\alpha\beta}$ saturates to a value of $\mu\sigma_{\alpha\alpha}$. Thus,

$$\hat{\sigma}_{\alpha\beta} = \frac{G_s}{h}\mu v_\beta \qquad \text{if } \sigma_{\alpha\beta} < \mu\sigma_{\alpha\alpha} \tag{6.23}$$

and

$$\hat{\sigma}_{\alpha\beta} = 0 \qquad \text{if } \sigma_{\alpha\beta} = \mu\sigma_{\alpha\alpha} \tag{6.24}$$

Such an assumption is consistent with Coulomb's law of friction.

The material model used to analyze axisymmetric extrusion is that due to Anand (1982) for Fe–0.05 carbon steel in a temperature range of 1173–1573 K and strain rate range of 1.4×10^{-4} to 2.3×10^{-2}/s. For the friction model, the parameters $G_s/h = 1.268 \times 10^5$ MPa/m.

The FEM analysis reported here uses a piecewise quadratic description of velocities over triangular elements. Time integrations for both the BEM and FEM programs are carried out by an explicit Euler-type method with automatic step control.

The geometry of axisymmetric extrusion considered here is shown in figure 6.3, and the numerical results are presented in figures 6.10–6.13. Figure 6.10 shows the steady-state distributions of normalized axial stress, σ_{ZZ}/S_0, for three different piston velocities in the absence of friction, where S_0 is a material constant. In particular, the axis of the workpiece is chosen to be the z axis and the stress distributions are shown for material points in the deformed configuration that initially had the same relative radial position ($R/a = 0$ and $R/a = 1$) in the billet. The maximum residual tensile stress in an extruded workpiece is of crucial significance in design since this is the primary potential source for crack initiation and growth. It is seen from figure 6.10 that the rate dependence of σ_{ZZ}/S_0 is quite significant. As the piston velocity is tripled from 0.254 to 0.762 mm/s, the BEM analysis predicts a change of 17.75 percent in the maximum axial tensile stress in the workpiece. The faster the billet is forced through the die, the less time there is for stresses to relax at material points in the workpiece as they move through the die. Consequently, the maximum axial tensile stress upon exit from the die increases substantially with the speed of extrusion.

Another important feature of elasto-viscoplastic analysis is that, following a peak value, the magnitude of σ_{ZZ} decreases as a function of z in most of the billet that has passed through the die. This is a result of stress relaxation in the workpiece after it is deformed.

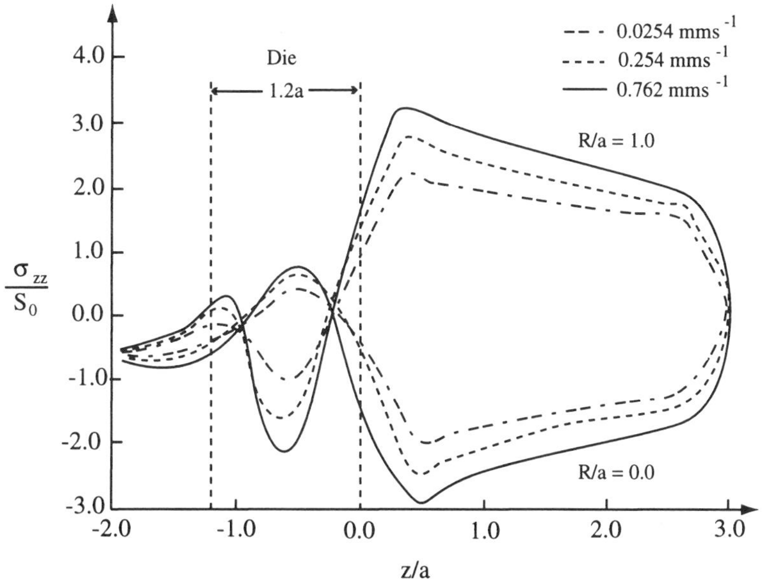

Figure 6.10. Steady-state distribution of σ_{ZZ} at different piston velocities for two values of R ($\mu = 0$).

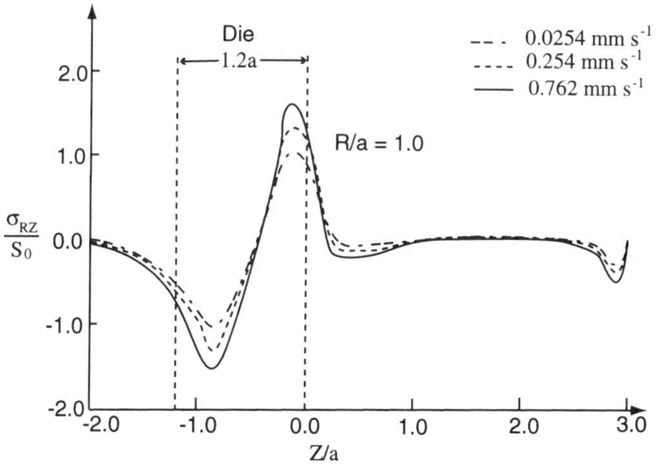

Figure 6.11. Steady-state distribution of σ_{RZ} as functions of piston velocity ($\mu = 0$).

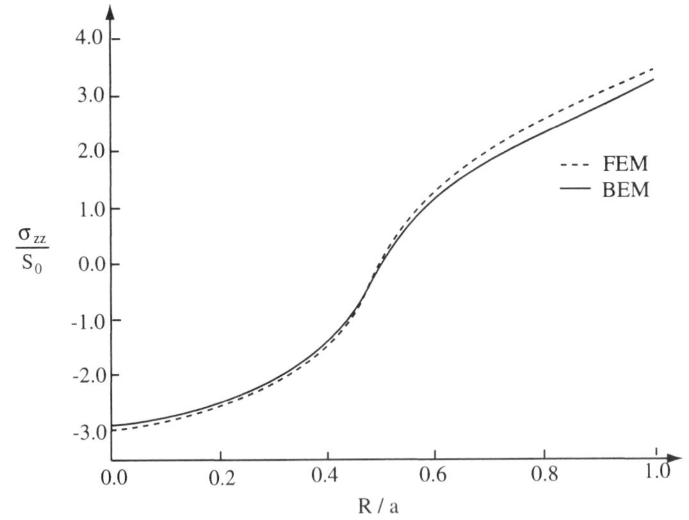

Figure 6.12. Residual stress distribution across a transverse section (piston velocity = 0.762 mm/s, $\mu = 0$).

The results for the steady-state distributions of the shearing stress σ_{RZ} are shown in figure 6.11. It is shown that there is a significant variation of shearing stress in the die region. Residual axial stress distribution over a cross-section at $z = 0.375a$ from the die exit is shown in figure 6.12. It should be noted that these residual stress distributions must be self-equilibrating and, for the residual stresses obtained from the current BEM analysis, the error in satisfying equilibrium is about 8 percent for σ_{ZZ} at $z = 0.375a$ from the die exit. At the die exit, the equilibrium of the residual σ_{ZZ} is satisfied with a 4 percent error and, at $z = 2a$ the error in equilibrium drops to less than 1 percent.

For comparison, the finite element code NIKE2D (Hallquist 1986) is used for analyzing the axisymmetric extrusion problem with no friction. NIKE2D cannot handle the state variable material model of Anand (1982). Accordingly, the stress–strain curve for a tension test at a strain rate of 0.5×10^{-2}/s is provided as the input material model to NIKE2D. The residual σ_{ZZ} at $z = 0.375$ obtained from NIKE2D is also shown in figure 6.12. The distributions obtained from the BEM analysis compare well with those obtained from the FEM analysis.

Figure 6.13 shows the effects of friction at the die–workpiece interface on the axial stress. Three different values of friction coefficients are chosen for a piston velocity of 0.762 mm/s. The presence of friction increases the peak value of the axial stress. Increasing the friction coefficient from $\mu = 0.0$

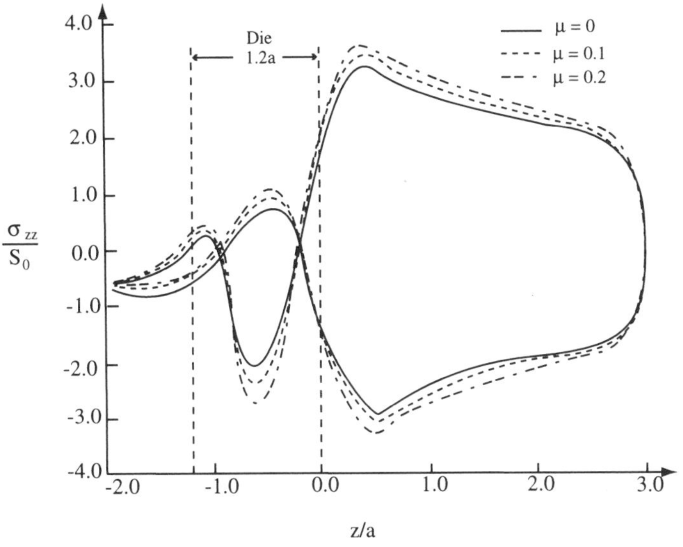

Figure 6.13. Steady-state distribution of σ_{ZZ} as functions of friction coefficients for two values of R (piston velocity = 0.762 mm/s).

to $\mu = 0.2$ raises the peak value of σ_{ZZ}/S_0 from 3.25 to 3.63. The spatial position of the peak remains almost unchanged. It should be noted here that the friction model with constant coefficient of friction, as used in the present analysis, is approximate and much work needs to be done on accurate and realistic modeling of friction in forming processes. The present analysis does, however, show the effects of friction on the metal extrusion processes.

6.5 Design Sensitivity and Optimization Issues

Issues regarding determination of design sensitivities of axisymmetric forming processes and their design optimization can be handled exactly the same way as discussed for planar forming problems. Section 5.5 illustrates concurrent preform and process design for planar forming processes. A similar procedure can also be used for axisymmetric forming processes with axisymmetric equations for analysis and design sensitivity studies.

References

Anand, L. (1982). "Constitutive Equations for the Rate-dependent Deformation of Metals," *J. Eng. Mater. Technol., ASME*, **104**, 12–17.

Banerjee, P. K. and Butterfield, R. (1981). *Boundary Element Methods in Engineering Science*. McGraw-Hill, London.

Bathe, K. J. and Chaudhary, A. (1985). "A Solution Method for Planar and Axisymmetric Contact Problems," *Int. J. Num. Meth. Eng.*, **21**, 65–88.

Bui, H. D. (1978). "Some Remarks About the Formulation of Three-dimensional Thermoelastic Problems by Integral Equations," *Int. J. Solids Structures*, **14**, 935–939.

Carter, W. T., Jr. and Lee, D. (1986). "Further Analysis of Axisymmetric Upsetting," *J. Eng. Ind., ASME*, **108**, 193–204.

Chandra, A. and Saigal, S. (1991). "A Boundary Element Analysis of the Axisymmetric Extrusion Processes," *Int. J. Nonlinear Mech.*, **26**, 1–13.

Chandra, A. and Srivastava, R. (1991). "A Boundary Element Analysis of Axisymmetric Upsetting," *Math. Comp. Modelling*, **15**, 81–92.

Chen, C. C. and Kobayashi, S. (1978). "Rigid Plastic Finite Element Analysis of Ring Compression," *Applications of Numerical Methods to Forming Processes, ASME*, AMD-**28**, 63–174.

Hallquist, J. O. (1986). "NIKE2D—A Vectorized Implicit Finite Deformation Finite Element Code for Analyzing the Static and Dynamic Response of 2-D Solids with Interactive Rezoning and Graphics," Lawrence Livermore National Laboratory, Livermore, Calif.

Mukherjee, S. (1982). *Boundary Element Methods in Creep and Fracture*. Elsevier Applied Science, Barking, Essex, U.K.

Mukherjee, S. and Chandra, A. (1987). "Nonlinear Solid Mechanics," *Boundary Element Methods in Mechanics* (ed. D. E. Beskos), pp. 285–331. North-Holland, Amsterdam.

Rajiyah, H. and Mukherjee, S. (1987). "Boundary Element Analysis of Inelastic Axisymmetric Problems with Large Strains and Rotations," *Int. J. Solids Structures*, **23**, 1679–1698.

Rizzo, F. J. and Shippy, D. J. (1968). "A Formulation and Solution Procedure for the General Nonhomogeneous Elastic Inclusion Problems," *Int. J. Solids Structures*, **4**, 1161–1179.

7

Solidification Processes

This chapter is concerned with the analysis and design of solidification of pure metals. In the first part of the chapter, a direct analysis is presented for the motion of the solid–liquid freezing interface and the time-dependent temperature field. An iterative implicit algorithm has been developed for this purpose using the boundary element method (BEM) with time-dependent Green's functions and convolution integrals. Emphasis is given to two-dimensional examples. The second part of the chapter provides a methodology for the solution of an inverse design Stefan problem. A method for controlling the fluxes at the freezing front and its velocity is demonstrated. The BEM, in conjunction with a sequential least squares technique, is used to solve this ill-posed problem, which has important technological applications. The accuracy of the method is illustrated through one-dimensional numerical examples.

7.1 Introduction

Problems of solidification of pure substances share the characteristic of an isothermal moving interface (freezing front). The freezing front motion and fluxes must be calculated as part of the solution of the phase change boundary value problem. Heat conduction is assumed in both solid and liquid phases and all thermal properties are considered temperature independent.

The flux discontinuity at any point of the interface is related to its normal velocity by the equation balancing the rate of heat flow with the energy rate required to create a fresh amount of solid per unit time (Stefan condition).

A solidification problem is considered direct when the temperature or the flux on the fixed boundary of a solidifying body, with given material properties, is prescribed.

There is an extensive literature on the above and related "Stefan" problems. The methods used to solve these problems can be categorized (Crank 1984) into analytical, front-tracking, front-fixing, and fixed-domain methods. The existing analytical solutions are primarily for one-dimensional problems (Crank 1984) and two-dimensional wedge-shaped spaces (Rathjen and Jiji 1971, Budhia and Kreith 1973).

Front-tracking methods involve finite differences or finite elements on a fixed grid (Lazaridis 1969, Rao and Sastri 1984) or on a variable space grid (Murray and Landis 1959), or the use of adaptive meshes (Bonnerot and Janet 1977, Lynch 1982, Albert and O'Neill 1986, Zabaras and Ruan 1990). An alternative formulation includes front-fixing methods (Crank and Gupta 1975), where the moving front is fixed by a suitable choice of space coordinates. In the fixed-domain methods the problem is formulated in such a way that the interface condition becomes implicit in a new form of the equations, which applies over the whole of a fixed domain (Voller and Cross 1981, Ralph and Bathe 1982, Roose and Storrer 1984, Hsiao 1985).

Integral formulations for one-dimensional problems have been applied by several authors (e.g., Chuang and Szekely 1972, Banerjee and Shaw 1982, O'Neill 1983, Sadegh et al. 1985, Heinlein et al. 1986). O'Neill (1983) gave a general integral formulation for quasi-static phase change problems, while Zabaras and Mukherjee (1987, 1994) extended this work to transient problems. Similar work has also been reported by Sadegh et al. (1985). Hong et al. (1984) have solved two-dimensional solidification problems by updating the position of the interface at each time step while keeping the interface location fixed during the calculation of the temperature field at each time step.

This first part of this chapter is concerned with a BEM formulation of two-dimensional direct solidification problems. The general integral equations are presented together with their numerical implementation. Special emphasis is given to key issues such as the accurate calculation of singular integrals, the iterative technique, and the calculation of interface motion. Numerical results for some sample two-dimensional solidification problems are presented and discussed. A detailed analysis of this problem is available in Zabaras and Mukherjee (1987).

Design Stefan problems, where the temperature fluxes and velocity are prescribed on the freezing front, while the temperature and the flux on the fixed boundary of the domain of interest are unknown and must be determined by the analysis, are also discussed in this chapter. They were first introduced in one dimension by Zabaras et al. (1988), who extended Beck's (1985) sensitivity analysis to problems with phase changes using a BEM analysis. By controlling the freezing interface fluxes and velocities during solidification, the cast structure can be controlled and made more uniform

(Zabaras et al. 1988, Zabaras and Mukherjee 1994, Flemings 1974). Further discussion and finite element results of such inverse design Stefan problems are given in publications by Zabaras and coworkers (Zabaras and Ruan 1989, Zabaras 1990, Ruan and Zabaras 1991, Zabaras et al. 1992). In these papers, the work reported in Zabaras et al. (1988) is extended to two-dimensional problems. Smoothing in time and space is introduced in the sense used by Zabaras et al. (1992). The BEM analysis allows easy and accurate calculation of the sensitivity coefficients and provides certain other advantages by permitting direct calculation of the surface fluxes. Typical one-dimensional results for the inverse problem are reported and discussed at the end of this chapter.

7.2 Direct Analysis of Solidification

This section begins with the governing differential equations for the problem. This is followed by an integral formulation. Next, numerical implementation of the integral formulation is discussed, with special attention paid to the issues of evaluation of integrals, modeling of corners, and obtaining a discretized (matrix) formulation.

7.2.1 Governing Differential Equations

A liquid at an initially uniform temperature T_i (equal to or above the melting point T_m) is assumed to occupy a region with a fixed boundary ∂B_0. At time $t > 0$, the boundary ∂B_0 is cooled to a temperature lower than the melting temperature T_m. Solidification starts all around ∂B_0 and proceeds inward. The interface at some time t is denoted by ∂B_I (figure 7.1)

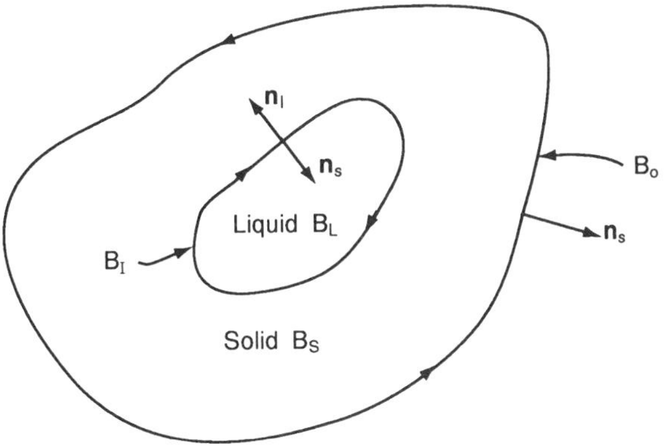

Figure 7.1. Geometry of the solidification problem (from Zabaras and Mukherjee 1987).

The governing differential equations, in the absence of heat sources and with constant material parameters, are (Crank 1984)

$$\frac{\partial T_s}{\partial t}(X,t) = k_s \nabla^2 T_s(X,t), \qquad X \in B_S \tag{7.1}$$

$$\frac{\partial T_l}{\partial t}(X,t) = k_l \nabla^2 T_l(X,t), \qquad X \in B_L \tag{7.2}$$

where, for example, $T_s(X,t)$ is the temperature at the point $X \in B_S$ at time t and the rest of the notation is clear from figure 7.1.

The thermal diffusivity k_s of the solid phase is equal to $K_s/\rho_s c_s$ in terms of its conductivity K_s, density ρ_s, and specific heat c_s, respectively. Similar notation is used for k_l. The boundary, initial, and freezing interface conditions are given as

$$T(X,t) = T_0(X,t), \qquad X \in \partial B_{0_1} \tag{7.3}$$

$$K_s \frac{\partial T}{\partial n} \equiv q(X,t) = q_0(X,t), \qquad X \in \partial B_{0_2} \tag{7.4}$$

$$T(X,t) = T_m, \qquad X \in \partial B_I \tag{7.5}$$

$$K_s \frac{\partial T_s}{\partial n_s} - K_l \frac{\partial T_l}{\partial n_s} = \rho_s L \frac{\partial V_n}{\partial t}, \qquad X \in \partial B_I \tag{7.6}$$

$$B_S = 0 \quad \text{at} \quad t = 0 \tag{7.7}$$

$$T(X,0) \equiv T_i = \text{const}, \qquad X \in B_L(0) \tag{7.8}$$

where T_0 is a prescribed temperature history on the part ∂B_{0_1} of ∂B_0 and q_0 is the prescribed flux on the remaining part ∂B_{0_2} of ∂B_0, T_m is the melting point of the solid, and L is the latent heat of fusion. Further, V_n is the normal velocity of the solidification front at a point on ∂B_I.

To simplify the calculations, it is assumed that T_i is constant throughout $B_L(0)$. With new simplified notations for fluxes, one can write the freezing interface normal velocity in the form

$$V_n = \frac{1}{\rho_s L} q_{ms} - \frac{1}{\rho_s L} q_{ml} \tag{7.9}$$

where

$$q_{ms} = K_s \frac{\partial T_s}{\partial n_s} \quad \text{and} \quad q_{ml} = K_l \frac{\partial T_l}{\partial n_s} \quad \text{for} \quad X \in \partial B_i$$

A direct solidification problem is defined as one of solving equations (7.1–7.9) for the interface normal velocity V_n, and the temperature field. This can be achieved by solving the integral equations as presented next.

7.2.2 *Integral Formulation*

One can write the following integral equations (Zabaras and Mukherjee 1987). For the solid phase, with $P \in \partial B_s \equiv \partial B_0 \cup \partial B_I$,

$$
\begin{aligned}
c(P)\,T(P,t) = \int_0^t dt_0 \int_{\partial B_0} \Big[& G_s(P,t;Q,t_0)\,q_0(Q,t_0) \\
& - k_s \frac{\partial G_s}{\partial n_s}(P,t;Q,t_0)\,T_0(Q,t_0) \Big] dS_Q \\
+ \int_0^t dt_0 \int_{\partial B_I} \Big[& G_s(P,t;Q,t_0)\,q_{ms}(Q,t_0) \\
& - k_s \frac{\partial G_s}{\partial n_s}(P,t;Q,t_0)\,T_m + G_s(P,t;Q,t_0)\,T_m V_n(Q,t_0) \Big] dS_Q
\end{aligned}
\tag{7.10}
$$

with the temperature T on the left-hand side of the above equation being equal to T_m and T_0, respectively, for the cases $P \in \partial B_I$ and $P \in \partial B_0$. These equations are called (7.10*a*) and (7.10*b*), respectively, for ease of reference.

For the liquid phase, with $P \in \partial B_I$,

$$
\begin{aligned}
c(P)(T_m - T_i) = \int_0^t dt_0 \int_{\partial B_I} \Big[& G_l(P,t;Q,t_0)(-q_{ml}(Q,t_0)) \\
& - k_l \frac{\partial G_l}{\partial n_l}(P,t;Q,t_0)(T_m - T_i) \\
& + G_l(P,t;Q,t_0)(T_m - T_i)(-V_n(Q,t_0)) \Big] dS_Q
\end{aligned}
\tag{7.11}
$$

with the Green's functions G_s and G_l defined as

$$
G(p,t;q,t_0) = \frac{\exp\left(-\dfrac{r^2}{4k(t-t_0)}\right)}{[4\pi k(t-t_0)]^{m/2}}
\tag{7.12}
$$

where m is the dimension of the problem and $c(P)$ in equations (7.10) and (7.4) is specified as in Zabaras and Mukherjee (1987).

A major simplification arises in equation (7.11) if $T_i = T_m$.

7.2.3 Numerical Implementation

The solution strategy consists of the use of suitable shape functions, in space as well as in time, for the unknowns of the problem, and marching forward in time. The solid–liquid interface, of course, is part of the solution and must be updated continuously in time. Convolution-type integrals must be calculated over a variable domain all the way from the initial zero time.

The boundaries ∂B_0 and ∂B_I are divided into N_1 and N_2 (at time zero) linear straight segments. The freezing interface mesh is considered, in general, a function of time, in order to account for the movement of the freezing front. Omitting indications of source and field points, the discretized forms of equations (7.10a), (7.10b), and (7.11) are as follows.

For the solid phase, with $P \in \partial B_I$,

$$c(P)\,T_m(t_F) = \sum_{f=1}^{F} \int_{t_{f-1}}^{t_f} dt_0 \sum_{k=1}^{N_1} \int_{\partial B_{0k}} \left(G_s q_0 - k_s \frac{\partial G_s}{\partial n_s} T_0 \right) dS_Q$$
$$+ \int_{f=1}^{F} \int_{t_{f-1}}^{t_f} dt_0 \sum_{k=1}^{N_2(t_0)} \int_{\partial B_{Ik}} \left(G_s(q_{ms} + T_m V_n) - k_s \frac{\partial G_s}{\partial n_s} T_m \right) dS_Q \tag{7.13}$$

and, with $P \in \partial B_0$,

$$c(P)\,T_0(P, t_F) = \sum_{f=1}^{F} \int_{t_{f-1}}^{t_f} dt_0 \sum_{k=1}^{N_1} \int_{\partial B_{0k}} \left(G_s q_0 - k_s \frac{\partial G_s}{\partial n_s} T_0 \right) dS_Q$$
$$+ \sum_{f=1}^{F} \int_{t_{f-1}}^{t_f} dt_0 \sum_{k=1}^{N_2(t_0)} \int_{\partial B_{Ik}} \left(G_s(q_{ms} + T_m V_n) - k_s \frac{\partial G_s}{\partial n_s} T_m \right) dS_Q \tag{7.14}$$

For the liquid phase, with $P \in \partial B_I$,

$$c(P)(T_m(t_F) - T_i) = \sum_{f=1}^{F} \int_{t_{f-1}}^{t_f} dt_0 \sum_{k=1}^{N_2(t_0)} \int_{\partial B_{Ik}} \left(G_l(-q_{ml} - (T_m - T_i) V_n) \right.$$
$$\left. - k_l \frac{\partial G_l}{\partial n_l} (T_m - T_i) \right) dS_Q \tag{7.15}$$

Note that the interface velocity and position enter the above equations in an explicit as well as in an implicit and nonlinear manner through the Green's functions.

Linear shape functions in space and in time are chosen here. Specifically, for the time step (t_{f-1}, t_f), and for a straight boundary element on ∂B_I with nodes 1 and 2 denoting the start and end of the element, the flux q_{ms}, for example, can be written as

$$q_{ms} = \phi_1 \psi_1 q_{ms_1}^{f-1} + \phi_2 \psi_1 q_{ms_2}^{f-1} + \phi_1 \psi_2 q_{ms_1}^{f} + \phi_2 \psi_2 q_{ms_2}^{f} \tag{7.16}$$

where the spatial and time shape functions ϕ_i and ψ_i are given as

$$\phi_1 = 1 - \frac{s}{\Delta s}, \qquad \phi_2 = \frac{s}{\Delta s} \tag{7.17}$$

$$\psi_1 = \frac{t_f - t_0}{\Delta t_f}, \qquad \psi_2 = \frac{t_0 - t_{f-1}}{\Delta t_f} \tag{7.18}$$

with $\Delta t_f = t_f - t_{f-1}$ and $q_{ms_i}^f$ denoting the nodal flux at node i at time t_f. The distance s is measured along the element of length $\Delta s(t)$, starting at node 1. Expressions similar to equation (7.17) are also valid for q_{ml}, V_n, and q_0.

The integrals which appear in equations (7.13–7.15) have one of the following forms:

$$I_{i1} = k \int_{t_{f-1}}^{t_f} dt_0 \int_{\partial B_k} \psi_i \frac{\partial G}{\partial n} ds \tag{7.19}$$

$$I_{i2} = k \int_{t_{f-1}}^{t_f} dt_0 \int_{\partial B_k} \frac{s}{\Delta s} \phi_i \frac{\partial G}{\partial n} ds \tag{7.20}$$

$$I_{i3} = \int_{t_{f-1}}^{t_f} dt_0 \int_{\partial B_k} \psi_i G \, ds \tag{7.21}$$

$$I_{i4} = \int_{t_{f-1}}^{t_f} dt_0 \int_{\partial B_k} \frac{s}{\Delta s} \psi_i G \, ds \tag{7.22}$$

for $i = 1, 2$ and ∂B_k an element on the stationary boundary ∂B_0 or on the moving interface ∂B_I. The source point of reference in the above integrals can lie on ∂B_0 or ∂B_I.

7.2.4 *Evaluation of Integrals*

1. Integration over ∂B_0 with source point on ∂B_0. This case includes integrals similar to those for non-phase-change heat conduction problems (see Brebbia et al. 1984).
2. Integration on ∂B_0 with source point on ∂B_I. With estimation of the position of the source point, a Gaussian integration in space and in time can be effective (no singular integrals appear in this category). The interface position at time t_F can be estimated by assuming that the interface moves during the interval (t_{F-1}, t_F) with the velocity it has at time t_{F-1}.
3. Integration over ∂B_I with source point on ∂B_0.
   (i) Case $t_f = t_F$: Using an estimate of the position of the interface during the interval (t_{F-1}, t_F), simple Gaussian integration can be effective.
   (ii) Case $t_f \neq t_F$: Considering that the interface is moving during the interval (t_{f-1}, t_f) with the constant known velocity $(V_f + V_{f-1})/2$, Gaussian integration can be used. Here, V_f and V_{f-1} are the velocities at the end and beginning of the time interval.
4. Integration over ∂B_I with source point on ∂B_I.
   (i) Case $t_f \neq t_F$: This case is very similar to the cases 2 and 3 above.
   (ii) $t_f = t_F$: If the source point $P \notin \partial B_{I_k}$, similar ideas to those above can be applied. The singular case $P \in \partial B_{I_k}$ requires special care. Splitting, for example (7.21), into two parts I'_{i3} and I''_{i3},

$$I_{i3} = \int_{t_{F-1}}^{t_F^*} dt_0 \int_{\partial B_{Ik}} \psi_i G\, dS + \int_{t_F^*}^{t_F} dt_0 \int_{\partial B_{Ik}} \psi_i G\, dS$$

where

$$t_F^* = t_{F-1} + \tfrac{3}{4}(t_F - t_{F-1})$$

the first integral I'_{i3} is nonsingular and can be evaluated as before. The singular integral I''_{i3} is obtained as follows. Since the interval $t_F - t_F^*$ is small, it is assumed that the order of the spatial and time integrals can be reversed in this case even though the interface element moves a little during this time interval. This assumption permits I''_{i3} to be evaluated analytically, as is done for the integrals in case 1 above. The shape functions in I''_{i3} are defined over the entire interval (t_{F-1}, t_F). The final expressions for these integrals are given in Zabaras and Mukherjee (1987).

7.2.5 *Modeling of Corners*

Single corner nodes on the interface ∂B_I. A length-weighted average normal $\boldsymbol{n}$ at the corner node i is defined as

$$\boldsymbol{n} = \frac{l_{i-1}\boldsymbol{n}_{i-1} + l_i\boldsymbol{n}_i}{l_{i-1} + l_i} \tag{7.23}$$

in terms of the lengths l_{i-1} and l_i and unit normals $\boldsymbol{n}_{i-1}$ and $\boldsymbol{n}_i$ of contiguous elements. The normal velocity V_n at node i is now assigned along the unit vector $\boldsymbol{n}/|\boldsymbol{n}|$.

Double corner nodes at physical corners on ∂B_I. Let $V_{n(i)}$ and $V_{n(i+1)}$ be the normal velocities at the (physically same) nodes i and $i+1$. To avoid singular matrices, we assume that the physical corner has a unique velocity of magnitude V_n in the average normal direction $\boldsymbol{n}$ defined by equation (7.23) with l_i and $\boldsymbol{n}_i$ replaced by l_{i+1} and $\boldsymbol{n}_{i+1}$, respectively.

An independent relation between $V_{n(i)}$ and $V_{n(i+1)}$ results as

$$V_{n(i)}(\boldsymbol{n} \cdot \boldsymbol{n}_{i+1}) - V_{n(i+1)}(\boldsymbol{n} \cdot \boldsymbol{n}_{i-1}) = 0 \tag{7.24}$$

The tangential motion of the freezing interface modes does not come into the physics of the problem. However, it has to be specified artificially by the analyst so that the proper mesh is always preserved on ∂B_I. Further discussion of the importance of the tangential motion of interface nodes is given in Zabaras and Ruan (1990).

Double corner nodes at physical corners on ∂B_0. An interesting situation arises at a physical corner on ∂B_0 if the temperature T is prescribed on both elements meeting there. Again, a singular matrix can be avoided in this situation by including an extra equation. The unknown flux at one of the double nodes is replaced by the flux obtained by backward differences from the prescribed temperatures on contiguous elements at these nodes.

7.2.6 *Matrix Formulation*

Consider the case when T_0 is prescribed all over ∂B_0. Equations (7.9), (7.13), and (7.14) contain the unknown nodal boundary vectors $\boldsymbol{q}_{ms}^F$, $\boldsymbol{q}_{ml}^F$, and $\boldsymbol{q}_0^F$ at time t_F. In matrix form we can write them as

$$\begin{bmatrix} [A]\alpha & [A]\beta & [\Delta] \\ [B]\alpha & [B]\beta & [E] \\ [\Gamma]\gamma & [\Gamma]\delta & [\emptyset] \end{bmatrix} \begin{bmatrix} \boldsymbol{q}_{ms}^F \\ \boldsymbol{q}_{ml}^F \\ \boldsymbol{q}_0^F \end{bmatrix} = \begin{bmatrix} \varepsilon \\ \zeta \\ \eta \end{bmatrix} \tag{7.25}$$

where

$$\alpha = 1 + \frac{T_m}{\rho L} \tag{7.26}$$

$$\beta = -\frac{T_m}{\rho L} \tag{7.27}$$

$$\gamma = -\frac{(T_m - T_i)}{\rho L} \tag{7.28}$$

$$\delta = -1 + \frac{T_m - T_i}{\rho L} \tag{7.29}$$

and $[\emptyset]$ denotes the zero matrix. The matrices $[A]$, $[B]$, $[\Gamma]$, $[\Delta]$, and $[E]$ contain calculations over the last time step (t_{F-1} to t_F). The vectors on the right side of (7.25) are known from previous calculations (time zero to t_{F-1}) and the applied boundary conditions, and they are not given explicitly here.

For the case $T_m = T_i$, a reduced form of equation (7.25) (with $\boldsymbol{q}_{ml} = 0$) has been used. For the superheated case ($T_i > T_m$), equation (7.25) has been combined with

$$\frac{1}{\rho L}(\boldsymbol{q}_{ms}^F - \boldsymbol{q}_{ml}^F) = \boldsymbol{V}_{\text{pr}} \tag{7.30}$$

where $\boldsymbol{V}_{\text{pr}}$ is the predicted velocity vector in the time interval t_{F-1} to t_F.

Equations (7.25)–(7.30) have been solved in a least squares sense for the flux vectors $\boldsymbol{q}_{ms}$, $\boldsymbol{q}_{ml}$, and $\boldsymbol{q}_0$. Equation (7.30) is introduced in order to avoid numerical divergence that appears at early times, when $\partial B_0 = \partial B_{0_2}$ and when equation (7.25) is used alone. It is obvious that a proper scaling of equations (7.25) and (7.30) has to be used before their final solution.

A simple iterative procedure is adopted here. Before updating the geometry and continuation to the next time step, one must check whether the nodal positions on ∂B_I at time t_F, predicted at the end of a successful iteration, are such that numerical instabilities could appear at the next time step t_F to t_{F+1}. These instabilities usually occur when the freezing interface nodes come very close to each other. If this is the case, then remeshing must be performed by node removal or node rearrangement (i.e., by introducing nodal motion tangential to the interface).

The advantage of this BEM formulation, compared to the so-called domain methods and other front-tracking techniques, is that for the calculation of the temperature at points internal to the domain one does not have to calculate all the internal temperatures over the entire domain. The equation corresponding to internal temperature calculation is not given here, but it has a form similar to that of the boundary equations given earlier. Two disadvantages are also present in the BEM analysis. At first, as in other front-tracking techniques (Zabaras and Ruan 1990), one must assume at time zero the existence of some solid (Zabaras and Mukherjee 1987). The second major disadvantage is that the fluxes q_{ms} and q_{ml} appear separately in the analysis and not in the combined form that yields the normal interface velocity (see equation (7.9)). As a result, the direct calculation of both interface fluxes separately leads to progressively incorrect front motion which can eventually violate global energy conservation. For a weak form of the Stefan condition, an energy-conserving scheme, and more references on the subject, see Zabaras and Ruan (1990).

7.3 An Inverse (Design) Solidification Problem

This section begins with a definition of an inverse solidification problem. These problems are generally ill-posed and regularization methods are usually needed in order to make them tractable. Such methods are discussed. This is followed by a discussion of methods of accurate calculation of the sensitivity coefficients that are needed for inverse problems.

7.3.1 The Problem

Of concern here is an inverse (design) solidification problem that is defined as follows: "Given the thermal properties of the solid and liquid phases and the melting temperature, calculate the boundary flux/temperature on ∂B_0 that achieves a desired freezing front motion."

For solidification in a one-dimensional region $0 \leq x \leq l$, one must specify the interface fluxes $q_{ms}(t)$ and $q_{ml}(t)$ instead of specifying only the interface motion $V(t)$. In this case, Zabaras et al. (1988) have shown that two uncoupled inverse problems arise: one in the solid and another in the liquid phase. They derived an unstable analytical solution in the form of an infinite series involving the prescribed interface flux and velocity and their time derivatives.

The importance of the above and related design solidification problems lies in the fact that the quality of the solidifying crystals is directly related to the freezing interface fluxes and velocity rather than the applied cooling boundary conditions on $\partial\Omega_0$. These problems are ill-posed in the sense that their solution may not be unique and stable to small changes in the desired interface motion. Here a general methodology is presented for two-dimensional problems and some one-dimensional examples are discussed.

7.3.2 Future Information and Spatial Regularization Methods

The boundary element analysis prescribed earlier is a convenient tool for the analysis of the above design problem since the freezing interface position/motion is known a priori. Indeed, let us consider a boundary element discretization of ∂B_0 and ∂B_I and a time-stepping process. The main unknowns of the design problem are considered to be the nodal fluxes (or temperatures) all over ∂B_0. Let $\boldsymbol{q}_0^F$ denote the boundary nodal unknown fluxes on ∂B_0 at $t = t_F$, that is,

$$\boldsymbol{q}_0^F = \{q_{01}^F, q_{02}^F, \ldots, q_{0i}^F, \ldots, q_{0N_1}^F\}^{\mathrm{T}} \tag{7.31}$$

where N_1 is the number of variables to be estimated, and q_{0i}^F is the boundary heat flux at the ith boundary node at time t_F. The given interface motion/position is treated as a boundary condition on ∂B_I. Then, one can consider that the temperature field at any point inside the domain is a function of $\boldsymbol{q}_0^F$ (through the solution of a direct boundary value problem). Assume that the temperature distribution at time t_{F-1} is known and that $\boldsymbol{q}_{0^*}^F$ is an estimate of the vector of boundary nodal fluxes. Let the times $t_{F+i-1}, i = 1, 2, \ldots, r$, be future times with $r - 1$ denoting the number of future time steps and N_2 the number of nodes on ∂B_I. Then, following Beck et al. (1985), the vector $\boldsymbol{q}_0^{F+i-1}$ is temporally constrained to be given as $\boldsymbol{q}_0^{F+i-1} = \boldsymbol{q}_{0^*}^F$, $i = 1, 2, \ldots, r$. The temperatures at the N_2 nodal points in the solid–liquid interface at time t_{F+i-1} can be approximated using the following truncated Taylor series expansion:

$$T_k^{F+i-1} = \overset{*}{T}{}_k^{F+i-1} + \frac{\partial T_k^{F+i-1}}{\partial \boldsymbol{q}_0^F}(\boldsymbol{q}_0^F - \boldsymbol{q}_{0^*}^F)$$

$$k = 1, 2, \ldots, N_2, \qquad i = 1, 2, \ldots, r \tag{7.32}$$

Equation (7.32) can be written in a compact form as

$$\boldsymbol{T} = \overset{*}{\boldsymbol{T}} + \boldsymbol{S}(\boldsymbol{q}_0^F - \boldsymbol{q}_{0^*}^F) \tag{7.33}$$

where

$$\boldsymbol{T} = (\boldsymbol{T}_1, \boldsymbol{T}_2, \ldots, \boldsymbol{T}_i, \ldots, \boldsymbol{T}_r)^{\mathrm{T}}$$

with

$$\boldsymbol{T}_i = (T_1^{F+i-1}, T_2^{F+i-1}, \ldots, T_k^{F+i-1}, T_{N_2}^{F+i-1})^{\mathrm{T}} \tag{7.34}$$

and the sensitivity matrix $\boldsymbol{S}$ is defined as

$$\boldsymbol{S} = (\boldsymbol{S}_1, \boldsymbol{S}_2, \ldots, \boldsymbol{S}_i, \ldots, \boldsymbol{S}_r)^{\mathrm{T}}$$

where

$$\boldsymbol{S}_i = (\boldsymbol{S}_1^{F+i-1}, \boldsymbol{S}_2^{F+i-1}, \ldots, \boldsymbol{S}_k^{F+i-1}, \boldsymbol{S}_{N_2}^{F+i-1})^{\mathrm{T}} \tag{7.35}$$

with

$$\boldsymbol{S}_k^{F+i-1} = \frac{\partial T_k^{F+i-1}}{\partial \boldsymbol{q}_0^F} \tag{7.36}$$

The vector $\overset{*}{\boldsymbol{T}}$ has a form similar to the vector $\boldsymbol{T}$ and is calculated through a direct problem, using $\boldsymbol{q}_{0^*}^F$ and the known freezing front motion. More details on this will be given later.

The goal here is to calculate the optimum value of the vector $\boldsymbol{q}_0^F$ such that the error between the approximated temperatures T_k^{F+i-1} and the given interface temperature (T_m) is minimum, that is,

$$\min_{\boldsymbol{q}_0^F} \{(\boldsymbol{Y} - \boldsymbol{T})^{\mathrm{T}} \boldsymbol{W} (\boldsymbol{Y} - \boldsymbol{T}) + \alpha_0 (\boldsymbol{q}_0^F)^{\mathrm{T}} \boldsymbol{W}_0 \boldsymbol{q}_0^F + \alpha_1 (\boldsymbol{H}\boldsymbol{q}_0^F)^{\mathrm{T}} \boldsymbol{W}_1 (\boldsymbol{H}\boldsymbol{q}_0^F)\} \tag{7.37}$$

where α_0 and α_1 are regularization parameters with $\alpha_0 > 0, \alpha_1 > 0$; $\boldsymbol{W}$, $\boldsymbol{W}_0$, and $\boldsymbol{W}_1$ are optimization weighting matrices; and $\boldsymbol{H}$ is the first-order spatial regularization coefficient matrix. These matrices are discussed in detail in Zabaras et al. (1992). The vector $\boldsymbol{Y}$ is defined as

$$\boldsymbol{Y} = (\boldsymbol{Y}_1, \boldsymbol{Y}_2, \ldots, \boldsymbol{Y}_i, \ldots, \boldsymbol{Y}_r)^{\mathrm{T}} \qquad \text{where} \qquad \boldsymbol{Y}_i = (T_m, T_m, \ldots, T_m, \ldots, T_m)^{\mathrm{T}} \tag{7.38}$$

The second term in equation (7.37) has been added to keep the estimated boundary fluxes at finite values, and the last term is necessary to avoid large flux variation between adjacent nodal points (Tikhonov and Arsenin 1977). Performing the minimization, and after some manipulation,

$$\boldsymbol{q}_0^F = (\boldsymbol{S}^{\mathrm{T}} \boldsymbol{W} \boldsymbol{S} + \alpha_0 \boldsymbol{W}_0 + \alpha_0 \boldsymbol{H}^{\mathrm{T}} \boldsymbol{W}_1 \boldsymbol{H})^{-1} [\boldsymbol{S}^{\mathrm{T}} \boldsymbol{W} (\boldsymbol{Y} - \overset{*}{\boldsymbol{T}}) + (\boldsymbol{S}^{\mathrm{T}} \boldsymbol{W} \boldsymbol{S} \boldsymbol{q}_{0^*}^F] \tag{7.39}$$

A discussion on the selection of the regularization parameters is given in Zabaras et al. (1992). Using equation (7.39), the boundary nodal unknowns can be found. The temperature field can then be obtained by solving a direct boundary value problem. An iterative procedure must be performed due to the nonlinearity of the problem. For a related BEM analysis of an inverse elasticity problem, see Zabaras et al. (1989).

To evaluate $\overset{*}{\boldsymbol{T}}$ one must solve a direct problem with prescribed flux $\boldsymbol{q}_{0*}^F$ on ∂B_0 and known interface motion on ∂B_I. This is a slightly different direct problem from the one presented in the first part of this chapter.

The integral equation (7.9) and (7.13–7.15) provide $(N_1 + 3N_2)$ equations that can be solved for the $(N_1 + 3N_2)$ unknowns that include $\boldsymbol{T}_0^F(N_1), \boldsymbol{q}_{ms}^F(N_2), \boldsymbol{q}_{ml}^F(N_2)$, and $\overset{*}{\boldsymbol{T}}(N_2)$. It must be emphasized again that it is the interface velocity [i.e., from equation (7.9) the interface flux discontinuity] rather than the individual interface fluxes, $\boldsymbol{q}_{ms}$ and $\boldsymbol{q}_{ml}$ that are required in solving this direct problem.

7.3.3 *Calculation of the Sensitivity Coefficients*

An easy and rather obvious way to calculate sensitivity coefficients is by finite difference approximations. For example, one can write

$$\frac{\partial T_k^{F+i-1}}{\partial q_{0a}^F} \approx \frac{\overset{**}{T}{}_k^{F+i-1} - \overset{*}{T}{}_k^{F+i-1}}{\lambda \overset{*}{q}{}_{0a}^F}$$

$$k = 1, 2, \ldots, N_2; \qquad i = 1, 2, \ldots, r; \qquad a = 1, 2, \ldots, N_1 \tag{7.40}$$

where the temperature $\overset{**}{\boldsymbol{T}}$ is calculated by solving the direct problem similar to $\overset{*}{\boldsymbol{T}}$ with the boundary condition $(1 + \lambda)\boldsymbol{q}_{0*}^F$ and λ equal to, say, 0.001.

An alternative, direct, and more accurate way of calculating these sensitivity coefficients using the BEM was presented by Zabaras et al. (1988). To demonstrate the calculation of the sensitivity coefficients at t_F, one writes equations (7.13)–(7.15) a matrix form as follows [see Brebbia et al. (1984) for notation]. Equation (7.13):

$$\begin{aligned} \boldsymbol{A}_F^F \boldsymbol{T}_0^F + \boldsymbol{AI}_F^F \boldsymbol{T}_m^F &= \boldsymbol{B}_F^F \boldsymbol{q}_0^F + \boldsymbol{BI}_F^F \boldsymbol{q}_{ms}^F + \boldsymbol{A}_F^{F-1} \boldsymbol{T}_0^{F-1} + \boldsymbol{AI}_F^{F-1} \boldsymbol{T}_m^{F-1} \\ &+ \boldsymbol{B}_F^{F-1} \boldsymbol{q}_0^{F-1} + \boldsymbol{BI}_F^{F-1} \boldsymbol{q}_{ms}^{F-1} + \cdots \end{aligned} \tag{7.41}$$

where all matrices $\boldsymbol{A}_F^J$ and $\boldsymbol{B}_F^J$, $J = F, \ldots, 1$ are of order $N_2 \times N_1$, and all matrices $\boldsymbol{AI}_F^J$ and $\boldsymbol{BI}_F^J$, $J = F, \ldots, 1$ are of order $N_2 \times N_2$. Similarly, equation (7.14) becomes

$$\begin{aligned} \boldsymbol{C}_F^F \boldsymbol{T}_0^F + \boldsymbol{CI}_F^F \boldsymbol{T}_m^F &= \boldsymbol{D}_F^F \boldsymbol{q}_0^F + \boldsymbol{DI}_F^F \boldsymbol{q}_{ms}^F + \boldsymbol{C}_F^{F-1} \boldsymbol{T}_0^{F-1} + \boldsymbol{CI}_F^{F-1} \boldsymbol{T}_m^{F-1} \\ &+ \boldsymbol{D}_F^{F-1} \boldsymbol{q}_0^{F-1} + \boldsymbol{DI}_F^{F-1} \boldsymbol{q}_{ms}^{F-1} + \cdots \end{aligned} \tag{7.42}$$

where all matrices $\boldsymbol{C}_F^J$ and $\boldsymbol{D}_F^J$, $J = F, \ldots, 1$ are of order $N_J \times N_1$, while $\boldsymbol{CI}_F^J$ and $\boldsymbol{DI}_F^J$ are of order $N_1 \times N_2$. Finally, equation (7.15) becomes

$$\boldsymbol{G}_F^F \boldsymbol{T}_m^F = \boldsymbol{E}_F^F \boldsymbol{q}_{ml}^F + \boldsymbol{G}_F^{F-1} \boldsymbol{T}_m^{F-1} + \boldsymbol{E}_F^{F-1} \boldsymbol{q}_{ml}^{F-1} + \cdots \tag{7.43}$$

where all matrices $\boldsymbol{G}_F^J$ and $\boldsymbol{E}_F^J$, $J = F, \ldots, 1$ are of order $N_2 \times N_2$.

Subscripts in the above matrices denote current time of reference, while superscripts denote the time interval during which the integration is carried out. $\boldsymbol{T}_0^J(N_1 \times 1)$ denotes all nodal temperatures on ∂B_0 at time t_J; $\boldsymbol{q}_0^J(N_1 \times 1)$ denotes all nodal fluxes at t_J; and $\boldsymbol{q}_{ms}^J$ and $\boldsymbol{q}_{ml}^J$ (both $N_2 \times 1$) denote the interface fluxes in the solid and liquid phases, respectively. Finally, $\boldsymbol{T}_m^J(N_2 \times 1)$ are the calculated temperatures at the moving front. Note that the above matrices can easily be calculated in an explicit manner since the freezing front motion is known a priori. One-dimensional calculations can be found in Zabaras et al. (1988).

Let the interface velocity be given. Suppose that one wants to find the sensitivity of $\boldsymbol{T}_m^F$ with respect to $\boldsymbol{q}_0^F$, that is, $\partial \boldsymbol{T}_m^F / \partial \boldsymbol{q}_0^F$. The equations (7.41)–(7.43) are written as follows:

$$\begin{bmatrix} \boldsymbol{A}_F^F & \boldsymbol{AI}_F^F & -\boldsymbol{BI}_F^F & 0 \\ \boldsymbol{C}_F^F & \boldsymbol{CI}_F^F & -\boldsymbol{DI}_F^F & 0 \\ 0 & \boldsymbol{G}_F^F & 0 & -\boldsymbol{E}_F^F \\ 0 & 0 & \boldsymbol{I} & -\boldsymbol{I} \end{bmatrix} \begin{Bmatrix} \boldsymbol{T}_0^F \\ \boldsymbol{T}_m^F \\ \boldsymbol{q}_{ms}^F \\ \boldsymbol{q}_{ml}^F \end{Bmatrix} = \begin{bmatrix} \boldsymbol{B}_F^F & 0 \\ \boldsymbol{D}_F^F & 0 \\ 0 & 0 \\ 0 & \rho L \boldsymbol{I} \end{bmatrix} \begin{Bmatrix} \boldsymbol{q}_0^F \\ \boldsymbol{V}^F \end{Bmatrix}$$

$$+ \text{known terms from calculations at earlier times} \tag{7.44}$$

where $\boldsymbol{I}$ is a unit diagonal matrix of order $N_2 \times N_2$. The sizes of the above quantities are, respectively, $(N_1 + 3N_2) \times (N_1 + 3N_2)$, $(N_1 + 3N_2) \times 1$, $(N_1 + 3N_2) \times (N_1 + N_2)$, and $(N_1 + N_2) \times 1$.

Equation (7.14) can be rewritten as

$$\begin{Bmatrix} \boldsymbol{T}_0^F \\ \boldsymbol{T}_m^F \\ \boldsymbol{q}_{ms}^F \\ \boldsymbol{q}_{ml}^F \end{Bmatrix} = [\boldsymbol{M}] \begin{Bmatrix} \boldsymbol{q}_0^F \\ \boldsymbol{V}^F \end{Bmatrix} + \text{known terms} \tag{7.45}$$

where $\boldsymbol{M}$ is a $(N_1 + 3N_2) \times (N_1 + N_2)$ matrix. The coefficients of this matrix contain the sensitivities of the variables on the left with respect to those on the right. For example, the sensitivity matrix

$$\left[\frac{\partial \boldsymbol{T}_m^F}{\partial \boldsymbol{q}_0^F} \right] = [\boldsymbol{N}] \tag{7.46}$$

where $\boldsymbol{N}$ is the submatrix of $\boldsymbol{M}$ that contains rows $(N_1 + 1)$ through $(N_1 + N_2)$ and columns 1 through N_1 of $\boldsymbol{M}$.

No one has yet investigated the merits of equation (7.45) relative to those of the finite difference equations (7.40). Note that to evaluate $\partial \boldsymbol{T}_m^J / \partial \boldsymbol{T}_0^F$, $J = F+1, \ldots, F+r-1$, one should write equations similar to equation (7.44) where the reference time is not t_F but $t_{F+i-1}, i = 2, \ldots, r$. In doing so, the boundary fluxes should be regularized in time such that

$$\boldsymbol{q}_0^{F+i-1} = \boldsymbol{q}_0^F, \qquad i = 1, 2, \ldots, r$$

Details and final expressions for one-dimensional problems can be found in Zabaras et al. (1988).

7.4 Numerical Examples

Numerical results, for both direct and inverse problems, are presented in this section.

7.4.1 Dimensionless Parameters

For the problems to be considered, the following dimensionless parameters have been used:

$$K_s = 1, \qquad \hat{K}_l = \frac{K_l}{K_s}, \qquad c_s = 1, \qquad \hat{c}_l = \frac{c_l}{c_s}, \qquad \hat{x} = \frac{x}{R}, \qquad \hat{y} = \frac{y}{R}$$

$$\tau = \frac{k_s t}{R^2}, \qquad \theta = \frac{T - T_m}{T_m - T_0}, \qquad L = \frac{1}{St}$$

where R is a characteristic length, and St is the Stefan number defined as

$$St = \frac{c_s(T_m - T_0)}{L}$$

7.4.2 The Direct Problem

The first example considered is that of a square 2×2 which is filled with liquid at the melting point $(\theta_i = 0)$. The surface is suddenly cooled to $\theta_0 = -1$.

Figure 7.2 shows the interface locations at various dimensionless times τ. Figures 7.3 and 7.4 show interface progression with time along the adiabatic $(x_1 = 1)$ and diagonal $(x_1 = x_2)$. The BEM results are here

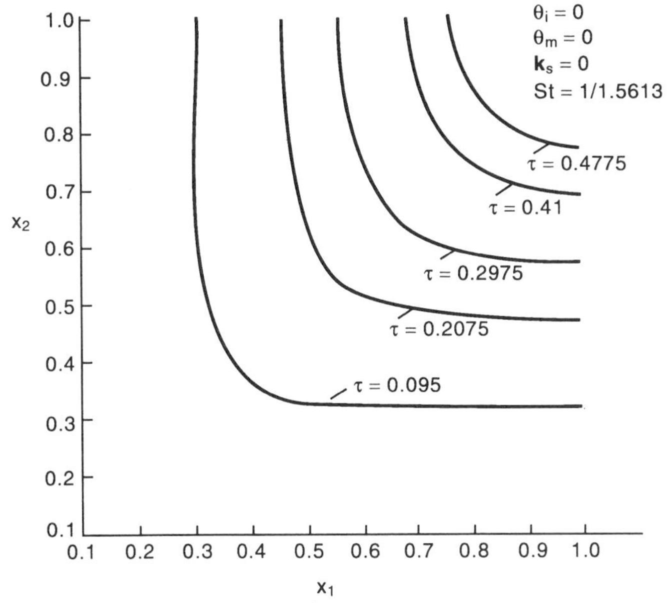

Figure 7.2. Interface motion in a square mold as a function of time (from Zabaras and Mukherjee 1987).

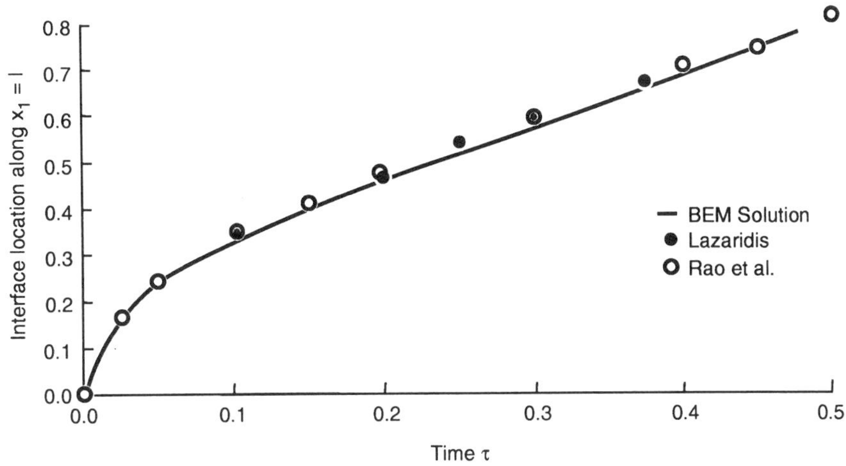

Figure 7.3. Interface motion in a square mold along the adiabatic $x_1 = 1$ as a function of time. Same situation as in figure 7.2 (from Zabaras and Mukherjee 1987).

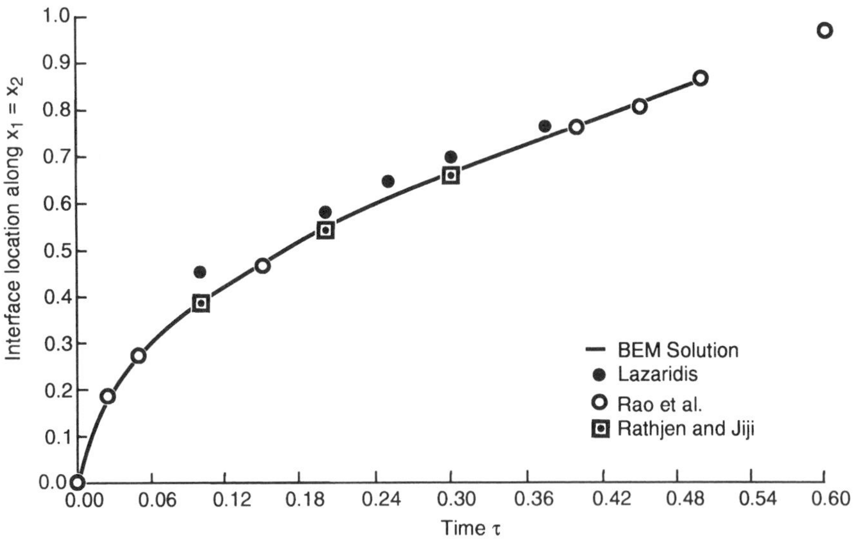

Figure 7.4. Interface motion in a square mold along the diagonal $x_2 = x_2$ as a function of time. Same situation as in figure 7.2 (from Zabaras and Mukherjee 1987).

compared in figure 7.3 with the implicit finite difference solution of Rao and Sastri (1984) and the work of Lazaridis (1969), and also in figure 7.4 with the semianalytical solution of Rathjen and Jiji (1971). These BEM solutions compare very well with the solutions from other numerical schemes as is seen in figures 7.3 and 7.4. Temperature calculations for some internal points are shown in figure 7.5.

The following example involves a square of 2×2 with

$$\hat{c}_l = 1, \qquad k_l = 1, \qquad \theta_0 = -1, \qquad \theta_i = 0.3 \quad \text{and} \quad St = 4$$

The same problem, has been analysed previously by Budhia and Kreith (1973), Comini et al. (1974), Ralph and Bathe (1982), and Zabaras and Ruan (1990).

Figure 7.6 shows the front position on the diagonal ($x_1 = x_2$) and figure 7.7 the temperature at the internal point $x_1 = x_2 = 0.5$. In figure 7.6 comparison is made with Ralph and Bathe (1982). The time step for the BEM is $\Delta t = 0.0225$ and that for the FEM is $\Delta t = 0.02$. A similar comparison is made in figure 7.7, where two time steps have been used. The FEM results of Ralph and Bathe (1982) show a big difference with a change in time step from 0.02 to 0.1. They have been found not to be in good agreement with the semianalytical solution of Rathjen and Jiji (1971) (see Zabaras and Ruan 1990).

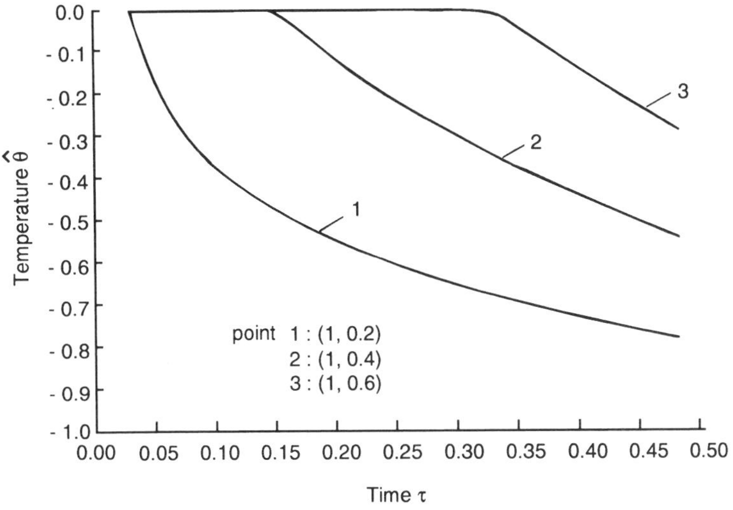

Figure 7.5. Temperature distribution at some internal points with respect to time during solidification in a square mold. Same situation as in figure 7.2 (from Zabaras and Mukherjee 1987).

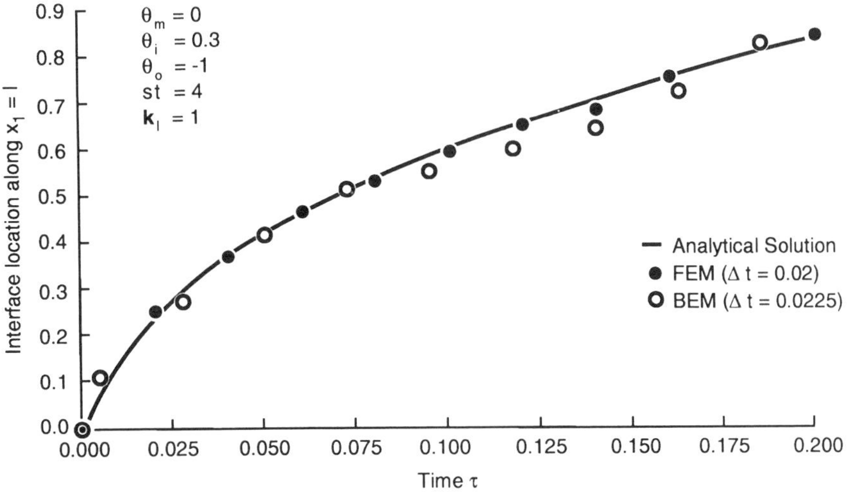

Figure 7.6. Interface motion in a square mold along the diagonal $x_1 = x_2$ as a function of time. Analytical solution from Budhia and Kreith (1973). FEM solution from Ralph and Bathe (1982) (from Zabaras and Mukherjee 1987).

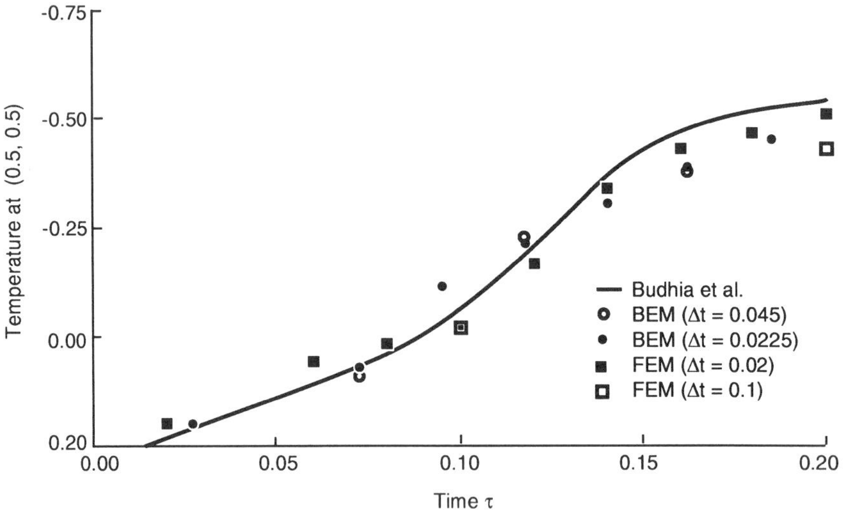

Figure 7.7. Temperature history of the point (0.5, 0.5) during the solidification in a square mold. Same situation as in figure 7.6. FEM solution from Ralph and Bathe (1982) (from Zabaras and Mukherjee 1987).

A similar example with

$$\hat{c}_l = 1, \qquad k_l = 1, \qquad \theta_0 = -1, \qquad \theta_i = 1 \quad \text{and} \quad St = 2$$

has been analyzed. Figure 7.8 shows the interface location at various times. Figures 7.9 and 7.10 compare the diagonal and asymptotic positions of the interface, respectively, with the results given by Rao and Sastri (1984). Figure 7.11 compares the temperature history at the center of the mold with Rao and Sastri (1984).

In all the above examples, the minimum time step was 0.0225 and the maximum was 0.1. The error tolerance parameter $\varepsilon_{\max} = 10^{-2}$. The equivalent heat capacity model has been used up to time 0.005 to initialize the BEM calculation. Double nodes have been considered both on ∂B_0 and ∂B_I.

7.4.3 *The Design Problem*

Consider solidification in a one-dimensional semi-infinite region $T_i = T_m = 0$, $K_s = 1$, $\rho_s = 1$, $c_s = 1$, $L = 1/2$, and $q_{ms} = 1$. Then $q_{ml} = 0$ and $V = 2$. The analytical solution of this problem is given as

$$q_{0s}(t) = e^{4t} \tag{7.47}$$

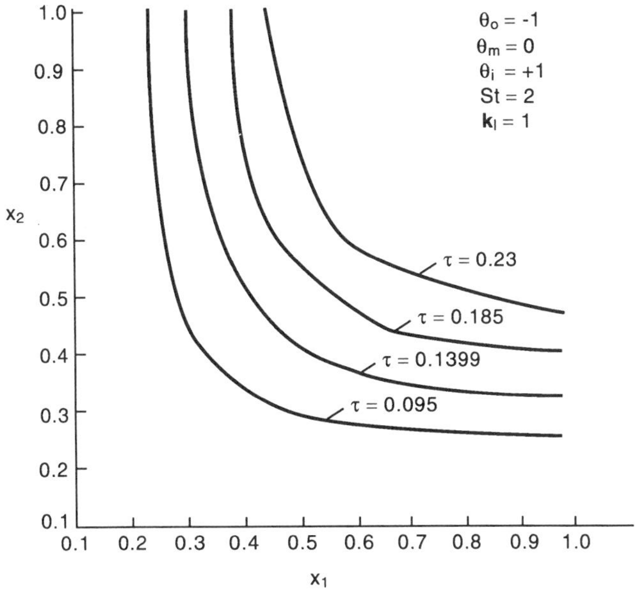

Figure 7.8. Interface motion in a square mold as a function of time (from Zabaras and Mukherjee 1987).

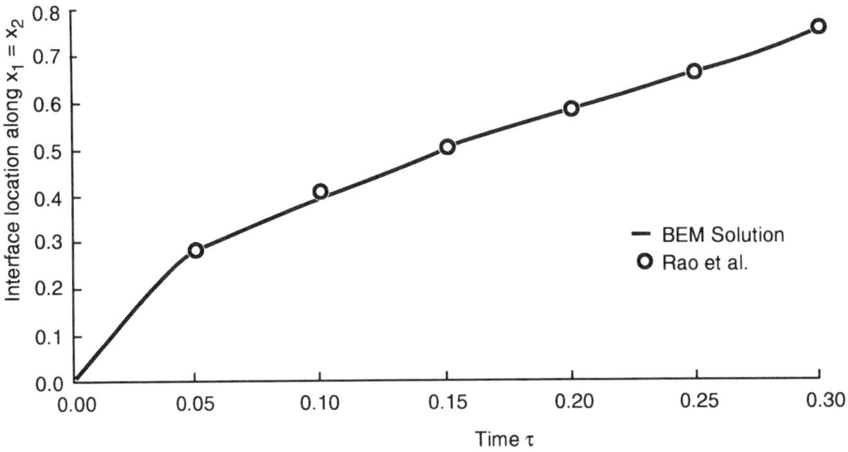

Figure 7.9. Interface motion in a square mold along the diagonal $x_1 = x_2$ as a function of time. Same situation as in Figure 7.8 (from Zabaras and Mukherjee, 1987).

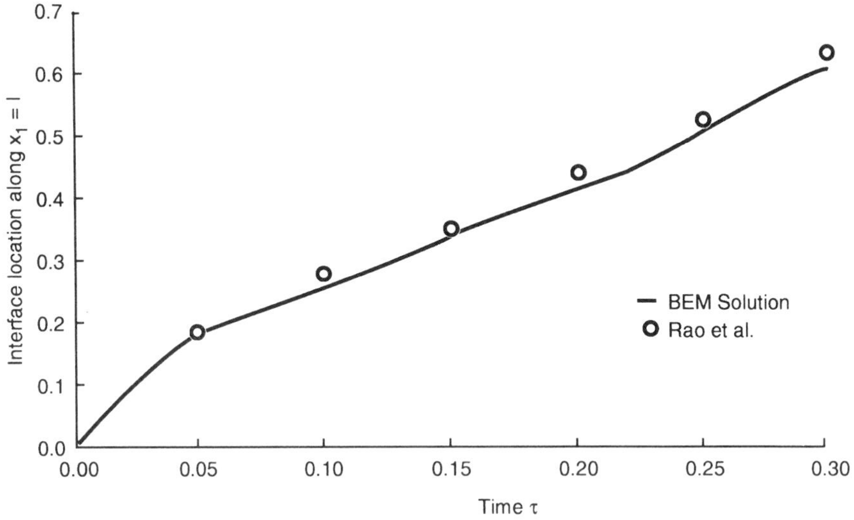

Figure 7.10. Interface location in a square mold along the adiabatic ($x_1 = 1$) as a function of time. Same situation as in figure 7.8 (from Zabaras and Mukherjee 1987).

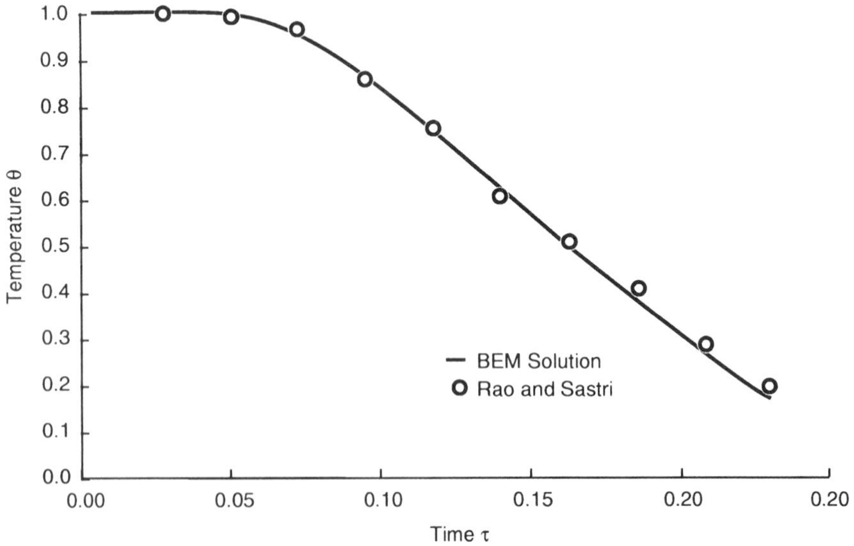

Figure 7.11. Temperature distribution at the center of a square mold as a function of time. Same situation as in figure 7.8 (from Zabaras and Mukherjee 1987).

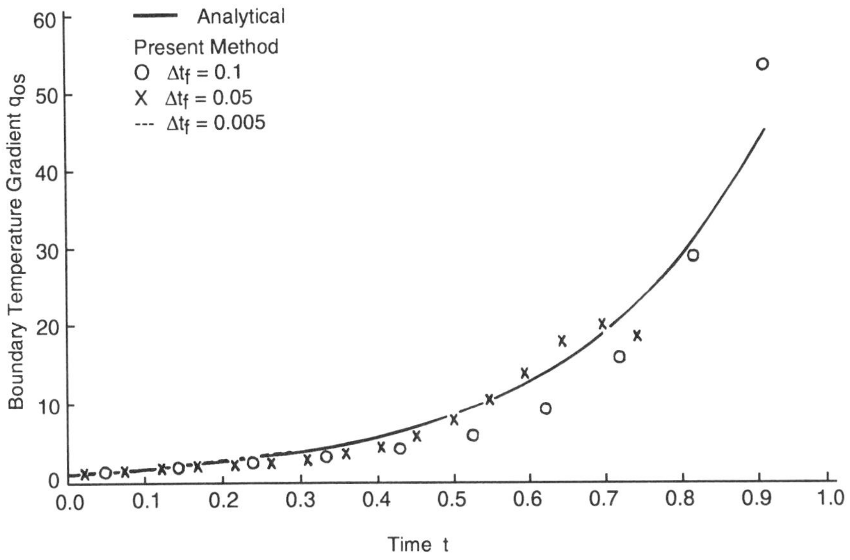

Figure 7.12. Boundary temperature q_{0s} as a function of time for different time steps (from Zabaras et al. 1988).

and

$$T_{0s}(t) = \tfrac{1}{2}(1 - e^{4t}) \tag{7.48}$$

No spatial regularization is involved in this one-dimensional problem. The sensitivity coefficients are calculated analytically in a way similar to that presented earlier. Figures 7.12 and 7.13 show plots of the flux $q_{0s}(t)$ together with the exact solutions given by equations (7.47) and (7.48) for three different time steps, $\Delta t_f = 0.1$, 0.05, and 0.005. One future time step ($r = 2$) has been used to stabilize the solution. As expected, when the interface moves away from the $x = 0$ boundary, oscillations or divergence from the analytical solution occur. In the above figures only the stable region has been plotted. As can be seen, the smaller the time step, the more accurate the numerical solution, but also the sooner it starts to diverge from the exact solution.

To test the algorithm in the liquid phase, the following case is considered here:

$$T_i = -1, \quad T_m = 0, \quad k_l = 1, \quad \hat{K}_l = 1, \quad \rho l = 1, \quad \hat{c}_l = 1, \quad L = 2, \quad l = 1$$

$$V = \frac{0.43}{\sqrt{t}}, \qquad h = 0.86\sqrt{t}, \qquad q_{ml} = -0.761\,78\,\frac{1}{\sqrt{t}}$$

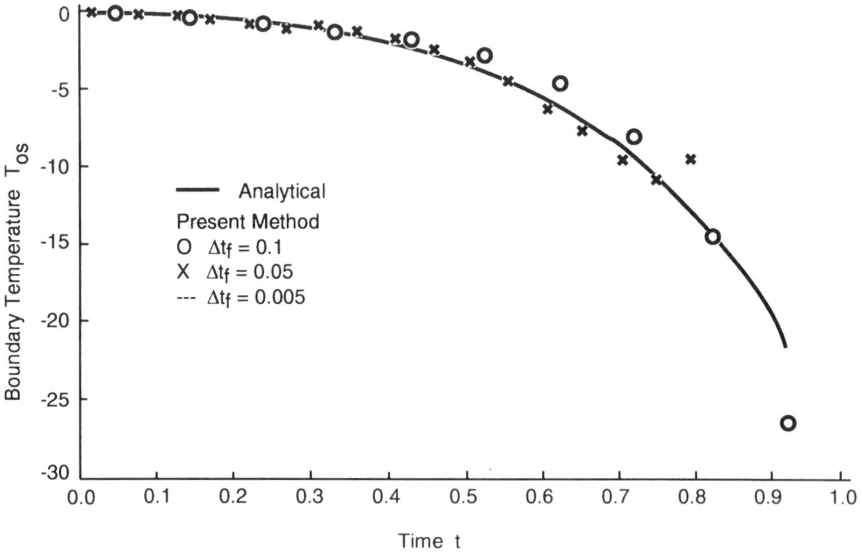

Figure 7.13. Boundary temperature T_{0s} as a function of time for different time steps (from Zabaras et al. 1988).

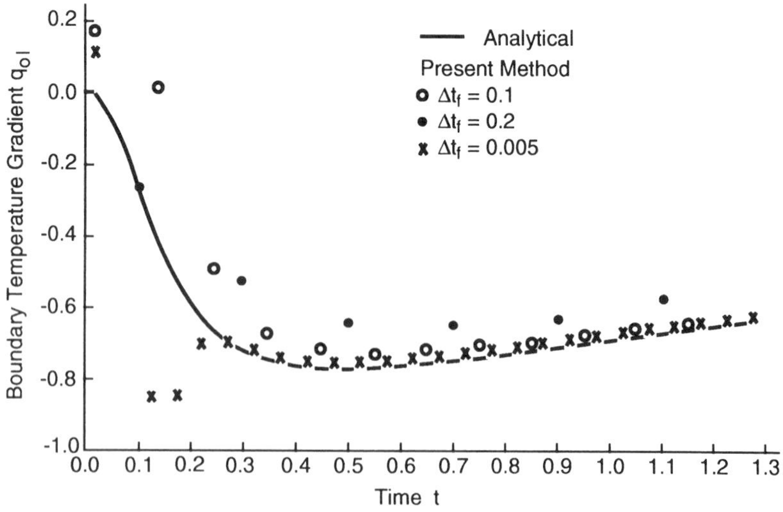

Figure 7.14. Boundary temperature gradient q_{0l} as a function of time for different time steps (from Zabaras et al. 1988).

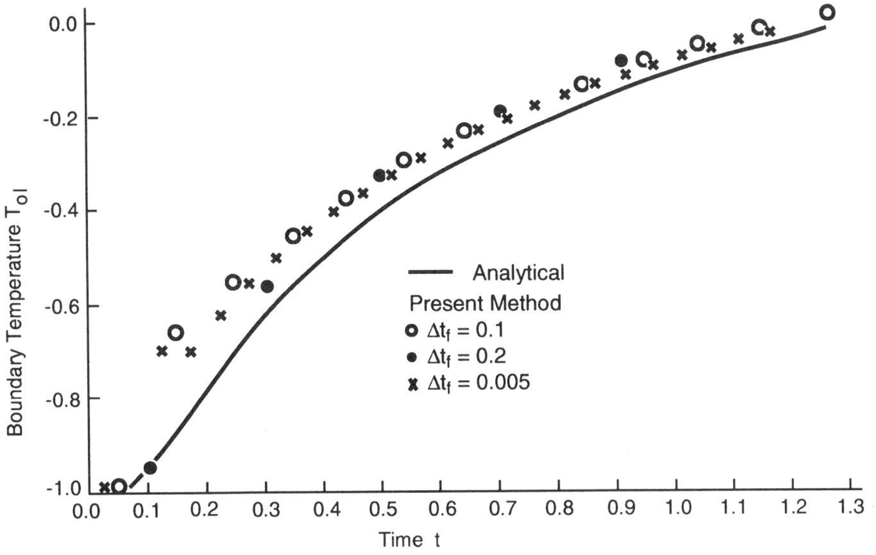

Figure 7.15. Boundary temperature T_{0l} as a function of time for different time steps (from Zabaras et al. 1988).

The above are approximations to the exact desired data that correspond to the analytical solution

$$q_{0l}(t) = 1.043\,7802 \frac{1}{\sqrt{t}}\, e^{-1/4t} \tag{7.49}$$

$$T_{0l}(t) = 1 + 1.850\,017 \text{ erfc} \frac{1}{2\sqrt{t}} \tag{7.50}$$

(see Zabaras 1990).

Figures 7.14 and 7.15 show the calculated temperature gradient and temperature at $x = l$. As expected, the solution is inaccurate (still stable) at early times, while very accurate and stable later. Note that if the exact desired data are used ($V = 0.432\,756/\sqrt{t}$, $q_{ms} = 0$), the estimated q_{0l} and T_{0l} are in very good agreement with the exact solutions (7.49) and (7.50).

References

Albert, M. R. and O'Neill, K. (1986). "Moving Boundary-Moving Mesh Analysis of Phase Change Using Finite Elements with Transfinite Mappings," *Int. J. Num. Meth. Eng.*, **23**, 591–607.

Banerjee, P. K. and Shaw, R. P. (1982). "Boundary Element Formulation for Melting and Solidification Problems," *Developments in Boundary Element Methods—2* (ed. P. K. Banerjee and R. P. Shaw), pp. 1–18. Elsevier Science, Barking, Essex, U.K.

Beck, J. V., Blackwell, B., and St. Clair, C. R. (1985). *Inverse Heat Conduction, Ill-posed Problems*. Wiley-Interscience, New York.

Bonnerot, R. and Janet, P. (1977). "Numerical Computation of the Free Boundary for the Two-dimensional Stefan Problem by Space-Time Finite Elements," *J. Comput. Phys.*, **25**, 163–181.

Brebbia, C. A., Telles, J. C., and Wrobel, L. C. (1984). *Boundary* Element *Techniques—Theory and Applications in Engineering*. Springer-Verlag, Berlin.

Budhia, H. and Kreith, F. (1973). "Heat Transfer with Melting or Freezing in a Wedge," *Int. J. Heat Mass Transfer*, **16**, 195–211.

Chuang, Y. K. and Szekely, J. (1972). "On the Use of Green's Functions for Solving Melting and Solidification Problems," *Int. J. Heat Mass Transfer*, **15**, 1171–1174.

Comini, G., Del Guidice, S., Lewis, R. W., and Zienkiewicz, O. C. (1974). "Finite Element Solution to Non-linear Heat Conduction Problems with Special Reference to Phase Change," *Int. J. Num. Meth. Eng.*, **8**, 613–624.

Crank, J. (1984). *Free and Moving Boundary Problems*. Clarendon Press, Oxford.

Crank, J. and Gupta, R. (1975). "Isotherm Migration Method in Two Dimensions," *Int. J. Heat Mass Transfer*, **18**, 1101–1117.

Flemings, M. C. (1974). *Solidification Processing*. McGraw-Hill, New York.

Heinlein, M., Mukherjee, S., and Richmond, O. (1986). "A Boundary Element Method Analysis of Temperature Fields and Stresses During Solidification," *Acta Mech.*, **59**, 58–81.

Hong, C. P., Umeda, T., and Kimura, Y. (1984). "Numerical Models for Casting Solidification: Part II. Application of the Boundary Element Method to Solidification Problems," *Metall. Trans. B*, **15B**, 101–107.

Hsiao, J. S. (1985). "An Efficient Algorithm for Finite-Difference Analysis of Heat Transfer with Melting and Solidification," *Num. Heat Transfer*, **8**, 653–666.

Lazaridis, A. (1969). "A Numerical Solution of the Solidification (or Melting) Problem in Multidimensional Space," Doctoral dissertation, Columbia University.

Lynch, R. D. (1982). "Unified Approach to Simulation on Deforming Elements with Application to Phase Change Problems," *J. Comput. Phys.*, **47**, 387–411.

Murray, W. D. and Landis, F. (1959). "Numerical and Machine Solutions of Transient Heat-Conduction Problems Involving Melting or Freezing," *Trans. ASME(c), J. Heat Transfer*, **81**, 106–112.

O'Neill, K. (1983). "Boundary Integral Equation Solution of Moving Boundary Phase Change Problems," *Int. J. Num. Meth. Eng.*, **19**, 1825–1850.

Ralph, W. and Bathe, K.-J. (1982). "An Efficient Algorithm for Analysis of Nonlinear Heat Transfer with Phase Changes," *Int. J. Num. Meth. Eng.*, **18**, 119–134.

Rao, P. and Sastri, V. M. K. (1984). "Efficient Numerical Method for Two-dimensional Phase Change Problems," *Int. J. Heat Mass Transfer*, **27** (11), 2077–2084.

Rathjen, K. A. and Jiji, L. M. (1971). "Heat Conduction with Melting or Freezing in a Corner," *Trans. ASME, J. Heat Transfer*, **93**, 101–109.

Roose, J. and Storrer, W. O. (1984). "Modelization of Phase Changes by Fictitious Heat Flow," *Int. J. Num. Meth. Eng.*, **20**, 217–225.

Ruan, Y. and Zabaras, N. (1991). "An Inverse Finite Element Technique to Determine the Change of Phase Interface Location in Two-Dimensional Melting Problems," *Commun. Appl. Num. Meth.*, **7**, 325–338.

Sadegh, A., Jiji, L. M., and Weinbaum, S. (1985). "Boundary Integral Equation Technique with Application to Freezing around a Buried Pipe," ASME paper presented at the Winter Annual Meeting, Miami Beach, Florida, 17–21 November.

Tikhonov, A. N. and Arsenin, V. Y. (1977). *Solution of Ill-posed Problems*. V.H. Winston, Washington, D.C.

Voller, V. R. and Cross, M. (1981). "Accurate Solutions of Moving Boundary Problems Using the Enthalpy Method," *Int. J. Heat Mass Transfer*, **24**, 545–556.

Zabaras, N. (1990). "Inverse Finite Element Techniques for the Analysis of Solidification Processes," *Int. J. Num. Meth. Eng.*, **29**, 1569–1587.

Zabaras, N., Morellas, V., and Schnur, D. (1989). "Spatially Regularized Solution of Inverse Elasticity Problems Using the Boundary Element Method," *Commun. Appl. Num. Meth.*, **5**, 547–553.

Zabaras, N. and Mukherjee, S. (1987). "An Analysis of Solidification Problems by the Boundary Element Method," *Int. J. Num. Meth. Eng.*, **10**, 1879–1900.

Zabaras, N., Mukherjee, S., and Richmond, O. (1988). "An Analysis of Inverse Heat Transfer Problems with Phase Changes Using an Integral Method," *ASME, J. Heat Transfer*, **110**, 554–561.

Zabaras, N. and Mukherjee, S. (1994). "Solidification Problems by the Boundary Element Method," *Int. J. Solids Structures*, **31**, 1829–1846.

Zabaras, N. and Ruan, Y. (1989). "A Deforming FEM Analysis of Inverse Stefan Problems," *Int. J. Num. Meth. Eng.*, **28**, 295–313.

Zabaras, N. and Ruan, Y. (1990). "Moving and Deforming Finite Element Simulation of Two-dimensional Stefan Problems," *Commun. Appl. Num. Meth.*, **6**, 495–506.

Zabaras, N., Ruan, Y., and Richmond, O. (1992). "On the Design of Two-dimensional Stefan Processes with Desired Freezing Front Motions," *Num. Heat Transfer*, **218**, 307–325.

8

Machining Processes

Elevated temperatures generated in machining operations significantly influence the chip formation mechanics, the process efficiency, and the surface quality of the machined part. Accordingly, a BEM approach is developed here to analyze the thermal aspects of machining processes. Particular attention is given to modeling of the tool–chip, chip–workpiece, and tool–workpiece interfaces. An exact expression for matching the boundary conditions across these interfaces is developed to avoid any iterations. A direct differentiation approach (DDA) is used to determine the sensitivities of temperature and flux distributions with respect to various design parameters.

The numerical results obtained by the BEM are first verified against existing analytical and FEM results. The temperature and flux fields for various machining conditions, along with their sensitivities, are presented next. The situations of progressive flank and crater wear of the tool with continued machining are also considered, and their effects on thermal fields are investigated. The BEM is found to be very robust and efficient for this class of steady-state conduction–convection problems. The application of DDA with BEM allows efficient determination of design sensitivities and avoids strongly singular kernels. This approach also provides a new avenue toward efficient optimization of the thermal aspects of machining processes.

8.1 Introduction

In machining operations, the elevated temperature fields generated by the severe inelastic deformations in the shear plane, along with the frictional conditions at the chip–tool interfaces, play crucial roles in determining the chip formation mechanics. Accordingly, the elevated temperature fields have significant effects on the surface quality of the finished product and the process efficiency. Tool life, as limited by its wear, depends to a large extent on the temperature in the vicinity of the cutting edge, which in turn imposes a practical limit on the rate of material removal. Excessive temperatures may lead to various types of surface damage. The shear zone temperatures in metal cutting influence the deformation processes, the occurrence of instabilities, and the behavior of free machining inclusions (Lemaire and Backofen 1972, Von Turkovich 1972, Ramalingam et al. 1977).

There exist various analytical (Loewen and Shaw 1954, Bhattacharyya 1984, Shaw 1984) and finite element analyses (Tay et al. 1974, Muraka et al. 1979, Stevensen et al. 1983, Dawson and Malkin 1984) of heat conduction with moving or stationary heat sources, together with kinematic, geometric, and energetic aspects of the metal cutting process. The widely used analytical model of Loewen and Shaw (1954) for orthogonal machining is based on the superposition of two planar heat sources, one at the shear plane and the other at the chip–tool interface. At each location, two temperature solutions are obtained, one for each side of the planar heat source, with a fraction of the total heat going to one side and the rest going to the other. At the shear plane, the temperature solution for the workpiece side is obtained by approximating the shear plane as a band heat source moving on the surface of a stationary semi-infinite solid at the shear velocity inclined at the shear angle to the cutting velocity. The remainder of the shearing energy not entering the workpiece is assumed to cause uniform heating of the chip. The partitioning of the total shearing energy between the workpiece and the chip may be obtained by equating the temperature along the shear plane from the workpiece side to that from the chip side. Thus, the average shear plane temperature may be determined by substituting the appropriate portion of the total shearing energy into either temperature solution. A similar procedure is also used to calculate the temperature rise at the chip–tool interface.

In the analytical models, it is often assumed that the chip is formed instantaneously at the shear plane, so that a uniform planar heat source and velocity discontinuity may be assumed to exist there. The second deformation zone has usually been neglected, and the chip–tool frictional heat source is typically assumed to be uniform.

The finite element analyses of Tay et al. (1974) and Stevensen et al. (1983) account for the primary and secondary zones arising from the fact that plastic deformation takes place over substantial zones both around the shear plane and the rake face of the tool. Dawson and Malkin (1984) have modified the heat transfer model of Loewen and Shaw (1954) for shear plane temperatures. Instead of moving the band heat source along the shear plane (relative to the workpiece), they move it at the cutting velocity, directly into the workpiece material to be removed ahead of the shear plane. Thus, preheated material directly ahead of the shear plane is removed. Accordingly, part of the heat entering the workpiece at the shear plane is subsequently removed by convection before it can be conducted downward below the path of the advancing cutting edge. Dawson and Malkin (1984) also consider the energy carried off by the chip due to heat convection from the workpiece across the shear plane. Muraka et al. (1979) have investigated the influence of several process variables, such as flank wear rate, coolant water, and others, on the temperature distributions in orthogonal machining using the finite element method.

The boundary element method is another powerful general-purpose method (Banerjee and Butterfield 1981, Mukherjee 1982, Brebbia et al. 1984, Beskos 1987). It is far more tolerant of aspect ratio degradation than the FEM and can yield secondary variables as accurate as the primary ones. The temperature distributions in machining processes vary sharply in the vicinity of the cutting zones. This requires very careful refinements in the FEM mesh. In the BEM, however, the internal equations are applied pointwise. Thus, sharp temperature gradients over the domain may be easily captured. In metal cutting operations, the crucial quantities are typically on the boundary, and BEM provides an accurate and efficient means for obtaining them. Recently, the BEM has been applied to several steady-state and transient heat conduction problems including moving boundary phase change problems (O'Neill 1983, Curran et al. 1986, Fleuries and Predeleanu 1987, Zabaras and Mukherjee 1987). Tanaka et al. (1986) have also obtained mixed boundary element solutions of steady-state convection diffusion problems in three dimensions and found the accuracy of the BEM solutions compared to exact solutions to be almost independent of the Peclet number. The BEM solutions were also unconditionally stable in space. These features make the BEM superior to domain-type numerical techniques, which have a criterion for numerical stability and whose accuracy depends to some extent on the Peclet number. Recently, Chan and Chandra (1991a) have also performed a boundary element analysis of steady-state metal cutting operations. Special attention was paid to the interface conditions at the tool–workpiece, tool–chip, and chip–workpiece interfaces, and a complete heat transfer model of steady-state turning has been obtained by matching the boundary conditions across the interfaces.

Typically, optimal designs of metal cutting operations must be carried out by nonlinear programming methods. Such algorithms require repeated iterations on the design variables, which may contain shape parameters, process parameters, and material parameters. Even for a very simple metal cutting process, such a procedure can be extremely computer intensive. A crucial ingredient for obtaining successful and economical solutions to such optimization problems is the accurate determination of design sensitivities.

A large amount of literature also exists on evaluation of design sensitivity coefficients, particularly in linear problems of solid mechanics. In a chapter such as this, it is very difficult to acknowledge all the worthwhile contributions in this field. Instead, the reader is referred to a comprehensive book by Haug et al. (1986). More recently, Tsay and Arora (1988) used FEM analysis to obtain design sensitivities in nonlinear structures with history-dependent effects, and Mukherjee and Chandra (1989, 1991) obtained BEM formulations for design sensitivities in problems involving material as well as geometric nonlinearities.

In general, two methods emerge as the most powerful ones for the determination of design sensitivities. These are the direct differentiation approach (DDA) and the adjoint structure approach (ASA). The DDA typically starts from a variational equation like the principle of virtual work (e.g., Tsay and Arora 1988) or from boundary integral equations (Mukherjee and Chandra 1989, 1991). Such an equation is differentiated with respect to the design variables, and the resulting equations are solved in order to obtain the sensitivities. The ASA, on the other hand, defines adjoint structures whose solutions permit explicit evaluation of the sensitivity coefficients (e.g., Haug et al. 1986).

The DDA, in conjunction with the BEM, provides an extremely elegant approach toward determination of design sensitivities. The differentiation procedure developed by Barone and Yang (1988) and by Mukherjee and Chandra (1989, 1991) does not increase the singularity of the relevant kernels. Thus, while for two-dimensional problems one usually starts with kernels that are $\ln(r)$ and $1/r$ singular (r being the distance between a source and a field point), the differentiated kernels are regular and $1/r$ singular, respectively.

Recently, Saigal et al. (1989) developed a BEM strategy for sensitivity analysis of linear elasticity problems. There, the BEM equations are discretized first. Appropriate modes are then used on the discretized version to avoid direct numerical evaluation of kernels with $1/r$ singularity. Finally, the discretized BEM equations are used again for indirect evaluation of singular kernels arising in the sensitivity equations. In the present work, the approaches of Mukherjee and Chandra (1989, 1991) and Rice and Mukherjee (1990) are followed. Appropriate modes are first used to modify the BEM equations, and the modified BEM equations of steady-state conduction–convection are then differentiated with respect to design variables. Very recently, Chandra and Chan (1992) developed a design sensitivity formulation for the conduction–convection equation using the direct differentiation approach. Numerical results are compared with analytical solutions for uniform compression problems. The numerical results obtained from the BEM agree very well with those obtained analytically.

This chapter begins with a BEM formulation for steady-state conduction–convection problems suitable for analyzing machining processes. A description of the numerical implementation for planar problems follows. Particular attention is paid to modeling of the boundary conditions at the tool–chip, chip–workpiece, and workpiece–tool interfaces. An exact expression for matching is developed to satisfy the matching interface conditions without any iterations. Numerical results for BEM analyses of the thermal aspects of machining processes are presented. The results obtained by the BEM are compared to existing analytical and FEM results with regard to accuracy and efficiency.

The issues relating to design sensitivities are addressed next. A BEM formulation is developed for determining the sensitivities of the thermal fields with respect to various geometric, material, and process parameters. Numerical implementations of the BEM sensitivity formulation are discussed. The BEM sensitivity algorithm is first applied to the case of uniform compression of an initially square domain, for which an analytical solution exists. The BEM results are compared to the analytical results with regard to accuracy. Various machining situations, such as nonuniform chip thickness, gradual nose wear of the tool, and gradual flank and crater wear are also considered. For these cases, the design sensitivity results obtained from BEM are compared to those obtained from finite difference schemes.

8.2 Boundary Element Formulation

For metal cutting processes, various tool force measurements and cinephotographs confirm that the machining process is essentially steady for a continuous strip (Tay et al. 1974, Stevenson et al. 1983, Bhattacharyya 1984, Shaw 1984). The grain sizes of the work material and the tool material are also quite small compared to the sizes of the deformation zones. Hence, the workpiece, the tool, and the chip may be treated as continua that are homogeneous and isotropic. It is also reasonable to assume that thermal conductivity, specific heat, and density remain constant over the operating range of a typical metal cutting process (Tay et al. 1974, Muraka et al. 1979, Stevenson et al. 1983, Bhattacharyya 1984, Dawson and Malkin 1984, Shaw 1984). Hence, the governing equation for temperature distributions in steady-state turning operations may be expressed as

$$\rho c v_i^{(s)} \frac{\partial T}{\partial x_i} = k \frac{\partial^2 T}{\partial x_i^2} \qquad \text{in } \Omega \tag{8.1}$$

where T is the temperature, k is the thermal conductivity, ρ is the density, c is the heat capacity, and v_i^s is the scanning velocity. Here, the convention that repeated indices represent summation over the two directions is used. The boundary conditions are

$$T = \overline{T} \qquad \text{on } \Gamma_T \tag{8.2}$$

and

$$k\frac{\partial T}{\partial x_i}n_i = \overline{q} \qquad \Gamma_q \tag{8.3}$$

Equation (8.1) applies to an Eulerian reference frame that remains spatially fixed while material flows through it. The convective term represents the energy transported by the material as it moves through the reference frame. The surface flux $\overline{q}$ includes a contribution $q^{(c)}$ from convection cooling losses which may be written as

$$q^{(c)} = h(T - T_\infty) \tag{8.4}$$

Let us also consider the adjoint equation

$$-\rho c v_i^{(s)}\frac{\partial G}{\partial x_i} = k\frac{\partial^2 G}{\partial x_i^2} + \delta[x_i(q) - x_i(p)] \tag{8.5}$$

where p is a source point and q is the field point in the domain. P and Q represent a source point and a field point, respectively, on the boundary. Applying the divergence theorem, an integral representation of the governing equations may be obtained (Tanaka et al. 1986) as

$$T(p) = -k\int_\Gamma \left(\frac{\partial G(p,Q)}{\partial n_Q}T(Q) - G(p,Q)q^{(n)}(Q)\right)d\Gamma$$

$$-\rho c\int_\Gamma G(p,Q)T(Q)v_i^{(s)}n_i(Q)\,d\Gamma + \rho c\int_\Omega G(p,q)T(q)v_{i,i}^{(s)}(q)\,d\Omega \tag{8.6}$$

and

$$q^{(n)}(Q) = \frac{\partial T(Q)}{\partial n_Q}$$

Here, a comma denotes field point differentiation. The domain integral in equation (8.6) vanishes if the scanning velocity is constant. For metal cutting operations, the scanning velocity is equal to the cutting speed, which is typically constant for a particular operation. Hence, the domain integral in equation (8.6) need not be considered for present purposes. In turning operations, the scanning velocities for the tool, the workpiece, and the chip are different from each other. This, however, may be handled easily by considering three separate regions. As discussed in a later section, the complete temperature distribution in the tool, the workpiece, and the chip may then be obtained by appropriate matching.

The fundamental solution $G(p, q)$ (Carslaw and Jaeger 1986, Tanaka et al. 1986) is given as

$$G(p, q) = \frac{1}{2\pi k} \exp\left(\frac{v_i^{(s)}[x_i(q) - x_i(p)]}{2\kappa} \right) K_0\left(\frac{vr}{2\kappa} \right) \quad \text{in 2D} \tag{8.7}$$

Here, v is the magnitude of the scanning velocity vector $v_i^{(s)}$, r is the distance between the source point and the field point, and K_0 is the modified Bessel function of the second kind of order zero.

A boundary integral equation for the steady-state conduction–convection problem may now be obtained by taking the limit as p tends to P. This gives

$$\begin{aligned} C(P)T(P) = &-k \int_\Gamma \left(\frac{\partial G(P, Q)}{\partial n_Q} T(Q) - G(P, Q) q^{(n)}(Q) \right) d\Gamma \\ &- \rho c \int_\Gamma G(P, Q) T(Q) v_i^{(s)} n_i(Q) \, d\Gamma \end{aligned} \tag{8.8}$$

The coefficient C, in general, depends on the local geometry at P. If the boundary is locally smooth at P, $C = \frac{1}{2}$. Otherwise, it may be evaluated indirectly (Banerjee and Butterfield 1981, Mukherjee 1982, Brebbia et al. 1984).

8.2.1 Numerical Implementation

Numerical implementation of the BEM equations (8.6–8.8) for the conduction–convection problem is discussed in this section. The first step is the discretization of the boundary of the two-dimensional domain into boundary elements. A discretized version of the boundary integral equation (8.8) may be written (Banerjee and Butterfield 1981, Mukherjee 1982, Brebbia et al. 1984) as

$$C(P_M)T(P_M) = -k\sum_{i=1}^{N_s}\int_{\Delta s_i}\left(\frac{\partial G(P_M,Q)}{\partial n_Q}T(Q) - G(P_M,Q)q^{(n)}(Q)\right)dS_Q$$
$$-\rho c\sum_{i=1}^{N_s}\int_{\Delta s_i} G(P_M,Q)T(Q)v_l^{(s)}n_l(Q)\,dS_Q \tag{8.9}$$

where the boundary of the domain Γ is divided into N_s boundary segments and $T(P_M)$ represents the temperature at a point P that coincides with the node M.

A suitable shape function must now be chosen for the variation of temperature and flux over the boundary elements Δs_i. In the present work, both temperature and flux are assumed to be linear over individual boundary elements. Hence, equation (8.9) may be written as

$$C(P_M)T(P_M) = -\sum_{i=1}^{N_s}\int_{\Delta s_i}\left(k\frac{\partial G(P_M,Q)}{\partial n_Q} + \rho c G(P_M,Q)v_l^{(s)}n_l(Q)\right)dS_{\hat{Q}}$$
$$+k\sum_{i=1}^{N_s}\int_{\Delta\sigma_i} G(P_M,Q)[\psi_1 q_1^{(n)} + \psi_2 q_2^{(n)}]\,dS_Q \tag{8.10}$$

where the shape functions are

$$\psi_1 = \tfrac{1}{2}(1-\eta) \qquad \text{and} \qquad \psi_2 = \tfrac{1}{2}(1+\eta) \tag{8.11}$$

and η is the dimensionless local coordinate over individual boundary segments. T_1, T_2, $q_1^{(n)}$, and $q_2^{(n)}$ are the nodal quantities of the ith boundary segment.

Defining

$$\alpha_{Mj}^{\gamma} = c_M\delta_{Mj} + \int_{\Delta s_j}\psi_\gamma\left(k\frac{\partial G(P_M,Q)}{\partial n_Q} + \rho c G(P_M,Q)\,v_l^{(s)}n_l\right)dS_Q \tag{8.12}$$

and

$$b_{Mj}^{\gamma} = -k\int_{\Delta s_j}\psi_\gamma G(P_M,Q)\,dS_Q \tag{8.13}$$

equation (8.10) may now be expressed as

$$\sum_{j=1}^{N_s} A_{ij}T_j + \sum_{j=1}^{N_s} B_{ij}q_j^{(n)} = 0 \tag{8.14}$$

where N_s is the total number of boundary nodes and each nodal coefficient A_{ij} is equal to the sum of a_{ij}^2 of element $(j-1)$ and a_{ij}^1 of element (j) for an anticlockwise numbering system. The same procedure also applies to B_{ij}.

Integrals of kernels over elements in equations (8.12)–(8.13) must be obtained carefully. In the present work, the diagonal elements of $G(P_M, Q)$ that are $\log(r)$ singular may be evaluated numerically through an algorithm for improper integrals (Press et al. 1986). The proper combination of the constant $C(P_M)$ and the diagonal elements of $\partial G(P_M, Q)/\partial n_Q$ is evaluated indirectly by applying isothermal boundary conditions over the entire boundary (Banerjee and Butterfield 1981, Brebbia et al. 1984).

At each location over the entire boundary of the domain, one of T, $q^{(n)}$, or a combination of T and $q^{(n)}$ (see equations (8.2)–(8.4)) is prescribed for a well-posed problem. Equation (8.14) may be rearranged as

$$\sum_{j=1}^{N_s} \widetilde{A}_{ij} Y_j^{(u)} + \sum_{j=1}^{N_s} \widetilde{B}_{ij} Y_j^{(k)} = 0 \tag{8.15}$$

The matrix coefficients $\widetilde{A}_{ij}$ and $\widetilde{B}_{ij}$ in equation (8.15) are

$$\widetilde{A}_{ij} = \begin{cases} A_{ij}, & \text{for } q_j^{(n)} \text{ specified} \\ B_{ij}, & \text{for } T_j \text{ specified} \\ A_{ij} + B_{ij}h, & \text{for convective heat loss} \end{cases} \tag{8.16}$$

$$\widetilde{B}_{ij} = \begin{cases} B_{ij}, & \text{for } q_j^{(n)} \text{ specified} \\ A_{ij}, & \text{for } T_j \text{ specified} \\ -B_{ij}h, & \text{for convective heat loss} \end{cases} \tag{8.17}$$

and the column vectors $Y_j^{(u)}$ and $Y_j^{(k)}$ are

$$Y_{(u)_j} = \begin{cases} T_j, & \text{for } q_j^{(n)} \text{ specified, or for} \\ & \text{convective heat loss} \\ q_j^{(n)}, & \text{for } T_j \text{ specified} \end{cases} \tag{8.18}$$

$$Y_{(k)_j} = \begin{cases} q_j^{(n)}, & \text{for } q_j^{(n)} \text{ specified} \\ T_j, & \text{for } T_j \text{ specified} \\ T_\infty, & \text{for convective heat loss} \end{cases} \tag{8.19}$$

Equation (8.15) can now be used to solve for the unknown temperature and flux. Once T and $q^{(n)}$ have been obtained over the entire boundary, the internal equation (8.6) may be used to obtain temperature and flux at any internal point.

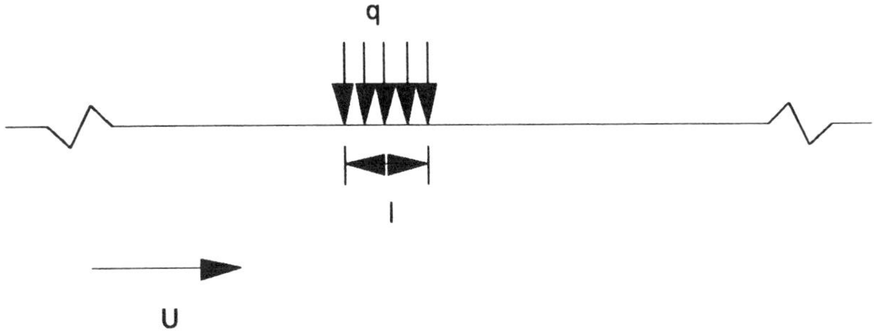

Figure 8.1. Schematic diagram of the Jaeger solution.

8.2.2 Verification of the Conduction–Convection Algorithm

The BEM formulation is first applied to calculate the surface temperature of a semi-infinite domain with surface heating over a finite region. This corresponds to the well-known Jaeger solution (Jaeger 1942). A schematic design of the Jaeger problem is given in figure 8.1. Introducing the dimensionless variables,

$$\tilde{x}_i = \frac{x_i}{l} \qquad \text{and} \qquad \tilde{T} = \frac{T - T_\infty}{q^{l/k}} \tag{8.20}$$

the governing equation in dimensionless form is

$$\frac{\partial \tilde{T}}{\partial \tilde{x}_1} = \frac{1}{Pe}\left(\frac{\partial^2 \tilde{T}}{\partial \tilde{x}_1^2} + \frac{\partial^2 \tilde{T}}{\partial \tilde{x}_2^2}\right) \tag{8.21}$$

$$\tilde{x}_2 = 0, \qquad \frac{\partial \tilde{T}}{\partial \tilde{x}_2} = \begin{cases} -1, & 0 \le \tilde{x}_1 \le 1 \\ 0, & \text{otherwise} \end{cases} \tag{8.22}$$

$$|\tilde{x}_1|, \qquad \tilde{x}_2 \to \infty, \qquad \tilde{T} = 0 \tag{8.23}$$

where

$$Pe = \frac{Ul}{\kappa} \tag{8.24}$$

The surface temperatures underneath the surface heat flux for four different Peclet numbers are plotted in figure 8.2. It can be observed that the BEM solutions compare very well with the corresponding Jaeger solutions, with a maximum discrepancy of 10 percent over a range of Peclet numbers from 4.5 to 45. The BEM solutions have also been compared with

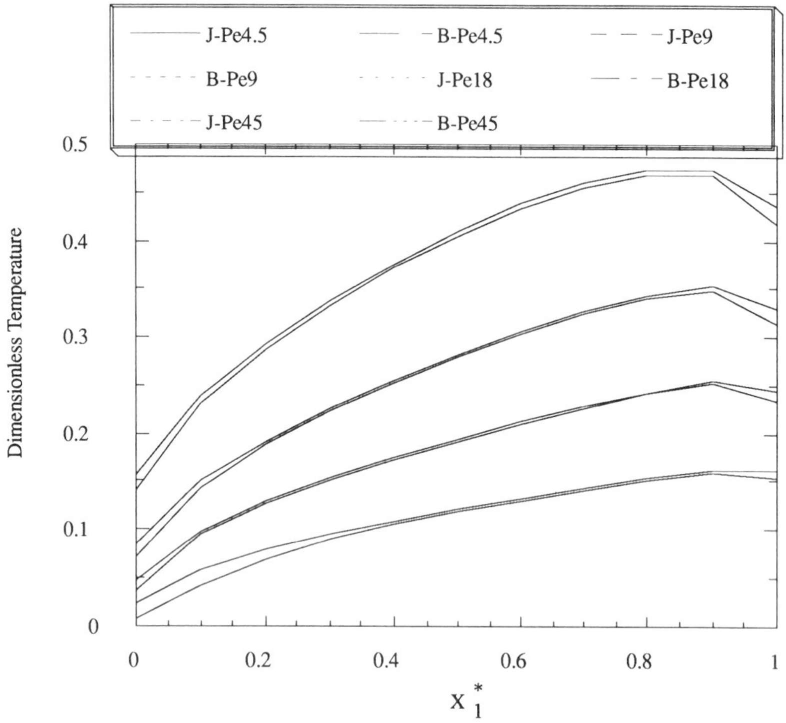

Figure 8.2. The Jaeger solution—comparison between the BEM and the analytical method (J = Jaeger; B = BEM).

other analytical solutions, such as the thermal entrance length of the slug flow problem. The BEM solutions correlated well with the analytical solutions.

8.3 Modeling of Machining Processes

In this section, the BEM formulation is used to model a steady-state metal cutting process. Typically, the tool is a large-angled wedge that is driven into the workpiece to remove a thin layer, the chip. A schematic diagram of the process is sketched in figure 8.3. As the tool is driven into the workpiece, the material undergoes a severe plastic deformation along the shear plane. As the chip forms, it diverts and slides across the tool face. There are two main sources of heat generation: (1) the heat generated by the plastic deformation in the shear plane—the primary zone; and (2) the frictional heating and plastic deformation as the chip slides over the tool face—the secondary zone. The heat transfer involved here is conduction and convection of the heat generated into the tool, the chip, and the workpiece.

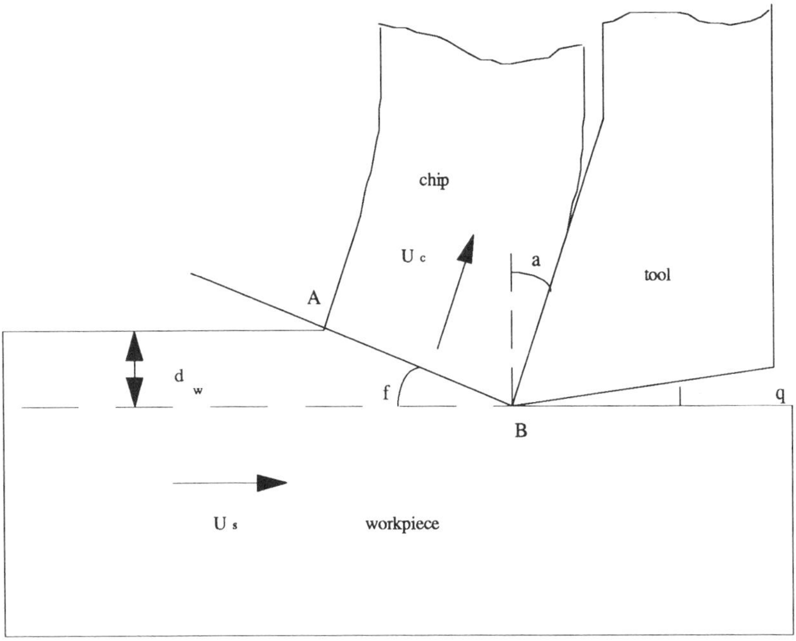

Figure 8.3. Schematic diagram of metal cutting process.

In a metal cutting operation, the velocities associated with the tool, the workpiece, and the chip are quite different. By fixing the reference frame to the tool, it may be considered stationary. The workpiece, with respect to such a reference frame, moves at the cutting velocity (scanning velocity). The chip moves in a different direction with a velocity related to the scanning and the shear plane angle. Consequently, the BEM algorithm is applied to each region separately. For oblique cutting, the chip velocity may also depend on the tool angles. Consequently, the previously developed BEM algorithm is applied separately to each of the regions. By matching the boundary conditions at the workpiece–chip interface and the chip–tool interface, a complete solution for the metal cutting problem may then be obtained.

8.3.1 Mathematical Formulation

Here, orthogonal machining is considered and a two-dimensional analysis of heat transfer is performed. To model the heat transfer efficiently, the tool, the chip, and the workpiece are formulated separately. Furthermore, a coordinate system unique to each region is defined for the purpose of better boundary element representation. The coordinate systems for the three regions are defined in figure 8.4. The mathematical formulations for the

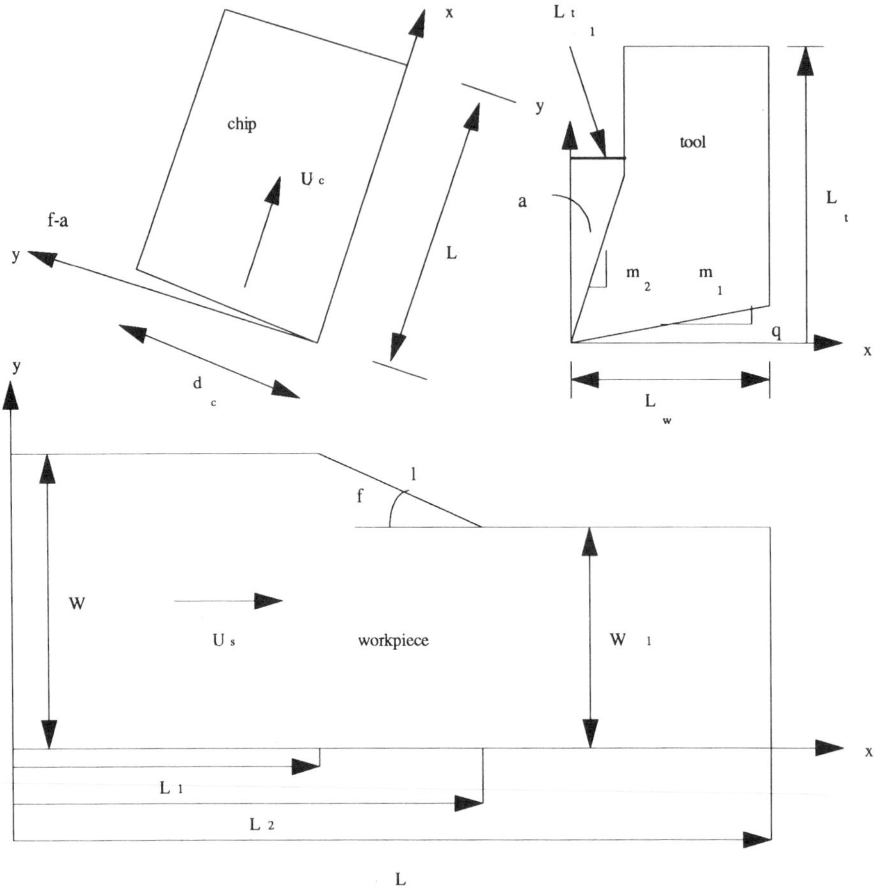

Figure 8.4. Coordinate system for each region.

conduction–convection heat transfer within each region are now presented in their dimensionless forms. The relevant scales are l as the length scale, $q'l/k_w$ as the temperature scale, and q' as the heat flux scale.

8.3.1.1 Within the workpiece

$$\frac{\partial \widetilde{T}_w}{\partial \widetilde{X}} = \frac{1}{Pe_w}\left(\frac{\partial^2 \widetilde{T}_w}{\partial \widetilde{X}^2} + \frac{\partial \widetilde{T}_w}{\partial \widetilde{Y}^2}\right) \qquad \text{in } \widetilde{\Omega}_w \tag{8.25}$$

$$Pe_w = \frac{v_s l}{\kappa_w} \tag{8.26}$$

The coordinate system is chosen so that the scanning velocity aligns with the X_w direction. The boundary conditions appropriate for the present region are

$$\tilde{X} = 0; \qquad \tilde{T}_w = 0 \tag{8.27}$$

$$\tilde{X} = \tilde{L}; \qquad \frac{\partial \tilde{T}_w}{\partial \tilde{X}} = 0 \tag{8.28}$$

$$\tilde{Y} = 0; \qquad \tilde{T}_w = 0 \tag{8.29}$$

$$\tilde{Y} = \tilde{W}, \qquad 0 \le \tilde{X} \le \tilde{L}_1; \qquad \frac{\partial \tilde{T}_w}{\partial \tilde{Y}} = -Nu_w \tilde{T} \tag{8.30}$$

$$\tilde{Y} = \tilde{W}, \qquad \tilde{L}_2 \le \tilde{X} \le \tilde{L}; \qquad \frac{\partial \tilde{T}_w}{\partial \tilde{Y}} = -Nu_w \tilde{T}_w \tag{8.31}$$

$$Nu_w = \frac{h_w l}{k_w} \tag{8.32}$$

The first boundary condition, equation (8.27), means that the incoming materials are at the ambient temperature. Equation (8.28) implies that beyond a certain distance downstream the heat transfer in the X_w direction is negligible. Equation (8.29) means that far from the source the material is not affected. Equations (8.30) and (8.31) represent the convective heat loss to the ambient. The boundary condition along the shear plane has to match with that of the chip and will be discussed later.

8.3.1.2 Within the chip

The governing equation within the chip is

$$\frac{\partial \tilde{T}_c}{\partial \tilde{x}_c} = \frac{1}{Pe_c}\left(\frac{\partial^2 \tilde{T}_c}{\partial \tilde{x}_c^2} + \frac{\partial^2 \tilde{T}_c}{\partial \tilde{y}_c^2}\right) \qquad \text{in } \tilde{\Omega}_c \tag{8.33}$$

$$Pe_c = \frac{V_c l}{\kappa_c} = Pe_w \frac{d_w \kappa_w}{d_c \kappa_c} \tag{8.34}$$

Once again, the coordinate system is chosen so that the motion of the chip is aligned with the x_c direction. It should be pointed out that the chip usually curls up. However, if the thickness of the chip is much smaller than the radius of curvature of the curling, the Cartesian coordinate form of equations can be used to approximate the problem. The boundary conditions appropriate in this region are

$$\tilde{x}_c = \tilde{L}_\infty; \qquad \frac{\partial \tilde{T}_c}{\partial \tilde{x}_c} = 0 \tag{8.35}$$

$$\tilde{y}_c = 0; \qquad \tilde{l}_c \le \tilde{x}_c \le \tilde{L}_\infty; \qquad \frac{\partial \tilde{T}_c}{\partial \tilde{y}_c} = Nu_c \tilde{T}_c \tag{8.36}$$

$$\tilde{y}_c = \tilde{d}_c, \qquad \tilde{d}_c \tan(\phi - \alpha) \le \tilde{x}_c \le \tilde{L}_\infty; \qquad -\frac{\partial \tilde{T}_c}{\partial \tilde{y}_c} = Nu_c \tilde{T} \tag{8.37}$$

$$Nu_c = \frac{h_c l}{k_c} \tag{8.38}$$

The first boundary condition, equation (8.35), means that at a certain distance from the shear plane the heat loss through the chip is negligible. Equations (8.36)–(8.37) represent convective heat loss to the ambient. The boundary condition at $x_c = 0$ is a matching condition along the shear plane, and that at $y_c = 0$ and $0 < x_c < l_c$ is a matching condition along the chip–tool interface; they will be discussed later.

8.3.1.3 *Within the tool*

With respect to the reference frame chosen, the tool is stationary. Consequently, the heat transfer within this region is pure conduction. The governing equation is

$$\frac{\partial^2 \tilde{T}_t}{\partial \tilde{x}_t^2} + \frac{\partial^2 \tilde{T}_t}{\partial \tilde{y}_t^2} = 0 \qquad \text{in } \tilde{\Omega}_t \tag{8.39}$$

The boundary conditions appropriate in this region are

$$\tilde{y}_t = m_1 \tilde{x}_t, \qquad 0 \le \tilde{x}_t \le \tilde{L}_w; \qquad \frac{\partial \tilde{T}_t}{\partial \tilde{n}_t} = -Nu_t \tilde{T}_t \tag{8.40}$$

$$\tilde{x}_t = \tilde{L}_w, \qquad m_1 \tilde{L}_w \le \tilde{y}_t \le \tilde{L}_t; \qquad \frac{\partial \tilde{T}_t}{\partial \tilde{x}_t} = -Nu_t \tilde{T}_t \tag{8.41}$$

$$\tilde{y}_t = \tilde{L}_t, \qquad \frac{\partial \tilde{T}_t}{\partial \tilde{y}_t} = -Nu_p \tilde{T}_t \tag{8.42}$$

$$\tilde{x}_t = \tilde{L}_{t_1}, \qquad m_2 \tilde{L}_{t_1} \le \tilde{y}_t \le \tilde{L}_t; \qquad \frac{\partial \tilde{T}_t}{\partial \tilde{x}_t} = Nu_t \tilde{T}_t \tag{8.43}$$

$$\tilde{y}_t = m_2 \tilde{x}_t, \qquad \frac{\tilde{L}_{t_1}}{2} \le \tilde{x}_t \le \tilde{L}_{t_1}; \qquad \frac{\partial \tilde{T}_t}{\partial \tilde{n}_t} = -Nu_t \tilde{T}_t \tag{8.44}$$

$$Nu_t = \frac{h_t l}{k_t}, \qquad Nu_p = \frac{h_p l}{k_p} \tag{8.45}$$

The first and second boundary conditions (8.40, 8.41) represent the convective heat loss to the ambient. Equation (8.42) represents a convective heat loss model for the heat flux going into the tool holder. Equations (8.43) and (8.44) represent the convective heat loss to the ambient. The contact length between the tool and the chip depends strongly on the cutting condition and the material properties of the tool and the workpiece. It has also been observed by Trent (1984, chapter 9) that alloying elements have a significant influence on the chip–tool contact area. It has been observed over a wide range of experiments (Levy et al. 1976, Trent 1984) that the contact length is of the order of the length of the shear plane. In this analysis, the contact length is assumed to be 0.75 times the length of the shear plane. It should be noted, however, that the proposed scheme of analysis can handle any specified contact length that is not dependent on any particular assumption.

8.3.1.4 Matching boundary conditions

The matching boundary conditions along the shear plane and the chip–tool interface are presented now. Along the shear plane, often referred to as the primary deformation zone, heat is generated as a result of the large plastic deformation. Assuming that the rate of work in the deformation zone is converted entirely into heat, the heat generation can be calculated by the following equation (Tay et al. 1974, Stevenson et al. 1983, Dawson and Malkin 1984):

$$q' \widetilde{\delta}_{sp} = \widetilde{\tau} \cdot \widetilde{\dot{\gamma}} \tag{8.46}$$

where q' is the volumetric heat generation and $\widetilde{\tau}$ is the yield stress; $\widetilde{\dot{\gamma}}$ is the shear strain rate, which can be calculated from

$$\widetilde{\dot{\gamma}} = \frac{C_s \widetilde{V}_{\text{slide}}}{\widetilde{d}_w} = \frac{C_s}{\widetilde{d}_w} \frac{\cos \alpha}{\cos(\phi - \alpha)} \tag{8.47}$$

where C_s is an empirical constant; V_{slide} is the sliding velocity, which is related to the scanning velocity v_s, the rake angle α, and the shear plane angle ϕ; δ_{sp} is a Dirac delta function situated along the shear plane. Consequently, the heat generated is assumed to be concentrated along the shear plane. Applying an energy balance along the shear plane, the matching boundary condition can be obtained:

$$\widetilde{q}^{(n)}_{\text{workpiece}} + \widetilde{q}^{(n)}_{\text{chip}} = 1 \tag{8.48}$$

Furthermore, the continuity of temperature is also prescribed:

$$\widetilde{T}_w = \widetilde{T}_c \tag{8.49}$$

The secondary deformation zone is located along the chip–tool interface. Heat is generated by plastic deformation and frictional heating. The actual heat generation in the secondary zone depends on the exact cutting conditions. Based on the observations of Trent (1984, chapters 6 and 9) and Tay et al. (1974) for nonabrasive continuous chips and medium cutting speeds, the total heat generation due to frictional heating and plastic deformation in the secondary zone may reasonably be assumed to be between 0.20 and 0.35 times that in the primary zone. In our analysis, we have assumed the total heat generation in the secondary zone to be 0.25 times that in the primary zone. The actual distribution needs to be determined experimentally, and the proposed analysis scheme is capable of handling any specified spatial heat generation profile. Here, both effects are modeled by a total volumetric heat generation, $q_s' \delta_{ct}$, concentrated along the chip–tool interface. Once again, applying energy balance, the matching condition along the chip–tool interface is

$$\tilde{q}^{(n)}_{\text{chip}} + \tilde{q}^{(n)}_{\text{tool}} = \frac{q_s'}{q'} \tag{8.50}$$

and

$$\tilde{T}_c = \tilde{T}_t \tag{8.51}$$

Equations (8.25)–(8.51) form the mathematical model of the heat transfer within the tool, the chip, and the workpiece during metal cutting.

8.3.2 Matching Scheme

As mentioned before, the BEM algorithm is first applied to solve the heat transfer problem in each region separately. By matching the boundary conditions along the shear plane and the chip–tool interface, a complete solution may then be obtained. An exact expression can be derived to satisfy the matching conditions based on a guess solution. This matching scheme will now be presented.

Applying the algorithm developed in section 8.2, three matrix equations of the form of equations (8.15)–(8.19) can be derived. This equation can be interpreted as a relation between all the unknown $Y_j^{(n)}$ and the specified conditions. It is interesting, however, to note that the matrix coefficients depend only on the geometry, the fundamental solution, and the shape functions. They do not depend on either the temperature or the heat flux. Along the matching interfaces, typically, temperatures will be prescribed and fluxes will need to be evaluated. The change in the flux at the jth node due to a change in the temperature at the lth node can be obtained by taking the derivative of equation (8.15) with respect to T_l:

$$\sum_{j=1}^{N_s} \widetilde{B}_{ij} \frac{\partial q_j^{(n)}}{\partial T_l} = -\widetilde{A}_{il} \tag{8.52}$$

Furthermore, all the derivatives are independent of temperature and flux. Consequently, an exact linear equation can be derived as

$$q_j^{(n)} - q_j^{(n)^{(0)}} = \sum_{l=l_1}^{l_2} [T_l - T_l^{(0)}] \frac{\partial q_j^{(n)}}{\partial T_l} \tag{8.53}$$

Here, $q_j^{(n)^{(0)}}$ is the solution of equation (8.17) based on an arbitrarily prescribed $T_l^{(0)}$. It should be noted that, depending on the desired heat flux condition, the correct temperature T_l can be determined exactly from equation (8.53) without recourse to any iterative scheme. Here, (l_1, l_2) denotes the range of the boundary nodes over which matching is required. A similar strategy has also been used by Zabaras et al. (1988) for inverse heat conduction problems with phase changes.

The matching scheme is illustrated by considering two regions—the workpiece and the chip. A schematic diagram of the shear plane defining the numbering of the matching nodes is depicted in figure 8.5. Two equations similar to equation (8.53) for the workpiece and the chip can be derived:

$$\widetilde{q}_{w_j}^{(n)} - \widetilde{q}_{w_j}^{(n)^{(0)}} = \sum_{l=l_1}^{l_2} [\widetilde{T}_{w_l} - \widetilde{T}_{w_l}^{(0)}] \frac{\partial \widetilde{q}_{w_j}^{(n)}}{\partial \widetilde{T}_{w_l}}, \qquad j = l_1, \ldots, l_2 \tag{8.54}$$

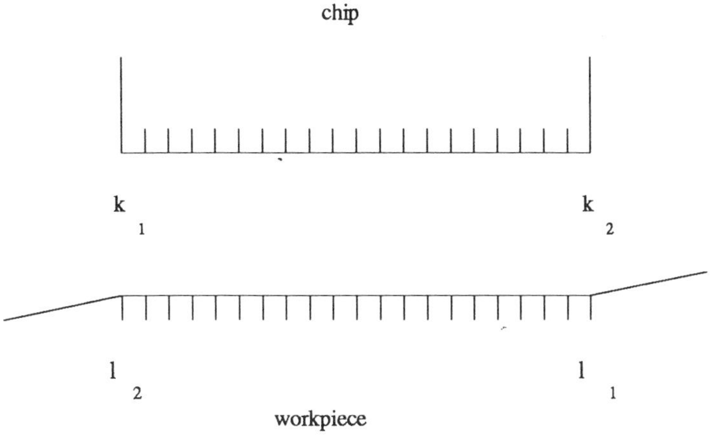

Figure 8.5. Schematic diagram of matching conditions for two regions.

and

$$\tilde{q}_{c_j}^{(n)} - \tilde{q}_{c_j}^{(n)^{(0)}} = \sum_{l=k_1}^{k_2} [\tilde{T}_{c_l} - \tilde{T}_{c_l}^{(0)}] \frac{\partial \tilde{q}_{c_j}^{(n)}}{\partial \tilde{T}_{c_l}}, \qquad j = k_1, \ldots, k_2 \tag{8.55}$$

Without loss of generality, it may be assumed that the starting temperatures are

$$\tilde{T}_{c_{j-1+k_1}}^{(0)} = \tilde{T}_{w_{l_2-j+1}}^{(0)} = 1, \qquad j = 1, \ldots, (k_2 - k_1 + 1) \tag{8.56}$$

The matching conditions, (8.48, 8.49), can be written as

$$\tilde{q}_{c_{j-1+k_1}}^{(n)} + \tilde{q}_{w_{l_2-j+1}}^{(n)} = 1, \qquad j = 1, \ldots, (k_2 - k_1 + 1) \tag{8.57}$$

and

$$\tilde{T}_{c_{j-1+k_1}} = \tilde{T}_{w_{l_2-j+1}} = 1, \qquad j = 1, \ldots, (k_2 - k_1 + 1) \tag{8.58}$$

Adding equations (8.54) and (8.55) via (8.56) and substituting for the matching condition equations (8.57) and (8.58), one has

$$1 - \tilde{q}_{c_{j-1+k_1}}^{(n)^{(0)}} - \tilde{q}_{w_{l_2-j+1}}^{(n)^{(0)}} = \sum_{l=1}^{k_2-k_1+1} \left(\frac{\partial \tilde{q}_{c_{j-1+k_1}}^{(n)}}{\partial \tilde{T}_{c_{l-1+k_1}}} + \frac{\partial \tilde{q}_{w_{l_2-j+1}}^{(n)}}{\partial \tilde{T}_{w_{l_2-j+1}}} \right) [\tilde{T}_{c_{l-1+k_1}} - \tilde{T}_{c_{j-1+k_1}}^{(0)}] \tag{8.59}$$

Equation (8.59) can be used to calculate the exact temperature at the appropriate interfaces.

The matching procedure for three regions will now be presented. The derivation is very similar. A schematic diagram defining the numbering system of the three regions is presented in figure 8.6. Equation (8.54) for the workpiece is still applicable. Equation (8.55) needs to be modified to include the additional chip–tool interface:

$$\tilde{q}_{c_j}^{(n)} - \tilde{q}_{c_j}^{(n)^{(0)}} = \sum_{l=k_1}^{k_2} [\tilde{T}_{c_l} - \tilde{T}_{c_l}^{(0)}] \frac{\partial \tilde{q}_{c_j}^{(n)}}{\partial \tilde{T}_{c_l}} + \sum_{l=k_3}^{k_4} [\tilde{T}_{c_l} - \tilde{T}_{c_l}^{(0)}] \frac{\partial \tilde{q}_{c_j}^{(n)}}{\partial \tilde{T}_{c_l}}$$

$$j = k_1, \ldots, k_2 \quad \text{and} \quad k_3, \ldots, k_4 \tag{8.60}$$

Here, $(k_1, \ldots, k_2)$ denotes the shear plane and $(k_3, \ldots, k_4)$ denotes the chip–tool interface. The first summation represents the contribution from the temperature change along the shear plane, while the second summation

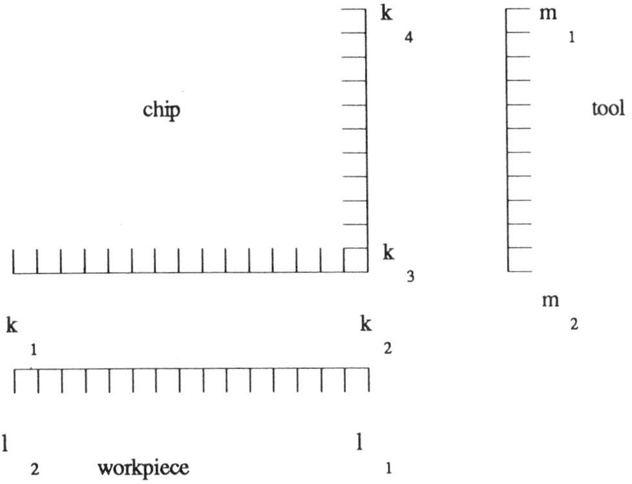

Figure 8.6. Schematic diagram of matching conditions for three regions.

represents the contribution from the temperature change along the chip–tool interface. Similarly, the equation applicable to the tool region is

$$\tilde{q}_{t_j}^{(n)} - \tilde{q}_{t_j}^{(n)^{(0)}} = \sum_{l=m_1}^{m_2} [\tilde{T}_{t_l} - \tilde{T}_{t_l}^{(0)}] \frac{\partial \tilde{q}_{t_j}^{(n)}}{\partial \tilde{T}_{t_l}}, \qquad j = m_1, \ldots, m_2 \qquad (8.61)$$

Here, $(m_1, \ldots, m_2)$ denotes the nodes located along the chip–tool interface.

The matching boundary conditions along the shear plane are given by equations (8.57) and (8.58). The matching boundary conditions along the chip–tool interface are

$$\tilde{q}_{c_{j-1+k_3}}^{(n)} + \tilde{q}_{t_{m_2-j+1}}^{(n)} = \frac{q_s'}{q'}, \qquad j = 1, \ldots, (k_4 - k_3 + 1) \qquad (8.62)$$

$$\tilde{T}_{c_{j-1+k_3}} = \tilde{T}_{t_{m_2-j+1}}, \qquad j = 1, \ldots, (k_4 - k_3 + 1) \qquad (8.63)$$

Using the matching condition equations (8.57), (8.58), (8.62), and (8.63), equations (8.54), (8.60), and (8.61) can be combined:

For $j = k_1, \ldots, k_2$,

$$1 - \tilde{q}_{c_j}^{(n)^{(0)}} - \tilde{q}_{w_{l_2-j+k_1}}^{(n)^{(0)}} = \sum_{l=k_1}^{k_2} \left(\frac{\partial \tilde{q}_{c_j}^{(n)}}{\partial \tilde{T}_{c_l}} + \frac{\partial \tilde{q}_{w_{l_2-j+k_1}}^{(n)}}{\partial \tilde{T}_{w_{l_2-j+k_1}}} \right) [\tilde{T}_{c_l} - \tilde{T}_{c_l}^{(0)}]$$
$$+ \sum_{l=k_3}^{k_4} \left(\frac{\partial \tilde{q}_{c_j}^{(n)}}{\partial \tilde{T}_{c_l}} \right) [\tilde{T}_{c_l} - \tilde{T}_{c_l}^{(0)}] \qquad (8.64)$$

For $j = k_3, \ldots, k_4$,

$$\frac{q_s'}{q'} - \tilde{q}_{c_j}^{(n)^{(0)}} - \tilde{q}_{w_{m_2-j+k_3}}^{(n)^{(0)}} = \sum_{l=k_1}^{k_2} \frac{\partial \tilde{q}_{c_j}^{(n)}}{\partial \tilde{T}_{c_l}} [\tilde{T}_{c_l} - \tilde{T}_{c_l}^{(0)}]$$

$$+ \sum_{l=k_3}^{k_4} \left(\frac{\partial \tilde{q}_{c_j}^{(n)}}{\partial \tilde{T}_{c_l}} + \frac{\partial \tilde{q}_{t_{m_2-j+k_3}}^{(n)}}{\partial \tilde{T}_{t_{m_2-l+k_3}}} \right) [\tilde{T}_{c_l} - \tilde{T}_{c_l}^{(0)}] \tag{8.65}$$

Equations (8.64) and (8.65) form a system of linear algebraic equations for the matching temperature along the shear plane and the chip–tool interface.

8.4 Results from BEM Analyses

As described in section 8.2, the BEM algorithm is first verified against the well-known Jaeger solutions (Jaeger 1942) for surface temperature of a semi-infinite domain with surface heating over a finite region. Steady-state turning operations are considered next. The particular processing conditions considered (Tay et al. 1974, Stevenson et al. 1983, Dawson and Malkin 1984) are tabulated in table 8.1. The geometric dimensions for these calculations are tabulated in table 8.2. In all the following cases, the thermophysical properties are assumed to be the same in all three regions. This assumption does not change the physics of the problem but simplifies the numerical implementation. These calculations provide an indication of the speed and efficiency of the method. The boundaries of the workpiece, the chip, and the tool are discretized into 62, 47, and 50 segments, respectively. The segments are divided in such a way that fine scales are used to resolve the high-gradient region. The calculation was done on a Sun Microsystem 3/260. The CPU time for all cases was approximately 0.4 second.

The dimensionless surface temperatures $[Pe(T - T_\infty)k_w/q' l]$ along the primary shear plane for several different Peclet numbers ranging from 1 to 40 are plotted in figure 8.7a. The abscissa is the dimensionless distance along the shear plane (see figure 8.3) measured from $A(x = 0)$ to $B(x = 1)$. It can

Table 8.1. Processing conditions

Workpiece speed, v_s	0.5–2.5 m/s
Chip speed, v_c	0.23–1.15 m/s
Shear angle, ϕ	30 degrees
Rake angle, α	20 degrees
Clearance angle, θ	6 degrees
Pe	1, 10, 20, 30, 40
Nu	0.003

Table 8.2. Dimensions in different regions

Workpiece:	
L^*	12.0
L_1^*	2.0
W^*	5.0
Tool:	
L_w^*	6.0
L_t^*	10.0
Chip:	
l_c^*	$\cos(\phi - \alpha)$
L_∞^*	10.0

be observed that the temperature increases gradually, attains a maximum, and then decreases. For $Pe = 1$ the maximum occurs at $x = 0.64$. As the Peclet number increases, the location of the maximum shifts to the right, and for $Pe = 40$ the maximum occurs at $x = 0.91$. Its magnitude also changes from 0.637 to 0.051 as the Peclet number goes from 1 to 40. The comparisons between the present BEM single-region results and those obtained by Dawson and Malkin (1984) using the FEM are presented in figure 8.7b. The Peclet number is 4.5 and the shear angle is 30 degrees. For this case, the maximum discrepancy between the two is 15 percent. The BEM results obtained by matching all three regions are also plotted in figure 8.7b. It can be observed that the interface temperature is considerably lower because not all the heat generated in the primary zone is conducted into the workpiece. The dimensionless temperatures along the chip–tool interface for five different Peclet numbers are plotted in figure 8.8. The abscissa is the dimensionless distance measured from $B(x = 0)$ to $C(x = 1)$ along the tool–chip interface (see figure 8.3). For $Pe = 1$ the maximum dimensionless temperature is 0.598 and occurs at $x = 0.1$. As the Peclet number increases, the temperature peak shifts to the right, and for $Pe = 40$ a temperature peak of 0.087 occurs at $x = 0.5$. The maximum temperature is believed to be responsible for the crater wear in the tool.

Figure 8.9 presents the dimensionless heat flux along the shear plane going into the workpiece ($q^{(n)}_{\text{workpiece}}/q'$) for various Peclet numbers. The dimensionless heat flux is scaled by the heat generation in the shear plane. Conservation of energy requires that the sum of the fluxes going into the workpiece and the chip along the shear plane be equal to 1. It is interesting to note that in some parts of the shear plane the heat flux going to the workpiece is greater than 1. This means that, in addition to the heat generated within the shear plane, heat is conducted to the workpiece from the chip. This effect is more noticeable near the trailing edge (point B in figure 8.3). This is expected, since the secondary contact region (between the tool and

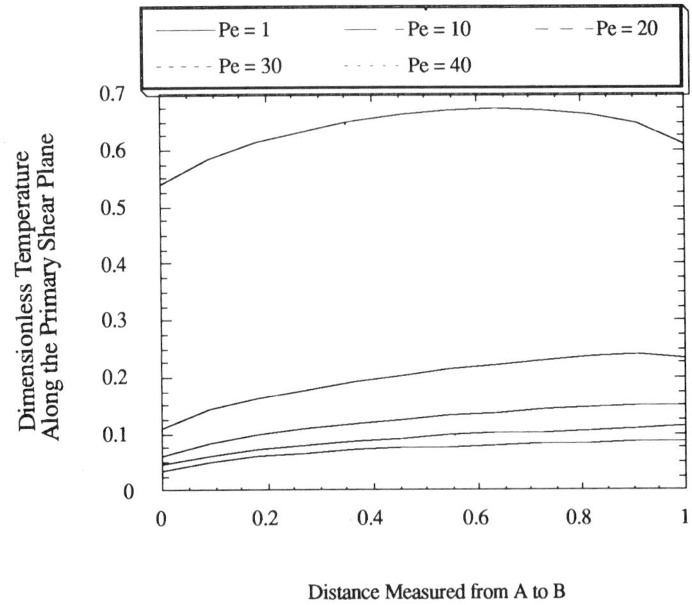

(a)

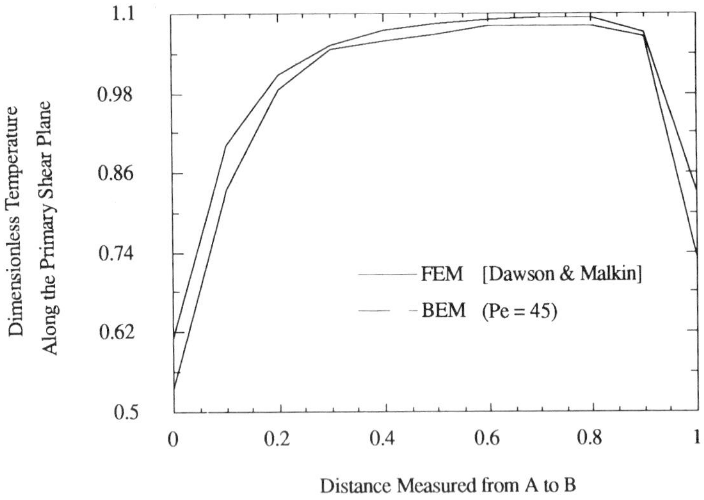

(b)

Figure 8.7. Dimensionless temperature along the primary shear plane: (a) for different Peclet numbers and (b) comparisons of BEM and FEM results for $Pe = 4.5$.

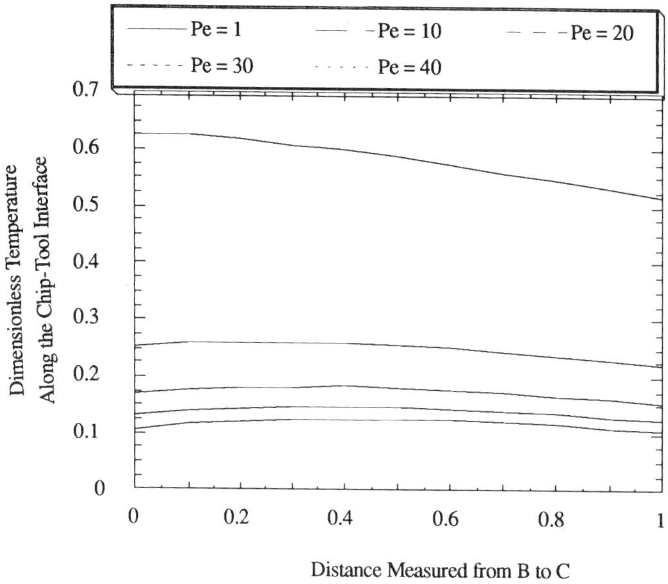

Figure 8.8. Dimensionless temperature along the contact surface for different Peclet numbers (ϕ = 30 degrees, α = 20 degrees, θ = 5 degrees).

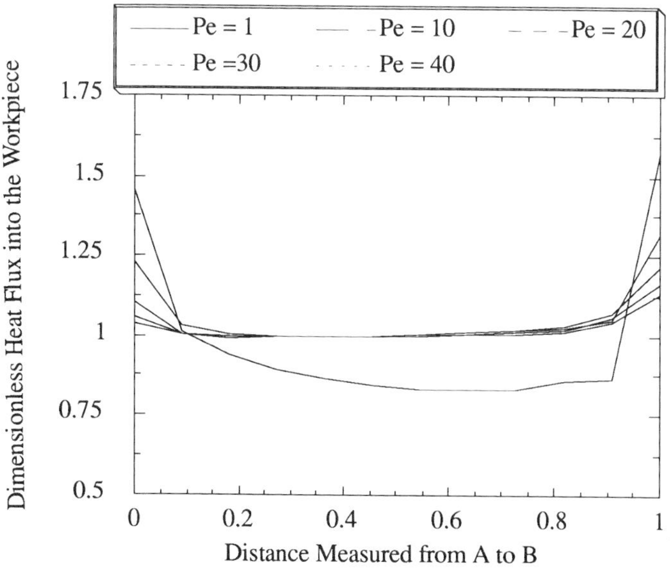

Figure 8.9. Dimensionless heat flux in the workpiece along the shear plane for different Peclet numbers (ϕ = 30 degrees, α = 20 degrees, θ = 5 degrees).

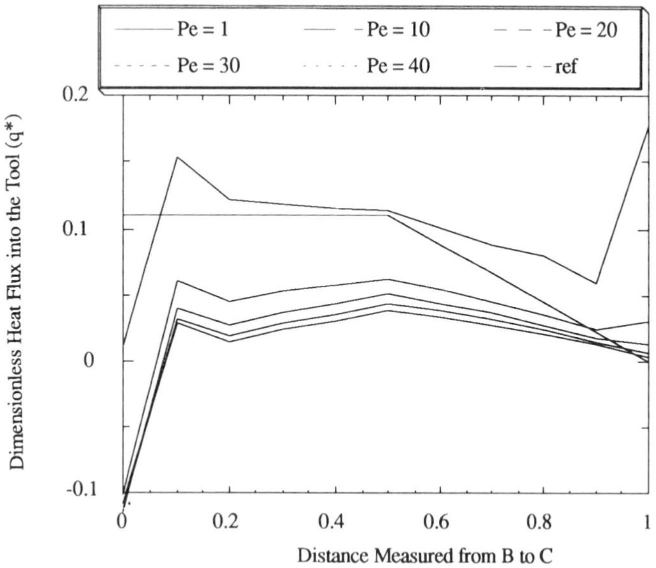

Figure 8.10. Dimensionless heat flux in the tool along the contact surface for different Peclet numbers (ϕ = 30 degrees, α = 20 degrees, θ = 5 degrees).

the chip) is right next to it. The heat generated in the chip–tool contact region is dissipated into the chip and then to the workpiece. It is observed that this effect decreases as the Peclet number (i.e., the cutting speed) increases. As the cutting speed increases, the heat convected by the chip also increases. Consequently, less heat is dissipated into the workpiece.

The dimensionless heat fluxes along the chip–tool interface (B–C) going into the tool are plotted in figure 8.10. As mentioned before, it has been assumed (Tay et al. 1974, Levy et al. 1976, Stevenson et al. 1983) that the total heat generation due to frictional heating and plastic deformation in the secondary zone is one-quarter of that in the primary zone. The dashed curve labeled "ref" shows the total heat generation in the secondary zone, which is the sum of the heat fluxes going into the tool and the chip. It is assumed to be constant ($q_s'/q' = 2/9$) between $x = 0$ and $x = 0.5$ and to decrease linearly to 0 between $x = 0.5$ and $x = 1$. Figure 8.10 also presents the proportion of the total flux going into the tool and the chip for different Peclet numbers. The vertical distance between an appropriate curve and the horizontal line for $\tilde{q} = 0$ represents the magnitude of the flux going into the tool, while the vertical distance between the appropriate curve and the "ref" curve represents that going into the chip at any value of x. In all the cases considered, it has been observed that more heat is dissipated into the chip than into the tool.

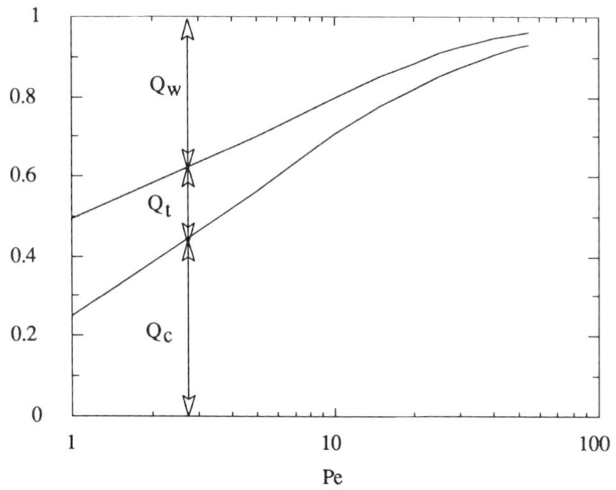

Figure 8.11. Distribution of the total heat generation dissipated into the chip, the tool, and the workpiece. Q_c, Q_t, and Q_w are the fractions for the chip, the tool, and the workpiece, respectively.

The variation of the distribution of the total heat flux (primary and secondary) dissipated into the chip, the tool, and the workpiece with respect to the Peclet number is plotted in figure 8.11. At $Pe = 1$, about 25 percent of the total heat generated is carried into the chip, while about 50 percent goes into the workpiece. As the Peclet number increases, the proportion of the heat carried by the chip increases rapidly. For $Pe = 50$, about 90 percent of the heat generated dissipates through the chip, while only 4 percent goes to the workpiece.

A schematic diagram for a machining process with flank wear is shown in figure 8.12. Figures 8.13 and 8.14 show the temperature and flux fields, respectively, along the primary shear plane, the secondary chip–tool interface, and the tool–workpiece or flank-wear interface. The processing conditions are shown in table 8.3. The dimensionless interface temperatures along the primary shear plane (A–B) and the flank-wear region (B–D) are

Table 8.3. Processing conditions

Shear angle, ϕ	5 degrees
Rake angle, α	20 degrees
Clearance angle, θ	6 degrees
Flank land	0.4 shear plane length
Pe	18
Nu	0.003

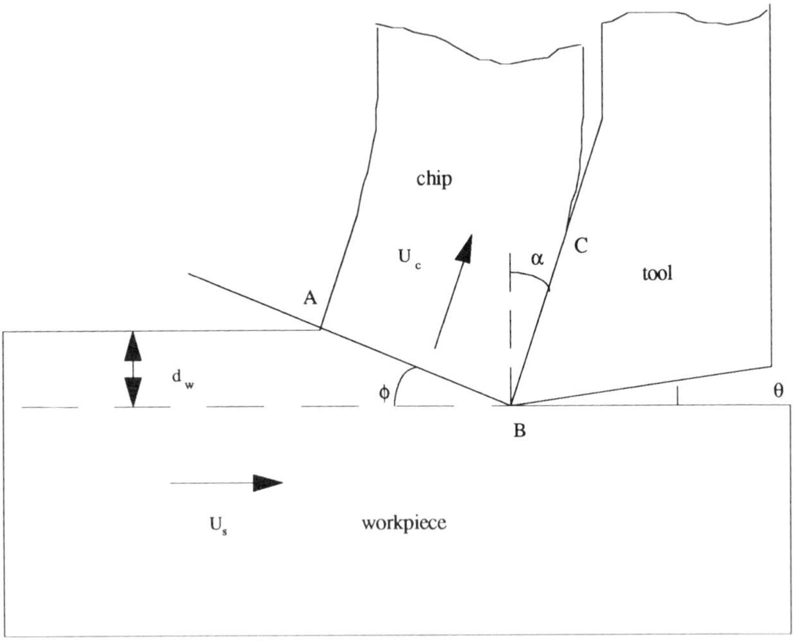

Figure 8.12. Schematic diagram of orthogonal cutting.

plotted in figure 8.13a. The temperature increases to a maximum of $\widetilde{T}(\widetilde{T} = Pe(T - T_\infty)k_w/q_l') = 0.21$ near point B and then decreases gradually to $\widetilde{T} = 0.17$ at point D. Figure 8.13b shows the variation of $\widetilde{T}$ along the chip–tool interface (C–B). $\widetilde{T}$ reaches a peak value of 0.24 at about the midpoint of CB. These temperature distributions are as expected for machining problems and, qualitatively, compare well with the earlier results of Dawson and Malkin (1984) and Chan and Chandra (1991a).

Figure 8.14 shows the dimensionless heat flux ($\widetilde{q} = q^{(n)}/q'$) along the workpiece–chip, workpiece–tool, and chip–tool interfaces. Conservation of energy requires that the heat generation be balanced by the fluxes. It is interesting to note that, in some parts of the shear plane, the heat flux going to the workpiece is greater than 1. This implies that, in addition to the heat generated within the shear plane, heat is conducted to the workpiece from the chip. Similar effects are also observed along the chip–tool interface near points C and B. Along the workpiece–tool interface, $\widetilde{q}$ is negative. This implies heat conduction from the tool to the workpiece.

8.5 BEM Sensitivity Formulation

As discussed earlier, the governing equation for temperature distributions in each of the regions (tool, chip, and workpiece) in steady-state machining

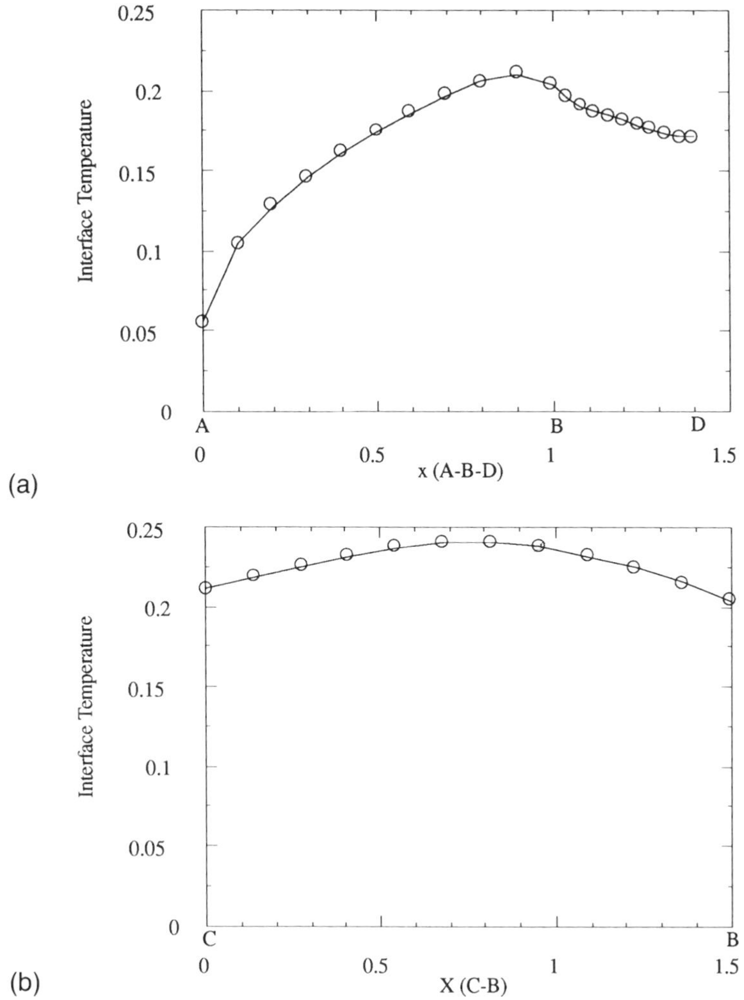

Figure 8.13. Temperature sensitivities at the interfaces for two different flank lengths: (a) workpiece–chip (A–B) and workpiece–tool (B–D) interfaces and (b) chip–tool interface (C–B).

operations may be expressed (Chan and Chandra 1991c) as

$$-k\frac{\partial^2 T}{\partial x_i^2} + \rho c v_i^{(s)} \frac{\partial T}{\partial x_i} = 0 \qquad \text{in } B \tag{8.66}$$

Here, $v_i^{(s)}$ is the convective velocity in that region and $\kappa = k/\rho c$ is the thermal diffusivity. The boundary conditions may be expressed as

$$T = \overline{T} \qquad \text{on } \partial B_T \tag{8.67a}$$

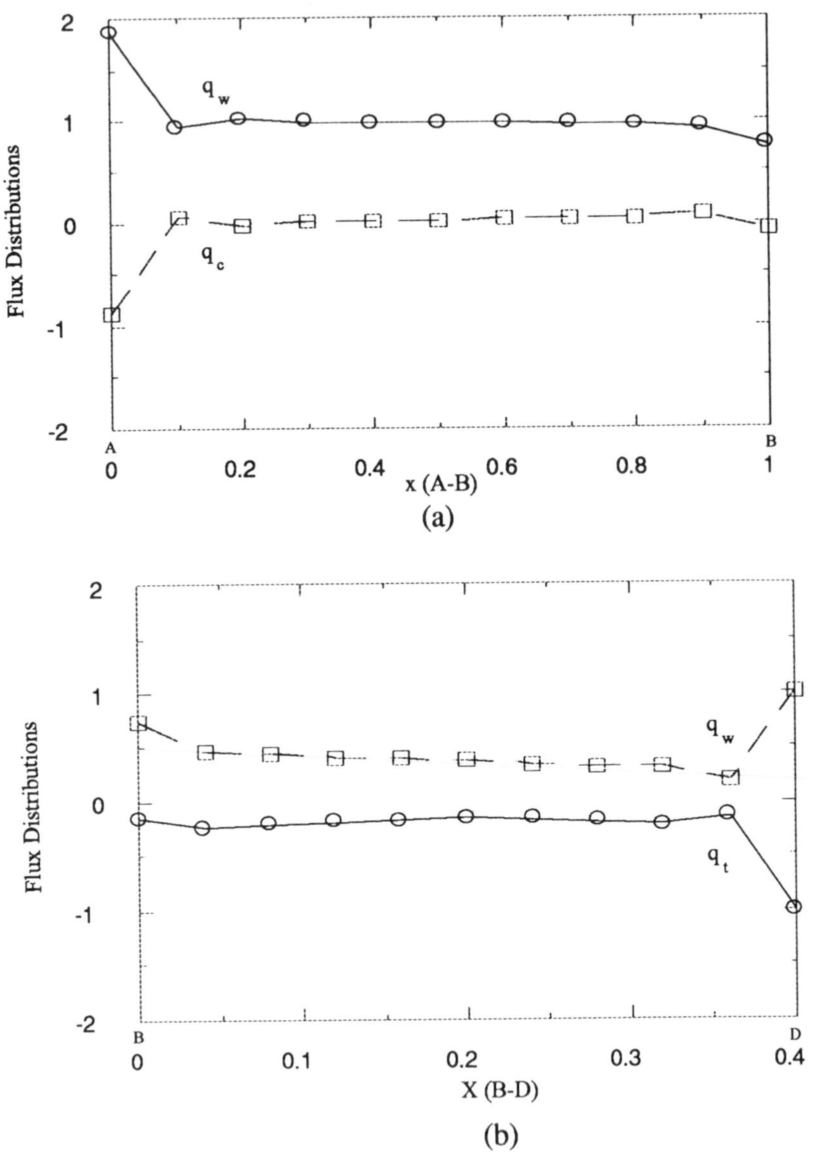

Figure 8.14. Flux sensitivities at the interfaces for two different flank lengths: (a) workpiece–chip (A–B) interface, (b) workpiece–tool (B–D) interface, and (c) chip–tool (C–B) interface.

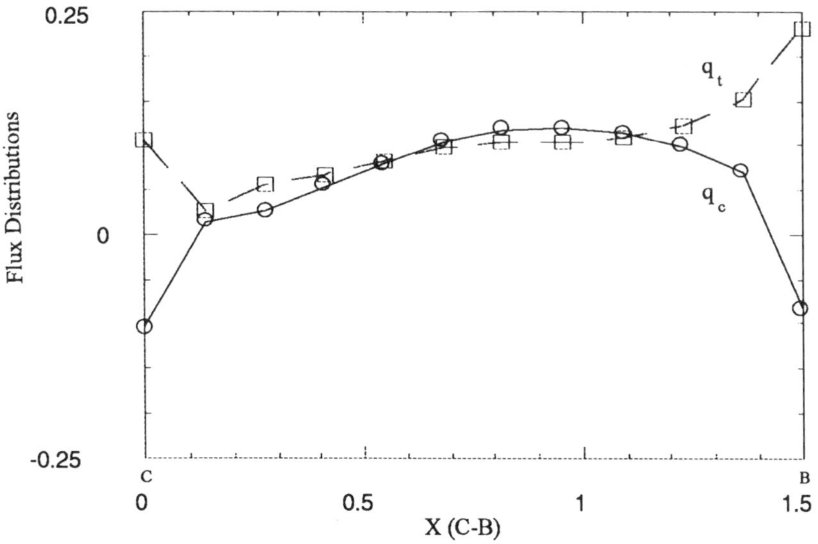

(c)

Figure 8.14—*continued*

and

$$k\frac{\partial T}{\partial x_i}n_i = \overline{q} \qquad \text{on } \partial B_q \tag{8.67b}$$

or

$$k\frac{\partial T}{\partial x_i}n_i = h(T - T_\infty) \qquad \text{on } \partial B_h \tag{8.67c}$$

Here, T represents the temperature field, q represents the flux, k is the thermal conductivity of the material, h is the convection coefficient, T_∞ represents the ambient temperature, x_i represents the spatial coordinates, and n_i are the components of the outward normal on the boundary of the domain. Equation (8.66) applies to an Eulerian reference frame that remains spatially fixed while the material flows through it. The convective term represents the energy transported by the material as it moves through the reference frame. Following Tanaka et al. (1986) and Chan and Chandra (1991a), an integral equation for an internal point may be written as

$$T(p) = \int_{\partial B}\left[-\left(k\frac{\partial G(p,Q)}{\partial n_Q} + \rho c G(p,Q)v_i^{(s)}n_i(Q)\right)T(Q) + kG(p,Q)q^{(n)}(Q)\right]dS(Q) \tag{8.68a}$$

when

$$v_{i,i}^{(s)} = 0 \tag{8.68b}$$

Here, a comma denotes field point differentiation; p and q represent a source point and a field point, respectively, in the domain; and P and Q represent a source point and a field point on the boundary. $G(p, q)$ is the Green's function for the steady-state conduction–convection equation (8.66). For two- or three-dimensional applications, appropriate versions of the fundamental solution $G(p, q)$ should be used. These are available in several references (Tanaka et al. 1986, Chan and Chandra 1991a). In the present work, the numerical implementation of the BEM formulation is done in two-dimensional situations only. Hence, two-dimensional versions of the above equations are presented in this section. The BEM formulation, however, is also valid for three-dimensional situations.

A boundary integral equation for the steady-state conduction–convection problem may now be obtained as $p \to P$ (Chan and Chandra 1991a). This gives

$$C(P)\,T(P) = \int_{\partial B}\left[-\left(k\frac{\partial G(P,Q)}{\partial n_Q} + \rho c G(P,Q)\,v_i^{(s)} n_i(Q)\right)T(Q) + kG(P,Q)\,q^{(n)}(Q)\right]dS(Q) \tag{8.69}$$

The coefficient $C(P)$, called the "corner tensor," depends upon the local geometry at P. If the boundary is locally smooth at P, $C = \frac{1}{2}$. Otherwise, it may be evaluated indirectly (Tanaka et al. 1986, Chan and Chandra 1991a) as shown below.

By applying a unit temperature field all over the boundary of the body, it may be shown that

$$C(P) = \int_{\partial B} -\left(k\frac{\partial G(P,Q)}{\partial n_Q} + \rho c G(P,Q)\,v_i^{(s)} n_i(Q)\right)dS(Q) \tag{8.70}$$

Substituting equation (8.70) into equation (8.69), one gets

$$0 = \int_{\partial B}\left[-\left(k\frac{\partial G(P,Q)}{\partial n_Q} + \rho c G(P,Q)\,v_i^{(s)} n_i(Q)\right)[T(Q) - T(P)] + kG(P,Q)\,q^{(n)}(Q)\right]dS(Q) \tag{8.71}$$

In order to determine the design sensitivities of temperature and flux with respect to any design parameter, equation (8.71) may be differentiated with respect to the particular parameter of interest. Barone and Yang (1988) and Mukherjee and Chandra (1989, 1991) have investigated shape optimization and assumed at this stage that the shape of a body is determined by a finite-dimensional vector with components b_i and that the shape changes occur continuously. The design parameters b_i, however, need not be restricted to shape variables. Process parameters like the scanning velocity, material parameters like thermal diffusivity, or a combined parameter like the Peclet number may also be chosen as the design parameter of interest. It is also possible to choose the specified boundary conditions as design parameters. Differentiating equation (8.71) with respect to a typical b_i (designated here as a shape parameter, b), one gets (Chandra and Chan 1992)

$$
\begin{aligned}
0 = \int_{\partial B} \Bigg[&- \left(k \frac{\partial G(b,P,Q)}{\partial n_Q} + \rho c G(b,P,Q) v_i^{(s)}(b) n_i(b,Q) \right) \\
&\times \left(\overset{*}{T}(b,Q) - \overset{*}{T}(b,P) \right) + kG(b,P,Q) \overset{*}{q}^{(n)}(b,Q) \Bigg] dS(b,Q) \\
+ \int_{\partial B} \Bigg[&- \left(k \frac{\partial \overset{*}{G}(b,P,Q)}{\partial n_Q} + \rho c \overset{*}{G}(b,P,Q) v_i^{(s)}(b) n_i(b,Q) \right. \\
&\left. + \rho c G(b,P,Q) v_i^{(s)}(b) \overset{*}{n}_i(b,Q) \right) [T(b,Q) - T(b,P)] \\
&+ k\overset{*}{G}(b,P,Q) q^{(n)}(b,Q) \Bigg] dS(b,Q) \\
+ \int_{\partial B} \Bigg[&- \left(k \frac{\partial G(b,P,Q)}{\partial n_Q} + \rho c G(b,P,Q) v_i^{(s)}(b) n_i(b,Q) \right) \\
&\times [T(b,Q) - T(b,P)] + kG(b,P,Q) q^{(n)}(b,Q) \Bigg] d\overset{*}{S}(b,Q) \qquad (8.72)
\end{aligned}
$$

where a superposed asterisk (*) denotes a derivative with respect to b. Other types of parameters such as process parameters (e.g., cutting velocity) and material parameters (e.g., thermal conductivity, specific heat) may also be chosen as design variables. In such cases, appropriate sensitivity equations including the requisite additional terms may be derived easily following the same procedure outlined above.

The kernels $\overset{*}{G}(b, P, Q)$ and $[\partial\overset{*}{G}(b, P, Q)]/\partial n_Q$ may be expressed (Barone and Yang 1988, Mukherjee and Chandra 1989, 1991) as

$$\overset{*}{G}(b, P, Q) = G_{,k}(b, P, Q)[\overset{*}{x}_k(Q) - \overset{*}{x}_k(P)] \tag{8.73}$$

$$\frac{\partial\overset{*}{G}(b, P, Q)}{\partial n_Q} = \frac{\partial G_{,k}(b, P, Q)}{\partial n_Q}[\overset{*}{x}_k(Q) - \overset{*}{x}_k(P)] + \overset{*}{n}_k(Q)G_{,k}(b, P, Q) \tag{8.74}$$

and

$$d\overset{*}{S}(b, Q) = \left(\frac{\dfrac{\partial}{\partial b}\left[\left| \dfrac{\partial}{\partial \eta}(x_k e_k) \right| \right]}{\left| \dfrac{\partial}{\partial \eta}(x_k e_k) \right|} \right) dS(b, Q) \tag{8.75}$$

where b is a shape parameter and e_k represents a component of a unit vector. The quantity $[\overset{*}{x}_k(Q) - \overset{*}{x}_k(P)] \sim O(r)$. Hence, when b is a shape parameter, $\overset{*}{G}$ and $\partial\overset{*}{G}/\partial n_Q$ are regular and $1/r$ singular, respectively, for two-dimensional applications. When b is not a shape parameter, $d\overset{*}{S}$ is zero. It can be seen that differentiation with respect to a nonspatial variable does not affect the order of singularity in either G or $\partial G/\partial n_Q$.

Once the standard BEM analysis is performed, the temperature $T(b, Q)$ and the flux $q^{(n)}(b, Q)$ are known everywhere on the boundary. From the given boundary conditions, half of the quantities of $\overset{*}{T}(b, Q)$ and $\overset{*}{q}^{(n)}(b, Q)$ are known for a well-posed problem. Hence, equations (8.72)–(8.75) may be used to solve for the unknown temperature sensitivities and flux sensitivities on the boundary in terms of the known ones.

A sensitivity equation for an internal point may now be obtained by differentiating equation (8.68*a*) with respect to b. This gives,

$$\begin{aligned}
\overset{*}{T}(b,p) = &\int_{\partial B}\Bigg[-\left(k\frac{\partial G(b,p,Q)}{\partial n_Q} + \rho c G(b,p,Q) v_i^{(s)}(b) n_i(b,Q) \right) \overset{*}{T}(b,Q) \\
&+ kG(b,p,Q)\overset{*}{q}^{(n)}(b,Q) \Bigg] dS(b,Q) \\
&+ \int_{\partial B}\Bigg[-\Bigg(k\frac{\partial \overset{*}{G}(b,p,Q)}{\partial n_Q} + \rho c \overset{*}{G}(b,p,Q) v_i^{(s)}(b) n_i(b,Q) \\
&+ \rho c G(b,p,Q) v_i^{(s)}(b) \overset{*}{n}_i(b,Q) \Bigg) T(b,Q) \\
&+ k\overset{*}{G}(b,p,Q) q^{(n)}(b,Q) \Bigg] dS(b,Q) \\
&+ \int_{\partial B}\Bigg[-\left(k\frac{\partial G(b,p,Q)}{\partial n_Q} + \rho c G(b,p,Q) v_i^{(s)}(b) n_i(b,Q) \right) T(b,Q) \\
&+ kG(b,p,Q) q^{(n)}(b,Q) \Bigg] d\overset{*}{S}(b,Q)
\end{aligned} \tag{8.76}$$

It should be noted here that equation (8.76) requires only algebraic evaluations for the determination of $\overset{*}{T}(b, p)$. The kernels, in this case, do not become singular, since Q lies on the boundary while p is strictly an interior point.

Numerical implementation of the BEM equations (8.69, 8.72–8.75) for heat transfer and design sensitivities in conduction–convection problems is discussed in this section. In the present work, two-dimensional problems are considered.

The first step is the discretization of the boundary of the two-dimensional domain into boundary elements. A discretized version of the standard boundary integral equation may be written as

$$
\begin{aligned}
C(P_M)T(P_M) = -\sum_{i=1}^{N_s}\int_{\Delta S_i}\bigg(k\frac{\partial G(P_M, Q)}{\partial n_Q}T(Q) \\
- kG(P_M, Q)q^{(n)}(Q)\bigg)dS(b, Q) \\
- \sum_{i=1}^{N_s}\int_{\Delta S_i}\rho c G(P_M, Q)T(Q)v_l^{(s)}n_l(Q)\,dS(b, Q) \qquad (8.77)
\end{aligned}
$$

A discretized BEM sensitivity equation may also be written as

$$
\begin{aligned}
0 = \sum_{i=1}^{N_s}\int_{\Delta s_i}\bigg[-\bigg(k\frac{\partial G(b, P_M, Q)}{\partial n_Q} + \rho c G(b, P, Q)v_l^{(s)}(b, Q)n_l(b, Q)\bigg) \\
\times[\overset{*}{T}(b, Q) - \overset{*}{T}(b, P_M)] + kG(b, P_M, Q)\overset{*}{q}^{(n)}(b, Q)\bigg]dS(Q) \\
+ \sum_{i=1}^{N_s}\int_{\Delta s_i}\bigg[-\bigg(k\frac{\partial \overset{*}{G}(b, P_M, Q)}{\partial n_Q} + \rho c\overset{*}{G}(b, P, Q)v_l^{(s)}(b)n_l(b, Q) \\
+ \rho c G(b, P_M, Q)v_l^{(s)}(b)\overset{*}{n}_l(b, Q)\bigg)[T(b, Q) - T(b, P)] \\
+ k\overset{*}{G}(b, P_M, Q)q^{(n)}(b, Q)\bigg]dS(b, Q) \\
+ \sum_{i=1}^{N_s}\int_{\Delta s_i}\bigg[-\bigg(k\frac{\partial G(b, P_M, Q)}{\partial n_Q} + \rho c G(b, P_M, Q)v_l^{(s)}(b)n_l(b, Q)\bigg) \\
\times[T(b, Q) - T(b, P_M)] + kG(b, P_M, Q)q^{(n)}(b, Q)\bigg]d\overset{*}{S}(b, Q) \qquad (8.78)
\end{aligned}
$$

where the boundary of the domain ∂B is divided into N_s boundary segments and $T(P_M)$ represents temperature at a point P that coincides with node M.

Suitable shape functions must now be chosen for the variation of temperature, temperature sensitivity, flux, and flux sensitivity over each boundary element Δs_i. In the present work, each of the above quantities is assumed to vary linearly over individual boundary elements. A matrix equation for the standard BEM equation (8.77) may be derived as (details in Chan and Chandra 1991b)

$$\sum_{j=1}^{N_s} A_{ij} T_j = \sum_{j=1}^{N_s} [b_{ij-1}^{(2)} q_j^{(n)} + b_{ij}^{(1)} q_j'^{(n)}] \tag{8.79}$$

Here, $q_j^{(n)}$ is the flux at the jth node in the normal direction of the element just before the node, while $q_j'^{(n)}$ is the flux at the jth node in the normal direction of the element just after the node. When the surface is smooth at the jth node, $q_j^{(n)}$ and $q_j'^{(n)}$ are equal to one another. By splitting up the normal flux, a geometrical corner can be handled properly (Chan and Chandra 1991b). A_{ij} represents an element of the assembled matrix that multiplies T_j, where $b_{ij-1}^{(2)}$ and $b_{ij}^{(1)}$ refer to the contributions from the second node of the $(j-1)$th segment and the first node of the jth segment, respectively, for the matrix multiplying the normal flux. This allows proper modeling of the jump in the normal flux across a geometric corner.

Similarly, a matrix equation for the design sensitivity calculations may be derived as

$$\sum_{j=1}^{N_s} A_{ij} \overset{*}{T}_j - \sum_{j=1}^{N_s} [b_{ij-1}^{(2)} \overset{*}{q}_j^{(n)} + b_{ij}^{(1)} \overset{*}{q}_j'^{(n)}] = -\sum_{j=1}^{N_s} \overset{*}{A}_{ij} T_j + \sum_{j=1}^{N_s} [\overset{*}{b}_{ij-1}^{(2)} q_j^{(n)} + \overset{*}{b}_{ij}^{(1)} q_j'^{(n)}] \tag{8.80}$$

It is clear from equation (8.80) that equation (8.79) should be solved first. Once temperature and flux are known everywhere on the boundary, equation (8.80) can be solved for the design sensitivities. It should be noted that the coefficient matrices A_{ij}, $b_{ij-1}^{(2)}$, and $b_{ij}^{(1)}$ appearing in equation (8.80) are exactly the same as those in equation (8.79).

At each location over the entire boundary of the domain, either $T(\overset{*}{T})$ or $q^{(n)}(\overset{*}{q}^{(n)})$ (or a combination of the two) is prescribed for a well-posed problem. Equation (8.79) may now be solved for the unknown T and $q^{(n)}$ in terms of the known ones. Using these results, equation (8.80) may be solved for the unknown sensitivities $\overset{*}{T}$ and $\overset{*}{q}^{(n)}$ (or $\overset{*}{q}'^{(n)}$ as appropriate). Typically, the temperature or the flux sensitivity is zero when the temperature or the flux is specified respectively at any particular point on the boundary.

It should be noted that the BEM sensitivity formulation can also accommodate any arbitrarily specified $\overset{*}{T}$ or $\overset{*}{q}$ on any portion of the boundary. Once $\overset{*}{T}$ and $\overset{*}{q}$ have been obtained over the entire boundary, the internal equation (8.76) may be used to obtain the temperature sensitivity at any internal point. A derivative form of the equation (8.76) may also be used to determine the flux sensitivity at any internal point.

It is also important to note here that the matrices A_{ij} and B_{ij} in equation (8.78) depend on the reference configuration only and do not depend on the choice of the design parameter b. All the effects for a particular choice of b are incorporated through the right-hand-side vector F_i in equation (8.78). This makes the BEM formulation very efficient when sensitivities with respect to a (relatively) large number of design variables are sought for a particular reference configuration. For the small increase in additional costs due to additional evaluations of the right-hand side, sensitivities with respect to several design variables may be obtained. Using parallel processing features, such design sensitivities may also be tracked simultaneously.

8.6 Sensitivities of Machining Processes

In this section, the BEM sensitivity formulation is used to model a steady-state machining process. A schematic diagram of the process is sketched in figure 8.3. Typically, the tool is a large-angled wedge that is driven into the workpiece to remove a thin layer, the chip. As the tool is driven into the workpiece, the material undergoes a severe plastic deformation along the shear plane. As the chip forms, it diverts and slides across the tool face. The tool also wears out as the machining continues. There are three main sources of heat generation: (1) the heat generated by the plastic deformation in the shear plane; (2) the frictional heating and plastic deformation as the chip slides over the tool face; and (3) the frictional heating and plastic deformation due to the flank wear of the tool at the tool–workpiece interface. The heat transfer involved here is conduction and convection of the heat generated into the tool, the chip, and the workpiece.

In a machining operation, the velocities associated with the tool, the workpiece, and the chip are quite different. Fixing the reference frame to the tool, it may be considered stationary. The workpiece, with respect to such a reference frame, moves at the cutting velocity (scanning velocity). The chip moves in a different direction with a velocity related to the scanning and the shear plane angle. Therefore, the algorithm is applied to each region separately. For oblique cutting, the BEM algorithm is applied separately to each of the regions because the chip velocity may also depend on the tool angles. By matching the boundary conditions at the workpiece–chip interface, the workpiece–tool interface, and the chip–tool interface, a complete solution for the machining problem may then be obtained. Orthogonal machining is considered here, and two-dimensional analyses of heat transfer and their sensitivities are performed. In order to model the heat transfer efficiently, the tool, the chip, and the workpiece are formulated separately. Furthermore, a coordinate system unique to each region is defined in order to better represent the boundary elements. The coordinate systems for the three regions are defined in figure 8.12.

8.6.1 Matching Boundary Conditions for Sensitivity Calculations

The matching boundary conditions along the shear plane, the chip–tool interface, and the tool–workpiece interface are now presented. Along the shear plane, often referred to as the primary deformation zone, heat is generated as a result of the large plastic deformation. Assuming that the rate of work in the deformation zone is converted entirely into heat, the heat generation can be calculated by the following equation (Tay et al. 1974, Stevenson et al. 1983, Dawson and Malkin 1984):

$$q'\,\delta_{sp} = \tau\dot{\gamma} \tag{8.81}$$

where q' is the volumetric heat generation, τ is the yield stress, and $\dot{\gamma}$ is the shear strain rate which can be calculated from

$$\dot{\gamma} = \frac{CV_{\text{slide}}}{d_w} = \frac{C}{d_w}\,\frac{v_s \cos\alpha}{\cos(\phi-\alpha)} \tag{8.82}$$

where C is an empirical constant; V_{slide} is the sliding velocity, related to the scanning velocity v_s, the rake angle α, and the shear plane angle ϕ; and δ_{sp} is a Dirac delta function situated along the shear plane. Consequently, the heat generated is assumed to be concentrated along the shear plane. Applying an energy balance along the shear plane, the matching boundary condition can be obtained:

$$q^{(n)}_{\text{workpiece}} + q^{(n)}_{\text{chip}} = q' \tag{8.83}$$

Furthermore, the continuity of temperature is also prescribed,

$$T_w = T_c \tag{8.84}$$

The matching boundary conditions for the sensitivities can be obtained by differentiating the above equations:

$$\overset{*}{q}{}^{(n)}_{\text{workpiece}} + \overset{*}{q}{}^{(n)}_{\text{chip}} = \overset{*}{q}{}' \tag{8.85}$$

$$\overset{*}{T}_w = \overset{*}{T}_c \tag{8.86}$$

If the heat generation in the shear plane is unaffected by the design parameter b, then $\overset{*}{q}{}'$ is equal to zero.

The secondary deformation zone is located along the chip–tool interface. Heat is generated by plastic deformation and frictional heating. Here, both effects are modeled by a total volumetric heat generation $q_s' \delta_{ct}$ concentrated along the chip–tool interface. Once again, applying the energy balance, the matching condition along the chip–tool interface is

$$q^{(n)}_{\text{chip}} + q^{(n)}_{\text{tool}} = q_s' \tag{8.87}$$

and

$$T_c = T_t \tag{8.88}$$

The matching conditions for the sensitivities are

$$\overset{*}{q}{}^{(n)}_c + \overset{*}{q}{}^{(n)}_{\text{tool}} = \overset{*}{q}{}_s' \tag{8.89}$$

$$\overset{*}{T}_c = \overset{*}{T}_t \tag{8.90}$$

For crater wear, the matching boundary conditions are given by equations (8.87)–(8.90).

In the case of flank wear, there is a third deformation zone located along the workpiece–tool interface. Heat is generated by plastic deformation and frictional heating. Here, both effects are modeled by a total volumetric heat generation $q_F' \delta_{wt}$ concentrated along the workpiece–tool interface. The matching boundary conditions are

$$q^{(n)}_w + q^{(n)}_t = q_F' \tag{8.91}$$

$$T_w = T_t \tag{8.92}$$

and

$$\overset{*}{q}{}^{(n)}_w + \overset{*}{q}{}^{(n)}_t = \overset{*}{q}{}_F' \tag{8.93}$$

$$\overset{*}{T}_w = \overset{*}{T}_t \tag{8.94}$$

8.6.2 Matching Scheme for the Sensitivity Problem

As mentioned before, the BEM algorithm is first applied to solve the heat transfer in each region separately. By matching the sensitivity boundary conditions along the shear plane, the chip–tool interface, and the workpiece–tool interface, a complete solution may then be obtained. Starting from an initial solution, an exact expression can be derived to satisfy the matching conditions without iterations. A detailed description can be found in Chan and Chandra (1991c). This matching scheme will now be presented briefly.

Applying the BEM sensitivity algorithm to each region, the following matrix equations may be obtained (Chan and Chandra 1991a,b):

$$\sum_{j=1}^{N_s} A_{ij}\overset{*}{T}_j = \sum_{j=1}^{N_s} [b^{(2)}_{ij-1}\overset{*}{q}{}_j^{(n)} + b^{(1)}_{ij}\overset{*}{q}{}_j'^{(n)}] \tag{8.95}$$

A switching process is applied by keeping the unknown quantities on the left-hand side and the known quantities on the right-hand side (Chan and Chandra 1991b).

Along a matching interface, the heat flux is treated as an unknown and the temperature is treated as a known variable. Taking the derivative of the matrix equation with respect to a node temperature along an interface ($\overset{*}{\gamma}{}_j^{(n)}$ is the unknown sensitivity variable at the jth node after switching),

$$\sum_{j=1}^{N_s} \widetilde{A}_{ij} \frac{\partial \overset{*}{Y}{}_j^{(u)}}{\partial \overset{*}{T}_l} = \widetilde{B}_{il} \tag{8.96}$$

All the derivatives are independent of temperature and flux. Consequently, an exact linear equation can be derived as

$$\overset{*}{q}{}_j^{(n)} - \overset{*}{q}{}_j^{(n)^{(0)}} = \sum_{l=l_1}^{l_2} [\overset{*}{T}_l - \overset{*}{T}{}_l^{(0)}] \frac{\partial \overset{*}{q}{}_j^{(n)}}{\partial \overset{*}{T}_l} \tag{8.97}$$

Here, $\overset{*}{q}{}_j^{(n)^{(0)}}$ is the solution of equation (8.95) based on an arbitrarily prescribed $\overset{*}{T}{}_l^{(0)}$. On a node along the matching interface, two equations (one from each region) may be derived. Using the two matching boundary conditions (e.g., equations 8.83–8.84), the heat flux may be eliminated to obtain an equation with temperature being the unknown. Applying this to each node along the interface, a matrix equation can be obtained for the matching interface temperatures:

For the workpiece–chip interface, $l_1 \le l \le l_2$,

$$\overset{*}{q}' - \overset{*}{q}^{(n)^{(0)}}_{w_l} - \overset{*}{q}^{(n)^{(0)}}_{c_{k_3-(l-l_2)}} = \sum_{j=l_1}^{l_2} \left(\frac{\partial \overset{*}{q}^{(n)}_{w_l}}{\partial \overset{*}{T}_{w_j}} + \frac{\partial \overset{*}{q}_{c_{k_3-(l-l_2)}}}{\partial \overset{*}{T}_{c_{k_3-(j-l_2)}}} \right) [\overset{*}{T}_{w_j} - \overset{*}{T}^{(0)}_{w_j}]$$

$$+ \sum_{j=l_2+1}^{l_3} \frac{\partial \overset{*}{q}^{(n)}_{w_l}}{\partial \overset{*}{T}_{w_j}} [\overset{*}{T}_{w_j} - \overset{*}{T}^{(0)}_{w_j}] + \sum_{j=k_1}^{k_2-1} \frac{\partial \overset{*}{q}_{c_{k_3-(l-l_2)}}}{\partial \overset{*}{T}_{c_j}} [\overset{*}{T}_{c_j} - \overset{*}{T}^{(0)}_{c_j}] \tag{8.98}$$

For the workpiece–tool interface, $l_2 < l \le l_3$,

$$\overset{*}{q}'_F - \overset{*}{q}^{(n)^{(0)}}_{w_l} - \overset{*}{q}^{(n)^{(0)}}_{t_{m_2-(l-l_2)}} = \sum_{j=l_1}^{l_2-1} \frac{\partial \overset{*}{q}^{(n)}_{w_l}}{\partial \overset{*}{T}_{w_j}} [\overset{*}{T}_{w_j} - \overset{*}{T}^{(0)}_{w_j}]$$

$$+ \sum_{j=l_2}^{l_3} \frac{\partial \overset{*}{q}^{(n)}_{w_l}}{\partial \overset{*}{T}_{w_j}} + \left(\frac{\partial \overset{*}{q}_{t_{m_2-(l-l_2)}}}{\partial \overset{*}{T}_{t_{m_2-(j-l_2)}}} \right) [\overset{*}{T}_{w_j} - \overset{*}{T}^{(0)}_{w_j}]$$

$$+ \sum_{j=k_1}^{k_2-1} \frac{\partial \overset{*}{q}_{t_{m_2-(l-l_2)}}}{\partial \overset{*}{T}_{t_{m_2-(j-k_1)}}} [\overset{*}{T}_{c_j} - \overset{*}{T}^{(0)}_{c_j}] \tag{8.99}$$

For the chip–tool interface, $k_1 \le k < k_2$,

$$\overset{*}{q}'_s - \overset{*}{q}^{(n)^{(0)}}_{c_k} - \overset{*}{q}^{(n)^{(0)}}_{t_{m_2-(k-k_1)}} = \sum_{j=l_1}^{l_2-1} \frac{\partial \overset{*}{q}^{(n)}_{c_k}}{\partial \overset{*}{T}_{w_{k_3-(j-l_1)}}} [\overset{*}{T}_{w_j} - \overset{*}{T}^{(0)}_{w_j}]$$

$$+ \left(\frac{\partial \overset{*}{q}_{c_k}}{\partial \overset{*}{T}_{c_{k_2}}} + \frac{\partial \overset{*}{q}_{t_{m_2-(k-k_2)}}}{\partial \overset{*}{T}_{t_{m_2}}} \right) [\overset{*}{T}_{w_{l_2}} - \overset{*}{T}^{(0)}_{w_{l_2}}]$$

$$+ \sum_{j=l_2+1}^{l_3} \frac{\partial \overset{*}{q}_{t_{m_2-(k-k_1)}}}{\partial \overset{*}{T}_{t_{m_2-(j-l_2)}}} [\overset{*}{T}_{w_j} - \overset{*}{T}^{(0)}_{w_j}]$$

$$+ \sum_{j=k_1}^{k_2-1} \left(\frac{\partial \overset{*}{q}_{c_k}}{\partial \overset{*}{T}_{c_j}} + \frac{\partial \overset{*}{q}_{t_{m_2-(l-l_2)}}}{\partial \overset{*}{T}_{t_{m_2-(j-k_1)}}} \right) [\overset{*}{T}_{c_j} - \overset{*}{T}^{(0)}_{c_j}] \tag{8.100}$$

At the node where all three regions meet,

$$\overset{*}{q} + \overset{*}{q}'_s + \overset{*}{q}'_F - \overset{*}{q}^{(n)^{(0)}}_{w_{l_2}} - \overset{*}{q}'^{(n)^{(0)}}_{w_{l_2}} - \overset{*}{q}^{(n)^{(0)}}_{c_{k_2}} - \overset{*}{q}'^{(n)^{(0)}}_{c_{k_2}} - \overset{*}{q}^{(n)^{(0)}}_{t_{m_2}} - \overset{*}{q}'^{(n)^{(0)}}_{t_{m_2}}$$

$$= \sum_{j=l_1}^{l_2-2} \left(\frac{\partial \overset{*}{q}^{(n)}_{w_{l_2}}}{\partial \overset{*}{T}_{w_j}} + \frac{\partial \overset{*}{q}^{(n)}_{c_{k_2}}}{\partial \overset{*}{T}_{c_{k_3-(j+l_1)}}} \right) [\overset{*}{T}_{w_j} - \overset{*}{T}^{(0)}_{w_j}]$$

$$+ \left(\frac{\partial \overset{*}{q}^{(n)}_{w_{l_2}}}{\partial \overset{*}{T}_{w_{l_2-1}}} + \frac{\partial \overset{*}{q}'^{(n)}_{w_{l_2}}}{\partial \overset{*}{T}_{w_{l_2-1}}} + \frac{\partial \overset{*}{q}^{(n)}_{c_{k_2}}}{\partial \overset{*}{T}_{c_{k_2+1}}} + \frac{\partial \overset{*}{q}'^{(n)}_{c_{k_2}}}{\partial \overset{*}{T}_{c_{k_2+1}}} \right) [\overset{*}{T}_{w_{l_2-1}} - \overset{*}{T}^{(0)}_{w_{l_2-1}}]$$

$$+\left(\frac{\partial \overset{*}{q}{}_{w_{l_2}}^{(n)}}{\partial \overset{*}{T}_{w_{l_2}}}+\frac{\partial \overset{*}{q}{}_{w_{l_2}}^{\prime(n)}}{\partial \overset{*}{T}_{w_{l_2}}}+\frac{\partial \overset{*}{q}{}_{c_{k_2}}^{(n)}}{\partial \overset{*}{T}_{c_{k_2}}}+\frac{\partial \overset{*}{q}{}_{c_{k_2}}^{\prime(n)}}{\partial \overset{*}{T}_{c_{k_2}}}+\frac{\partial \overset{*}{q}{}_{t_{m_2}}^{(n)}}{\partial \overset{*}{T}_{t_{m_2}}}+\frac{\partial \overset{*}{q}{}_{t_{m_2}}^{\prime(n)}}{\partial \overset{*}{T}_{t_{m_2}}}\right)[\overset{*}{T}_{w_{l_2}}-\overset{*}{T}{}_{w_{l_2}}^{(0)}]$$

$$+\left(\frac{\partial \overset{*}{q}{}_{w_{l_2}}^{(n)}}{\partial \overset{*}{T}_{w_{l_2+1}}}+\frac{\partial \overset{*}{q}{}_{w_{l_2}}^{\prime(n)}}{\partial \overset{*}{T}_{w_{l_2-1}}}+\frac{\partial \overset{*}{q}{}_{t_{m_2}}^{(n)}}{\partial \overset{*}{T}_{t_{m_2-1}}}+\frac{\partial \overset{*}{q}{}_{t_{m_2}}^{\prime(n)}}{\partial \overset{*}{T}_{t_{m_2-1}}}\right)[\overset{*}{T}_{w_{l_2+1}}-\overset{*}{T}{}_{w_{l_2+1}}^{(0)}]$$

$$+\sum_{j=l_2+2}^{l_3}\left(\frac{\partial \overset{*}{q}{}_{w_{l_2}}^{(n)}}{\partial \overset{*}{T}_{w_j}}+\frac{\partial \overset{*}{q}{}_{t_{m_2}}^{(n)}}{\partial \overset{*}{T}_{t_{m_2-(j+l_2)}}}\right)[\overset{*}{T}_{w_j}-\overset{*}{T}{}_{w_j}^{(0)}]$$

$$+\sum_{j=k_1}^{k_2-2}\left(\frac{\partial \overset{*}{q}{}_{t_{m_2}}^{(n)}}{\partial \overset{*}{T}_{t_{m_3-(j-k_1)}}}+\frac{\partial \overset{*}{q}{}_{c_{k_2}}^{(n)}}{\partial \overset{*}{T}_{c_j}}\right)[\overset{*}{T}_{c_j}-\overset{*}{T}{}_{c_j}^{(0)}]$$

$$+\left(\frac{\partial \overset{*}{q}{}_{c_{k_2}}^{(n)}}{\partial \overset{*}{T}_{c_{k_2-1}}}+\frac{\partial \overset{*}{q}{}_{c_{k_2}}^{\prime(n)}}{\partial \overset{*}{T}_{c_{k_2-1}}}+\frac{\partial \overset{*}{q}{}_{t_{m_2}}^{(n)}}{\partial \overset{*}{T}_{t_{m_2+1}}}+\frac{\partial \overset{*}{q}{}_{t_{m_2}}^{\prime(n)}}{\partial \overset{*}{T}_{t_{m_2+1}}}\right)[\overset{*}{T}_{c_{k_2-1}}-\overset{*}{T}{}_{c_{k_2-1}}^{(0)}] \tag{8.101}$$

The solution procedure may be started by assuming initial interface temperature sensitivities, $\overset{*}{T}{}_l^{(0)}$, to be unity. Then, $\overset{*}{q}{}_j^{(n)^{(0)}}$ in each zone may be obtained from the standard BEM equation (8.94). The exact interface temperature sensitivities may then be determined from equations (8.98–8.101). The BEM equation is used once more to determine the interface flux sensitivities along with other unknowns.

8.7 Results from BEM Sensitivity Analysis

It is assumed here that the boundary ∂B of a simply connected region may be decomposed as

$$\partial B = \partial B_{\text{no-opt}} \cup \partial B_{\text{opt}}$$

where ∂B_{opt} is the portion of the boundary being varied or optimized. Once again, steady-state turning operations are considered here. The design sensitivities of temperature and flux fields are first considered in single regions, and two example problems involving parabolic variation of shear plane geometry and gradual nose wear of a cutting tool are presented.

In metal cutting processes, the chip thickness is typically assumed to be uniform. In many situations (Trent 1984), however, the chip thickness is not uniform. Hence, the shear angle may vary across the depth of cut, and the shear plane will also be curved. Similar situations also arise in plunge grinding. To investigate the implications of the uniform chip thickness assumption, the shear plane geometry is varied as a parabola (see figure 8.15) represented as

$$x' = \eta$$

$$y' = b(\eta - 0.5)(\eta + 0.5), \qquad -0.5 \leq \eta \leq 0.5 \tag{8.102}$$

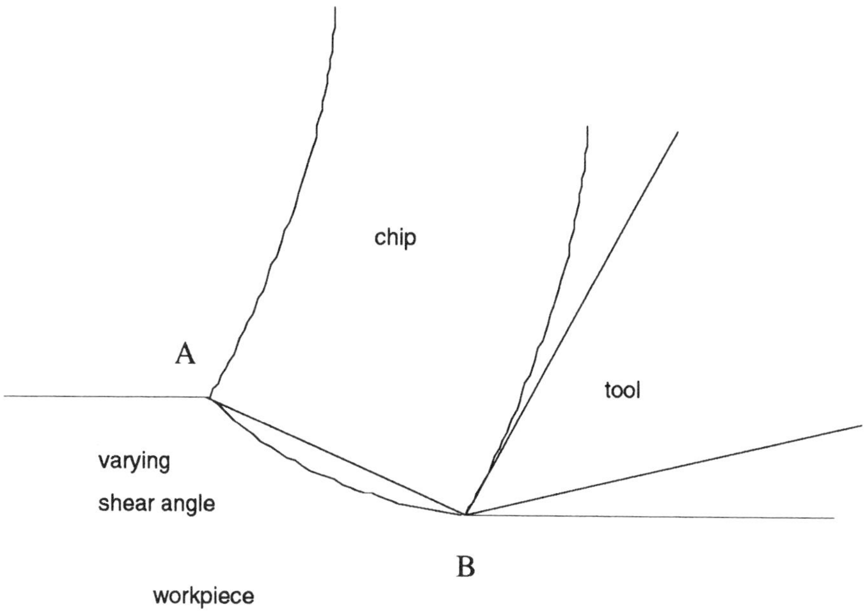

Figure 8.15. Schematic diagram of shear angle variations in machining problems.

where x' and y' are local tangential and normal coordinates, respectively, at the shear plane. Here, b is a parameter determining the maximum normal distance between the nominal and the perturbed parabolic shape. For example, the maximum normal distance will be $0.25b$ if the nominal shape is a straight line ($b = 0$).

The nominal configuration is a straight line ($b = 0$). Figure 8.16 shows the temperature and its sensitivity along a shear plane (in the nominal configuration) for $Pe = 10$ and 30 at a shear angle of 45 degrees. The abscissa is the dimensionless distance along the shear plane (see figure 8.15) measured from $A(x = 0)$ to $B(x = 1)$. The sensitivity of the temperature decreases with the Peclet number and, as Pe goes from 10 to 30, the maximum temperature sensitivity goes from -0.06 to -0.025. The position of the maximum also shifts toward point A (in figure 8.15).

The wear of the tool nose with machining is considered next. The nose curve is parametrized (design parameter b represents the tool nose radius, $b = 0$ corresponds to a sharp tool) as

$$x' = \eta$$

$$y' = a\left[b\exp\left(-\frac{1}{b}|\eta|\right) + |\eta|\right], \qquad -1.0 \leq \eta \leq 1.0 \qquad (8.103)$$

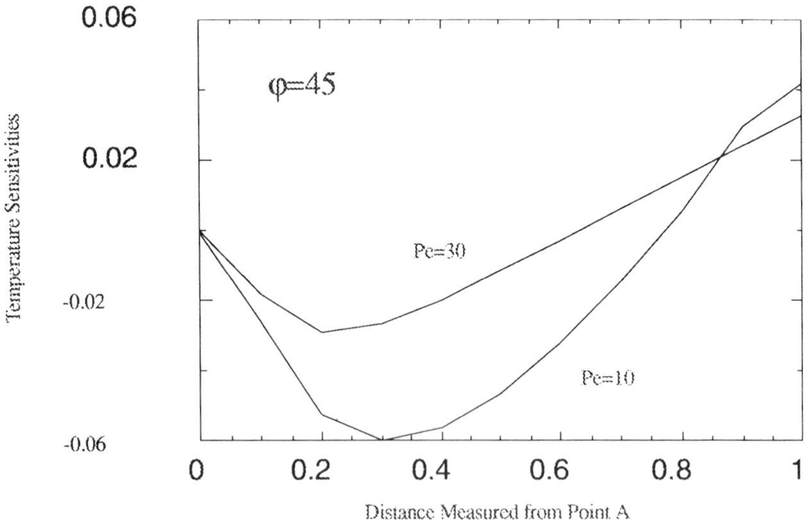

Figure 8.16. Temperature sensitivities along the shear plane due to shear angle variations.

where x' and y' are the local tangential and normal coordinates, respectively, at the tool nose. The parameter a is the cotangent of half of the included angle in the tool wedge, and b represents the amount of blunting of the tool nose. A perfectly sharp tool is represented by setting $b = 0$, and the blunting of the nose is represented by increasing b. Figure 8.17 shows a schematic diagram of the tool wear. Figure 8.18 shows the temperature sensitivity along the chip–tool interface. The abscissa is the distance measured from point B in figure 8.17. In this case, the nominal configuration is of crucial importance. The temperature sensitivity increases significantly as the tool wears out. The maximum temperature sensitivity occurs at point B and goes from 0.021 to 0.129 as b goes from 0.01 to 0.02. This also corroborates the physical observations in real-life machining operations.

As the machining process continues, the length of the flank land gradually increases. As the flank land develops, the tool is fed in the direction perpendicular to the cutting velocity. In the present work, the flank wear is modeled by seven different regions, as shown schematically in figure 8.19. Here, b is the design variable representing the amount of wear perpendicular to the flank land and x' and y' are, respectively, the local tangential and normal coordinates. In our convention, the origin for a region is set at the starting-point for that region (e.g., the origin for the region a–b is at point "a").

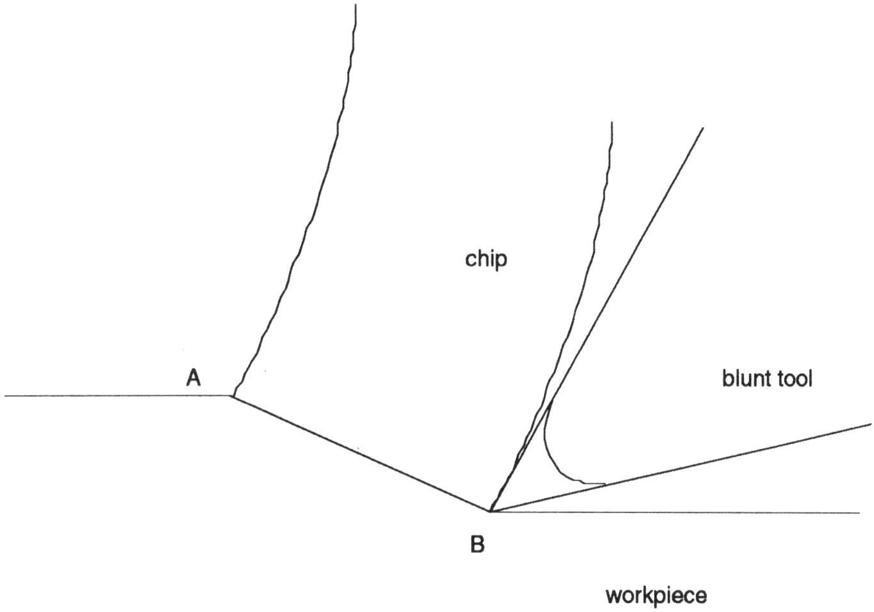

Figure 8.17. Schematic diagram of gradual nose wear of a cutting tool in machining problems.

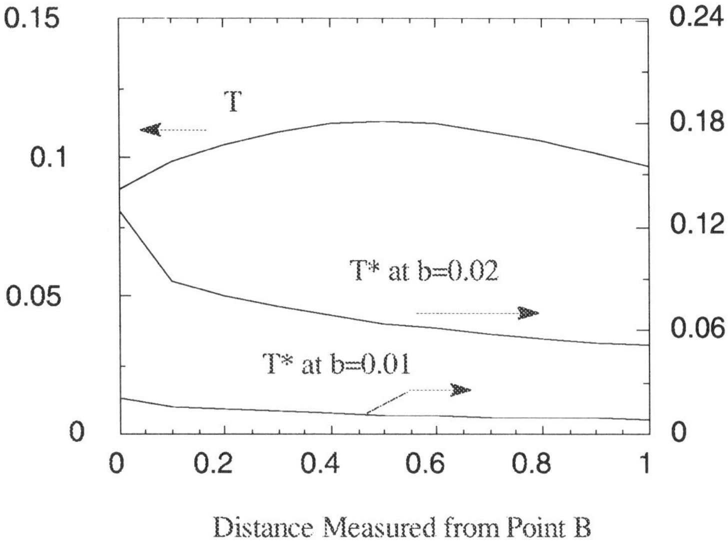

Figure 8.18. Temperature sensitivities along the chip–tool interface due to gradual nose wear.

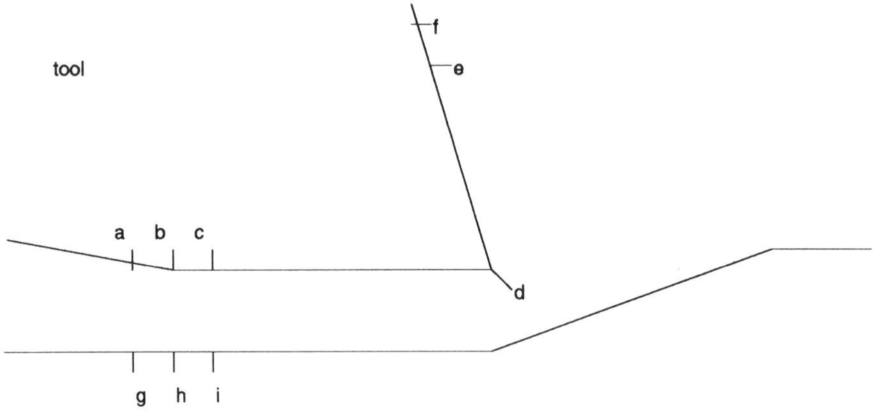

Figure 8.19. Schematic diagram of flank wear in machining.

Region 1 (*a–b*):

$$x' = \left(\Delta s_{ab} - \frac{b}{\sin\theta}\right)\frac{\eta}{\Delta s_{ab}}, \qquad 0 \leq \eta \leq \Delta s_{ab}$$
$$y' = 0 \tag{8.104}$$

Region 2 (*b–c*):

$$x' = -b\cot\left((\theta - \alpha)\frac{\Delta s_{bc} - \eta}{\Delta s_{bc}} + \alpha\right), \qquad 0 \leq \eta \leq \Delta s_{bc}$$
$$y' = b \tag{8.105}$$

Region 3 (*c–d*):

$$x' = \eta - b\tan\alpha, \qquad 0 \leq \eta \leq \Delta s_{cd}$$
$$y' = b \tag{8.106}$$

Region 4 (*d–e*):

$$x' = \eta - \frac{b}{\cos\alpha}, \qquad 0 \leq \eta \leq \Delta s_{de}$$
$$y' = 0 \tag{8.107}$$

Region 5 (*e–f*):

$$x' = \frac{\Delta s_{ef} - (b/\cos\alpha)}{\Delta s_{ef}}(s_{ef} + \eta), \qquad 0 \leq \eta \leq \Delta s_{ef}$$
$$y' = 0 \tag{8.108}$$

Region 6 (g–h):

$$x' = \frac{\Delta s_{gh} - b[\tan(\pi/2 - \theta) - \tan\alpha]}{\Delta s_{gh}}\eta, \qquad 0 \le \eta \le \Delta s_{gh}$$
$$y' = 0 \tag{8.109}$$

Region 7 (h–i):

$$x' = \frac{\Delta s_{hi} - b[\tan(\pi/2 - \theta) - \tan\alpha]}{\Delta s_{hi}}(\Delta s_{hi} - \eta), \qquad 0 \le \eta \le \Delta s_{hi}$$
$$y' = 0 \tag{8.110}$$

Figure 8.19 shows the temperature sensitivities along the interfaces with variations in the flank length. Figure 8.20a shows the temperature sensitivities along the workpiece–chip interface (A–B) and the workpiece–tool interface (B–D). It may be observed that the temperature sensitivity is negative for both cases of $L_{FW} = 0.4$ and 0.5 (L_{FW} = flank length/length of the shear plane). As L_{FW} increases from 0.4 to 0.5, the magnitude of the temperature sensitivity along the interfaces ABD decreases. Assuming that the flank-wear rate is predominantly governed by the flank temperature, this will be expected as transition from the primary flank-wear region to the secondary flank-wear takes place. Figure 8.20b shows the temperature sensitivities at the chip–tool interface (C–B) for $L_{FW} = 0.4$ and 0.5. A similar trend is also observed at the chip–tool interface.

Figure 8.21 shows the flux sensitivities at the interfaces for $L_{FW} = 0.4$ and 0.5. Figure 8.21a shows the sensitivity of the flux going into the workpiece along the workpiece–chip interface. The flux sensitivity is negative in both cases, and its magnitude reduces with increase in flank length. It should be noted that flux balance must be maintained and the sensitivity of the flux going into the chip along the interface A–B is positive with equal magnitude. The sensitivity of the flux going into the workpiece along the workpiece–tool interface B–D is shown in figure 8.21b. It is observed, as expected, that the flux sensitivity along B–D is much higher than that at other interfaces. This is due to the fact that the flank length B–D is being altered. Figure 8.20c shows the sensitivity of the flux going into the chip at the chip–tool interface C–B.

As the machining process continues, the size of the crater at the tool–chip interface gradually increases. In the present work, the crater is modeled as an arc of a circle (shown schematically in figure 8.22), and crater wear is modeled by six regions. It is assumed that the crater is small so that the chip is basically undeformed. Here, R is the design variable representing the radius of the arc, h is the perpendicular distance from the center of the crater

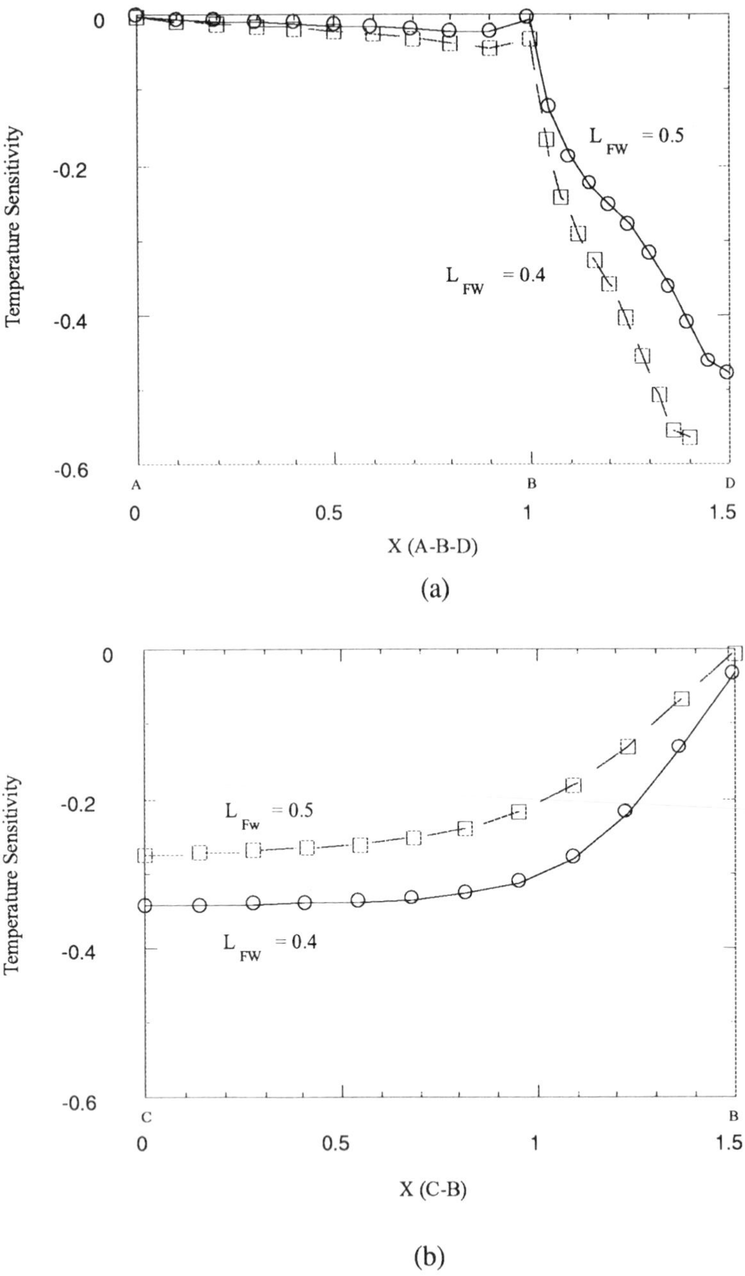

Figure 8.20. Temperature sensitivities at the interfaces for two different flank lengths: (a) workpiece–chip (A–B) and workpiece–tool (B–D) interfaces and (b) xhip–tool interface (C–B).

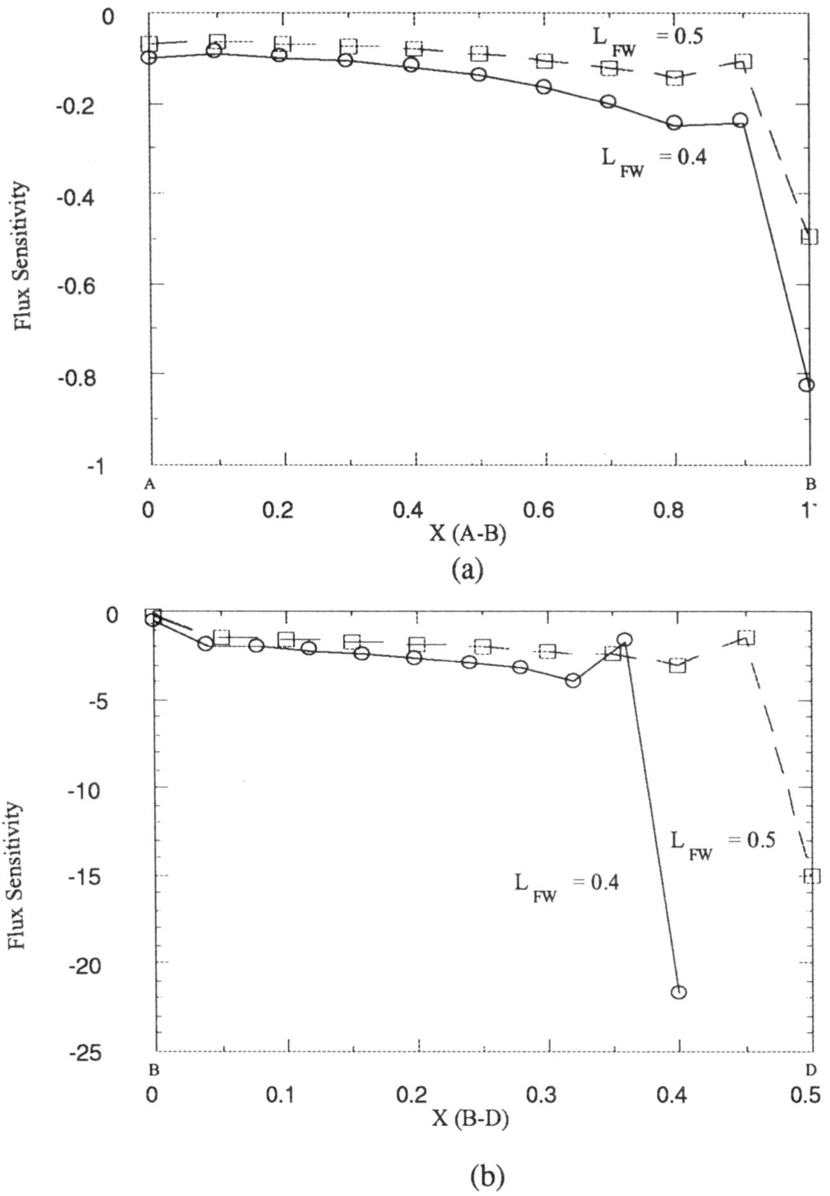

Figure 8.21. Flux sensitivities at the interfaces for two different flank lengths: (a) workpiece–chip (*A–B*) interface, (b) workpiece–tool (B–D) interface, and (c) chip–tool (*C–B*) interface.

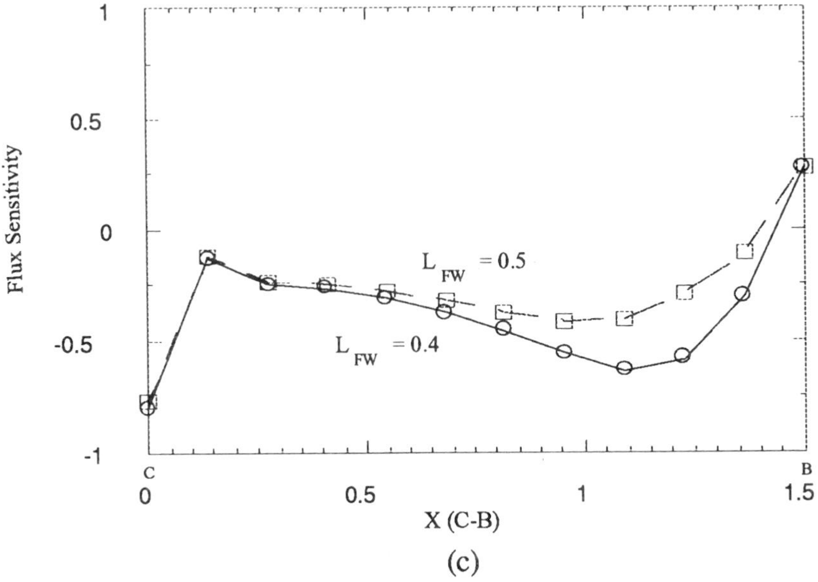

Figure 8.21—*continued*

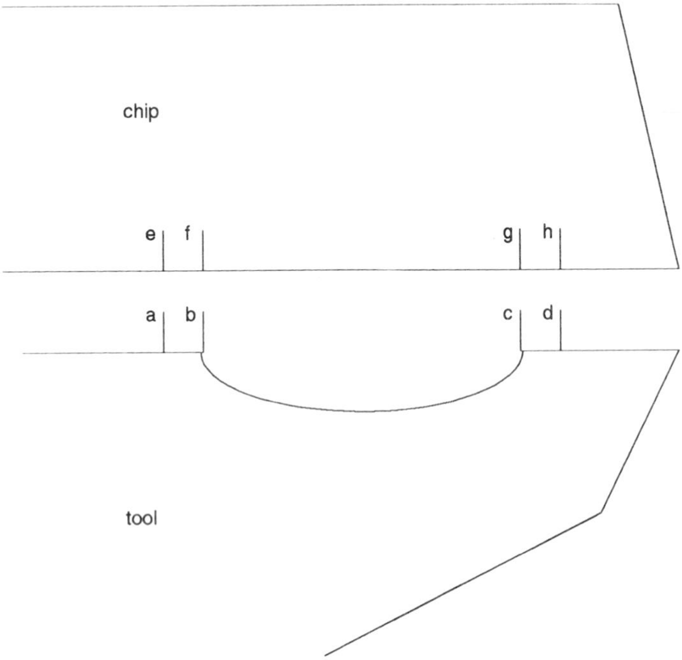

Figure 8.22. Schematic diagram of crater wear in machining.

to the original rake face of the tool, and x' and y' are, respectively, the local tangential and normal coordinates. In our convention, the origin for a region is set at the starting-point for that region (e.g., the origin for the region *a–b* is at point "*a*"). The only exception is region 2, where the origin is set at the center of the crater.

Region 1 (*a–b*):

$$x' = -\frac{\eta}{\Delta s_{ab}} R \sin\left[\cos^{-1}\left(\frac{h}{R}\right)\right], \qquad 0 \le \eta \le \Delta s_{ab}$$
$$y' = 0 \tag{8.111}$$

Region 2 (*b–c*):

$$x' = R \sin\left[\eta \cos^{-1}\left(\frac{h}{R}\right)\right]$$
$$y' = R \cos\left[\eta \cos^{-1}\left(\frac{h}{R}\right)\right]$$

where

$$\eta = \frac{\beta}{\cos^{-1}(h/R_0)} \qquad \text{and} \qquad \beta = \tan^{-1}\left(\frac{x'_0}{y'_0}\right) \tag{8.112}$$

Here, the subscripted zero denotes a reference crater configuration.

Region 3 (*c–d*):

$$x' = \left(1 - \frac{\eta}{\Delta s_{cd}}\right) R \sin\left[\cos^{-1}\left(\frac{h}{R}\right)\right], \qquad 0 \le \eta \le \Delta s_{cd}$$
$$y' = 0 \tag{8.113}$$

Region 4 (*e–f*):

$$x' = -\frac{\eta}{\Delta s_{ef}} R \sin\left[\cos^{-1}\left(\frac{h}{R}\right)\right], \qquad 0 \le \eta \le \Delta s_{ef}$$
$$y' = 0 \tag{8.114}$$

Region 5 (*f–g*):

$$x' = 2\frac{\eta}{\Delta s_{fg}} R \cos^{-1}\left(\frac{h}{R}\right) - R \sin\left[\cos^{-1}\left(\frac{h}{R}\right)\right], \qquad 0 \le \eta \le \Delta s_{fg}$$
$$y' = 0 \tag{8.115}$$

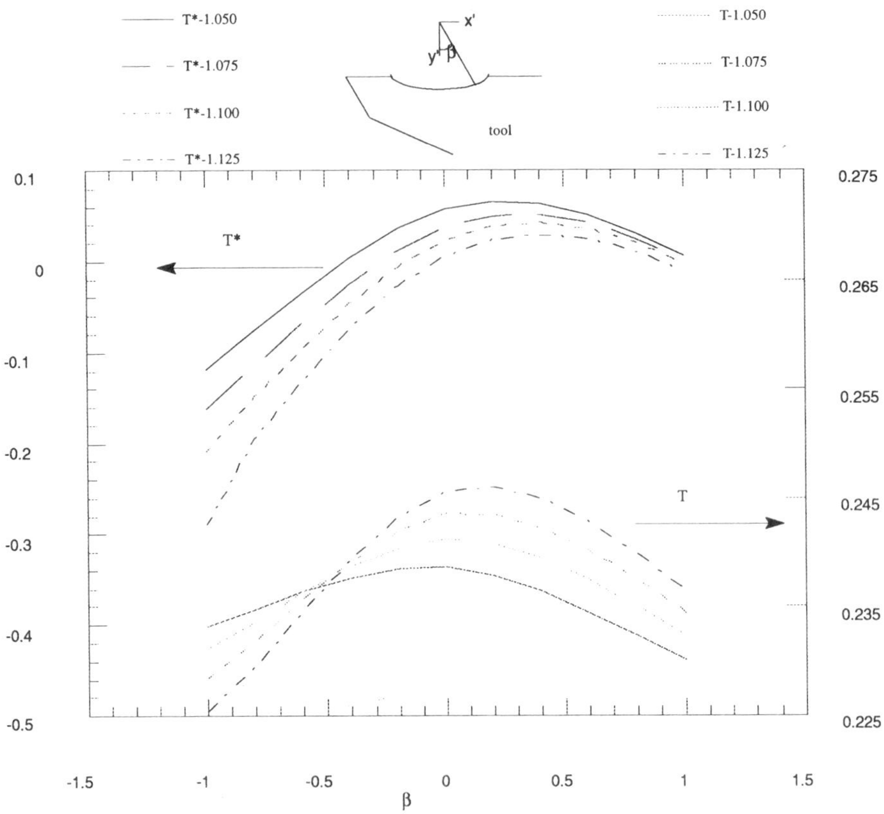

Figure 8.23. Temperature and temperature sensitivities along the crater for different crater sizes.

Region 6 (g–h):

$$x' = \left(1 - \frac{\eta}{\Delta s_{gh}}\right)\left\{2R\cos^{-1}\left(\frac{h}{R}\right) - R\sin\left[\cos^{-1}\left(\frac{h}{R}\right)\right]\right\}, \qquad 0 \leq \eta \leq \Delta s_{gh}$$

$$y' = 0 \tag{8.116}$$

Figure 8.23 shows the temperature sensitivities along the chip–tool interface with variations in the size of the crater. These calculations, are started from a normalized (with respect to length of the shear plane) crater depth of 0.05. Since this crater size is quite small, one expects to see the transition from primary crater to secondary crater with progressive wear of the tool. It may be observed that the temperature sensitivities decrease with the increase in the size of the crater. This is expected at the transition from primary to secondary craters. The temperature sensitivity at the leading edge of the crater is negative while the sensitivities at the center and the trailing edge of the crater are positive. Accordingly, one would expect higher rates of crater wear at the center and the trailing edge. This observation also correlates well with experimental observation.

8.8 Discussion and Conclusion

Boundary element method (BEM) formulations for analyses and design sensitivity studies of the thermal aspects of steady-state machining processes are presented here. The BEM analysis algorithm is first applied to obtain the surface temperature in a semi-infinite domain with surface heating over a finite region, and the results are compared to the well-known Jaeger solutions. The BEM approach is then applied to steady-state machining operations. The analysis of the heat transfer involves applications of the BEM separately to the workpiece, the chip, and the tool because of different velocities in each of these regions. A complete heat transfer model is then obtained by matching the boundary conditions across the interfaces. An exact algebraic system of equations is derived to satisfy the matching boundary conditions. This obviates the need for any iterations.

Numerical solutions of heat transfer during steady-state turning under realistic processing conditions are also presented. It is observed that the temperature along the shear plane increases with the cutting speed. The maximum temperature on the chip–tool interface approximately doubles when the Peclet number varies from 10 to 50. It is also observed that more heat is conducted into the workpiece than into the chip across the shear plane. Heat conducted into the chip across the chip–tool interface is, in turn, greater than the heat conducted into the tool. It is also found that the portion of the total heat flux going into the chip increases rapidly with the Peclet number.

A boundary element method (BEM) formulation for the determination of design sensitivities in steady-state conduction–convection problems is developed through a direct differentiation approach (DDA). This does not increase the singularity of the kernels and does retain the accuracy advantages of the BEM. The BEM formulation is based on the fundamental solution of the full adjoint equation. Accordingly, the convection term is modeled with higher precision than that obtained by upwinding in finite difference. Consequently, the BEM solution is stable regardless of the Peclet number and does not show any false diffusion. Irregular boundaries may be handled easily by the BEM. It also has the advantage of using a smaller amount of core memory since only the boundary of the domain needs to be discretized. It is also important to note that the kernels G and $\partial G/\partial n_Q$ in equation (8.78) depend on the reference configuration only and do not depend on the choice of design parameter b. All the effects for a particular choice of b are incorporated through the right-hand-side vector. Accordingly, for a relatively small increase in additional cost due to evaluation of right-hand-side vectors, sensitivities with respect to several design variables may be obtained simultaneously. This makes the BEM sensitivities algorithm very efficient.

The BEM is also applied to several machining problems, and the design sensitivities of temperature and flux fields are obtained with respect to several geometric and process parameters. The analysis and the sensitivity calculations involve separate applications of the BEM to the workpiece, the tool, and the chip regions. A complete set of results for the machining process is obtained by matching the boundary conditions across the interfaces. Exact algebraic systems of equations (8.95, 8.96) are derived to satisfy the matching boundary conditions. This obviates the need for any iteration.

References

Banerjee, P. K. and Butterfield, R. (1981). *Boundary Element Methods in Engineering Science*. McGraw-Hill, London.

Barone, M. R. and Yang, R.-J. (1988). "Boundary Integral Equations for Recovery of Design Sensitivities in Shape Optimization," *AIAA J.*, **26**, 589–594.

Beskos, D. E. (ed.) (1987). *Boundary Element Methods in Mechanics*. North-Holland, Amsterdam.

Bhattacharyya, A. (1984). *Metal Cutting Theory and Practice*. Central Book Publisher, Calcutta.

Brebbia, C. A., Telles, J. C. F., and Worbel, L. C. (1984). *Boundary Element Techniques Theory and Applications in Engineering*. Springer-Verlag, Berlin.

Carslaw, H. S. and Jaeger, J. C. (1986). *Conduction of Heat in Solids*, 2d ed. Clarendon Press, Oxford.

Chan, C. L. and Chandra, A. (1991a). "A Boundary Element Analysis of Temperature Distributions in Metal Cutting," *ASME J. Eng. Ind.*, **113**, 311–319.

Chan, C. L. and Chandra, A. (1991b). "An Algorithm for Handling Corners in the Boundary Element Method: Application to Conduction–Convection Equations," *Appl. Math. Modelling*, **15**, 244–255.

Chan, C. L. and Chandra, A. (1991c). "A BEM Approach to Thermal Aspects of Machining Processes and Their Design Sensitivities," *Appl. Math. Modelling*, **15**, 562–575.

Chandra, A. and Chan, C. L. (1992). "A BEM Formulation for Design Sensitivities in Steady-State Conduction–Convection Problems," *ASME J. Appl. Mech.*, **59**, 182–190.

Curran, D. A. S., Lewis, B. A., and Cross, M. (1986). "A Boundary Element Method for the Solutions of the Transient Diffusion Equation in Two Dimensions," *Appl. Math. Modelling*, **10**, 107–113.

Dawson, P. R. and Malkin, S. (1984). "Inclined Moving Heat Source Model for Calculating Metal Cutting Temperatures," *ASME J. Eng. Ind.*, **106**, 179–186.

Fleuries, J. and Predeleanu, M. (1987). "On the Use of Coupled Fundamental Solutions in BEM for Thermoelastic Problems," *Eng. Anal.*, **4**, 70–74.

Haug, E. J., Choi, K. K., and Komkov, V. (1986). *Design Sensitivity Analysis of Structural Systems*. Academic Press, New York.

Jaeger, J. C. (1942). "Moving Sources of Heat and the Temperature at Sliding Contacts," *Proc. Roy. Soc. New South Wales*, **76**, 203–224.

Lemaire, J. C. and Backofen, W. A. (1972). "Adiabatic Instability of the Orthogonal Cutting of Steel," *AIME Metall. Trans.*, **3**, 477–481.

Levy, E. K., Tsai, C. L., and Groover, M. P. (1976). "Analytical Investigation of the Effect of Tool Wear on the Temperature Variations in a Metal Cutting Tool," *ASME J. Eng. Ind.*, **98**, 251–257.

Loewen, E. G. and Shaw, M. C. (1954). "On the Analysis of Cutting Tool Temperatures," *Trans. ASME*, **76**, 217–221.

Mukherjee, S. (1982). *Boundary Element Methods in Creep and Fracture*. Elsevier Applied Science, Barking, Essex, U.K.

Mukherjee, S. and Chandra, A. (1989). "A Boundary Element Formulation for Design Sensitivities in Materially Nonlinear Problems," *Acta Mech.*, **78**, 243–253.

Mukherjee, S. and Chandra, A. (1991). "A Boundary Element Formulation for Design Sensitivities in Problems Involving Both Geometric and Material Nonlinearities," *Math. and Computer Modelling*, **15**, 245–255.

Muraka, P. D., Barrow, G., and Hinduja, S. (1979). "Influence of the Process Variables on the Temperature Distribution in Orthogonal Machining Using the Finite Element Method," *Int. J. Mech. Sci.*, **21**, 445–456.

O'Neill, K. (1983). "Boundary Integral Equation Solution of Moving Boundary Phase Change Problems," *Int. J. Num. Meth. Eng.*, **19**, 1825–1850.

Press, W. H., Flannery, B. P., Teukolsky, S. A., and Vetterling, W. T. (1986). *Numerical Recipes*. Cambridge University Press, New York.

Ramalingam, S., Basu, K., and Malkin, S. (1977). "Deformation Index of MnS Inclusions in Resulferized and Leaded Steels," *Mater. Sci. Eng.*, **29**, 117–121.

Rice, J. S. and Mukherjee, S. (1990). "Design Sensitivity Coefficients for Axisymmetric Elasticity Problems by Boundary Element Methods," *Eng. Anal. Boundary Elements*, **7**, 13–20.

Saigal, S., Borgaard, J. T., and Kane, J. H. (1989). "Boundary Element Implicit Differentiation Equations for Design Sensitivities in Axisymmetric Structures," *Int. J. Solids Structures*, **25**, 527–538.

Shaw, M. C. (1984). *Metal Cutting Principles*. Clarendon Press, Oxford.

Stevenson, M. G., Wright, P. K., and Chow, J. G. (1983). "Further Developments in Applying the Finite Element Method to the Calculation of Temperature Distributions in Machining and Comparisons with Experiment," *ASME J. Eng. Ind.*, **105**, 149–154.

Tanaka, Y., Honma, T., and Kaji, I. (1986). "On Mixed Boundary Element Solutions of Convection–Diffusion Problems in Three Dimensions," *Appl. Math. Modelling*, **10**, 170–175.

Tay, A. E., Stevenson, M. G., and DeVahl Davis, G. (1974). "Using the Finite

Element Method to Determine Temperature Distributions in Orthogonal Machining," *Proc. Inst. Mech. Eng.*, **188** (55), 627–638.

Trent, E. M. (1984). *Metal Cutting*. Butterworths, London.

Tsay, J.-J. and Arora, J. S. (1988). *Design Sensitivity Analysis of Nonlinear Structures with History Dependent Effects*, Technical Report No. ODL-88.4, University of Iowa, Iowa City.

Von Turkovich, B. F. (1972). "On a Class of Thermo-Mechanical Processes During Rapid Plastic Deformation (with Special Reference to Metal Cutting)," *Ann. CIRP*, **21**, 15–16.

Zabaras, N. and Mukherjee, S. (1987). "An Analysis of Solidification Problems by the Boundary Element Method," *Int. J. Num. Meth. Eng.*, **24**, 1879–1900.

Zabaras, N., Mukherjee, S., and Richmond, O. (1988). "An Analysis of Inverse Heat Transfer Problems with Phase Changes Using an Integral Method," *J. Heat Transfer*, **110**, 554-561.

9

Integral Equations for Ceramic Grinding Processes

Strength degradation of the finished part due to surface and subsurface damage is a critical issue in grinding of ceramic materials. In this chapter, the chip formation process is modeled as a two-dimensional system of radial and lateral cracks, and an integral equation approach is developed to investigate the interactions of these radial and lateral cracks with various distributions of planar microcracks. The effects in the vicinity of the free surface are modeled explicitly. The amplification and shielding effects on stress intensity factors due to various microcrack distributions are studied in detail. Numerical results are presented for crack systems representing single-grit as well as multi-grit grinding, and the implications of crack interactions on the strength degradation of finished products are discussed. Effective elastic properties for workpieces containing microcrack distributions are obtained and their ramifications on the grinding process are investigated. Some benevolent microcrack distributions that will facilitate chip formation as well as suppress surface and subsurface damage evolution are also presented.

The integral equation approach is then extended to situations involving interactions among rigid lines and cracks near a free surface that is representative of grinding short-fiber reinforced composites. The results from integral equation analyses are also utilized to develop numerical fundamental solutions for analyzing elastic components with clusters of microdefects.

9.1 Introduction

Over the last decade, interest in grinding of ceramics has grown substantially with the widespread use of ceramic components in numerous engineering applications. The ceramic components offer many advantages over their metallic counterparts in terms of improved performance and better efficiency.

Compared to metals, however, ceramics are also much more susceptible to surface and subsurface damage during grinding, and their subsequent behaviors under loads are much less forgiving to surface and subsurface damage induced by grinding. Precision ceramic components require strict adherence to required geometry, tolerance, and finish while maintaining the desired level of strength, and finishing operations such as grinding may constitute a significant portion (typically 25–50 percent) of the total cost of such components. Thus, the economic feasibility and competitiveness of high-performance ceramic components depend in no small way on the effectiveness and efficiency of the particular machining technique being used to finish these components. Surface and subsurface damage to the workpiece constitutes a major constraint against lowering of grinding costs by operating at faster stock removal rates, and this is especially critical for grinding of ceramic materials (Malkin and Ritter 1989).

There exists an extensive body of literature on grinding (e.g., Komanduri and Maas 1985, King and Hahn 1986, Malkin 1989). Two main approaches, the "machining" approach and the "indentation fracture mechanics" approach, may be identified from this body of research. Each of these approaches provides important insights into the grinding process and its associated damage evolution.

The machining approach typically involves the measurements of grinding forces and specific energy coupled with microscopic observations of the surface morphology and grinding detritus. Such investigations on metals have shown that as an abrasive grain passes through the grinding zone, material is initially plowed aside at shallow grain depths of cut, followed by a plastic flow mode of chip formation after the grain penetrates to a critical depth. At a later stage, the chip formation occurs through elastic or elasto-plastic fracture. For ceramic materials, the fracture process plays a predominant role in chip formation. In a series of experiments, Huerta and Malkin (1976a,b) observed plowing similar to that found in metals at the beginning of cuts made in glass by individual abrasive grains. It was observed that fracture became the predominant chip formation mechanism at larger grain depths of cut when lateral cracking could occur. The specific grinding energy for chip formation through fracture was also observed (Huerta and Malkin 1976a,b) to be an order of magnitude lower than that for a plastic flow mode of chip formation. Unfortunately, however, the finished surface is highly fragmented and the strength of the part is lower when fracture is the predominant mode of chip formation. Higher fracture toughnesses are typically observed for finished parts produced by a plastic flow mode of chip formation.

The strength degradation of the finished part due to induced surface and subsurface damage is a very important issue in grinding. The indentation fracture mechanics approach assumes that the damage produced by grinding can be modeled as an idealized crack system produced by a sharp indenter. For ceramic materials, the plastic flow regime is very small. Accordingly, a fracture mechanics approach may also be pursued to describe the material removal process. Moreover, except for a small zone directly under the abrasive grain where severe heating may occur, the fracture process in ceramic materials may be assumed to be elastic. Thus, for ceramic grinding, an indentation fracture mechanics approach offers an avenue for describing both the material removal process and its influence on strength degradation.

Such an approach based on the elastic fracture mechanics is pursued in the present work. The abrasive–workpiece interactions for ceramic grinding are likened to localized small-scale indentation events. As a first attempt, the state of the material in the vicinity of the indentation is assumed to be isothermal. The temperature rise in this region due to grinding action is assumed to be uniform, and appropriate elastic parameters at the elevated temperature are used. In reality, the radial and lateral cracks involved in single-grit scratch tests are three-dimensional in geometry. As a first attempt, a two-dimensional model is used in the present analysis. The actual process of grinding always involves a large number of grits simultaneously indenting the workpiece in very close proximity, and the cracks are well developed. Restricting ourselves to planar interactions, a two-dimensional model can be justified for the well-developed cracks observed in actual grinding involving multiple grits.

This chapter starts with a review of the literature on the indentation fracture mechanics approach for modeling ceramic grinding processes. The chip formation process is modeled as a system of median or radial cracks and lateral cracks. The effect of the small plastic zone directly under the indenter is accounted for by incorporating appropriate residual stresses (Lawn et al. 1980) in the elastic fracture analysis. An integral equation formulation is then developed to investigate the interactions of these radial and lateral cracks with various distributions of microcracks. These microcracks may preexist in a ceramic workpiece, or they may be initiated ahead of the grinding zone. The vicinity of the free surface to the crack systems is included explicitly in the proposed crack interaction models. The amplification and shielding effects on stress singularities are studied in detail. The interactions among the radial and the lateral crack systems are also considered. Effective elastic properties are obtained for workpieces containing microcrack distributions, and the effects of variations in these elastic properties on the grinding process are investigated. In addition to investigations of the single-grit scratch tests, effects of multiple indentations in the vicinity of each other and the interactions of their associated crack systems are also investigated to obtain a realistic picture for multipoint grinding operations.

Numerical results for radial and lateral crack systems and their interactions with various microcrack distributions are presented, and their implications for the ceramic grinding process are discussed. Some benevolent microcrack distributions that will facilitate chip formation and suppress surface and subsurface damage evolution are also obtained, and their effects are discussed.

Strength degradation in grinding ceramic composites is investigated next. Short-fiber reinforcements are modeled as rigid lines. It is also interesting to note that, for a given configuration, the interactions between two cracks and between a crack and a rigid line bound the range of all possible interaction effects.

Finally, local integral equation analysis schemes capable of detailed representations of the microfeatures of a problem are integrated with a macro-scale BEM technique capable of handling complex finite geometries and realistic boundary conditions. The micro-scale effects are introduced into the macro-scale BEM analysis through an augmented fundamental solution obtained from an integral equation representation of the micro-scale features. The proposed hybrid micro–macro BEM formulation allows decomposition of the complete problem into two subproblems, one residing entirely at the micro-level and the other at the macro-level. This allows for investigations of the effects of the microstructural attributes while retaining the macro-scale geometric features and actual boundary conditions for the component or structure under consideration. As a first attempt, elastic fracture mechanics problems with interacting cracks at close spacings are considered. The numerical results obtained from the hybrid BEM analysis establish the accuracy and effectiveness of the proposed micro–macro computational scheme for this class of problems. The proposed micro–macro BEM formulation can easily be extended to investigate the effects of other microfeatures (e.g., interfaces, short or continuous fiber reinforcements, voids, and inclusions, in the context of linear elasticity) on macroscopic failure modes.

9.2 Background of Strength Degradation in Ceramic Grinding

The mechanical properties of ceramic materials are quite different from those of metals. They typically have much higher hardness and are brittle. For glass or glass-ceramics, many investigators present evidence indicating that the material removal occurs by brittle fracture (Fielden and Rubenstein 1969, Vaidyanathan and Finnie 1972). It is also well known, however, that ceramics such as glass may be made to flow plastically without fracture at room temperature under conditions of large hydrostatic pressure (e.g., Marsh 1963). Permanent flow under such conditions has been observed in hardness indentations and in plowed grooves and flow-chips produced by slowly dragging a pointed tool across a glass surface (McClintock and Argon 1966). In grinding, the rake angles may be negative by 60 degrees or more (Komanduri 1971) and may produce a favorable stress condition for plastic flow of glass at the cutting edge of an abrasive grain. Huerta and Malkin (1976a,b) carried out several grinding experiments on glass using vitreous-bonded silicon carbide wheels and bronze-bonded diamond wheels. The specific grinding energy for glass is observed to follow trends similar to those for metals. The specific grinding energy for glass increases as the undeformed chip thickness is decreased either by slowing the workpiece velocity or by reducing the downfeed. The specific grinding energy is also found to increase with the softening temperature of glass. In diamond grinding of high-density polycrystalline alumina and other ceramics (Hockey 1972, Spur et al. 1985, Subramanian and Keat 1985, Inasaki 1986), both flow and fracture have been observed, although the material removal occurs primarily by fracture.

It has been observed that the specific grinding energy needed in the fracture mode is about an order of magnitude lower than that needed for the plastic flow mode of chip formation. Accordingly, a fracture mode is preferred from the perspective of efficiency. Unfortunately, the finished surface obtained through a fracture mode of chip formation is also highly fragmented, and the strength after grinding is significantly lower than that obtained by a plastic flow mode of chip formation.

The characterization of strength after grinding requires a detailed knowledge about the evolution of surface and subsurface damage. Busch and Prins (1972) performed experiments using a single diamond point on Coors AD96 alumina and observed the wear of the diamond particle and the damage inflicted on the ceramic workpiece using electron microscopy. Various researchers have followed the indentation fracture mechanics approach, which models the abrasive–workpiece interactions in grinding of ceramics as localized small-scale indentation events, to characterize the fracture pattern in grinding. Lawn and Swain (1975) investigated the microfracture beneath point indentations in brittle solids. Evans (1979) examined the effects of lateral fracture mechanisms on the abrasive wear of ceramics. Hagan (1979) studied the nucleation of median and lateral cracks around Vickers indentations in soda-lime glass. Kirchner et al. (1979) and Kirchner (1984) also investigated the mechanisms of fragmentation and damage penetration for point and line contact loadings. Several researchers (e.g., Lawn et al. 1980, Marshall et al. 1982, Evans 1984, Marshall 1984) have investigated the radial and the lateral crack systems for various indenters and associated failure from surface flaws. The radial cracks are initiated during loading, but may propagate during unloading due to the residual stresses. The lateral cracks are found to initiate and propagate during unloading only. Larchuk et al. (1985) also identified crushing as a mechanism for material removal during abrasive machining, which leads to mixed mode fracture. Indentation fracture mechanics has also been applied to analyze the effects of solid particle erosion at normal incidence on the strength of soda-lime glass and sintered alumina (Ritter et al. 1984, 1985). Some preliminary results obtained through this approach correlate well with the experimental results obtained by Evans and Marshall (1981) and Koepke and Stokes (1979). In a very interesting experimental investigation by Marshall et al. (1983) using acoustic wave scattering, strength measurements, and post-failure fractography on machining-induced cracks, it was observed that the residual stresses significantly affect the crack response. Two components of residual stresses, a crack wedging force due to the plastic zone beneath the strength-controlling machining groove and a compressive surface layer due to adjacent grooves, were identified. The wedging force typically dominates and causes stable equilibrium crack extension during a breaking test.

It is evident from the literature that the indentation by a sharp indenter may be modeled as two principal crack systems, in the median or radial direction and in the lateral direction. It is also observed that both types of cracks are affected by the residual stresses due to the plastic zone. To date, however, these crack systems have been modeled as isolated cracks. For grinding applications, both types of cracks are present in the vicinity of each other. The radial cracks represent induced surface damage. The lateral cracks may contribute to chipping. A buried lateral crack also represents subsurface damage. To improve the surface integrity of the finished workpiece, the designer attempts to avoid the surface and subsurface damage. In terms of radial and lateral crack systems, this implies minimizing the length of the radial cracks while developing highly curved lateral cracks to induce chipping. This will add to the depth of cut and also minimize surface damage. Accordingly, the interactions of the radial and the lateral crack systems become a critical issue. Moreover, the stress and deformation fields during grinding may induce microcracks in the workpiece, and their interactions with the major crack systems (radial and lateral) should also be incorporated in the analysis.

9.3 Indentation Fracture Mechanics Model for Monolithic Ceramics

Here, the elasto-plastic indentation problem is modeled as an elastic problem with the appropriate residual stresses from the plastic zone (Lawn et al. 1980, Marshall et al. 1982). In a series of experiments on ceramic specimens in which the surfaces were either progressively polished or annealed, Petrovic et al. (1976) obtained a strength recovery commensurate with the true critical stress intensity factor. This demonstrates the central role of the plastic deformation zone about the indentation as a source of residual stresses. Using an elastic model augmented by the residual stress field, the principle of superposition may be utilized while capturing the effects of the plastic deformation zone. Using the above assumptions, the chip formation process, as well as the damage evolution in ceramic grinding, may be viewed as an elastic fracture mechanics problem involving crack systems with appropriate distributions. The stress intensity factors and propagations of these cracks will then be investigated to gain insight into the chip formation and damage evolution characteristics of ceramic grinding processes.

A problem involving N cracks in a linear elastic solid may then be represented by a superposition of N problems, each involving a single crack loaded by unknown tractions that are induced on a given crack line by other cracks and by the remote loading (e.g., Chudnovsky et al. 1987). Kachanov and Montagut (1986) and Montagut and Kachanov (1988) also used a pseudo traction or force formulation approach based on the superposition technique and the ideas of self-consistency applied to the average tractions on individual cracks. Such an approach provides an approximate solution, but it is efficient.

It is also well known that the problem of interactions among many cracks can be reduced to a set of integral equations through a displacement formulation. Dundurs and Sendeckyj (1965), Atkinson (1972a,b), and Erdogan and Gupta (1975) studied the problem of a crack near a circular inclusion using the point dislocation solution due to Dundurs and Mura (1964). However, their treatise is not directly applicable for a crack at an arbitrary orientation near a free surface. Hills and Ashelby (1980) obtained the stress intensity factor (SIF) for a fatigue wear problem using an integral equation approach and a pitting model for rolling contact fatigue. Keer et al. (1982) and Keer and Bryant (1983) also used an integral formulation to analyze the interactions of a Hertzian contact with surface-breaking cracks. Erdogan et al. (1973), Li and Hills (1990), Nowell and Hills (1987), and Hills and Nowell (1989) also investigated problems involving subsurface cracks and obtained very accurate results.

9.3.1 An Integral Equation Formulation for Grinding of Monolithic Ceramics

In the present work, an integral equation formulation is developed to analyze grinding of ceramics. The grinding process is modeled as a moving indenter problem with its associated radial and lateral crack systems and, as a first attempt, only steady-state conditions are considered. As discussed before,

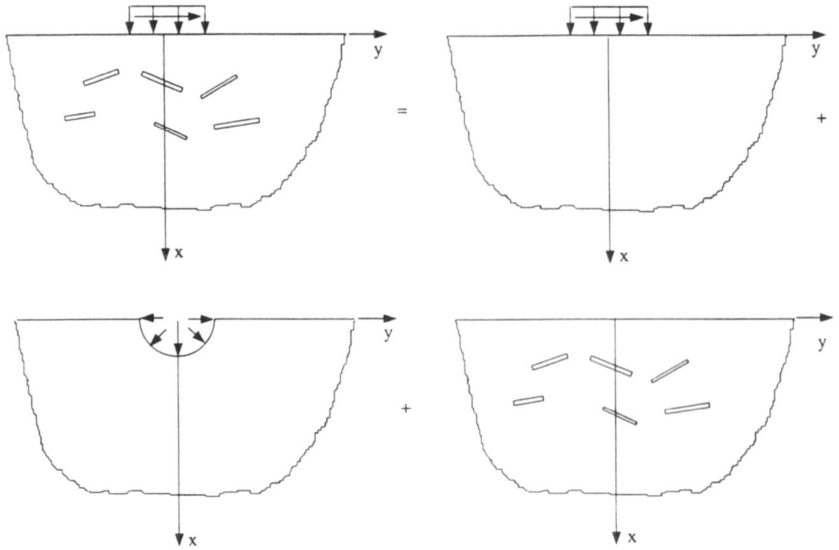

Figure 9.1. Grinding as superposition of three problems.

the physics of the problem requires that the formulation be valid for closely spaced cracks with strong interactions. In grinding, the cracks occur very near the free surface. Accordingly, the influence of the free surface must be incorporated. This requires that the tractions be zero on the crack faces, as well as on the free surface. Accordingly, existing full-plane crack solutions are not directly applicable.

Following the integral equation approach of Hu and Chandra (1993a), the cracks are modeled as continuous distributions of dislocations. The well-known solution for a point dislocation in an elastic half-plane (Dundurs and Mura 1964, Dundurs and Sendeckyj 1965) is utilized as the fundamental solution in the present analysis. Such a fundamental solution also incorporates the effects of the free surface. As shown in figure 9.1, the grinding problem is viewed as the superposition of three problems:

1. The first problem consists of the elastic half-plane without any crack under the specified loads. The solution for this problem can easily be obtained by integrating the well-known Flamant solution (Timoshenko and Goodier 1973) over the desired loading distribution.
2. The effects of the plastic zone under the indenter and the associated material mismatch (Lawn et al. 1980, Samuel et al. 1988) are addressed next. The residual stresses due to the plastic zone provide the driving force for both the radial and the lateral crack systems. It

is usually assumed that the external compressive loads in a grinding operation provide no contribution to the stress intensity factors for median (or radial) flat cracks. This, however, is not the case for slightly opened cracks (Madenci 1991). It has also been observed that the tensile stresses induced by compression can significantly affect the stress intensity factors and the crack growth patterns during grinding. Accordingly, the integral equation approach due to Hu and Chandra (1993a) is adapted in the present study to incorporate such effects.

3. The singular stress fields due to the presence of the cracks and their interactions are considered next. It also should be noted here that the stress fields associated with the first and the second problems are regular. The solutions to the third problem alone contain the singular behaviors at crack tips. The crack surfaces in the third problem are subject to distributed dislocations and the free surfaces are traction free.

Such an approach based on an integral equation formulation is capable of modeling the radial and lateral crack systems along with their interactions. The effects of the free surface are incorporated explicitly. The evolution of these crack systems may now be investigated to gain insight into the chip formation, as well as surface and subsurface damage evolution characteristics, in ceramic grinding processes.

Figure 9.2a shows a system of interacting cracks with arbitrary orientations in the vicinity of a free surface. As shown in figure 9.2b, an indentation process may be represented by a special set of such radial and lateral cracks, along with the surrounding microcracks. A grinding process may then be modeled as a moving indenter problem. It is important to note here that, even when the cracks are loaded only by normal tractions (mode I loading), the interactions among cracks can induce both mode I and mode II deformations. Accordingly, both climb dislocations $b_s^{(i)}$ $(t^{(i)})$ and glide

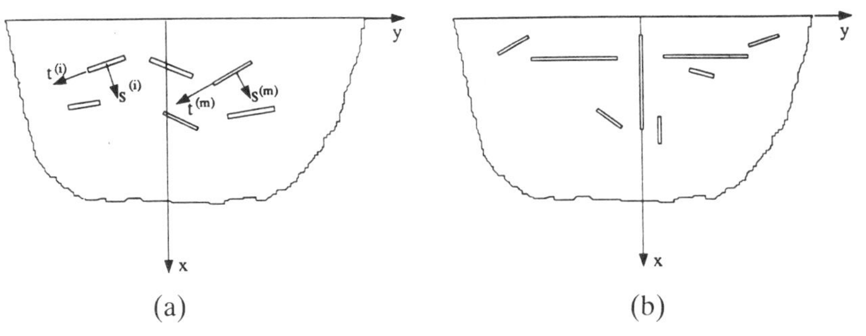

Figure 9.2. Schematic diagrams of (a) system of interacting cracks, (b) system of radial and lateral cracks in grinding along with surrounding microcracks.

dislocations $b_t^{(i)}(t^{(i)})$ are considered on the crack surfaces. Each crack is then represented as continuous distributions of these dislocations. The requirement that the crack surfaces be traction free results in the following integral equations:

$$\frac{G}{\pi(\kappa+1)}\sum_{i=1}^{N}\left(\int_{-a^{(i)}}^{a^{(i)}} K_{11}(t^{(m)},t^{(i)})\,b_t^{(i)}(t^{(i)})\,dt^{(i)} + \int_{-a^{(i)}}^{a^{(i)}} K_{12}(t^{(m)},t^{(i)})\,b_s^{(i)}(t^{(i)})\,dt^{(i)}\right) + \sigma_{ss}^{(m0)}(t^{(m)}) = 0$$

$$-a^{(m)} < t^{(m)} < a^{(m)}, \qquad m = 1,\ldots,N \tag{9.1a}$$

and

$$\frac{G}{\pi(\kappa+1)}\sum_{i=1}^{N}\left(\int_{-a^{(i)}}^{a^{(i)}} K_{21}(t^{(m)},t^{(i)})\,b_t^{(i)}(t^{(i)})\,dt^{(i)} + \int_{-a^{(i)}}^{a^{(i)}} K_{22}(t^{(m)},t^{(i)})\,b_s^{(i)}(t^{(i)})\,dt^{(i)}\right) + \sigma_{ts}^{(m0)}(t^{(m)}) = 0$$

$$-a^{(m)} < t^{(m)} < a^{(m)}, \qquad m = 1,\ldots,N \tag{9.1b}$$

where $\kappa = (3-4\nu)$ for plane strain and $\kappa = (3-\nu)/(1+\nu)$ for plane stress; G is the shear modulus and ν is the Poisson ratio; and N is the number of cracks in the system. Here, $(t^{(i)}, s^{(i)})$ is a pair of local tangential–normal coordinates with their origin at the center of the ith crack, $(-a^{(m)}, a^{(m)})$ is the interval occupied by the mth crack along $t^{(m)}$, and $\sigma_{ss}^{(m0)}$ and $\sigma_{ts}^{(m0)}$ are the stresses under the given loading at the supposed location of the mth crack in a half-plane, but for an elastic body without any cracks. Accordingly, $\sigma_{ss}^{(m0)}$ and $\sigma_{ts}^{(m0)}$ are easily obtainable. The kernels K_{11} through K_{22} are deduced from the elastic solutions for half-plane problems due to point dislocations (Dundurs and Mura 1964, Dundurs and Sendeckyj 1965). Thus, the effects of the free surface are explicitly incorporated through these kernels. The details of these kernels may be expressed as

$$\begin{aligned} K_{11}(t^{(m)},t^{(i)}) &= \cos^2\theta^{(m)}\,[\sin\theta^{(i)} G_{xxx}(t^{(i)},t^{(m)}) - \cos\theta^{(i)} G_{yxx}(t^{(i)},t^{(m)})] \\ &+ 2\sin\theta^{(m)}\cos\theta^{(m)}\,[\sin\theta^{(i)} G_{xxy}(t^{(i)},t^{(m)}) - \cos\theta^{(i)} G_{yxy}(t^{(i)},t^{(m)})] \\ &+ \sin^2\theta^{(m)}\,[\sin\theta^{(i)} G_{xyy}(t^{(i)},t^{(m)}) - \cos\theta^{(i)} G_{yyy}(t^{(i)},t^{(m)})] \end{aligned} \tag{9.2}$$

$$\begin{aligned} K_{12}(t^{(m)},t^{(i)}) &= \cos^2\theta^{(m)}\,[\cos\theta^{(i)} G_{xxx}(t^{(i)},t^{(m)}) + \sin\theta^{(i)} G_{yxx}(t^{(i)},t^{(m)})] \\ &+ 2\sin\theta^{(m)}\cos\theta^{(m)}\,[\cos\theta^{(i)} G_{xxy}(t^{(i)},t^{(m)}) + \sin\theta^{(i)} G_{yxy}(t^{(i)},t^{(m)})] \\ &+ \sin^2\theta^{(m)}\,[\cos\theta^{(i)} G_{xyy}(t^{(i)},t^{(m)}) + \sin\theta^{(i)} G_{yyy}(t^{(i)},t^{(m)})] \end{aligned} \tag{9.3}$$

$$K_{21}(t^{(m)}, t^{(i)}) = \sin\theta^{(m)}\cos\theta^{(m)}[\sin\theta^{(i)} G_{xxx}(t^{(i)}, t^{(m)}) - \cos\theta^{(i)} G_{yxx}(t^{(i)}, t^{(m)})] + [\sin^2\theta^{(m)} - \cos^2\theta^{(m)}][\sin\theta^{(i)} G_{xxy}(t^{(i)}, t^{(m)}) - \cos\theta^{(i)} G_{yxy}(t^{(i)}, t^{(m)})] - \sin\theta^{(m)}\cos\theta^{(m)}[\sin\theta^{(i)} G_{xyy}(t^{(i)}, t^{(m)}) - \cos\theta^{(i)} G_{yyy}(t^{(i)}, t^{(m)})] \quad (9.4)$$

$$K_{22}(t^{(m)}, t^{(i)}) = \sin\theta^{(m)}\cos\theta^{(m)}[\cos\theta^{(i)} G_{xxx}(t^{(i)}, t^{(m)}) + \sin\theta^{(i)} G_{yxx}(t^{(i)}, t^{(m)})] + [\sin^2\theta^{(m)} - \cos^2\theta^{(m)}][\cos\theta^{(i)} G_{xxy}(t^{(i)}, t^{(m)}) + \sin\theta^{(i)} G_{yxy}(t^{(i)}, t^{(m)})] - \sin\theta^{(m)}\cos\theta^{(m)}[\cos\theta^{(i)} G_{xyy}(t^{(i)}, t^{(m)}) + \sin\theta^{(i)} G_{yyy}(t^{(i)}, t^{(m)})] \quad (9.5)$$

$$K_{31} = \sin^2\theta^{(m)}[\sin\theta^{(i)} G_{xxx}(t^{(i)}, t^{(m)}) - \cos\theta^{(i)} G_{yxx}(t^{(i)}, t^{(m)})] - 2\sin\theta^{(m)}\cos\theta^{(m)}[\sin\theta^{(i)} G_{xxy}(t^{(i)}, t^{(m)}) - \cos\theta^{(i)} G_{yxy}(t^{(i)}, t^{(m)})] + \cos^2\theta^{(m)}[\sin\theta^{(i)} G_{xyy}(t^{(i)}, t^{(m)}) - \cos\theta^{(i)} G_{yyy}(t^{(i)}, t^{(m)})] \quad (9.6)$$

$$K_{32} = \sin^2\theta^{(m)}[\sin\theta^{(i)} G_{xxx}(t^{(i)}, t^{(m)}) + \sin\theta^{(i)} G_{yxx}(t^{(i)}, t^{(m)})] - 2\sin\theta^{(m)}\cos\theta^{(m)}[\cos\theta^{(i)} G_{xxy}(t^{(i)}, t^{(m)}) + \sin\theta^{(i)} G_{yxy}(t^{(i)}, t^{(m)})] + \cos^2\theta^{(m)}[\cos\theta^{(i)} G_{xyy}(t^{(i)}, t^{(m)}) + \sin\theta^{(i)} G_{yyy}(t^{(i)}, t^{(m)})] \quad (9.7)$$

where

$$G(t^{(i)}, t^{(m)}) = G(x, y; c) \quad (9.8)$$

$$c = x_c^{(i)} + t^{(i)}\sin\theta^{(i)} \quad (9.9)$$

$$x = x_c^{(m)} + t^{(m)}\sin\theta^{(m)} \quad (9.10)$$

$$y = t^{(i)}\cos\theta^{(i)} - t^{(m)}\cos\theta^{(m)} + y_c^{(m)} - y_c^{(i)} \quad (9.11)$$

Here, $(x_c^{(i)}, y_c^{(i)})$ are the coordinates of the center of the ith crack, $\theta^{(i)}$ is the angle between the ith crack and a horizontal line, and $G(x, y; c)$ is deduced from solutions given by Dundurs and Sendeckyj (1965) and can be expressed as

$$G_{xxx}(x, y; c) = -\frac{6y}{r_1^2} + \frac{6y}{r_2^2} + \frac{4y^3}{r_1^4} - \frac{4y^3}{r_2^4} - 2c\left(\frac{12cy}{r_2^4} - \frac{12(c+x)y}{r_2^4} - \frac{16cy^3}{r_2^6} + \frac{16(c+x)y^3}{r_2^6}\right) \quad (9.12)$$

$$G_{xyy}(x, y; c) = -\frac{2y}{r_1^2} + \frac{2y}{r_2^2} + \frac{4(-c+x)^2 y}{r_1^4} - \frac{4(c+x)^2 y}{r_2^4} - 2c\left(\frac{4cy}{r_2^4} - \frac{12(c+x)y}{r_2^4} - \frac{16c(c+x)^2 y}{r_2^6} + \frac{16(c+x)^3 y}{r_2^6}\right) \quad (9.13)$$

$$G_{xxy}(x, y; c) = -\left[-\frac{2(-c+x)}{r_1^2} + \frac{2(c+x)}{r_2^2} + \frac{4(-c+x)y^2}{r_1^4} - \frac{4(c+x)y^2}{r_2^4} \right.$$
$$- 2c\left(\frac{2}{r_2^2} + \frac{4c(c+x)}{r_2^4} - \frac{4(c+x)^2}{r_2^4} - \frac{4y^2}{r_2^4} - \frac{16c(c+x)y^2}{r_2^6} \right.$$
$$\left.\left. + \frac{16(c+x)^2y^2}{r_2^6} \right) \right] \tag{9.14}$$

$$G_{yxx}(x, y; c) = \frac{2(-c+x)}{r_1^2} - \frac{2(c+x)}{r_2^2} - \frac{4(-c+x)y^2}{r_1^4} + \frac{4(c+x)y^2}{r_2^4}$$
$$+ 2c\left(\frac{2}{r_2^2} - \frac{4c(c+x)}{r_2^4} + \frac{4(c+x)^2}{r_2^4} - \frac{4y^2}{r_2^4} + \frac{16c(c+x)y^2}{r_2^6} \right.$$
$$\left. - \frac{16(c+x)^2y^2}{r_2^6} \right) \tag{9.15}$$

$$G_{yyy}(x, y; c) = \frac{6(-c+x)}{r_1^2} - \frac{4(-c+x)^3}{r_1^4} - \frac{6(c+x)}{r_2^2} + \frac{4(c+x)^3}{r_2^4}$$
$$+ 2c\left(-\frac{2}{r_2^2} - \frac{12c(c+x)}{r_2^4} + \frac{16(c+x)^2}{r_2^4} + \frac{16c(c+x)^3}{r_2^6} \right.$$
$$\left. - \frac{16(c+x)^4}{r_2^6} \right) \tag{9.16}$$

$$G_{yxy}(x, y; c) = -\left[\frac{2y}{r_1^2} - \frac{2y}{r_2^2} - \frac{4(-c+x)^2y}{r_1^4} + \frac{4(c+x)^2y}{r_2^4} + 2c\left(-\frac{4cy}{r_2^4} \right.\right.$$
$$\left.\left. + \frac{4(c+x)y}{r_2^4} + \frac{16c(c+x)^2y}{r_2^6} - \frac{16(c+x)^3y}{r_2^6} \right) \right] \tag{9.17}$$

where

$$r_1^2 = (x-c)^2 + y^2 \tag{9.18}$$

$$r_2^2 = (x+c)^2 + y^2 \tag{9.19}$$

The components $b_t^{(i)}$ and $b_s^{(i)}$ of the dislocation density vector along the ith crack may be expressed as

$$b_t^{(i)}(t^{(i)}) = \frac{\partial}{\partial t^{(i)}}[u_t^{(i)}(t^{(i)}, 0^+) - u_t^{(i)}(t^{(i)}, 0^-)] = \frac{\partial \Delta_t^{(i)}(t^{(i)})}{\partial t^{(i)}} \tag{9.20a}$$

and

$$b_s^{(i)}(t^{(i)}) = \frac{\partial}{\partial t^{(i)}}[u_s^{(i)}(t^{(i)}, 0^+) - u_s^{(i)}(t^{(i)}, 0^-)] = \frac{\partial \Delta_s^{(i)}(t^{(i)})}{\partial t^{(i)}} \tag{9.20b}$$

where $\Delta_t^{(i)}$ and $\Delta_s^{(i)}$ denote the glide and climb displacements, respectively, for the ith crack surface.

The point dislocation solutions due to Dundurs and Mura (1964) and Dundurs and Sendeckyj (1965) result in Cauchy-type singularities of order $(1/t^{(i)})$ when $t^{(i)} = t^{(m)}$. The solution procedure needs to account for such singularities. Also, the $2N$ integral equations give rise to $2N$ additional constants of integration that must be evaluated to obtain a complete solution. The $2N$ additional equations for evaluating these $2N$ unknowns may be obtained from the requirement that the crack faces must meet at both ends of each crack. Thus, we obtain

$$\int_{-a^{(i)}}^{a^{(i)}} b_t^{(i)}(t^{(i)})\, dt^{(i)} = 0, \qquad i = 1, \ldots, N \tag{9.21a}$$

and

$$\int_{-a^{(i)}}^{a^{(i)}} b_s^{(i)}(t^{(i)})\, dt^{(i)} = 0, \qquad i = 1, \ldots, N \tag{9.21b}$$

Equations (9.1) through (9.21) may now be solved to determine the dislocation density vectors.

Once the dislocation density vectors for all cracks are obtained, the stress field for the half-plane containing N interacting cracks near a free surface can also be constructed by superposition of the same three problems described before. The stress components at any point in terms of a local coordinate system $(t^{(\alpha)}, s^{(\alpha)})$ may be written as

$$\sigma_{ss}^{(\alpha)}(t^{(\alpha)}) = \frac{G}{\pi(k+1)} \sum_{i=1}^{N} \left(\int_{-a^{(i)}}^{a^{(i)}} K_{11}(t^{(\alpha)}, t^{(i)})\, b_t^{(i)}(t^{(i)})\, dt^{(i)} + \int_{-a^{(i)}}^{a^{(i)}} K_{12}(t^{(\alpha)}, t^{(i)})\, b_s^{(i)}(t^{(i)})\, dt^{(i)} \right) + \sigma_{ss}^{(\alpha 0)}(t^{(\alpha)}) \tag{9.22a}$$

$$\sigma_{st}^{(\alpha)}(t^{(\alpha)}) = \frac{G}{\pi(k+1)} \sum_{i=1}^{N} \left(\int_{-a^{(i)}}^{a^{(i)}} K_{21}(t^{(\alpha)}, t^{(i)})\, b_t^{(i)}(t^{(i)})\, dt^{(i)} + \int_{-a^{(i)}}^{a^{(i)}} K_{22}(t^{(\alpha)}, t^{(i)})\, b_s^{(i)}(t^{(i)})\, dt^{(i)} \right) + \sigma_{st}^{(\alpha 0)}(t^{(\alpha)}) \tag{9.22b}$$

and

$$\sigma_{tt}^{(\alpha)}(t^{(\alpha)}) = \frac{G}{\pi(k+1)} \sum_{i=1}^{N} \left(\int_{-a^{(i)}}^{a^{(i)}} K_{31}(t^{(\alpha)}, t^{(i)}) \, b_t^{(i)}(t^{(i)}) \, dt^{(i)} + \int_{-a^{(i)}}^{a^{(i)}} K_{32}(t^{(\alpha)}, t^{(i)}) \, b_s^{(i)}(t^{(i)}) \, dt^{(i)} \right) + \sigma_{tt}^{(\alpha 0)}(t^{(\alpha)}) \tag{9.22c}$$

where $\sigma_{ss}^{(\alpha 0)}$, $\sigma_{st}^{(\alpha 0)}$, and $\sigma_{tt}^{(\alpha 0)}$, are the stress components under the prescribed loading for an uncracked body. The additional kernels K_{31} and K_{32} are given in equations (9.6) and (9.7).

9.3.2 *Numerical Solution Procedure*

The integral equations (9.1) for unknown dislocation functions $b_t^{(i)}$ and $b_s^{(i)}$ are of the first kind and are fully coupled. Accordingly, a numerical technique for solving these equations is adapted here. The interval of integration is first normalized by setting

$$t^{(m)} = a^{(m)} v^{(m)} \tag{9.23a}$$

and

$$t^{(i)} = a^{(i)} v^{(i)} \tag{9.23b}$$

The governing integral equations (9.1) may now be written as

$$\frac{G}{\pi(\kappa+1)} \sum_{i=1}^{N} \left(\int_{-1}^{1} \widetilde{K}_{11}(v^{(m)}, u^{(i)}) \, b_t^{(i)}(u^{(i)}) \, du^{(i)} + \int_{-1}^{1} \widetilde{K}_{12}(v^{(m)}, u^{(i)}) \, b_s^{(i)}(u^{(i)}) \, du^{(i)} \right) + \sigma_{ss}^{(m0)}(v^{(m)}) = 0$$

$$-1 < v^{(m)} < 1, \qquad m = 1, \ldots, N \tag{9.24a}$$

and

$$\frac{G}{\pi(\kappa+1)} \sum_{i=1}^{N} \left(\int_{-1}^{1} \widetilde{K}_{21}(v^{(m)}, u^{(i)}) \, b_t^{(i)}(u^{(i)}) \, du^{(i)} + \int_{-1}^{1} \widetilde{K}_{22}(v^{(m)}, u^{(i)}) \, b_s^{(i)}(u^{(i)}) \, du^{(i)} \right) + \sigma_{st}^{(m0)}(v^{(m)}) = 0$$

$$-1 < v^{(m)} < 1, \qquad m = 1, \ldots, N \tag{9.24b}$$

where the new coefficient has been absorbed into the modified kernels.

The variables $b_t^{(i)}$ and $b_s^{(i)}$ contain singularities of the order of $[u^{(i)}]^{1/2}$ at the end points ($u^{(i)} = \pm 1$) of the integration. In the present work, this singularity is abstracted by expressing the solutions in terms of regular functions $B_t^{(i)}(u^{(i)})$ and $B_s^{(i)}(u^{(i)})$ along with a weighting function containing the square root singularity. This may be defined as

$$b_t^{(i)}(u^{(i)}) = \frac{B_t^{(i)}(u^{(i)})}{(1 - u^{(i)^2})^{1/2}} \tag{9.25a}$$

and

$$b_s^{(i)}(u^{(i)}) = \frac{B_s^{(i)}(u^{(i)})}{(1 - u^{(i)^2})^{1/2}} \tag{9.25b}$$

where $B_t^{(i)}(u^{(i)})$ and $B_s^{(i)}(u^{(i)})$ are bounded functions in the closed interval of the ith crack. The presentation shown in equations (9.25) also allows us to use a Gauss–Chebychev quadrature scheme (Erdogan et al. 1973). The discretized versions of equations (9.24) may then be expressed as

$$\frac{G}{\pi(\kappa+1)} \sum_{i=1}^{N} \sum_{l=1}^{N} \left(\frac{\pi}{L} K_{11}(v^{(m,k)}, u^{(i,l)})\, B_t^{(i)}(u^{(i,l)}) + \frac{\pi}{L} K_{12}(v^{(m,k)}, u^{(i,l)})\, B_s^{(i)}(u^{(i,l)}) \right) + \sigma_{ss}^{(m0)}(v^{(m,k)}) = 0$$

$$k = 1, \ldots, L-1, \qquad m = 1, \ldots, N \tag{9.26a}$$

and

$$\frac{G}{\pi(\kappa+1)} \sum_{i=1}^{N} \sum_{l=1}^{N} \left(\frac{\pi}{L} K_{12}(v^{(m,k)}, u^{(i,l)})\, B_t^{(i)}(u^{(i,l)}) + \frac{\pi}{L} K_{22}(v^{(m,k)}, u^{(i,l)})\, B_s^{(i)}(u^{(i,l)}) \right) + \sigma_{ts}^{(m0)}(v^{(m,k)}) = 0$$

$$k = 1, \ldots, L-1, \qquad m = 1, \ldots, N \tag{9.26b}$$

where L is the number of integration points and

$$u^{(i,l)} = \cos\left(\frac{2l-1}{2L}\pi\right)$$

$$l = 1, \ldots, L, \qquad i = 1, \ldots, N \tag{9.27a}$$

$$v^{(m,k)} = \cos\left(\frac{k}{L}\pi\right)$$

$$k = 1, \ldots, L-1, \qquad m = 1, \ldots, N \tag{9.27b}$$

It should be noted here that the system of equations (9.26) contains $2N(L-1)$ equations and $2NL$ unknowns. The additional $2N$ equations can be obtained from the discretized versions of equations (9.21) ensuring continuity of crack opening shapes. These additional equations may be written as

$$\sum_{l=1}^{L} \frac{\pi}{L} B_t^{(i)}(u^{(i,l)}) = 0, \qquad i = 1, \ldots, N \tag{9.28a}$$

and

$$\sum_{l=1}^{L} \frac{\pi}{L} B_s^{(i)}(u^{(i,l)}) = 0, \qquad i = 1, \ldots, N \tag{9.28b}$$

Utilizing equations (9.26)–(9.28), the unknowns $B_t^{(i)}$ and $B_s^{(i)}$ can be determined at all collocation points.

The stress intensity factors may now be determined as

$$K_I^{(i)}(\pm a^{(i)}) = \pm 2\sqrt{\pi a^{(i)}} \frac{G}{(\kappa+1)} B_s^{(i)}(\pm 1) \tag{9.29a}$$

and

$$K_{II}^{(i)}(\pm a^{(i)}) = \pm 2\sqrt{\pi a^{(i)}} \frac{G}{(\kappa+1)} B_t^{(i)}(\pm 1) \tag{9.29b}$$

Here, $B_s^{(i)}(\pm 1)$ and $B_t^{(i)}(\pm 1)$ are obtained by extrapolating the values of $B_s^{(i)}$ and $B_t^{(i)}$, respectively, at the collocation points. In the present work, the interpolation functions proposed by Krenk (1975) are adopted. These may be expressed as

$$B_s^{(i)}(+1) = \frac{1}{L} \sum_{l=1}^{L} B_s^{(i)}(u^{(i,l)}) \left[\frac{\sin\left[\left(\frac{2L-1}{4L}\right)(2l-1)\pi\right]}{\sin\left[\left(\frac{2L-1}{4L}\right)\pi\right]} \right] \tag{9.30a}$$

and

$$B_s^{(i)}(-1) = \frac{1}{L} \sum_{l=1}^{L} B_s^{(i)}(u^{(i,L+1-l)}) \left[\frac{\sin\left[\left(\frac{2L-1}{4L}\right)(2l-1)\pi\right]}{\sin\left[\left(\frac{2L-1}{4L}\right)\pi\right]} \right] \tag{9.30b}$$

Similar interpolation schemes may also be used for determining $B_t^{(i)}(\pm 1)$.

Using a crack propagation criterion (e.g., Sih 1973), the direction of propagation for a crack under mixed mode I and mode II loading may also be obtained. Assuming crack propagation in the direction normal to that of the principal tensile stress, we obtain

$$K_I \sin\frac{\theta_p}{2}\cos\frac{\theta_p}{2} + K_{II}\left(3\cos^2\frac{\theta_p}{2} - 2\right) = 0 \tag{9.31}$$

where θ_p denotes the direction of crack propagation.

9.4 Determination of Effective Elastic Properties

The effective elastic properties of the cracked half-plane over a representative volume are investigated next. The stress–strain relationship for the cracked body may be written as

$$\langle\varepsilon\rangle = C^{\text{eff}}:\langle\sigma\rangle \tag{9.32}$$

where $\langle\varepsilon\rangle$ is the average strain over a representative volume, $\langle\sigma\rangle$ is the corresponding remotely applied average stress, and C^{eff} represents the effective material properties and is a fourth-rank tensor. The semicolon denotes the contraction between C^{eff} and $\langle\sigma\rangle$ over two indices. The average strain $\langle\varepsilon\rangle$ is commonly represented (Hashin 1983, 1988, Kachanov 1987) as

$$\langle\varepsilon\rangle = C^0:\langle\sigma\rangle + \frac{1}{2v}\sum_{i=1}^{N}\int_{s^{(i)}}\langle \boldsymbol{n}^{(i)}\Delta^{(i)} + \Delta^{(i)}\boldsymbol{n}^{(i)}\rangle\, ds^{(i)} \tag{9.33}$$

where C^0 is a tensor representing the elastic compliance tensor for the material without cracks and $\boldsymbol{n}^{(i)}$ denotes the unit normal vector to the ith crack with a surface $s^{(i)}$. The crack opening displacement vector across $s^{(i)}$ is represented by $\Delta^{(i)}$. It is important to note here that $\boldsymbol{n}^{(i)}\Delta^{(i)}$ forms a second-order tensor. For flat cracks, the unit normal vector remains constant along each of the cracks. Thus, we get

$$\langle\varepsilon\rangle = C^0:\langle\sigma\rangle + \frac{1}{2v}\sum_{i=1}^{N}(\boldsymbol{n}^{(i)}\langle\Delta^{(i)}\rangle + \langle\Delta^{(i)}\rangle\boldsymbol{n}^{(i)})\,s^{(i)} \tag{9.34}$$

where $\langle\Delta^{(i)}\rangle$ denotes the average of $\Delta^{(i)}$ over $s^{(i)}$.

The problem of determining effective elastic properties for the damaged or cracked body is then reduced to the determination of the average crack opening displacements for all cracks if σ can be related to the crack geometry. Here,

$$\Delta_t^{(i)}(t^{(i)}) = \int_{-a^{(i)}}^{t^{(i)}} b_t^{(i)}(t^{(i)})\,dt^{(i)} \tag{9.35a}$$

and

$$\Delta_s^{(i)}(t^{(i)}) = \int_{-a^{(i)}}^{t^{(i)}} b_s^{(i)}(t^{(i)})\,dt^{(i)} \tag{9.35b}$$

Using equations (9.35), the average $\Delta_t^{(i)}$ and $\Delta_s^{(i)}$ can readily be evaluated when $B_t^{(i)}$ and $B_s^{(i)}$ are known. The effective elastic properties may then be estimated.

9.4.1 Numerical Results for Monolithic Ceramics

Applications of the proposed integral equation formulations to investigations of surface and subsurface damage evolution during grinding of ceramic materials are presented in this section. The integral equation formulations are based on the fundamental solutions due to point dislocations in an elastic half-plane and can handle arbitrary orientations and distributions of cracks, along with arbitrary loading conditions. In our approach, the grinding process is modeled as a moving indenter problem and, as a first attempt, only steady-state characteristics are considered.

It has been observed by several researchers (Conway and Kirchner 1980, Lawn et al. 1980, Marshall et al. 1982) that a sharp indenter produces a combination of radial and lateral cracks in a ceramic body (as shown in figure 9.2b). For sharp indenters, it has also been observed that there exists a small plastic zone directly under the indenter. This plastic zone exerts a residual opening force on the accompanying cracks. Following the approach of Lawn et al. (1980) and Marshall et al. (1982), an equivalent elastic model for this elasto-plastic indentation problem is also developed by incorporating the residual stresses due to the plastic zone. Analyses are then carried out using uniform indenter loading, as well as nonuniform residual stresses due to the plastic zone. In a real grinding process, the resultant force distributions (due to indenter loading and residual stresses) on the crack faces are expected to be nonuniform. As a first attempt, the resultant driving force distributions are approximated as uniform in some of the examples. These cases are referred to as "uniform loading cases." Such approximations also allow us to gain detailed insights into the nature of the interactions. The "nonuniform loading cases" are also considered (see figure 9.5b) to better represent the actual situation during grinding.

In the present work, the lateral and radial cracks near a free surface are first investigated individually. The results obtained by the proposed technique for a simple radial crack near a free surface compare well with those reported by Erdogan et al. (1973). The combination of radial and lateral cracks is investigated next, and the interactions of these cracks with preexisting microcracks are studied in detail. In addition to investigating single-grit scratch tests, the effects of multiple indentations and their associated crack systems in the vicinity of each other are investigated to obtain a realistic description of damage evolution during multipoint grinding operations. For all of the cases, the specific problem geometries are shown as insets in the figures.

A lateral crack near a free surface subject to indentation loading is considered first. It has been observed that the residual stresses due to the plastic zone also provide crack driving forces upon unloading of the indenter. The normalized stress intensity factors (SIFs) in both mode I and mode II are shown in figure 9.3a for uniform loading. It is interesting to observe the existence of the direct shear mode (mode II) for this problem. The SIF in mode II is a direct consequence of the vicinity of the free surface and, as expected, the SIF in mode II drops to zero as the buried depth increases. The SIFs in both modes I and II tend to increase as the normalized buried depth decreases. This increase is significant when the normalized buried depth drops below 0.4. For uniform loading, as the normalized buried depth drops from 0.4 to 0.2, the SIF in mode I changes from 2.8 to 7.5 and the SIF in mode II changes from 1.2 to 3.2. As the buried depth is increased, the full-plane solution is recovered and the mode II SIF tends to vanish. The crack propagation direction is also of great interest, since the effective material removal rate may be controlled by varying the crack propagation direction. Figure 9.3b shows the variations in crack propagation directions with the buried depth. As the buried depth is reduced, the effect of the free surface becomes more pronounced, and the direction of propagation for a lateral crack rotates toward the normal to the free surface.

Interactions between two collinear lateral cracks near a free surface (the near tips are separated by 0.2) and subject to uniform loading are considered next. The variations of the normalized SIFs with changes in the buried depth are presented in figure 9.4. The full-plane solution (Tada et al. 1985) is recovered as the buried depth is increased. Comparisons with the SIFs in figure 9.3a reveal a shielding effect as the normalized buried depth drops below 0.5. As the normalized buried depth is increased, however, a stress amplification effect is observed. It is also interesting to note here that, at small buried depths, the normalized mode I SIF at the far tip becomes greater than that at the near tip. At a normalized buried depth of 0.8, the far-tip SIF is 1.5 and the near-tip SIF is 2.2. At a normalized buried depth of 0.2, however,

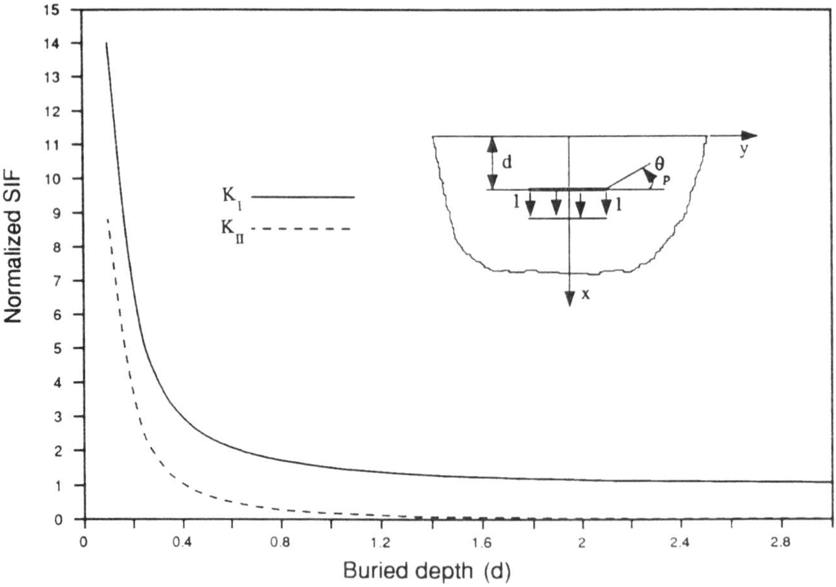

(a)

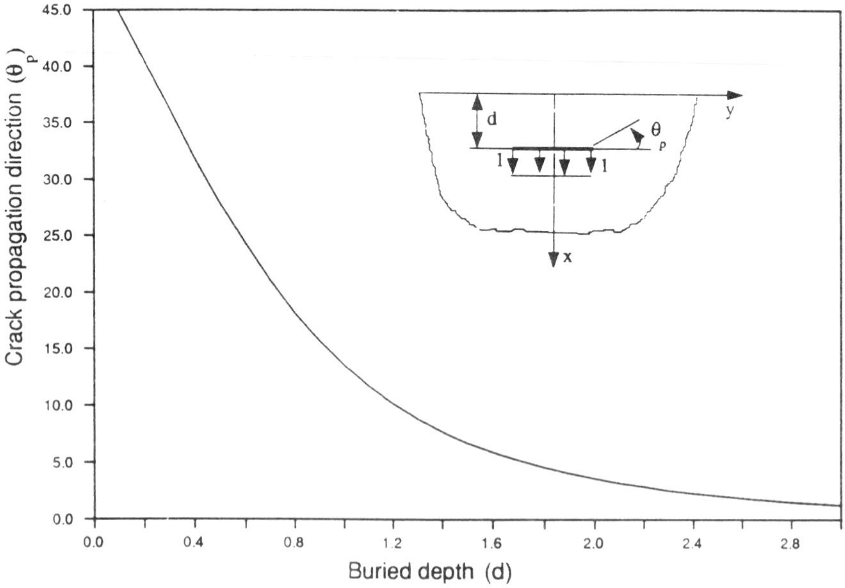

(b)

Figure 9.3. A lateral crack near a free surface: (a) SIFs under uniform loading; (b) crack propagation under uniform loading.

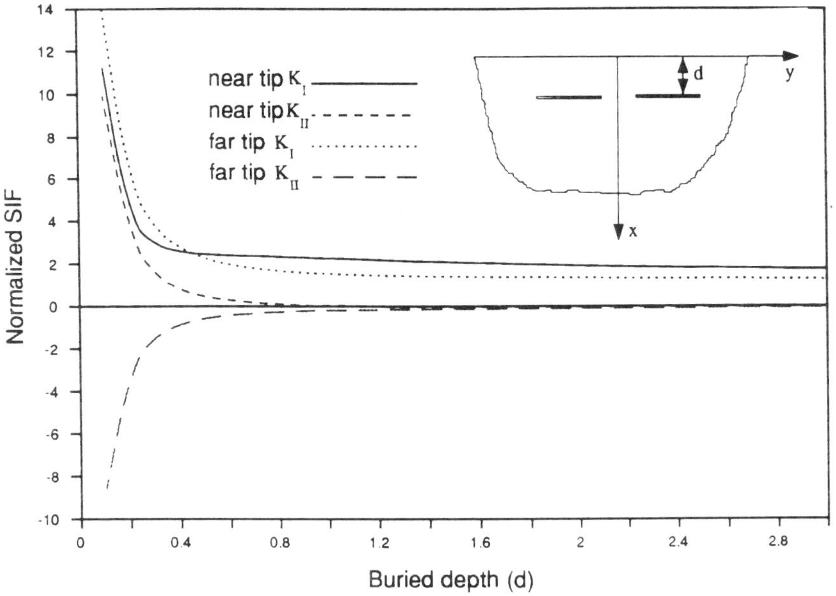

Figure 9.4. SIFs for two collinear lateral cracks near a free surface.

the far-tip mode I SIF becomes 6.7 while the near-tip mode I SIF becomes 4.2. Accordingly, interactions among preexisting lateral cracks and the lateral cracks due to grinding will tend to enhance lateral propagation of the preexisting cracks. This will tend to increase the effective chip thickness.

The radial cracks near a free surface subject to grinding loads are considered next. Figure 9.5 shows the normalized mode I SIFs for a single radial crack. Both uniform loading and nonuniform residual stresses due to the plastic zone are considered. Based on the observations by Lawn et al. (1980) and Marshall et al. (1983), the resultant loading due to nonuniform residual stresses is approximated by a distribution whose intensity doubles as the base of the plastic zone is approached from the lower tip of the radial crack. The mode II SIFs for both cases are zero. The mode I SIFs shown in figure 9.5 at different buried depths are almost identical to those reported by Erdogan et al. (1973) for a single radial crack near a free surface. Figure 9.6 shows the normalized SIFs for two radial cracks (separating distance of 0.2) near a free surface and subject to uniform loading. It is interesting to observe from figure 9.6 that there exist mode II SIFs for both the upper tip and the lower tip. The existence of these direct shear effects is a consequence of the crack interactions. Both mode I and mode II SIFs increase with decreasing buried depth. Comparing the mode I SIF for a single radial crack (figure 9.5) with those presented in figure 9.6, a significant shielding effect is observed for both upper and lower tips.

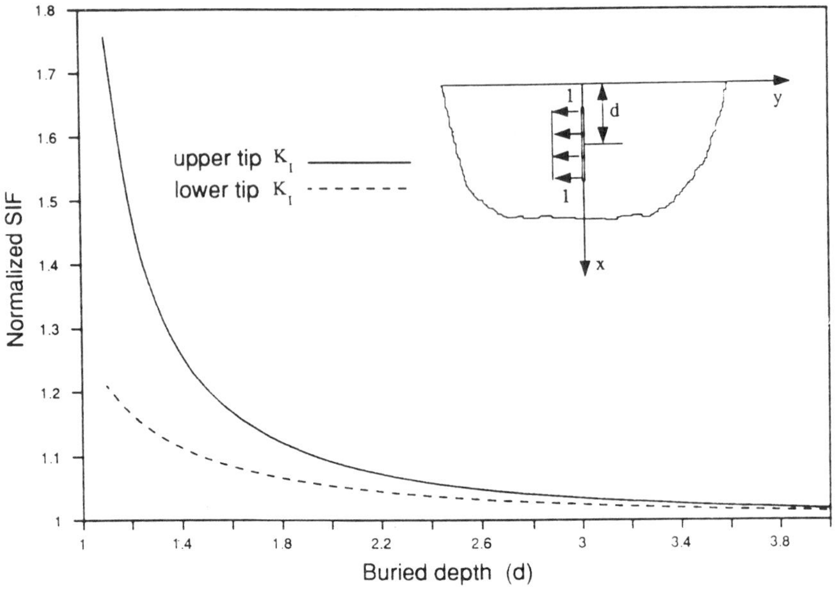

(a)

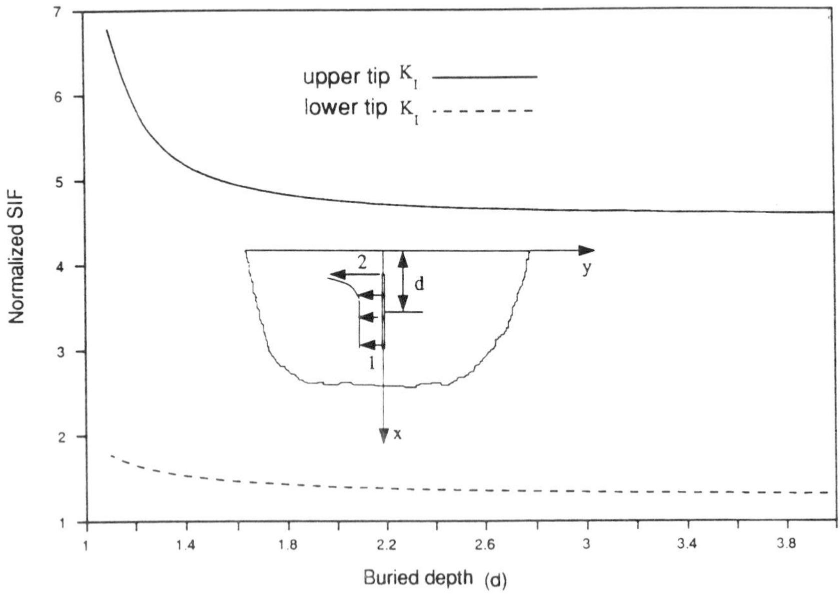

(b)

Figure 9.5. A radial crack near a free surface: (a) SIFs under uniform loading; (b) SIFs under nonuniform forces.

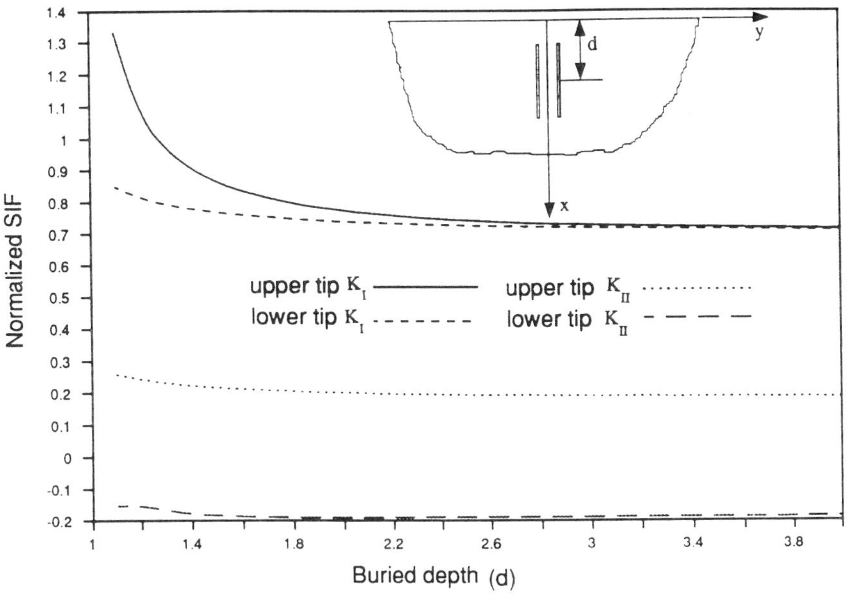

Figure 9.6. SIFs for two radial cracks near a free surface.

Interactions among two cracks with arbitrary relative orientations and subjected to uniform loading are considered in figure 9.7. The normalized buried depth is held constant at 0.4. The normalized distance between the crack centers is 2.2. One of the cracks is held lateral and the orientation of the other one is varied from −22 degrees to +22 degrees. The variations of the normalized SIFs in modes I and II for the lateral crack are shown in figure 9.7a. It may be observed that the changes in mode I SIFs with variations in θ are relatively small for both crack tips. Compared to the mode I SIF shown in figure 9.3a for a single lateral crack at a buried depth of 0.4, there is a slight amplification when θ lies between −22 and −4 degrees and a small shielding effect may be observed as θ increases from −4 to 22 degrees. The mode II SIF at the near tip also changes from −0.9 at $\theta = -22$ degrees to −0.45 at $\theta = -6$ degrees and again to −2 at $\theta = 22$ degrees. These variations are due to crack interactions. Figure 9.7b shows the SIFs for the crack with varying orientation. At $\theta = -22$ degrees, the near tip is close to the free surface and shows a relatively high mode I SIF of 7.2. The far tip moves near the free surface at $\theta = 22$ degrees and shows a mode I SIF of 7.0. The mode II SIFs at the near tip are also relatively large for all values of θ between −22 and 22 degrees. Figure 9.7c shows the variations in the direction of propagation of the lateral crack with changes in θ. As θ is varied from −22 to −7 degrees, the crack propagation direction changes from 27

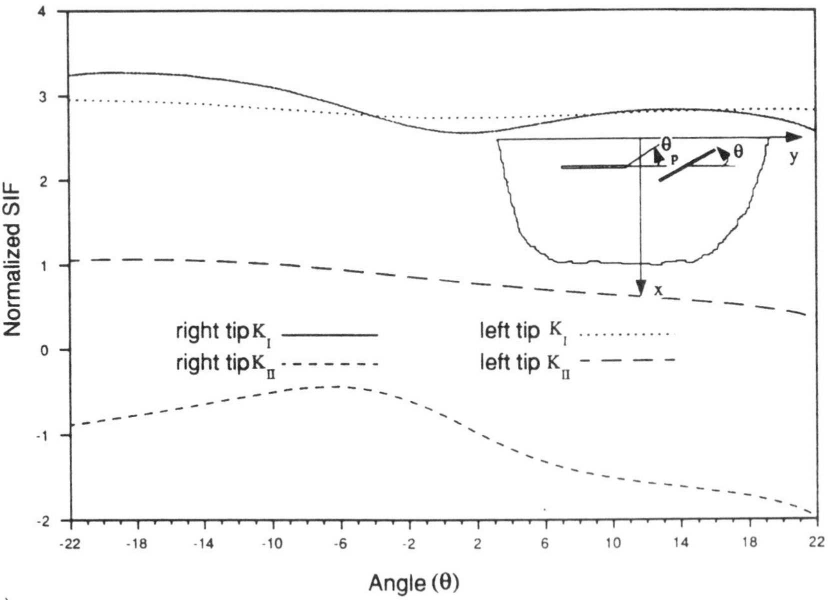

(a)

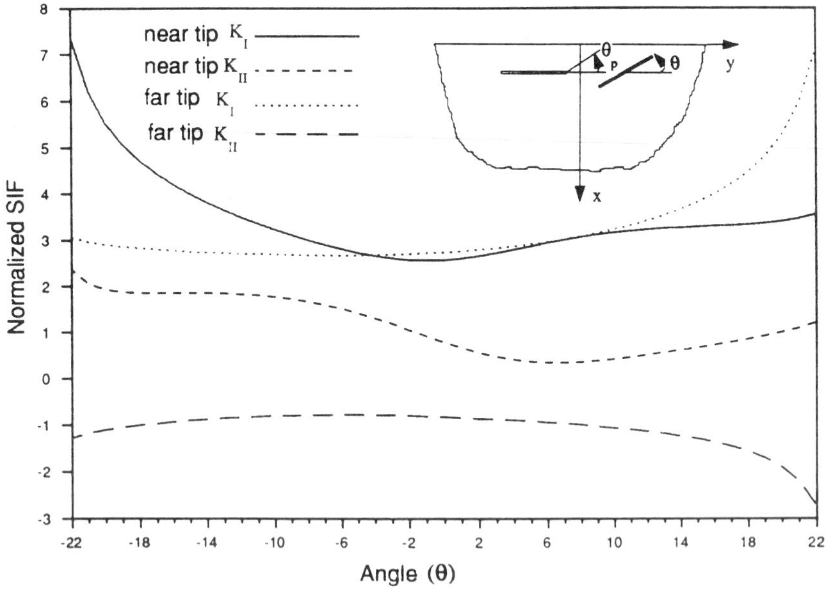

(b)

Figure 9.7. Two cracks near a free surface with arbitrary relative orientations: (a) variations in SIFs for the lateral crack with orientation of the accompanying crack; (b) variations in SIFs for the accompanying crack with its orientation; (c) variations in the direction of propagation for the lateral crack with orientation of the accompanying crack.

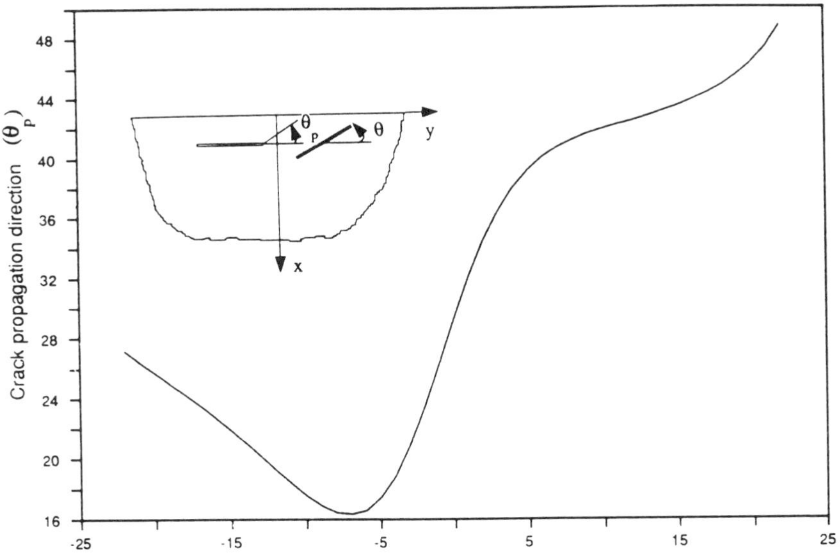

(c)

Figure 9.7—*continued*

to 16.5 degrees. However, as θ varies from -7 to 5 degrees, the crack propagation direction changes rapidly to 37 degrees and continues to 48.5 degrees at $\theta = 22$ degrees. It is evident here that the direction of propagation for the lateral crack can be controlled very effectively through small variations in orientation of the accompanying crack θ between -7 and 5 degrees. This provides an effective means for controlling effective material removal rates in ceramic grinding processes.

The interactions among a lateral crack and an array of four lateral microcracks are considered in figure 9.8. The buried depth of the lateral crack is 0.3. The buried depths of microcracks from top to bottom are 0.1, 0.2, 0.4, and 0.5. The separation of the near tips of the lateral crack and the microcracks is fixed at 0.1. An amplification effect may be observed for the near-tip mode I SIF, and this effect becomes more pronounced with the increase in size of the microcracks. The mode II SIF at the near tip, however, shows a shielding effect as the microcrack sizes are increased. Figure 9.9 considers a later situation for such an array of lateral cracks, when the primary crack has been extended to the centerline of the microcracks. The crack geometry is the same as in figure 9.8, except that the center of the microcracks is aligned with the tip of the lateral crack. It is interesting to observe the shielding effects for the mode I SIF at the near tip for this case. For this case, the centerlines of the microcracks were held fixed. Accordingly, the distance between the far tip of the primary crack and the near tips of

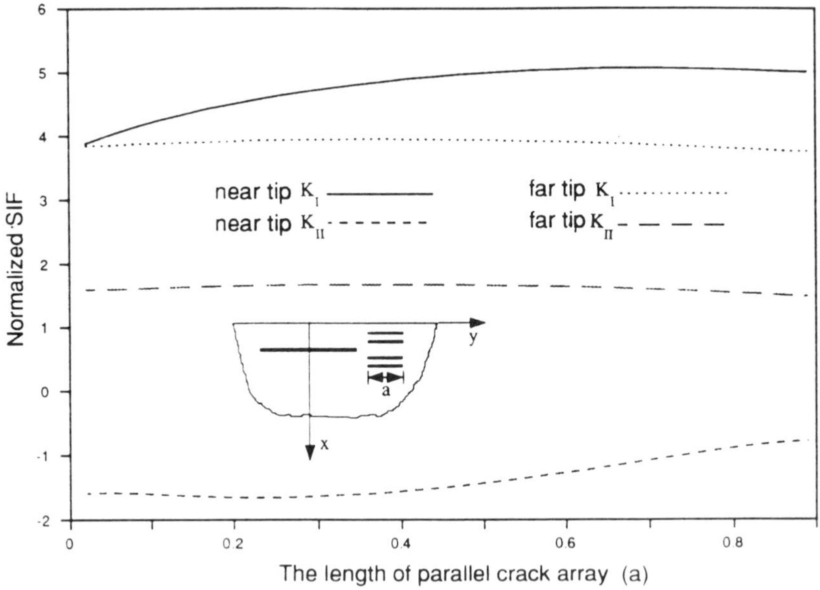

Figure 9.8. Interactions of a lateral crack with an array of lateral microcracks ahead of the crack tip.

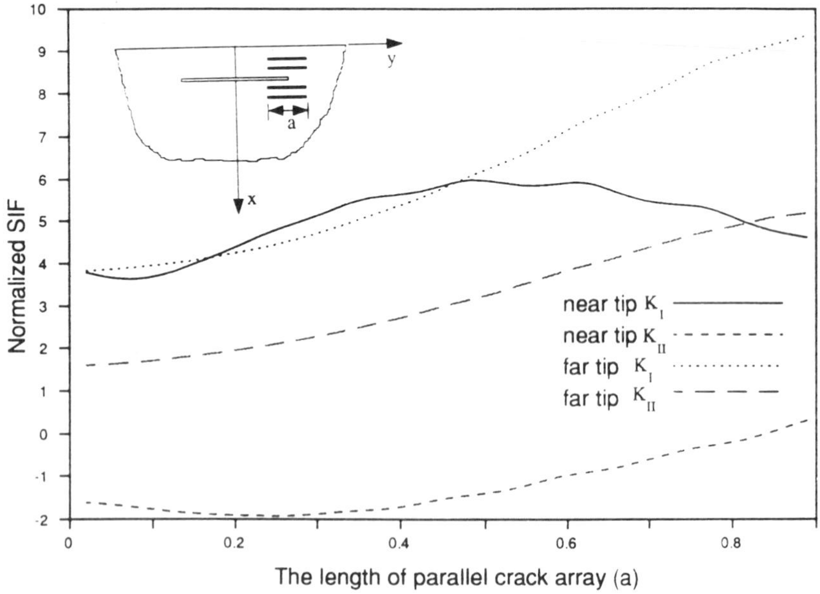

Figure 9.9. Interactions between an array of lateral microcracks with a lateral crack propagating through the array.

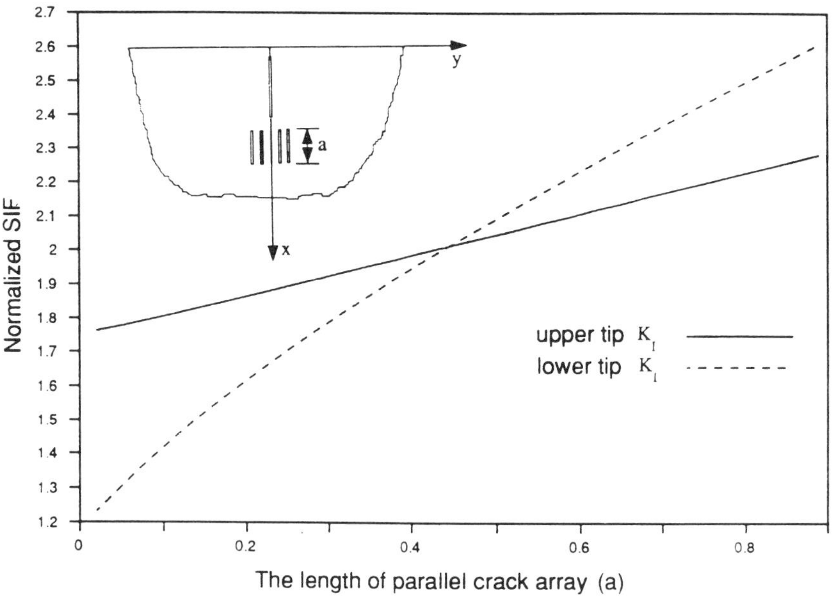

Figure 9.10. Interactions of a radial crack with an array of radial microcracks ahead of the crack tip.

the microcracks changed with the size of the microcrack. This produces the rapid rise in far-tip SIFs observed in figure 9.9.

The interactions among a radial crack and an array of four radial cracks are considered in figure 9.10. The center of the radial crack is located at a distance of 1.1 from the free surface. The separation of the tips of the radial crack from those of the microcracks is kept fixed at 0.1. The same crack system at a later stage, when the primary radial crack has been extended to the centerline of the microcracks, is considered in figure 9.11. The crack geometry is the same, except that the center of the microcracks is aligned with the tip of the radial crack. As the microcrack size is increased, an amplification effect is observed for the lower tip in figure 9.10; however, a rapid shielding effect is observed for the lower tip in figure 9.11. As the microcrack size varies from 0.025 to 0.2, the mode I SIF drops by half, from 1.2 to 0.6. This implies that a radial array of microcracks has the capability of arresting the growth of a radial crack as it attempts to propagate through the array.

A combination of radial and lateral cracks representative of a ceramic grinding situation is considered in figure 9.12. If brittle fracture is the predominant mechanism of chip formation, the buried depth will be relatively small. As more energy is expended in plastic deformation and the size of

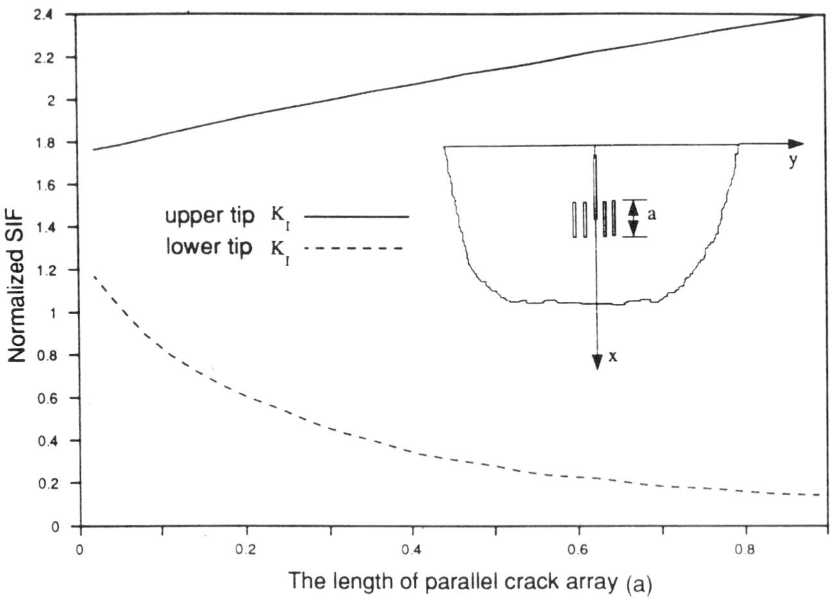

Figure 9.11. Interactions between an array of radial microcracks with a radial crack propagating through the array.

the plastic zone increases, the buried depth will also increase. It can be observed from figure 9.12a that the mode I SIF for the upper tip of the radial crack increases very rapidly as the buried depth decreases. As the buried depth decreases from 1.8 to 1.1, the mode I SIF varies from 2.4 to 5.7 for the upper tip. The lower tip mode I SIF, however, only varies from about 1.0 at a buried depth of 1.8 to 1.6 at a buried depth of 1.1. As seen in figure 9.12b, the mode I and mode II SIFs for both the near tip and the far tip of the lateral crack also increase rapidly in magnitude with decreasing buried depth.

A multipoint grinding situation is considered in figure 9.13. Two systems of radial and lateral cracks emanating from the ploughing action of two abrasive grits in the vicinity of each other are considered. The spacing between the crack systems is d. The buried depth of the lateral cracks is 0.5 and that of the center of the radial cracks is 1.4. The separation distance between the radial and lateral cracks is 0.2. Compared to single-point grinding, a slight shielding effect is observed for multipoint grinding. As the spacing between the two crack systems is reduced from 6.5 to 4.92, the mode I SIF at the near tip of the lateral crack "A" drops from 2.34 to 2.1 before rising again to 2.48 at a spacing of 4.6. The mode I SIF at the far tip remains almost constant. The mode II SIF at the near tip also remains constant when

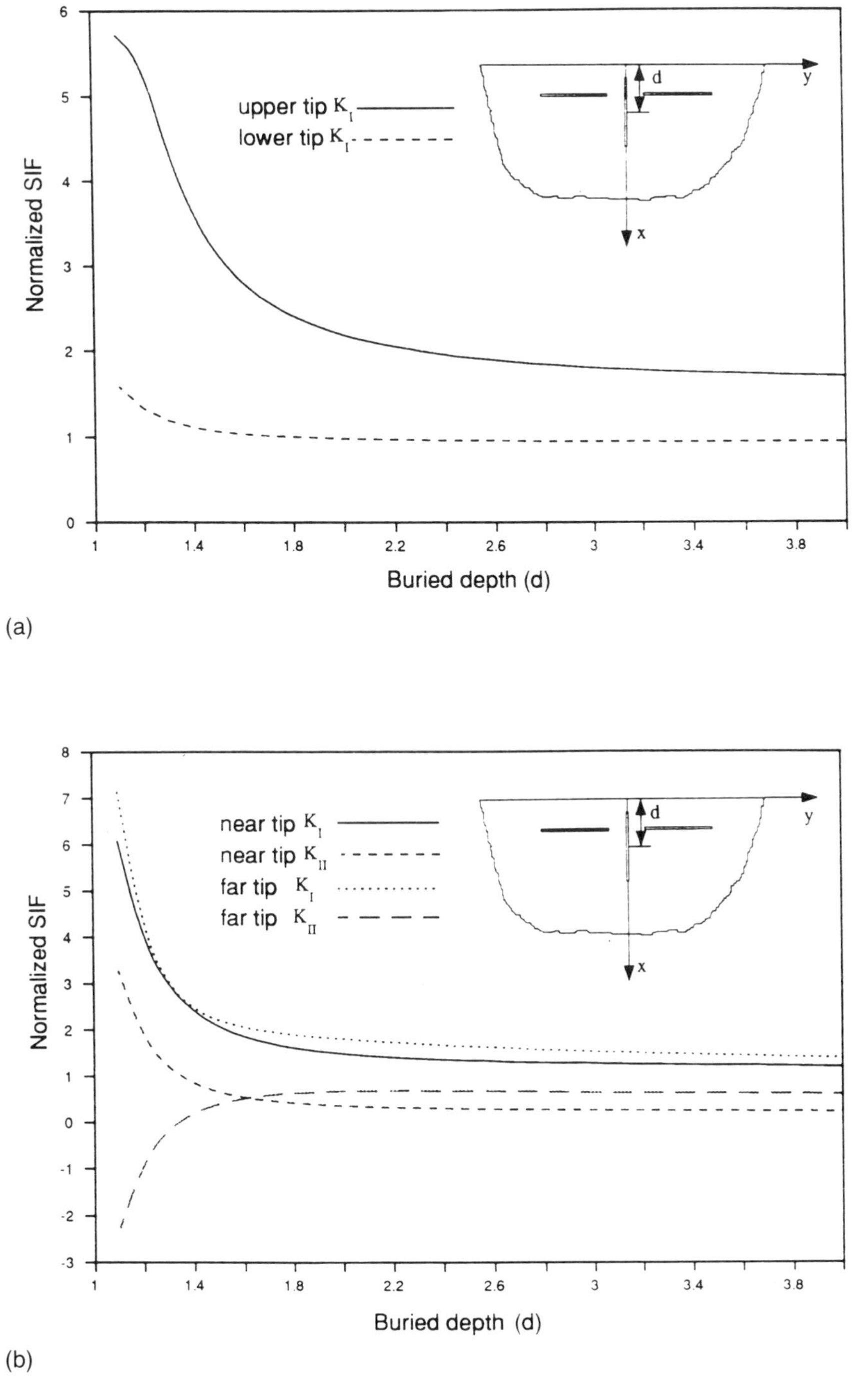

Figure 9.12. Representation of grinding as a system of radial and lateral cracks: (a) SIFs for the radial crack; (b) SIFs for the lateral crack.

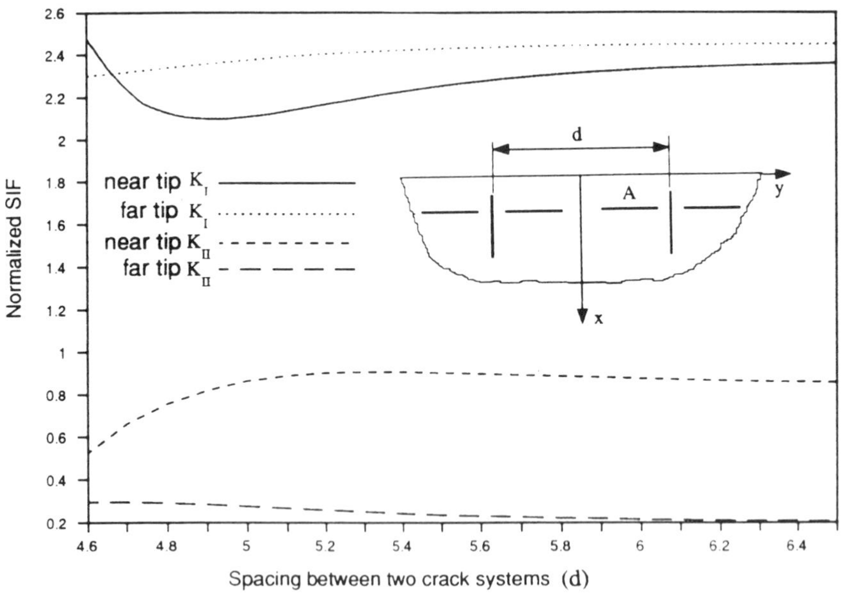

Figure 9.13. Interactions among two radial–lateral crack systems for multipoint grinding.

the spacing is greater than 5.7. As the spacing is reduced further, mode II SIF at the near tip drops to 0.52 at a spacing of 4.6. Similar shielding effects were also observed by Kirchner (1984) for multipoint grinding.

The effects of the microcrack distributions on the effective elastic properties are discussed next. For the purposes of grinding, only the properties of a thin layer (depth of the order of the depth of cut) of the workpiece is considered. Figure 9.14 shows the effects of an array of radial microcracks on the effective bulk modulus at a loading of $G/100$. It is observed that the effective normalized bulk modulus of the control layer of the workpiece decreases as the normalized spacing between the cracks (buried depth is taken to be 1.5) is reduced. It drops from 0.76 to 0.694 as the spacing is reduced from 4.0 to 2.0. The reduction in effective bulk modulus is more drastic as the spacing is further reduced, and at a spacing of 1.0 the effective normalized bulk modulus reduces to 0.61. Figure 9.15 shows the effects of an array of lateral cracks on the effective bulk modulus of the control layer of the workpiece at a loading of $G/100$. As the normalized spacing between lateral cracks (buried depth is assumed to be 1.0) is reduced from 6.5 to 3.25, the normalized effective bulk modulus reduces from 0.638 to 0.485. Upon further reduction of the crack spacing to 2.4, the bulk modulus reduces to 0.39. Thus, the effective elastic bulk modulus may be

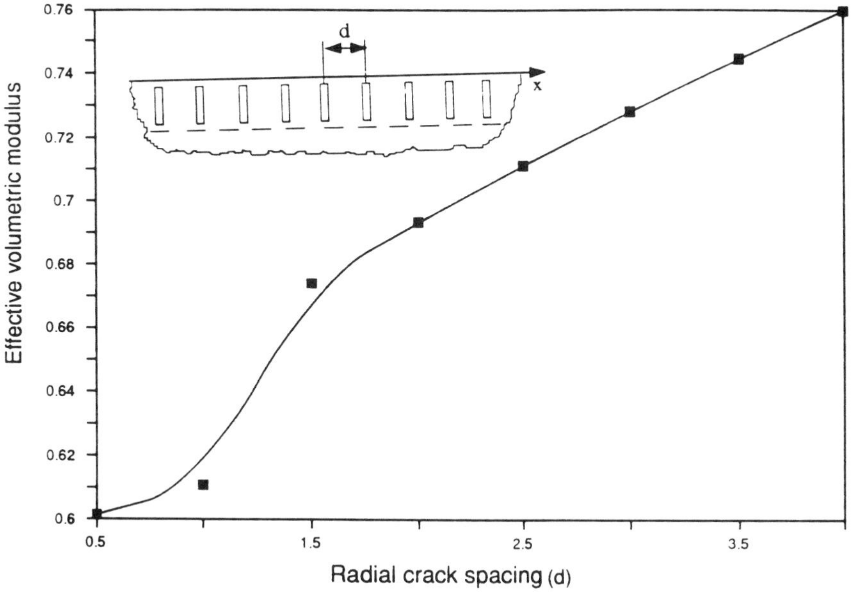

Figure 9.14. Variations in effective bulk modulus for an array of radial cracks.

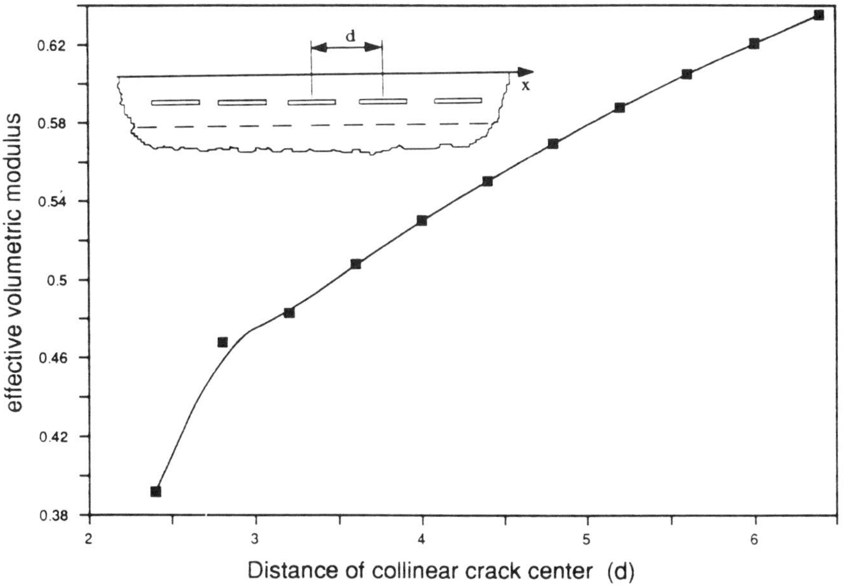

Figure 9.15. Variations in effective bulk modulus for an array of lateral cracks.

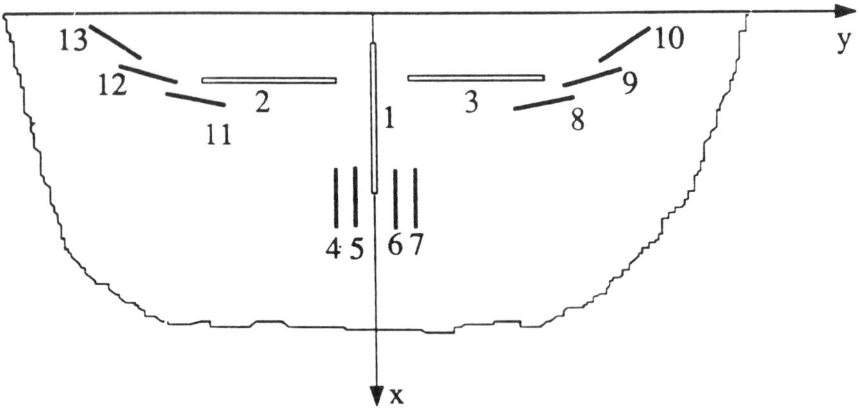

Figure 9.16. Schematic diagram of crack interactions induced by grinding, with chosen microcrack distributions.

reduced to only about 40 percent of that of the virgin material by introducing arrays of microcracks in the control layer near the free surface.

Based on the above observations, many preexisting microcrack distributions may be constructed to retard the growth of radial cracks while enhancing the growth of lateral cracks. The microcrack distribution may also be used to facilitate curving of the lateral cracks toward the free surface, thus enhancing the material removal rate. An example of such a microcrack distribution is shown in figure 9.16. It consists of an array of radial cracks at about the same buried depth as the length of the radial crack and another set of cracks with varying orientation. Crack 1 is the main radial crack and cracks 2 and 3 are the principal lateral cracks due to grinding. Cracks 4 through 7 consist of the radial array of microcracks, and cracks 8 through 10 and 11 through 13 are arranged at varying orientations. Details of the cracks are given in table 9.1. Significant shielding is observed at the lower tip of the principal radial crack. The mode I SIF at the lower tip of crack 1 is observed to be only 0.534. This is due to the presence of the radial array—the radial crack is arrested as it attempts to propagate through the array of radial microcracks. The mode I SIFs at the lower tips of the radial microcracks are observed to be around 1.3. Cracks 8, 9, and 10, on the other hand, significantly amplify the mode I SIF for the lateral crack to 2.89. The mode I SIFs for cracks 8, 9, and 10 are also very high (7.0, 5.6, and 2.8, respectively). Accordingly, these microcracks will tend to propagate and coalesce with the lateral crack, thus enhancing effective material removal. As observed in figures 9.14 and 9.15, it is also important to note that the microcrack distribution shown in figure 9.16 significantly reduces the elastic moduli of the top layer of the workpiece. This will also facilitate grinding with softer abrasives.

Table 9.1. Geometrical and stress intensity factors (SIFs) for crack interactions given in figure 9.16.

Crack number	Coordinates (center of crack)		Angle with free surface (degrees)	Length of crack	Mode I SIFs		Mode II SIFs	
	X	Y			Lower right hand	Upper left hand	Lower right hand	Upper left hand
1	1.4	0.0	90.0	2.0	0.534	4.320	0.000	0.000
3 (2)	0.5	1.2	0.0	2.0	2.823	2.892	−0.123	0.236
6 (5)	2.4	0.1	90.0	0.4	1.267	0.582	−0.001	−0.535
7 (4)	2.4	0.2	90.0	0.4	1.312	0.281	−0.506	−1.180
8 (11)	0.6	2.2	5.0	0.4	4.349	7.001	1.948	−3.786
9 (12)	0.5	2.5	10.0	0.4	5.876	5.614	−0.505	−1.953
10 (13)	0.3	2.8	15.0	0.4	4.461	2.792	−1.428	0.350

An indentation fracture mechanics approach is pursued in this section to gain insight into the evolution of surface and subsurface damage during grinding of monolithic ceramic materials. An integral equation approach based on the fundamental solutions due to point dislocations in an elastic half-plane is used to model the interactions among radial and lateral crack systems near a free surface.

The present study reveals several interesting situations of stress amplification and stress shielding for the radial and lateral cracks typically observed during ceramic grinding. It is also observed that the effective bulk modulus of the workpiece layer near the free surface can be reduced substantially by introducing appropriate distributions of microcracks. Based on these basic studies of simple radial and lateral crack systems, one can design a preexisting microcrack distribution in the workpiece to arrest the radial cracks while facilitating the propagation of lateral cracks and curving them upward toward the free surface. The same microcrack distribution may also be customized to significantly reduce the effective elastic properties of a thin layer (thickness of the order of depth of cut in grinding) near the free surface of the workpiece. This will allow grinding of hard (in virgin state) ceramic materials with relatively softer abrasives.

The success of such a grinding process, where microcrack distributions are intentionally introduced, depends, of course, on the ability to remove all such microcracks with the chip. Enlarging the plastic zone will facilitate nucleation of the lateral cracks at a greater buried depth and help in removing the microcracks introduced previously.

9.5 Grinding of Ceramic Composites

In recent years, various ceramic compounds and intermetallics, by virtue of possessing ordered superlatice structures and high activation energies for diffusion, have been found to be ideally suited for high-temperature strength and modulus retention. However, these ceramic compounds and intermatallics are generally very brittle at room temperature. In order to enhance the ductility at room temperature and to lower the brittle–ductile transition temperatures, various researchers have attempted to change the crystal structures using alloys. The alloying agents, however, introduce second-phase particles into the matrix. The networks of microdefects (voids or inclusions) introduced through alloying can significantly affect the final strength and characteristics of the products. Many intermetallics (e.g., Tial–Tizal) contain a lamellar microstructure, and the harder phase may be modeled as a rigid line compared to the softer phase. The rigid line approximation is also valid for representing chopped-fiber distributions in ceramic composites. In many material processing and structural applications, the interactions among these micro*defects (voids or inclusions) have been observed to be one of the important factors in determining the macro failure modes (e.g., Kachanov 1985, Horii and Nemat-Nasser 1986, Baeslack et al. 1988, Haritos et al. 1988, Jha et al. 1988).

For ceramics and intermetallics, this network of microdefects and their associated brittle fracture modes provides crucial avenues for material removal processes. As shown by Hu and Chandra (1993a), the apparent moduli of an ultrahard ceramic material may be reduced to less than 40 percent of the original values by inducing predetermined distributions of microcracks. This ability to control apparent moduli has very important ramifications for various material removal processes such as grinding and drilling. For grinding of ceramics, the surface and subsurface damage evolution is also significantly influenced by the interactions among cracks generated due to grinding and the preexisting microcracks (Hu and Chandra 1993a). For such material removal processes, the interactions occur near a free surface, and it has been observed that the free surface greatly modifies the nature of these interactions.

Two kinds of microdefects are typically observed in a material. The first is a microvoid, which can also be idealized as a microcrack. A crack is essentially a cut that transmits no traction across it but allows a displacement discontinuity. The other type of microdefect is a rigid inclusion. A rigid lamella may be modeled as the opposite of a crack, that is, it transmits tractions but prevents displacement discontinuity. Moreover, a rigid lamella can undergo only rigid-body motions and no deformations. Dundurs and Markenscoff (1989) also refer to these rigid lines as anticracks. In addition to representing harder inclusions, rigid lines may also be used to model the fibers or whiskers in a metal–matrix or ceramic–matrix composite material. Many advanced materials for cutting tools belong to this category, and interactions among rigid lines and microcracks near a free surface must be considered in order to model damage evolution in such applications. Many structural composite components may also require surface finishing, and interactions among rigid lines and cracks near a free surface need to be understood thoroughly in order to comprehend the extent of damage induced by such operations. Accordingly, the investigation of interactions among general systems of cracks and rigid lines near a free surface is the main thrust of this paper. An integral equation approach based on the fundamental fields due to point loads and point dislocations near a free surface is utilized for this purpose. A Gauss–Chebychev quadrature is used to reduce the integral equations to a linear system of equations consisting of the distribution of forces and Burger's dislocation vectors on the rigid lines and cracks, respectively. The proposed solution procedure also allows direct determination of the rigid-body rotations for the rigid lines.

This section starts with a review of the literature relevant to problems involving cracks and rigid lines, followed by a presentation of fundamental solutions due to point loads and point dislocations near a free surface. An integral equation formulation for general systems of cracks and rigid lines near a free surface is developed next. The numerical results obtained by the proposed technique are first verified against existing solutions for a single crack or a rigid line near a free surface. General systems of cracks and rigid lines in the vicinity of a free surface subject to general loading conditions are considered next, and the proposed technique is used to investigate the amplification and shielding effects in such systems. Implications of such interactions, or damage evolutions, during ceramic grinding are also discussed. It is also interesting to note that, for a given configuration, the interactions between two cracks and between a crack and a rigid line bound the range of all possible interaction effects.

The problems involving rigid lines, or anticracks, have also been widely investigated in the context of elasticity. An extensive review of elastic problems involving inclusions is given by Mura (1987, 1988). Initial work on rigid-line inclusions can be traced back to Muskhelishvili (1953) and Eshelby (1957, 1959). Since then, problems involving rigid-line inclusions have been investigated by several researchers (e.g., Atkinson 1973, Brussat and Westmann 1975, Hasebe et al. 1984) for isotropic elastic bodies. Chou and Wang (1983) and Wang et al. (1985, 1986) considered a rigid-line inclusion in an isotropic planar elastic body and derived analytical expressions for the stress field due to uniform remote loading. The same problem has also been considered by Ballarini (1987), using an integral transform method. Sendeckyj (1970) and Selvadurai (1980) investigated elastic-line inclusions. Atkinson (1973) also considered elastic-line inclusions, and a solution for a half-plane containing an anticrack perpendicular to the free surface is given. He obtained an asymptotic solution for the stress fields under the assumption that the inclusion is much harder than the matrix. Erdogan and Gupta (1972) studied the more general problem of bonded materials containing a flat inclusion that may be rigid or elastic with negligible bending rigidity. They formulated the problem as a system of singular integral equations that is solved by expanding the solutions in Chebychev polynomials. Dundurs and Markenscoff (1989) provide a direct Green's function formulation for rigid lines. Such a formulation is very suitable for solution by currently available numerical methods. They also provide pertinent fields for concentrated forces, dislocations, and couples applied on the line of the anticrack. Li and Ting (1989) investigated line inclusions embedded in an anisotropic infinite elastic medium subject to uniform remote loading. Stroh's (1958, 1962) formalism is used to obtain the displacement and stress fields. They consider both rigid and elastic inclusions. A pair of singular Fredholm integral equations of the second kind is derived for the difference in the stress on both surfaces of an elastic and anisotropic inclusion. If the relative rigidity of the matrix is small compared to that of the inclusion, Li and Ting (1989) showed that the governing equations can be decoupled. For such rigid-line cases, they also obtain asymptotic solutions for the traction and the rotation of the rigid line.

Most of the existing analyses including rigid lines, however, consider a single rigid-line inclusion. On the other hand, problems involving interacting cracks have been studied more extensively. Goldstein and Salganik (1974) studied brittle fracture of solids with arbitrary cracks. Utilizing the superposition technique and the ideas of self-consistency applied to the average tractions on individual cracks, Kachanov (1985, 1987) obtained approximate analytical solutions for the stress intensity factor due to interacting elastic cracks. Melin (1983) and Broberg (1987) investigated the directional stability of cracks using linear elastic fracture mechanics and observed that the interactions among cracks significantly influenced the directional stability and crack branching characteristics. Hori and Nemat-Nasser (1987) also used the method of Muskhelishvili to obtain a two-dimensional elasticity solution for interacting microcracks near the tip of a microcrack. Using a dislocation approach, various researchers have also reduced the problem of a linear elastic solid with interacting cracks to a system of integral equations (e.g., Chatterjee 1975, Erodogan and Gupta 1975, Lo 1978, Cotterell and Rice 1980, Melin 1986, Chudnovsky et al. 1987, Müller 1989).

The problem of a half-plane containing a crack has been studied by several researchers (e.g., Ioakimidis and Theocaris 1979, Keer et al. 1982, Keer and Bryant 1983, Chen 1984, Nowell and Hills 1987, Li and Hills 1990). Recently, Hu and Chandra (1993a) used an integral equation approach to investigate the interactions among cracks near a free surface and their implications for surface degradation during grinding of ceramics. A force-type integral equation, where the kernels contain only weak logarithmic singularities and the unknowns are Burgers' vectors on the crack faces, was developed by Cheung and Chen (1987) for a full-plane problem containing cracks. Zang and Gudmundson (1989, 1991) extended the approach to analyze half-plane and anisotropic problems. Brussat and Westmann (1975) investigated the interactions among collinear rigid-line inclusions subject to uniform remote loading. For this special case, they establish a direct correspondence between the Westergaard stress function for elastic crack problems and a stress function for rigid-line inclusion problems. A correspondence is also shown to exist between crack opening displacements and axial forces on the rigid inclusions.

9.5.1 Fundamental Fields due to Point Loads and Point Dislocations

An integral equation approach is pursued here to investigate the interactions among general systems of cracks and rigid lines near a free surface subject to general loading conditions. Cracks and rigid lines can be modeled as continuous spatial distributions of dislocations and tractions. Accordingly, the corresponding fundamental solutions due to a point load and a point dislocation are discussed in this section. Two-dimensional problems are considered here.

For a two-dimensional elastic half-plane ($x \geq 0$) subjected to a concentrated force with components p_x and p_y at the location $(\xi, 0)$, the solutions due to Dundurs and Hetenyi (1961) and Hetenyi and Dundurs (1962) at a point (x, y) may be written as

$$2Gu_{x,x} = \frac{1}{2\pi(\kappa+1)}[p_x H_1(x, y; \xi) + p_y H_2(x, y; \xi)] \tag{9.36a}$$

$$2Gu_{x,y} = \frac{1}{2\pi(\kappa+1)}[p_x H_3(x, y; \xi) + p_y H_4(x, y; \xi)] \tag{9.36b}$$

$$2Gu_{y,x} = \frac{1}{2\pi(\kappa+1)}[p_x H_5(x, y; \xi) + p_y H_6(x, y; \xi)] \tag{9.36c}$$

$$2Gu_{y,y} = \frac{1}{2\pi(\kappa+1)}[p_x H_7(x, y; \xi) + p_y H_8(x, y; \xi)] \tag{9.36d}$$

$$\sigma_{x,x} = \frac{1}{2\pi(\kappa+1)}[p_x H_9(x, y; \xi) + p_y H_{10}(x, y; \xi)] \tag{9.36e}$$

$$\sigma_{x,y} = \frac{1}{2\pi(\kappa+1)}[p_x H_{11}(x, y; \xi) + p_y H_{12}(x, y; \xi)] \tag{9.36f}$$

$$\sigma_{y,y} = \frac{1}{2\pi(\kappa+1)}[p_x H_{13}(x, y; \xi) + p_y H_{14}(x, y; \xi)] \tag{9.36g}$$

where G is the shear modulus. The constant $\kappa = (3 - 4\nu)$ for plane strain and $\kappa = (3 - \nu)/(1 + \nu)$ for plane stress, where ν is the Poisson's ratio. The details of the kernels H_1 through H_{14} may be obtained from Dundurs and Hetenyi (1961) and Hetenyi and Dundurs (1962) using MATHEMATICA (Wolfram 1991) and are expressed (Hu and Chandra 1993b) as

$$H_1(x, y; \xi) = 2(2-\kappa)\left(\frac{x_1}{r_1^2} + \frac{x_2}{r_2^2}\right) - 4\left(\frac{x_1^3}{r_1^4} + \frac{x_2^3}{r_2^4}\right)$$
$$- 4\xi\left(\frac{1}{r_2^2} - \frac{8x_2^2}{r_2^4} + \frac{8x_2^4}{r_2^6}\right) - 8\xi^2\left(\frac{3x_2}{r_2^4} - \frac{4x_2^3}{r_2^6}\right) \tag{9.37a}$$

$$H_2(x,y;\xi) = y\left[\frac{2}{r_1^2} - \frac{4x_1^2}{r_1^4} + \frac{\kappa^2+2\kappa-1}{r_2^2} - \frac{4\kappa x_2^2}{r_2^4}\right.$$
$$\left. + 8\xi\left(-\frac{(2-\kappa)x_2}{r_2^4} + \frac{4x_2^3}{r_2^6}\right) + 8\xi^2\left(\frac{1}{r_2^4} - \frac{4x_2^2}{r_2^6}\right)\right] \tag{9.37b}$$

$$H_3(x,y;\xi) = y\left[-\frac{2\kappa}{r_1^2} - \frac{4x_1^2}{r_1^4} - \frac{\kappa^2+1}{r_2^2} - \frac{4\kappa x_2^2}{r_2^4}\right.$$
$$\left. - 8\xi\left(\frac{x_2}{r_2^4} - 4\frac{x_2^3}{r_2^6}\right) - 8\xi^2\left(\frac{1}{r_2^4} - \frac{4x_2^2}{r_2^6}\right)\right] \tag{9.37c}$$

$$H_4(x,y;\xi) = \frac{2x_1}{r_1^2} - \frac{4x_1y^2}{r_1^4} + \frac{(\kappa^2+2\kappa-1)x_2}{r_2^2} - \frac{4\kappa x_2 y^2}{r_2^4}$$
$$- 4\xi\left(\frac{\kappa}{r_2^2} - \frac{2x_2^2}{r_2^4} - \frac{2\kappa y^2}{r_2^4} - \frac{8x_2^2y^2}{r_2^6}\right) + 8\xi^2\left(\frac{x_2}{r_2^4} - \frac{4x_2y^2}{r_2^6}\right) \tag{9.37d}$$

$$H_5(x,y;\xi) = y\left[2\kappa\left(\frac{1}{r_1^2} + \frac{1}{r_2^2}\right) - 4\kappa\left(\frac{x_1^2}{r_1^4} + \frac{x_2^2}{r_2^4}\right) + \frac{\kappa^2-1}{r_2^2}\right.$$
$$\left. + 4\xi\left(\frac{2(2+\kappa)x_2}{r_2^4} - \frac{8x_2^3}{r_2^6}\right) - 8\xi^2\left(\frac{1}{r_2^4} - \frac{4x_2^2}{r_2^6}\right)\right] \tag{9.37e}$$

$$H_6(x,y;\xi) = -\frac{4\kappa x_1^3}{r_1^4} + \frac{(3-\kappa^2)x_2}{r_2^2} - \frac{4x_2^3}{r_2^4}$$
$$- 4\xi\left(\frac{1}{r_2^2} - \frac{8x_2^2}{r_2^4} + \frac{8x_2^4}{r_2^6}\right) + 8\xi^2\left(-\frac{3x_2}{r_2^4} + \frac{4x_2^3}{r_2^6}\right) \tag{9.37f}$$

$$H_7(x,y;\xi) = \frac{2x_1}{r_1^2} - \frac{4x_1y^2}{r_1^4} + \frac{2\kappa x_2}{r_2^2} - \frac{4\kappa x_2 y^2}{r_2^4} - \frac{(\kappa-1)x_2}{r_2^2}$$
$$+ 4\xi\left(-\frac{\kappa}{r_2^2} + \frac{2\kappa y^2}{r_2^4} + \frac{2x_2^2}{r_2^4} - \frac{8x_2^2y^2}{r_2^6}\right) - 8\xi^2\left(\frac{x_2}{r_2^4} - \frac{4x_2y^2}{r_2^6}\right) \tag{9.37g}$$

$$H_8(x,y;\xi) = y\left[-\frac{2\kappa}{r_1^2} - \frac{2\kappa x_1}{r_1^4} - \frac{\kappa^2+1}{r_2^2} - \frac{4x_2^2}{r_2^4}\right.$$
$$\left. - 4\xi\left(-\frac{2x_2}{r_2^4} + \frac{8x_2^3}{r_2^6}\right) + 8\xi^2\left(-\frac{1}{r_2^4} + \frac{4x_2^2}{r_2^6}\right)\right] \tag{9.37h}$$

$$H_9(x,y;\xi) = -\frac{(\kappa-1)x_1}{r_1^2} - \frac{2(\kappa+3)x_1^3}{r_1^4} + \frac{(\kappa-1)x_2}{r_2^2} - \frac{2(3\kappa+1)x_2^3}{r_2^4} + 2\xi\left(-\frac{\kappa-1}{r_1^2} + \frac{2(\kappa+5)x_2^2}{r_2^4} - \frac{16x_2^4}{r_2^6}\right) - 8\xi^2\left(\frac{3x_2}{r_2^4} - \frac{4x_2^3}{r_2^6}\right) \tag{9.37i}$$

$$H_{10}(x,y;\xi) = y\left[\frac{\kappa-1}{r_1^2} - \frac{4x_1^2}{r_1^4} - \frac{\kappa-1}{r_2^2} - \frac{4\kappa x_2^2}{r_2^4} - 4\xi\left(-\frac{(\kappa-1)x_2}{r_2^4} - \frac{16x_2^3}{r_2^6}\right) + 8\xi^2\left(\frac{2}{r_2^4} - \frac{4x_2^2}{r_2^6}\right)\right] \tag{9.37j}$$

$$H_{11}(x,y;\xi) = y\left[-\frac{\kappa-1}{r_1^2} - \frac{4x_1^2}{r_1^4} + \frac{\kappa-1}{r_2^2} - \frac{4\kappa x_2^2}{r_2^4} + 4\xi\left(\frac{(3+\kappa)x_2}{r_2^4} - \frac{8x_2^3}{r_2^6}\right) - 8\xi^2\left(\frac{1}{r_2^4} - \frac{4x_2^2}{r_2^6}\right)\right] \tag{9.37k}$$

$$H_{12}(x,y;\xi) = -\frac{(\kappa+3)x_1}{r_1^2} + \frac{4x_1^3}{r_1^4} - \frac{(3\kappa+1)x_2}{r_2^2} + \frac{4\kappa x_2^3}{r_2^4} - 2\xi\left(-\frac{\kappa-1}{r_1^2} + \frac{2(\kappa-7)x_2^2}{r_2^4} + \frac{16x_2^4}{r_2^6}\right) + 8\xi^2\left(-\frac{3x_2}{r_2^4} + \frac{4x_2^3}{r_2^6}\right) \tag{9.37l}$$

$$H_{13}(x,y;\xi) = \frac{(\kappa-5)x_1}{r_1^2} + \frac{4x_1^3}{r_1^4} - \frac{(5\kappa-1)x_2}{r_2^2} + \frac{4\kappa x_2^3}{r_2^4} + 2\xi\left(\frac{\kappa+3}{r_2^2} - \frac{2(\kappa+9)x_2^2}{r_2^4} + \frac{16x_2^4}{r_2^6}\right) - 8\xi^2\left(-\frac{3x_2}{r_2^4} + \frac{4x_2^3}{r_2^6}\right) \tag{9.37m}$$

$$H_{14}(x,y;\xi) = y\left[-\frac{\kappa+3}{r_1^2} + \frac{4x_1^2}{r_1^4} - \frac{3\kappa+1}{r_2^2} + \frac{4\kappa x_2^2}{r_2^4} - 4\xi\left(\frac{(\kappa-5)x_2}{r_2^2} + \frac{8x_2^3}{r_2^6}\right) + 8\xi^2\left(-\frac{1}{r_2^4} + \frac{4x_2^2}{r_2^6}\right)\right] \tag{9.37n}$$

The fundamental solutions at (x, y) for a two-dimensional half-plane subjected to an edge dislocation with Burger's vector components b_x and b_y acting at a point $(\xi, 0)$ may also be expressed (Dundurs and Mura 1964) as

$$u_{x,x} = \frac{1}{2\pi(\kappa+1)} [b_x I_1(x, y; \xi) + b_y I_2(x, y; \xi)] \tag{9.38a}$$

$$u_{x,y} = \frac{1}{2\pi(\kappa+1)} [b_x I_3(x, y; \xi) + b_y I_4(x, y; \xi)] \tag{9.38b}$$

$$u_{y,x} = \frac{1}{2\pi(\kappa+1)} [b_x I_5(x, y; \xi) + b_y I_6(x, y; \xi)] \tag{9.38c}$$

$$u_{y,y} = \frac{1}{2\pi(\kappa+1)} [b_x I_7(x, y; \xi) + b_y I_8(x, y; \xi)] \tag{9.38d}$$

$$\sigma_{xx} = \frac{G}{\pi(\kappa+1)} [b_x I_9(x, y; \xi) + b_y I_{10}(x, y; \xi)] \tag{9.38e}$$

$$\sigma_{xy} = \frac{G}{\pi(\kappa+1)} [b_x I_{11}(x, y; \xi) + b_y I_{12}(x, y; \xi)] \tag{9.38f}$$

$$\sigma_{yy} = \frac{G}{\pi(\kappa+1)} [b_x I_{13}(x, y; \xi) + b_y I_{14}(x, y; \xi)] \tag{9.38g}$$

The details of the kernels I_1 through I_{14} may be expressed (Hu and Chandra 1993b) as

$$I_1(x, y; \xi) = y\Bigg[-\frac{\kappa-1}{r_1^2} - \frac{4x_1^2}{r_1^4} + \frac{\kappa-1}{r_2^2} + \frac{4x_2^2}{r_2^4} - 4\xi\left(\frac{(\kappa+3)x_2}{r_2^4} - \frac{8x_2^3}{r_2^6}\right) + 8\xi^2\left(\frac{2}{r_2^4} - \frac{4x_2^2}{r_2^6}\right)\Bigg] \tag{9.39a}$$

$$I_2(x, y; \xi) = \frac{(\kappa-5)x_1}{r_1^2} + \frac{4x_1^3}{r_1^4} - \frac{(\kappa-5)x_2}{r_2^2} - \frac{4x_2^3}{r_2^4} - 2\xi\left(\frac{\kappa-1}{r_2^2} + \frac{2(\kappa-7)x_2^2}{r_2^4} - \frac{16x_2^4}{r_2^6}\right) + 8\xi^2\left(\frac{3x_2}{r_2^4} - \frac{4x_2^3}{r_2^6}\right) \tag{9.39b}$$

$$I_3(x, y; \xi) = \frac{(\kappa+3)x_1}{r_1^2} - \frac{4x_1y^2}{r_1^4} - \frac{2(\kappa+2)x_2}{r_2^2} + \frac{4x_2y^2}{r_2^4}$$
$$- 2\xi\left(\frac{-\kappa+1}{r_2^2} + \frac{4x_2^2}{r_2^4} - \frac{2(\kappa-1)y^2}{r_2^4} - \frac{16x_2^2y^2}{r_2^6}\right)$$
$$+ 8\xi^2\left(\frac{x_2}{r_2^4} - \frac{4x_2y^2}{r_2^6}\right) \tag{9.39c}$$

$$I_4(x, y; \xi) = y\left[\frac{\kappa-1}{r_1^2} + \frac{4x_1^2}{r_1^4} - \frac{\kappa-1}{r_2^2} - \frac{4x_2^2}{r_2^4}\right.$$
$$\left. -4\xi\left(-\frac{(\kappa-1)x_2}{r_2^4} - \frac{8x_2^3}{r_2^6}\right) + 8\xi^2\left(\frac{1}{r_2^4} - \frac{4x_2^2}{r_2^6}\right)\right] \tag{9.39d}$$

$$I_5(x, y; \xi) = -\frac{(\kappa+3)x_1}{r_1^2} + \frac{4x_1^3}{r_1^4} + \frac{(\kappa+3)x_2}{r_2^2} - \frac{4x_2^3}{r_2^4}$$
$$- 2\xi\left(\frac{\kappa+3}{r_2^2} - \frac{2(\kappa+9)x_2^2}{r_2^4} + \frac{14x_2^2}{r_2^6}\right)$$
$$+ 8\xi^2\left(-\frac{3x_2}{r_2^4} + \frac{4x_2^3}{r_2^6}\right) \tag{9.39e}$$

$$I_6(x, y; \xi) = y\left[-\frac{\kappa+3}{r_1^2} + \frac{4x_1^2}{r_1^4} + \frac{\kappa+3}{r_2^2} - \frac{4x_2^2}{r_2^4}\right.$$
$$\left. -4\xi\left(\frac{(\kappa+3)x_2}{r_2^4} - \frac{8x_2^3}{r_2^6}\right) + 8\xi^2\left(\frac{1}{r_2^4} - \frac{4x_2^2}{r_2^6}\right)\right] \tag{9.39f}$$

$$I_7(x, y; \xi) = y\left[-\frac{\kappa-1}{r_1^2} + \frac{4x_1^2}{r_1^4} + \frac{\kappa-1}{r_2^2} - \frac{4x_2^2}{r_2^4}\right.$$
$$\left. -4\xi\left(\frac{(\kappa+3)x_2}{r_2^4} - \frac{8x_2^3}{r_2^6}\right) + 8\xi^2\left(-\frac{1}{r_2^4} + \frac{4x_2^2}{r_2^6}\right)\right] \tag{9.39g}$$

$$I_8(x, y; \xi) = \frac{(\kappa-1)x_1}{r_1^2} + \frac{4x_1y^2}{r_1^4} - \frac{(\kappa-1)x_2}{r_2^2} - \frac{4x_2y^2}{r_2^4}$$
$$- 2\xi\left(-\frac{\kappa-1}{r_2^2} + \frac{4x_2^2}{r_2^4} + \frac{2(\kappa-1)y^2}{r_2^4} - \frac{16x_2^2y^2}{r_2^6}\right)$$
$$+ 8\xi^2\left(\frac{x_2}{r_2^4} - \frac{4x_2y^2}{r_2^6}\right) \tag{9.39h}$$

$$I_9(x, y; \xi) = y\left[6\left(-\frac{1}{r_1^2} + \frac{1}{r_2^2}\right) + 4y^2\left(\frac{1}{r_1^4} - \frac{1}{r_2^4}\right)\right.$$
$$\left. - 8\xi\left(-\frac{3x_2}{r_2^4} + \frac{4x_2y^2}{r_2^6}\right) - 8\xi^2\left(\frac{3}{r_2^4} - \frac{4y^2}{r_2^6}\right)\right] \tag{9.39i}$$

$$I_{10}(x, y; \xi) = 2\left(\frac{x_1}{r_1^2} - \frac{x_2}{r_2^2}\right) - 4y^2\left(\frac{x_1}{r_1^4} + \frac{x_2}{r_2^4}\right)$$
$$+ 4\xi\left(\frac{1}{r_2^2} + \frac{2x_2^2}{r_2^4} - \frac{2y^2}{r_2^4} - \frac{8x_2^2y^2}{r_2^6}\right)$$
$$- 8\xi^2\left(\frac{x_2}{r_2^4} - \frac{4x_2y^2}{r_2^6}\right) \tag{9.39j}$$

$$I_{11}(x, y; \xi) = 2\left(\frac{x_1}{r_1^2} - \frac{x_2}{r_2^2}\right) - 4y^2\left(\frac{x_1}{r_1^4} + \frac{x_2}{r_2^4}\right)$$
$$+ 4\xi\left(-\frac{1}{r_2^2} + \frac{8x_2^2y^2}{r_2^6}\right) + 8\xi^2\left(\frac{x_2}{r_2^4} - \frac{4x_2y^2}{r_2^6}\right) \tag{9.39k}$$

$$I_{12}(x, y; \xi) = y\left[2\left(-\frac{1}{r_1^2} + \frac{1}{r_2^2}\right) + 4\left(\frac{x_1^2}{r_1^4} - \frac{x_2^2}{r_2^4}\right)\right.$$
$$\left. - 8\xi\left(\frac{x_2}{r_2^4} - \frac{4x_2^3}{r_2^6}\right) - 8\xi^2\left(-\frac{1}{r_2^4} + \frac{4x_2^2}{r_2^6}\right)\right] \tag{9.39l}$$

$$I_{13}(x, y; \xi) = y\left[2\left(-\frac{1}{r_1^2} + \frac{1}{r_2^2}\right) + 4\left(\frac{x_1^2}{r_1^4} - \frac{x_2^2}{r_2^4}\right)\right.$$
$$\left. - 8\xi\left(-\frac{3x_2}{r_2^4} + \frac{4x_2^3}{r_2^6}\right) - 8\xi^2\left(\frac{1}{r_2^4} - \frac{4x_2^2}{r_2^6}\right)\right] \tag{9.39m}$$

$$I_{14}(x, y; \xi) = 6\left(\frac{x_1}{r_1^2} - \frac{x_2}{r_2^2}\right) + 4\left(-\frac{x_1^3}{r_1^4} + \frac{x_2^3}{r_2^4}\right)$$
$$+ 4\xi\left(-\frac{1}{r_2^2} + \frac{8x_2^2}{r_2^4} - \frac{8x_2^4}{r_2^6}\right) + 8\xi^2\left(-\frac{3x_2}{r_2^4} + \frac{4x_2^3}{r_2^6}\right) \tag{9.39n}$$

where

$$r_1^2 = x_1^2 + y^2 \quad \text{and} \quad r_2^2 = x_2^2 + y^2 \tag{9.39o}$$

$$x_1 = x - \xi \quad \text{and} \quad x_2 = x + \xi \tag{9.39p}$$

It is important to notice here that the displacement gradient fields and the stress fields in equations (9.36)–(9.39) contain a singularity of order r as the distance r between the field point (x, y) and the source point $(\xi, 0)$ approaches zero. This order of singularity is the same as that for the full plane. The singular behavior for the point load case in a full plane has been discussed widely, particularly in the boundary integral equation literature (e.g., Banerjee and Butterfield 1981, Mukherjee 1982, Cruse 1988). Recently, Dundurs and Markenscoff (1989) also examined the singular behavior for a point dislocation in a full plane. A similar procedure is followed in the present work to represent the singular integrals for the half-plane in a Cauchy principal value sense.

9.5.2 An Integral Equation Formulation for General Crack–Anticrack Systems

It is assumed here that the crack, like any other void, cannot transmit any traction. Accordingly, the crack surface is required to be traction free. For the purpose of this work, any crack closure is neglected, and it is assumed that the crack remains open throughout the application of the external load. The anticrack, like any other rigid inclusion, can only admit rigid-body motions. A perfect bonding between the rigid line and the matrix is assumed here.

A general system containing M cracks and N anticracks embedded in arbitrary orientations in an elastic half-plane is considered here. We concentrate on the ith defect (crack or anticrack) and consider a local tangential–normal coordinate system with origin at the center of the defect, the normal direction denoted by $s^{(i)}$, and the tangential direction along the ith crack or anticrack denoted by $t^{(i)}$. The coordinates of the center of the ith defect are denoted by $(x_c^{(i)}, y_c^{(i)})$ and the angle between x and $t^{(i)}$ is denoted by $\theta^{(i)}$. This is shown schematically in figure 9.17. The occupancy of the ith defect is denoted as $-a^{(i)} < t^{(i)} < a^{(i)}$. The solution of such a system must meet the following requirements: (1) The stress field must satisfy equilibrium. (2) The free surface must be traction free. (3) Crack surfaces must be traction free. (4) Rigid lines, or anticracks, should undergo only rigid-body motions. These rigid-body motions should not cause any material separations (assuming perfect bonding).

Under the above assumptions, the boundary conditions for the cracks require that $(i = 1, 2, \ldots, M)$

$$\sigma_{ss}^{(i)} = \sigma_{ts}^{(i)} = 0 \qquad \text{for} \qquad -a^{(i)} < t^{(i)} < a^{(i)} \tag{9.40}$$

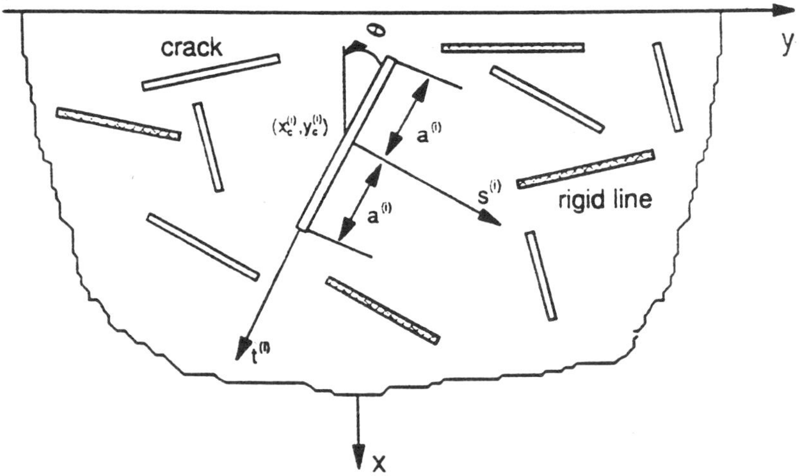

Figure 9.17. Schematic diagram of cracks and rigid lines near a free surface.

The boundary conditions for the rigid lines may also be expressed as ($j = 1, 2, \ldots, N$)

$$u_t^{(j)} = \chi^{(j)} - \omega^{(j)} s^{(j)}$$

$$u_s^{(j)} = \beta^{(j)} + \omega^{(j)} t^{(j)} \qquad \text{for} \qquad -a^{(j)} < t^{(j)} < a^{(j)} \tag{9.41}$$

where $\chi^{(j)}$ and $\beta^{(j)}$ are the rigid-body translations in the $t^{(j)}$ and $s^{(j)}$ directions, respectively, and $\omega^{(j)}$ is the rigid-body rotation in the $t^{(j)}$–$s^{(j)}$ plane.

The rigid-body displacements may be differentiated with respect to $t^{(j)}$, yielding the boundary conditions ($j = 1, 2, \ldots, N$)

$$\frac{\partial u_t^{(j)}}{\partial t^{(j)}} = 0 \qquad \text{and} \qquad \frac{\partial u_s^{(j)}}{\partial t^{(j)}} = \omega^{(j)} \qquad \text{for} \qquad -a^{(j)} < t^{(j)} < a^{(j)} \tag{9.42}$$

for anticracks, or rigid lines.

Let us now consider the effects of all cracks and anticracks on the mth crack. For consistency, the stress fields associated with different cracks and anticracks are transformed to the local tangential–normal ($t^{(m)}, s^{(m)}$) coordinate system for the mth crack. For example, the stress fields associated with the ith crack may be transformed as

$$\sigma_{ss}^{(m)} = \sigma_{tt}^{(i)} \sin^2 \theta^{(mi)} - 2\sigma_{ts}^{(i)} \sin \theta^{(mi)} \cos \theta^{(mi)} + \sigma_{ss}^{(i)} \cos^2 \theta^{(mi)} \tag{9.43a}$$

and

$$\sigma_{ts}^{(m)} = (\sigma_{ss}^{(i)} - \sigma_{tt}^{(i)}) \sin \theta^{(mi)} \cos \theta^{(mi)} + \sigma_{ts}^{(i)}(\cos^2 \theta^{(mi)} - \sin^2 \theta^{(mi)}) \tag{9.43b}$$

where $\theta^{(mi)}$ is the angle between axes $t^{(i)}$ and $t^{(m)}$.

The N anticracks are now represented by their corresponding distributed tractions and the M cracks are represented by their corresponding distributed dislocations. Summing the effects of all cracks and anticracks on the mth crack we get $(m = 1, 2, \ldots, M)$

$$\sum_{i=1}^{M} \int_{-a^{(i)}}^{a^{(i)}} [K_{11}t(t^{(m)}, t^{(i)}) b_t^{(i)}(t^{(i)}) + K_{12}(t^{(m)}, t^{(i)}) b_s^{(i)}(t^{(i)})]\, dt^{(i)}$$
$$+ \sum_{j=1}^{N} \int_{-a^{(j)}}^{a^{(j)}} [K_{13}(t^{(m)}, t^{(j)}) p_t^{(j)}(t^{(j)}) + K_{14}(t^{(m)}, t^{(j)}) p_s^{(j)}(t^{(j)})]\, dt^{(j)}$$
$$+ \sigma_{ss}^{(m0)}(t^{(m)}) = 0 \tag{9.44a}$$

and

$$\sum_{i=1}^{M} \int_{-a^{(i)}}^{a^{(i)}} [K_{21}t(t^{(m)}, t^{(i)}) b_t^{(i)}(t^{(i)}) + K_{22}(t^{(m)}, t^{(i)}) b_s^{(i)}(t^{(i)})]\, dt^{(i)}$$
$$+ \sum_{j=1}^{N} \int_{-a^{(j)}}^{a^{(j)}} [K_{23}(t^{(m)}, t^{(j)}) p_t^{(j)}(t^{(j)}) + K_{24}(t^{(m)}, t^{(j)}) p_s^{(j)}(t^{(j)})]\, dt^{(j)}$$
$$+ \sigma_{ts}^{(m0)}(t^{(m)}) = 0 \tag{9.44b}$$

where $b_t^{(i)}$, $b_s^{(i)}$, $p_t^{(j)}$, and $p_s^{(j)}$ represent the unknown dislocations and tractions on the ith crack and the jth rigid line, respectively; $\sigma_{ss}^{(m0)}$ and $\sigma_{ts}^{(m0)}$ are the stress components at the location of the mth crack but in the absence of all cracks and rigid lines. Here, the kernels K_{11} through K_{24} in equations (9.44) may be obtained by transforming the fundamental solution to an appropriate local coordinate system. The final results may be expressed as

$$K_{11}(t^{(m)}, t^{(i)}) = \frac{G}{\pi(\kappa+1)} \{\cos^2\theta^{(m)}[\sin\theta^{(i)} I_9(t^{(m)}, t^{(i)}) - \cos\theta^{(i)} I_{10}(t^{(m)}, t^{(i)})]$$
$$+ 2\sin\theta^{(m)}\cos\theta^{(m)}[\sin\theta^{(i)} I_{11}(t^{(m)}, t^{(i)}) - \cos\theta^{(i)} I_{12}(t^{(m)}, t^{(i)})]$$
$$+ \sin^2\theta^{(m)}[\sin\theta^{(i)} I_{13}(t^{(m)}, t^{(i)}) - \cos\theta^{(i)} I_{14}(t^{(m)}, t^{(i)})]\} \tag{9.45a}$$

$$K_{12}(t^{(m)}, t^{(i)}) = \frac{G}{\pi(\kappa+1)} \{\cos^2\theta^{(m)}[\cos\theta^{(i)} I_9(t^{(m)}, t^{(i)}) + \sin\theta^{(i)} I_{10}(t^{(m)}, t^{(i)})]$$
$$+ 2\sin\theta^{(m)}\cos\theta^{(m)}[\cos\theta^{(i)} I_{11}(t^{(m)}, t^{(i)}) + \sin\theta^{(i)} I_{12}(t^{(m)}, t^{(i)})]$$
$$+ \sin^2\theta^{(m)}[\cos\theta^{(i)} I_{13}(t^{(m)}, t^{(i)}) + \sin\theta^{(i)} I_{14}(t^{(m)}, t^{(i)})]\} \tag{9.45b}$$

$$K_{13}(t^{(m)}, t^{(j)}) = \frac{1}{2\pi(\kappa+1)} \{\cos^2\theta^{(m)}[\sin\theta^{(j)} H_9(t^{(m)}, t^{(j)})$$
$$- \cos\theta^{(j)} H_{10}(t^{(m)}, t^{(j)})] + 2\sin\theta^{(m)}\cos\theta^{(m)}[\sin\theta^{(j)} H_{11}(t^{(m)}, t^{(j)})$$
$$- \cos\theta^{(j)} H_{12}(t^{(m)}, t^{(j)})] + \sin^2\theta^{(m)}[\sin\theta^{(j)} H_{13}(t^{(m)}, t^{(j)})$$
$$- \cos\theta^{(j)} H_{14}(t^{(m)}, t^{(j)})]\} \tag{9.45c}$$

$$K_{14}(t^{(m)}, t^{(j)}) = \frac{1}{2\pi(\kappa+1)} \{\cos^2\theta^{(m)}[\cos\theta^{(j)} H_9(t^{(m)}, t^{(j)})$$
$$+ \sin\theta^{(j)} H_{10}(t^{(m)}, t^{(j)})] + 2\sin\theta^{(m)}\cos\theta^{(m)}[\cos\theta^{(j)} H_{11}(t^{(m)}, t^{(j)})$$
$$+ \sin\theta^{(j)} H_{12}(t^{(m)}, t^{(j)})] + \sin^2\theta^{(m)}[\cos\theta^{(j)} H_{13}(t^{(m)}, t^{(j)})$$
$$+ \sin\theta^{(j)} H_{14}(t^{(m)}, t^{(j)})]\} \tag{9.45d}$$

$$K_{21}(t^{(m)}, t^{(i)}) = \frac{G}{\pi(\kappa+1)} \{\sin\theta^{(m)}\cos\theta^{(m)}[\sin\theta^{(i)} I_9(t^{(m)}, t^{(i)})$$
$$- \cos\theta^{(i)} I_{10}(t^{(m)}, t^{(i)})] + [\sin^2\theta^{(m)} - \cos^2\theta^{(m)}][\sin\theta^{(i)} I_{11}(t^{(m)}, t^{(i)})$$
$$- \cos\theta^{(i)} I_{12}(t^{(m)}, t^{(i)})] + \sin\theta^{(m)}\cos\theta^{(m)}[\sin\theta^{(i)} I_{13}(t^{(m)}, t^{(i)})$$
$$- \cos\theta^{(i)} I_{14}(t^{(m)}, t^{(i)})]\} \tag{9.45e}$$

$$K_{22}(t^{(m)}, t^{(i)}) = \frac{G}{\pi(\kappa+1)} \{\sin\theta^{(m)}\cos\theta^{(m)}[\cos\theta^{(i)} I_9(t^{(m)}, t^{(i)})$$
$$+ \sin\theta^{(i)} I_{10}(t^{(m)}, t^{(i)})] + [\sin^2\theta^{(m)} - \cos^2\theta^{(m)}][\cos\theta^{(i)} I_{11}(t^{(m)}, t^{(i)})$$
$$+ \sin\theta^{(i)} I_{12}(t^{(m)}, t^{(i)})] - \sin\theta^{(m)}\cos\theta^{(m)}[\cos\theta^{(i)} I_{13}(t^{(m)}, t^{(i)})$$
$$+ \sin\theta^{(i)} I_{14}(t^{(m)}, t^{(i)})]\} \tag{9.45f}$$

$$K_{23}(t^{(m)}, t^{(j)}) = \frac{1}{2\pi(\kappa+1)} \{\sin\theta^{(m)}\cos\theta^{(m)}[\sin\theta^{(j)} H_9(t^{(m)}, t^{(j)})$$
$$- \cos\theta^{(j)} H_{10}(t^{(m)}, t^{(j)})] + [\sin^2\theta^{(m)} - \cos^2\theta^{(m)}][\sin\theta^{(j)} H_{11}(t^{(m)}, t^{(j)})$$
$$- \cos\theta^{(j)} H_{12}(t^{(m)}, t^{(j)})] - \sin\theta^{(m)}[\sin\theta^{(j)} H_{13}(t^{(m)}, t^{(j)})$$
$$- \cos\theta^{(j)} H_{14}(t^{(m)}, t^{(j)})]\} \tag{9.45g}$$

$$K_{24}(t^{(m)}, t^{(j)}) = \frac{1}{2\pi(\kappa+1)} \{\sin\theta^{(m)} \cos\theta^{(m)} [\cos\theta^{(j)} H_9(t^{(m)}, t^{(j)})$$
$$+ \sin\theta^{(j)} H_{10}(t^{(m)}, t^{(j)})] + [\sin^2\theta^{(m)} - \cos^2\theta^{(m)}] [\cos\theta^{(j)} H_{11}(t^{(m)}, t^{(j)})$$
$$+ \sin\theta^{(j)} H_{12}(t^{(m)}, t^{(j)})] - \sin\theta^{(m)} \cos\theta^{(m)} [\cos\theta^{(j)} H_{13}(t^{(m)}, t^{(j)})$$
$$+ \sin\theta^{(j)} H_{14}(t^{(m)}, t^{(j)})]\} \tag{9.45h}$$

where

$$I(t^{(m)}, t^{(i)}) \equiv I(x, y; \xi) \qquad \text{and} \qquad H(t^{(m)}, t^{(j)}) \equiv H(x, y; \xi) \tag{9.46}$$

with x, y, and ξ being substituted as ($k = i$ or j)

$$x = x_c^{(m)} + t^{(m)} \cos\theta^{(m)} \tag{9.47a}$$

$$y = t^{(k)} \sin\theta^{(k)} - t^{(m)} \sin\theta^{(m)} + y_c^{(m)} - y_c^{(k)} \tag{9.47b}$$

$$\xi = x_c^{(k)} + t^{(k)} \cos\theta^{(k)} \tag{9.47c}$$

The boundary conditions for anticracks, or rigid lines, are considered next. Utilizing the fact that the components of the displacement gradients transform as

$$u_{t,t}^{(m)} = u_{t,t}^{(j)} \cos^2\theta^{(mj)} + u_{t,s}^{(j)} \sin\theta^{(mj)} \cos\theta^{(mj)}$$
$$+ u_{s,t}^{(j)} \sin\theta^{(mj)} \cos\theta^{(mj)} + u_{s,s}^{(j)} \sin^2\theta^{(mj)} \tag{9.48a}$$

and

$$u_{s,t}^{(m)} = -u_{t,t}^{(j)} \sin\theta^{(mj)} \cos\theta^{(mj)} - u_{t,s}^{(j)} \sin^2\theta^{(mj)}$$
$$+ u_{s,t}^{(j)} \cos^2\theta^{(mj)} + u_{s,s}^{(j)} \sin\theta^{(mj)} \cos\theta^{(mj)} \tag{9.48b}$$

the effects of all cracks and rigid lines on the mth rigid line may be expressed as ($m = 1, 2, \ldots, N$)

$$\sum_{i=1}^{M} \int_{-a^{(i)}}^{a^{(i)}} [K_{31}t(t^{(m)}, t^{(i)}) b_t^{(i)}(t^{(i)}) + K_{32}(t^{(m)}, t^{(i)}) b_s{}^{(i)}(t^{(i)})] dt^{(i)}$$
$$+ \sum_{j=1}^{N} \int_{-a^{(j)}}^{a^{(j)}} [K_{33}(t^{(m)}, t^{(j)}) p_t^{(j)}(t^{(j)}) + K_{34}(t^{(m)}, t^{(j)}) p_s^{(j)}(t^{(j)})] dt^{(j)}$$
$$+ u_{tt}^{(m0)}(t^{(m)}) = 0 \tag{9.49a}$$

and

$$\sum_{i=1}^{M}\int_{-a^{(i)}}^{a^{(i)}}[K_{41}t(t^{(m)},t^{(i)})\,b_t^{(i)}(t^{(i)})+K_{42}(t^{(m)},t^{(i)})\,b_s^{(i)}(t^{(i)})]\,dt^{(i)}$$

$$+\sum_{j=1}^{N}\int_{-a^{(j)}}^{a^{(j)}}[K_{43}(t^{(m)},t^{(j)})\,p_t^{(j)}(t^{(j)})+K_{44}(t^{(m)},t^{(j)})\,p_s{}^{(j)}(t^{(j)})]\,dt^{(j)}$$

$$-\,\omega^{(m)}(t^{(m)})+u_{st}^{(m0)}(t^{(m)})=0 \qquad (9.49b)$$

where $u_{t,t}^{(m0)}$ and $u_{s,t}^{(m0)}$ are the displacement gradients at the location of the mth rigid line, but with no cracks or rigid lines present in the system, and $\omega^{(m)}$ is the rotation of the mth rigid line.

Here, the kernels K_{31} through K_{44} in equations (9.49) may also be obtained by transforming the fundamental solution to an appropriate local coordinate system. The final results may be expressed as

$$\begin{aligned}K_{31}(t^{(m)},t^{(i)}) = \frac{1}{2\pi(\kappa+1)}&\{\sin^2\theta^{(m)}[\sin\theta^{(i)}I_1(t^{(m)},t^{(i)})-\cos\theta^{(i)}I_2(t^{(m)},t^{(i)})]\\&-\sin\theta^{(m)}\cos\theta^{(m)}[\sin\theta^{(i)}(I_3(t^{(m)},t^{(i)})+I_5(t^{(m)},t^{(i)}))\\&-\cos\theta^{(i)}(I_4(t^{(m)},t^{(i)})+I_6(t^{(m)},t^{(i)}))]\\&+\cos^2\theta^{(m)}\,[\sin\theta^{(i)}I_7(t^{(m)},t^{(i)})-\cos\theta^{(i)}I_8(t^{(m)},t^{(i)})]\}\end{aligned} \qquad (9.50a)$$

$$\begin{aligned}K_{32}(t^{(m)},t^{(i)}) = \frac{1}{2\pi(\kappa+1)}&\{\sin^2\theta^{(m)}[\cos\theta^{(i)}I_1(t^{(m)},t^{(i)})+\sin\theta^{(i)}I_2(t^{(m)},t^{(i)})]\\&-\sin\theta^{(m)}\cos\theta^{(m)}[\cos\theta^{(i)}(I_3(t^{(m)},t^{(i)})+I_5(t^{(m)},t^{(i)}))\\&+\sin\theta^{(i)}(I_4(t^{(m)},t^{(i)})+I_6(t^{(m)},t^{(i)}))]\\&+\cos^2\theta^{(m)}\,[\cos\theta^{(i)}I_7(t^{(m)},t^{(i)})+\sin\theta^{(i)}I_8(t^{(m)},t^{(i)})]\}\end{aligned} \qquad (9.50b)$$

$$\begin{aligned}K_{33}(t^{(m)},t^{(j)}) = \frac{1}{4\pi G(\kappa+1)}&\{\sin^2\theta^{(m)}[\sin\theta^{(j)}H_1(t^{(m)},t^{(j)})\\&-\cos\theta^{(j)}H_2(t^{(m)},t^{(j)})]\\&-\sin\theta^{(m)}\cos\theta^{(m)}[\sin\theta^{(j)}(H_3(t^{(m)},t^{(j)})-H_5(t^{(m)},t^{(j)}))\\&-\cos\theta^{(j)}(H_4(t^{(m)},t^{(j)})+H_6(t^{(m)},t^{(j)}))]\\&+\cos^2\theta^{(m)}\,[\sin\theta^{(j)}H_7(t^{(m)},t^{(j)})-\cos\theta^{(j)}H_8(t^{(m)},t^{(j)})]\}\end{aligned} \qquad (9.50c)$$

$$
\begin{aligned}
K_{34}(t^{(m)}, t^{(j)}) = \frac{1}{4\pi G(\kappa+1)} & \{\sin^2 \theta^{(m)}[\cos \theta^{(j)} H_1(t^{(m)}, t^{(j)}) \\
& + \sin \theta^{(j)} H_2(t^{(m)}, t^{(j)})] \\
& - \sin \theta^{(m)} \cos \theta^{(m)}[\cos \theta^{(j)}(H_3(t^{(m)}, t^{(j)}) + H_5(t^{(m)}, t^{(j)})) \\
& + \sin \theta^{(j)}(H_4(t^{(m)}, t^{(j)}) + H_6(t^{(m)}, t^{(j)}))] \\
& + \cos^2 \theta^{(m)} \; [\cos \theta^{(j)} H_7(t^{(m)}, t^{(j)}) \\
& + \cos \theta^{(j)} H_8(t^{(m)}, t^{(j)})]\}
\end{aligned} \tag{9.50d}
$$

$$
\begin{aligned}
K_{41}(t^{(m)}, t^{(i)}) = \frac{1}{2\pi(\kappa+1)} & \{\sin \theta^{(m)} \cos \theta^{(m)}[\sin \theta^{(i)}(I_1(t^{(m)}, t^{(i)}) - I_7(t^{(m)}, t^{(i)})) \\
& - \cos \theta^{(i)}(I_2(t^{(m)}, t^{(i)}) \;\; - I_8(t^{(m)}, t^{(i)}))] \\
& + \sin^2 \theta^{(m)}[\sin \theta^{(i)} I_5(t^{(m)}, t^{(i)}) - \cos \theta^{(i)} I_6(t^{(m)}, t^{(i)})] \\
& - \cos^2 \theta^{(m)}[\sin \theta^{(i)} I_3(t^{(m)}, t^{(i)}) - \cos \theta^{(i)} I_4(t^{(m)}, t^{(i)})]\}
\end{aligned} \tag{9.50e}
$$

$$
\begin{aligned}
K_{42}(t^{(m)}, t^{(i)}) = \frac{1}{2\pi(\kappa+1)} & \{\sin \theta^{(m)} \cos \theta^{(m)}[\cos \theta^{(i)}(I_1(t^{(m)}, t^{(i)}) - I_7(t^{(m)}, t^{(i)})) \\
& + \sin \theta^{(i)}(I_2(t^{(m)}, t^{(i)}) \;\; - I_8(t^{(m)}, t^{(i)}))] \\
& + \sin^2 \theta^{(m)}[\cos \theta^{(i)} I_5(t^{(m)}, t^{(i)}) + \sin \theta^{(i)} I_6(t^{(m)}, t^{(i)})] \\
& - \cos^2 \theta^{(m)}[\cos \theta^{(i)} I_3(t^{(m)}, t^{(i)}) + \sin \theta^{(i)} I_4(t^{(m)}, t^{(i)})]\}
\end{aligned} \tag{9.50f}
$$

$$
\begin{aligned}
K_{43}(t^{(m)}, t^{(j)}) = \frac{1}{4\pi G(\kappa+1)} & \{\sin \theta^{(m)} \cos \theta^{(m)}[\sin \theta^{(i)}(H_1(t^{(m)}, t^{(j)}) \\
& - H_7(t^{(m)}, \;\; t^{(j)})) - \cos \theta^{(i)}(H_2(t^{(m)}, t^{(j)}) - H_8(t^{(m)}, t^{(j)}))] \\
& + \sin^2 \theta^{(m)}[\sin \theta^{(i)} H_5(t^{(m)}, t^{(j)}) - \cos \theta^{(i)} H_6(t^{(m)}, t^{(j)})] \\
& - \cos^2 \theta^{(m)}[\sin \theta^{(i)} H_3(t^{(m)}, t^{(j)}) - \cos \theta^{(i)} H_4(t^{(m)}, t^{(j)})]\}
\end{aligned} \tag{9.50g}
$$

$$
\begin{aligned}
K_{44}(t^{(m)}, t^{(j)}) = \frac{1}{4\pi G(\kappa+1)} & \{\sin \theta^{(m)} \cos \theta^{(m)}[\cos \theta^{(i)}(H_1(t^{(m)}, t^{(j)}) \\
& - H_7(t^{(m)}, t^{(j)})) + \sin \theta^{(i)}(H_2(t^{(m)}, t^{(j)}) - H_8(t^{(m)}, t^{(j)}))] \\
& + \sin^2 \theta^{(m)}[\cos \theta^{(i)} H_5(t^{(m)}, t^{(j)}) + \sin \theta^{(i)} H_6(t^{(m)}, t^{(j)})] \\
& - \cos^2 \theta^{(m)}[\cos \theta^{(i)} H_3(t^{(m)}, t^{(j)}) + \sin \theta^{(i)} H_4(t^{(m)}, t^{(j)})]\}
\end{aligned} \tag{9.50h}
$$

where

$$I(t^{(m)}, t^{(i)}) \equiv I(x, y; \xi) \qquad \text{and} \qquad H(t^{(m)}, t^{(j)}) \equiv H(x, y; \xi) \tag{9.51}$$

with x, y, and ξ also being transformed as two arguments, $t^{(m)}$ and $t^{(j)}$, according to equations (9.47). Here, equations (9.44) ensure traction-free conditions on crack surfaces, and equations (9.49) ensure compatible rigid-body motions for the rigid lines. The equilibrium conditions and the traction-free conditions at the free surface are automatically satisfied by virtue of the kernels K_{11} through K_{44}.

It can be observed that equations (9.44) and (9.49) give rise to $2(M+N)$ integral equations. In order to solve these equations completely for $b_t^{(i)}$, $b_s^{(i)}$, $p_t^{(j)}$, and $p_s^{(j)}$, however, we need to evaluate $(2M+2N)$ additional constants of integration. It is also important to notice that we need to evaluate the rotations $\omega^{(j)}$ for the N anticracks. This requires N additional equations, bringing the total number of additional equations required to $(2M+3N)$. The $(2M+3N)$ additional equations may be obtained by considering the continuity of the opening shapes of the cracks and the equilibrium conditions for the anticracks. The continuity requirements for crack opening shapes may be expressed as $(i = 1, 2, \ldots, M)$

$$\int_{-a^{(i)}}^{a^{(i)}} b_t^{(i)}(t^{(i)})\, dt^{(i)} = 0 \qquad \text{and} \qquad \int_{-a^{(i)}}^{a^{(i)}} b_s^{(i)}(t^{(i)})\, dt^{(i)} = 0 \tag{9.52}$$

Assuming that no external force or couple is applied directly on the anticrack, the equilibrium conditions for the anticracks may also be written as $(j = 1, 2, \ldots, N)$

$$\int_{-a^{(j)}}^{a^{(j)}} p_t^{(j)}(t^{(j)})\, dt^{(j)} = 0, \qquad \int_{-a^{(j)}}^{a^{(j)}} p_s^{(j)}(t^{(j)})\, dt^{(j)} = 0$$

and

$$\int_{-a^{(j)}}^{a^{(j)}} t^{(j)} p_s^{(j)}(t^{(j)})\, dt^{(j)} = 0 \tag{9.53}$$

Equations (9.44) and (9.49), along with the additional constraints of (9.52) and (9.53), now provide an integral equation representation of the interactions in a general system involving M cracks and N anticracks at arbitrary orientations. The above representation is based on the fundamental solutions due to a point load and a point dislocation in an elastic half-plane.

The integral equations described above can be solved numerically using a Gauss–Chebychev quadrature scheme. The numerical technique proposed by Erdogan et al. (1973) for singular integral equations is adapted here and generalized for systems involving cracks and rigid lines near a free surface, with particular attention given to modeling of rotations of rigid lines.

The discretized system contains $2(L-1)(M+N)$ equations (where L is the number of Gauss points on a crack or a rigid line, M is the number of cracks, and N is the number of rigid lines) but $2L(M+N)+N$ unknowns. The additional $(2M+3N)$ equations are obtained by requiring continuity of crack opening shapes and equilibrium for the rigid lines (Hu and Chandra 1993b). The interpolation function due to Krenk (1975) is used in the present work to evaluate the SIFs.

The performance of such a polynomial scheme for close spacings of interacting cracks and rigid lines is obviously an important issue in the present context. It is interesting to note that, in their analysis of singular integral equations, Theocaris and Ioakimidis (1977) show that the numerical scheme developed by Erdogan et al. (1973) requires only n quadrature points for accurate representations of polynomial functions of order $2n$. As the spacings between interacting cracks and rigid lines decrease, higher orders of polynomial functions are needed for accurate representations. However, this does not pose any major limitations. For a problem involving equal-length collinear cracks with a small tip separation of $0.01a$ ("a" being the half length of the crack), the present scheme with 30 quadrature points on a crack yielded SIF values within 0.1 percent of the analytical results of Erdogan (1962). Melin (1983), Li and Hills (1990), and Rubinstein (1990) also adapted a similar scheme to investigate kinked cracks. An alternative scheme using Fourier series expansion was developed by Hori and Nemat-Nasser (1987) for analyzing crack interactions at small spacings.

9.5.3 Numerical Results for Grinding of Ceramic Composites

In the grinding of ceramics, one attempts to retard the growth of radial cracks normal to the free surface. This improves the strength of the finished part after grinding. To improve the efficiency of the process and to increase the material removal rate, the designer may also facilitate the growth of lateral cracks parallel to the free surface. Based on our observations of interactions among cracks and rigid lines, a distribution of cracks and rigid lines may be constructed to improve the strength of the finished product, as well as to improve the material removal rate in the grinding of ceramics. Many real-life materials contain hard second-phase whiskers or lamellae that may be modeled as rigid lines. The proposed integral equation approach provides an avenue for investigating general systems of cracks and rigid lines near a free surface.

Applications of the proposed integral equation formulation to several example problems involving interactions of cracks and rigid lines near a free surface are presented in this section. The integral equation formulation is based on the fundamental solutions due to a point force and a point dislocation in an elastic half-plane and can handle arbitrary orientations and distributions of cracks and rigid lines, along with arbitrary loading conditions, in the context of elasticity. Thus, the proposed formulation is suitable for investigating the effects of microdefects in ceramics composites and intermetallics under general loading conditions.

The numerical results from the proposed integral equation formulation are first verified against existing solutions for special cases in the literature. Problems involving general systems of cracks and rigid lines near a free surface are addressed next, and the implications of these interactions for ceramic grinding processes are discussed.

In the following presentation, the Poisson's ratio (ν) is taken to be 0.25. If not otherwise specified, the reported stress intensity factors (SIFs) are always normalized with respect to those of a single crack in an infinite plane subject to the same loading. The singular intensities of tangential (P_t) and normal (P_s) tractions for rigid lines are also normalized with respect to the tangential traction singularity (P_0) for a rigid line in an infinite plane subject to the same magnitude of remote normal traction (Wang et al. 1985).

A rigid line normal to the free surface and subject to a remote tension is considered in figure 9.18. The results obtained from the present analysis agree very well with those obtained by Atkinson (1973). (Note that the lower- and upper-tip SIFs reported by Atkinson involve a transformation constant. To eliminate this constant, the ratio of the upper-tip SIF to the lower-tip SIF is used here.) It should be noted that the intensity of the tangential traction singularity at the lower tip [$P_t(+1)$] is stronger than that at the upper tip [$P_t(-1)$]. This is contrary to the variations of SIFs for a corresponding crack problem. As expected, the intensities at both tips tend to become equal as the buried depth is increased. Figure 9.19 shows a pressurized crack normal to the free surface. As the buried depth decreases, the free surface amplifies the upper tip SIF considerably. The mode I SIFs obtained from the present analysis also compare very well with those obtained by Erdogan et al. (1973). An inclined crack at a normalized buried depth of 1.1 and subject to a remote tension is considered in figure 9.20. The mode I SIFs reach a maximum when the crack is normal to the free surface, while the mode II SIFs attain their absolute maxima of 0.63 for the upper tip and 0.48 for the lower tip at $\theta = 50$ degrees and $\theta = 43$ degrees, respectively. As shown in figure 9.20, the predictions from the present analysis also agree very well with the results

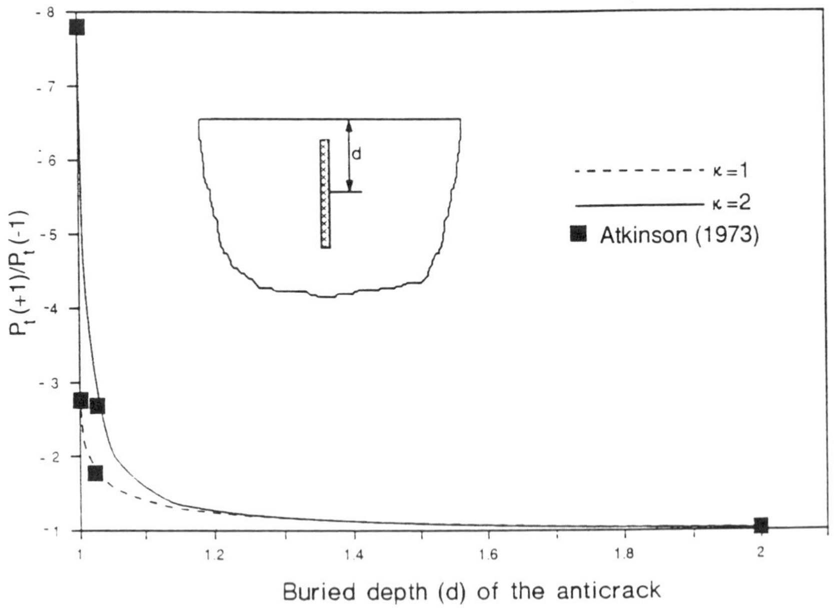

Figure 9.18. Variations in $P_t(+1)/P_t(-1)$ with the buried depth of a rigid line (anticrack).

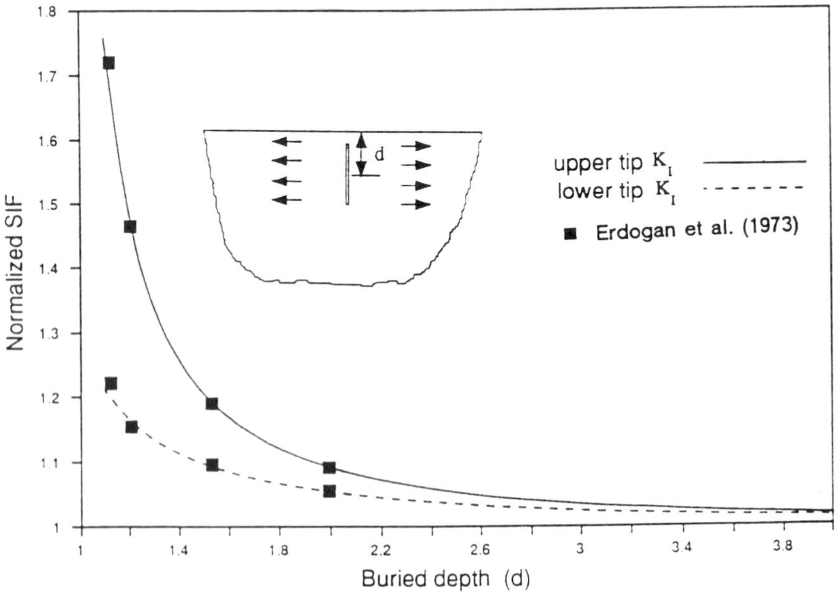

Figure 9.19. Variations in normalized SIF with the buried depth of a crack.

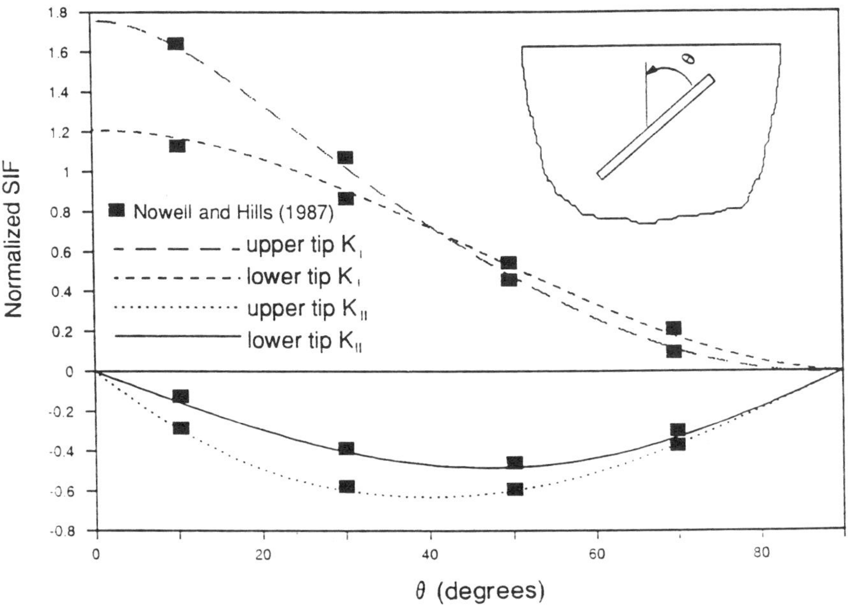

Figure 9.20. Variations in normalized SIF with the orientation of a crack.

obtained by Nowell and Hills (1987). An inclined rigid line (anticrack) at a normalized buried depth of 1.2 and subject to a remote tension is considered in figure 9.21. Figure 9.21a shows that the absolute maxima for normalized tangential traction intensity (P_t/P_0) occur for both upper and lower tips at $\theta = 0$. As shown in figure 9.21b, the free surface makes the normal traction intensities (P_s/P_0) at the upper and lower tips deviate significantly from each other. It is important to observe from figures 9.21a and 9.21b that, for $\theta = 60$ degrees, the singular intensities vanish for both tangential and normal tractions. Wang et al. (1985) show that there exists an orientation, for a rigid line in an infinite plane, for which the traction singularities at the tips of the rigid line disappear. Figures 9.21a and 9.21b establish that similar situations may also occur for a rigid line near a free surface. Figure 9.21c shows the variations in rigid-body rotations with the orientations of the rigid line. The maximum rotation of -0.02 radians (a negative sign implies clockwise rotation) occurs at $\theta = 45$ degrees.

The inset in figure 9.22 shows a crack (normalized buried depth = 0.5) between two parallel rigid lines (normalized buried depths of 0.4 and 0.6) subject to a remote loading, θ_0. This configuration produces significant mode I and mode II SIFs induced by interactions among the crack and the rigid lines. The mode I SIF (normalized by $\theta_0\sqrt{\pi a}$) reaches a maximum value of 0.7 at $b = 0.92$; the mode II SIF reaches a value of -0.5 at $b = 1.63$ ("a"

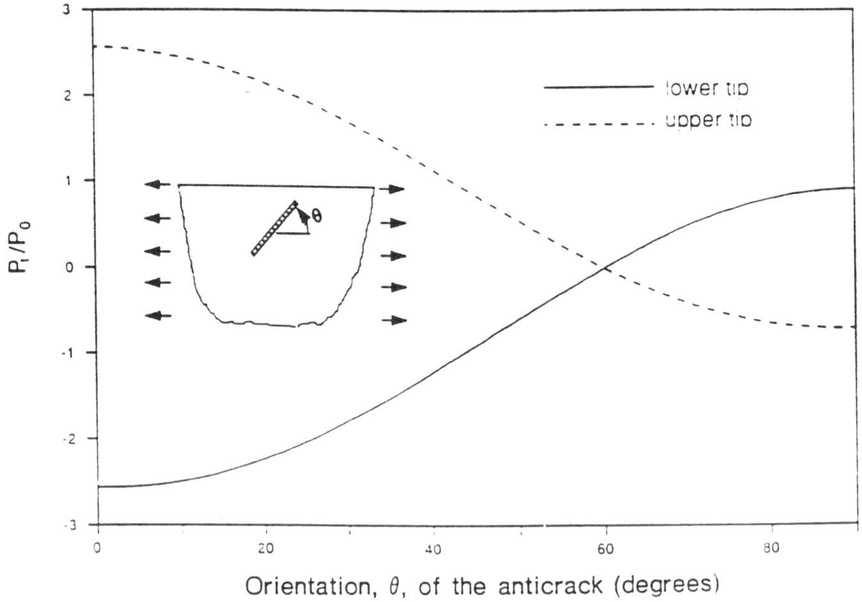

(a)

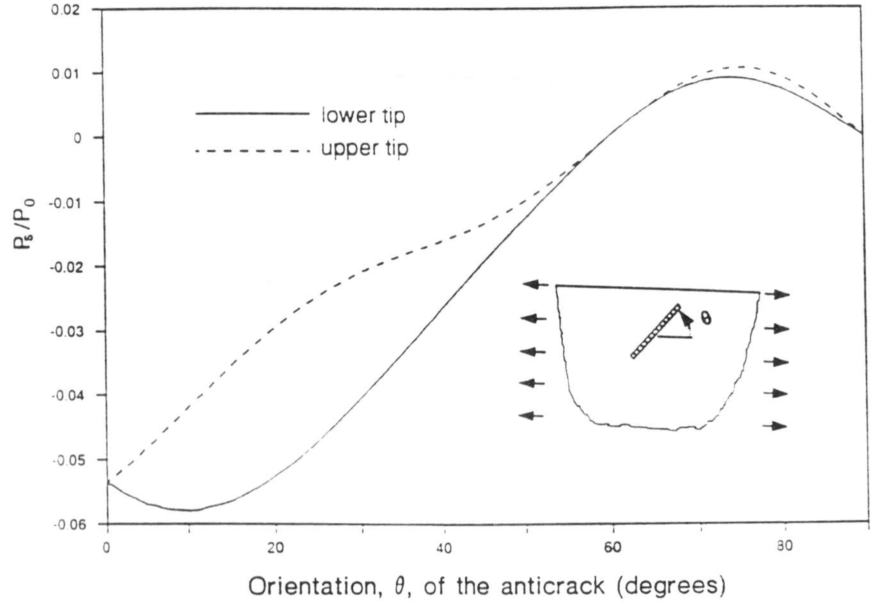

(b)

Figure 9.21. Orientations of a rigid line (anticrack): (a) with variations in P_t/P_0; (b) with variations in P_s/P_0; (c) with variations in the angular rotation of the line.

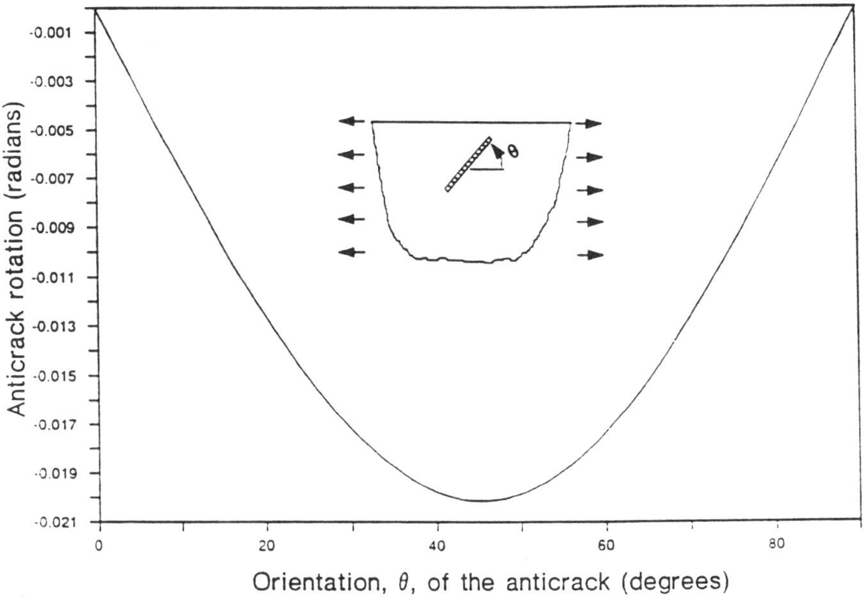

(c)

Figure 9.21—*continued*

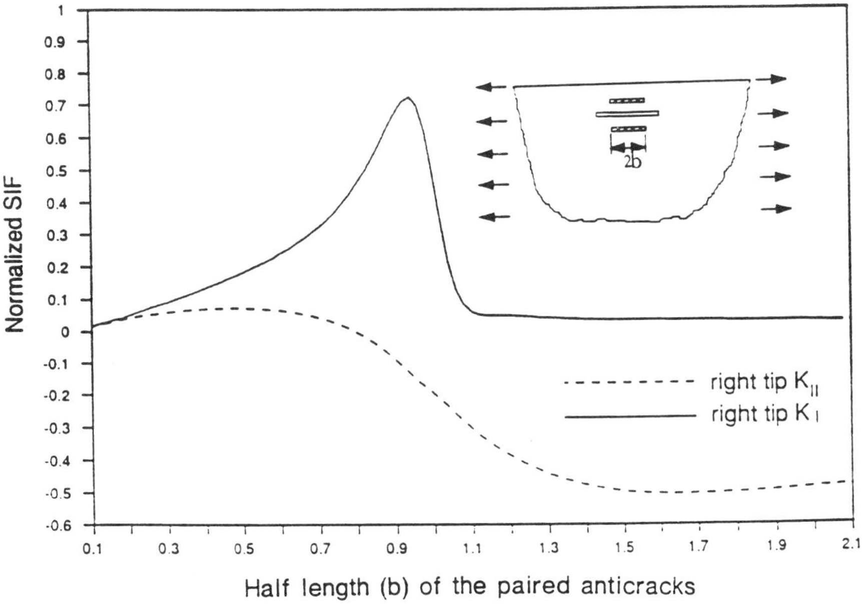

Figure 9.22. Coupling-induced stress singularities for a crack between two parallel rigid lines (anticracks).

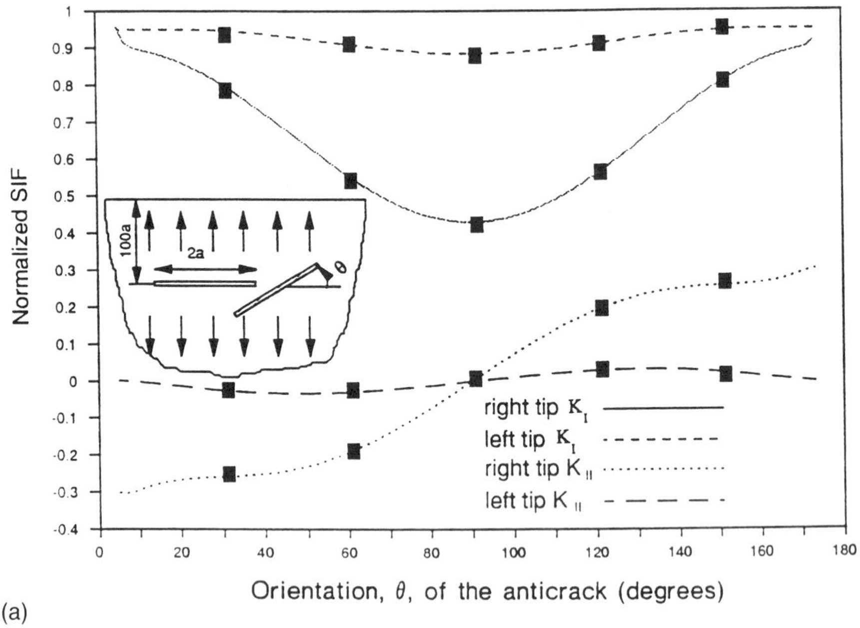

(a)

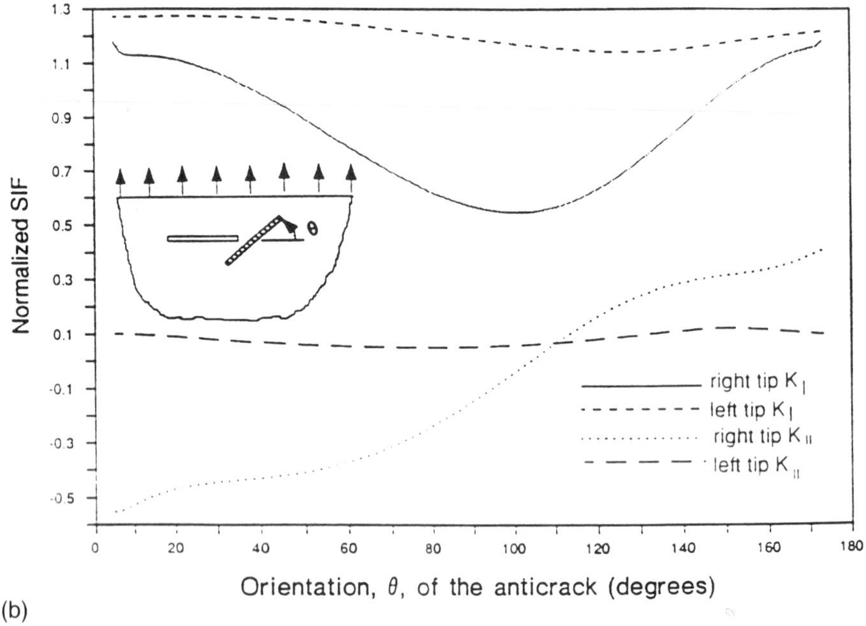

(b)

Figure 9.23. Interactions between a crack and a rigid line with arbitrary relative orientations: (a) verification against the full-plane solution; (b) variations in normalized SIF with θ; (c) variations in P_t/P_0 and P_s/P_0 with θ; (d) variations in rigid-body rotation of the rigid line with θ.

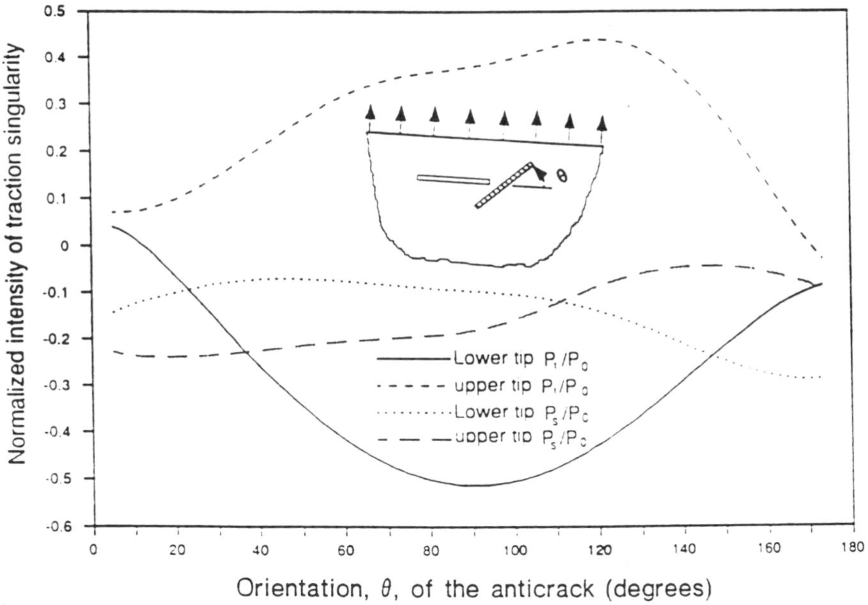

(c)

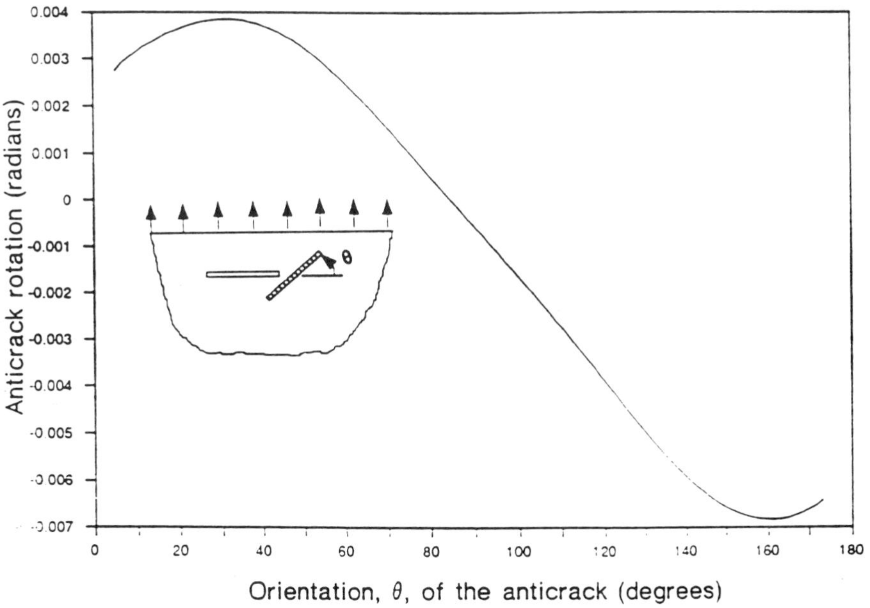

(d)

Figure 9.23—*continued*

is the half length of the crack, θ_0 the remote tension parallel to the crack, and "b" the ratio of lengths between the rigid lines and the crack). This is contrary to what is observed for cracks parallel to the direction of loading. In the present case, however, $u_{t,t}$ and $u_{s,t}$ [in equations (9.49)] will not necessarily be zero. This will produce nonzero p_s and p_t, which in turn will require nonzero b_s and b_t. Thus, nonzero SIFs may be induced at crack tips.

The interactions between a crack and rigid line with arbitrary relative orientations are considered next. The predictions from the present analysis at large buried depths are first verified against the full plane solution (Hu and Chandra 1993b). As shown in figure 9.23a, the results from the present analysis at a buried depth of 100a ("a" is the half length of the crack) matches very well with the full-plane solution. For the current example, the normalized distance between the centers of the crack and the rigid line is 1.2 (note that $0 < \theta < 180$ degrees—the crack and the rigid line are never allowed to overlap); the crack and rigid line are of equal length. The normalized SIFs for a normalized buried depth of 1.2 for the crack and rigid-line centers are presented in figure 9.23b. The normalized mode I SIF at the crack tip near the rigid line reaches a minimum value of 0.55 at $\theta = 100$ degrees. The normalized mode II SIF for the near tip varies noticeably from -0.55 at $\theta = 5$ degrees to 0.41 at $\theta = 173$ degrees and passes through zero at $\theta = 104$ degrees. Due to effects of the free surface, the position of the zero value deviates from $\theta = 90$ degrees. The far-tip SIFs are not significantly affected by variations in the orientation of the accompanying rigid line. Figure 9.23c shows the variations of normalized traction singularities for the rigid line with its orientation. Compared to the full-plane solutions (Hu and Chandra 1993b), the symmetry at the lower tip (about $\theta = 90$ degrees) is only slightly modified due to effects of the free surface. However, P_t/P_0 for the upper tip reaches a maximum of 0.44 at $\theta = 120$ degrees, which deviates considerably from the $\theta = 90$ degrees position of the maximum for the full plane. The variation in the rotations of the rigid line with its orientation is presented in figure 9.23d. The rotation reaches 0.0039 radians in the counterclockwise direction for $\theta = 32$ degrees. Then it reverses direction and reaches an extreme value of 0.0068 radians in the clockwise direction at $\theta = 160$ degrees.

Figure 9.24 shows the shielding effects between a crack and a rigid line perpendicular to each other. The crack is pressurized. The normalized buried depth of the lower tip of the crack is 1.2, and the rigid line is placed at a distance of 0.1 from this crack tip. Such pressurized cracks are typical in indentation problems. The process of grinding ceramics may also be viewed as a moving indentation problem, and such radial cracks are commonly observed in this process. To improve the strength of the finished workpiece,

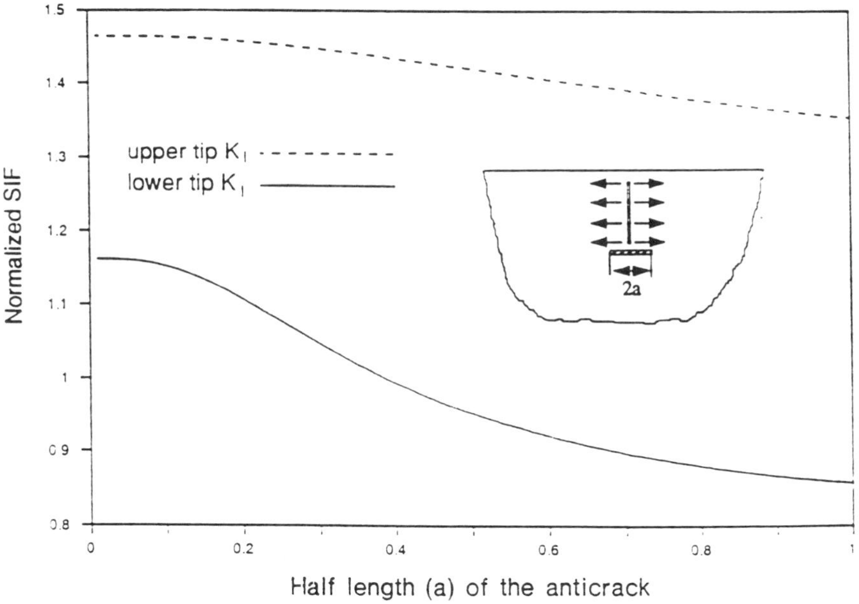

Figure 9.24. Shielding effects between a crack and a rigid line perpendicular to each other.

it is important to retard the growth of such cracks normal to the free surface. It is observed, here, that the arrangement shown in figure 9.24 makes the mode I SIF at the lower tip drop from 1.16 to 0.86 as the length of the rigid line is increased from 0.1 to 1.0. The arrangement shown in figure 9.25 has a buried crack depth of 1.2 and a buried depth of 2.2 for the centers of the rigid lines (the lower tip of the crack is aligned with the centers of the rigid lines). The rigid lines are spaced at a distance of 0.1 from each other. The crack is pressurized. As shown in figure 9.25, the mode I SIF at the lower tip rises from 1.16 to 1.28 as the half-length of the rigid line is increased to 0.26. This amplification is due to tip interactions that occur in the proximity of the crack and rigid-line tips at short lengths of the rigid line. As the half-length of the rigid line is increased to 1.0, the mode I SIF at the lower tip drops to 1.18. At the same time, the upper tip mode I SIF is also reduced from 1.46 to 1.28.

9.6 Micro-Scale Features in Macro-Scale Problems

The above analyses of ceramic grinding situations take advantage of the fact that the mechanics of grinding problems can be formulated as crack–crack

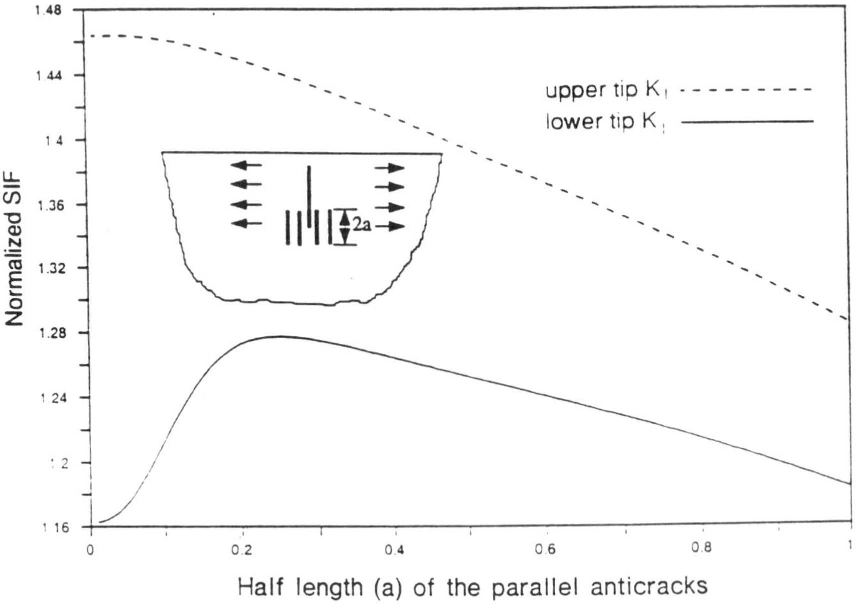

Figure 9.25. Interactions between a crack and a system of parallel rigid lines.

or crack–rigid line interaction problems near the free surfaces of semi-infinite bodies. Accordingly integral equation approaches have been pursued in previous sections to investigate these problems.

In recent years, performance and weight goals for the next generation of high-temperature applications have spurred the development of improved intermetallics and ceramic matrix composites capable of sustained operations in the 1000–1500°C range. These materials contain multiple phases and secondary-phase particles. To model real-life applications of these high-temperature components, one must incorporate general loading situations, finite and often complex geometries of particular components, and detailed representations of interacting inhomogeneities, along with their associated damage evolutions. Therein lies the fundamental difficulty in the analysis of these problems. Typically, the microdefects and their spacings are of the order of a few micrometers, while the overall dimensions of a component may range from a few centimeters to even a meter. Thus, the computational scheme is required, simultaneously, to provide a detailed representation of the underlying mechanics at two widely different scales: a local micro-scale ranging from 10–100 μm and a global macro-scale that may range from 10 to 1000 mm.

The computational techniques in existence today, such as the finite element method (e.g., Oden 1972, Gallagher 1975, Hughes 1987, Bathe 1982,

Zienkiewicz and Taylor 1991) and the boundary element method (e.g., Banerjee and Butterfield 1981, Mukherjee 1982, Cruse 1988, Lutz et al. 1992, Huang and Cruse 1994), are ideally suited for macro-scale analysis and can easily handle complex geometries along with general loading conditions. Thus, the defects are normally introduced as geometric entities in such macro-scale computational schemes. The resulting mixed boundary value problems essentially assume that the defect sizes are of the same order as the geometric dimensions of the body (e.g., Cruse and Polch 1986). In many cases, special quarter-point finite elements (e.g., Barsoum 1976, Yahia and Shephard 1985) or boundary elements (e.g., Crouch 1976, Kamel and Liaw 1991a,b, Raveendra and Banerjee 1992) are introduced to capture the r singularity at the tip of an elastic crack. In such cases, the crack size is limited by the level of discretization, and the technique becomes prohibitively expensive if the body contains a large number of microcracks. Moreover, such a technique is applicable only to isolated elastic cracks. These special elements cannot represent any effects due to crack interactions, and one must rely on numerical discretization to capture such effects. This poses significant difficulties whenever the cracks are closely packed. Besides, these special elements cannot be directly extended to model other types of defects, such as voids or inclusions (e.g., secondary phases and short or continuous fibers) that are commonly present in many real materials. An understanding of the evolution of such defects in a damage cluster is critically important for estimating the characteristics (e.g., strength and fracture toughness) of modern intermetallics and ceramic materials.

Several researchers (e.g., Tvergaard 1982, 1989a,b, 1990, Needleman 1987, Fleck et al. 1989, Needleman and Tvergaard 1991) have developed a unit cell approach (under assumptions of doubly periodic defect distributions) to investigate micro-scale issues such as void growth resulting in plastic instabilities, decohesion of hard particles from the matrix material, and so on, in the context of computational mechanics. Chandra and Tvergaard (1993) also utilized such a unit cell approach to investigate void nucleation and growth in plane strain extrusion processes. The unit cells, however, are by definition much smaller than the macroscopic dimensions of the body and exist essentially at a micro-scale of the same order as the defect sizes. Thus, the computations are carried out entirely at a micro-scale. Becker (1992) also carried out finite element analyses, entirely at a micro-scale, to investigate the effects of different crystallographic orientations in a polycrystalline material.

Thus, the computational techniques available today are capable of analyzing a problem at a macro-scale or at a micro-scale. However, they cannot bridge these two widely different scales in a single analysis. Banerjee and Henry (1992) introduced special boundary elements and Nakamura and

Suresh (1993) extended the unit cell analysis using FEM for modeling the effects of fiber packing in composite materials. However, much work is needed before one can directly investigate the effects of macro-scale design (geometric, loading, and boundary condition) considerations upon the evolutions of micro-scale defects, which essentially govern the strength and life of individual components in service.

Over the past few decades, analytical techniques have been used extensively to investigate various micromechanical phenomena. Several micromechanical models have been developed to study the behavior of materials containing various distributions of inhomogeneities, which may be reinforcements or defects. These approaches include self-consistent (Budiansky 1965, Hill 1965), differential (Roscoe 1952, Norris 1985), Mori-Tanaka (Taya and Chou 1981, Weng 1984), and generalized self-consistent (Christensen and Lo 1979, Huang et al. 1994) methods. Micromechanical models are particularly suited for the prediction of overall properties of composites, but they cannot accurately represent the local stress and deformation fields around each inhomogeneity. These local fields, however, have been demonstrated to be of extreme importance for defect initiation, growth, and coalescence (Becker et al. 1988).

Various researchers have attempted to investigate the interactions among microdefects in a damage cluster. Horii and Nemat-Nasser (1986) used a pseudo-traction approach for problems involving crack interactions. Hu and Chandra (1993a,b,c) modeled the microcrack as distributions of dislocations and the rigid lines as distributions of tractions in order to develop an integral equation approach for investigating interactions among cracks and rigid lines in a defect cluster. The approach of Hu and Chandra has also been extended to interactions of voids, cracks, and rigid lines (Hu et al. 1993a). The Gauss–Chebychev quadrature scheme (Erdogan et al. 1973) utilized by Hu and Chandra (1993a,b,c) requires only n quadrature points for accurate representations of polynomial functions of order $2n$. Using such a scheme, they were able to investigate defect interactions at very close spacings. For a problem involving equal-length collinear cracks with a small tip separation of only 1 percent of the half-length of the crack, the scheme of Hu and Chandra (1993b) with 30 quadrature points on a crack yields SIF values within 0.1 percent of the analytical results of Erdogan (1962). Hu et al. (1993a,b, 1994) also pursued a traction approach to investigate the interactions among bridged cracks and their implications on defect coalescence in various multiphase ceramic materials. Some of these have been discussed in previous sections (9.1–9.5).

The analytical and semianalytical investigations cited above can provide crucial insights into the behaviors of interacting micro-defects in a damage cluster. However, these analyses are mostly carried out under assumptions of infinite bodies or extremely simplified geometry and loading conditions, which severely restrict their applicability to real-life situations involving complex finite geometries and general loading conditions. Thus, on one hand, there are analytical models capable of yielding very accurate results for various micro-scale phenomena involving evolutions of micro-defects in a damage cluster, but for very simple geometries and loading situations. On the other hand, very powerful computational techniques have been developed to handle real-life macro-scale problems involving complex finite geometries and general loading conditions, yet it is very difficult to relate the effects of macro-scale parameters on the interactions and evolutions of micro-scale cracks, voids, and inclusions present in a real material.

Accordingly, the present section aims at integrating the local analysis schemes capable of detailed representations of the micro-scale phenomena with the macro-scale computational schemes capable of handling finite complex geometries and realistic boundary conditions. Unlike the finite element method (FEM), the boundary element method requires a fundamental solution that is the solution of the adjoint governing operator for a concentrated load subject to the homogeneous boundary conditions in an infinite domain. Over the years, this requirement has been the greatest hindrance for the application of the BEM technique to a wide class of problems, and various researchers have devised ways of circumventing this drawback. In the present work, however, this particular feature of the BEM provides an avenue for incorporating the insights gained from micro-scale investigations into the macro-scale computational technique. This allows us, in a single analysis, to investigate the effects of macro-scale variations in geometrical features and boundary conditions on the micro-scale evolutions of defects in damage clusters. The microphenomena are captured in the fundamental solution, and the conventional BEM technique is used to relate them to macro-scale parameters of the problem. Such an approach also allows sequential modeling of the microdefects and the macro-scale problem. Thus, it obviates the problem of proliferation of degrees of freedom that is commonly observed in computational efforts attempting to capture both micro- and macrofeatures of a problem.

This section starts with a brief exposition of some relevant analytical results. The micro-scale features of the problem are addressed next, and an integral equation technique is used to numerically construct the appropriate fundamental solution. As a first attempt, only microcracks and their interactions are considered in this paper. However, it should be emphasized that, following the work of Hu and Chandra (1993a,b,c) and Hu et al. (1993a,b, 1994), fundamental solutions can also be obtained for problems involving other microfeatures, such as voids, chopped fibers, and interfaces. The augmented fundamental solution is then incorporated into a direct BEM approach to solve the macro-scale elasticity problems. The numerical results obtained from the proposed hybrid micro–macro BEM formulation are first verified against existing results in the literature. The capabilities of the proposed scheme are investigated, and various salient features are discussed. Finally, a full BEM formulation is utilized to investigate chip formation during single-grit scratch tests.

9.6.1 Micro-Scale Fundamental Solutions

The fundamental solutions representing the microfeatures of the macroscopic problem are developed in this section. As a first attempt, only microcracks are considered and the macroscopic body is assumed to be elastic. A crack is essentially a cut that transmits no traction across it but allows a displacement discontinuity.

In the present section, the cracks are modeled as distributions of dislocations. As shown in figure 9.26, we consider an infinitely extended body containing M interacting microcracks. The components $b_t^{(i)}$ and $b_s^{(i)}$ of the dislocation density vector along the ith crack may be expressed as

$$b_t^{(i)}(t^{(i)}) = \frac{\partial}{\partial t^{(i)}}[u_t^{(i)}(t^{(i)}, 0^+) - u_t^{(i)}(t^{(i)}, 0^-)] = \frac{\partial \Delta_t^{(i)}(t^{(i)})}{\partial t^{(i)}} \tag{9.54a}$$

$$b_s^{(i)}(t^{(i)}) = \frac{\partial}{\partial t^{(i)}}[u_s^{(i)}(t^{(i)}, 0^+) - u_s^{(i)}(t^{(i)}, 0^-)] = \frac{\partial \Delta_s^{(i)}(t^{(i)})}{\partial t^{(i)}} \tag{9.54b}$$

where $\Delta_t^{(i)}$ and $\Delta_s^{(i)}$ denote the glide and climb dislocations, respectively, in the local tangential–normal $(t^{(i)}, s^{(i)})$ coordinate system for the ith microcrack, whose occupancy is denoted as $-a^{(i)} < t^{(i)} < a^{(i)}$.

Read (1953), Dundurs and Mura (1964), and Dundurs and Sendeckyj (1965) developed the fundamental solution at (x, y) for a two-dimensional infinite body subjected to an edge dislocation with Burgers' vector components b_x and b_y acting at a point $(\xi, 0)$. This fundamental solution due to an edge dislocation in an infinite plane may be expressed as

$$u_i = \frac{1}{2\pi(1-\nu)} H_{ij}(x, y; \xi)\, b_j \tag{9.55a}$$

$$u_{i,j} = \frac{1}{2\pi(\kappa+1)} I_{ijk}(x, y; \xi)\, b_k \tag{9.55b}$$

$$\sigma_{ij} = \frac{G}{\pi(\kappa+1)} J_{ijk}(x, y; \xi)\, b_k \tag{9.55c}$$

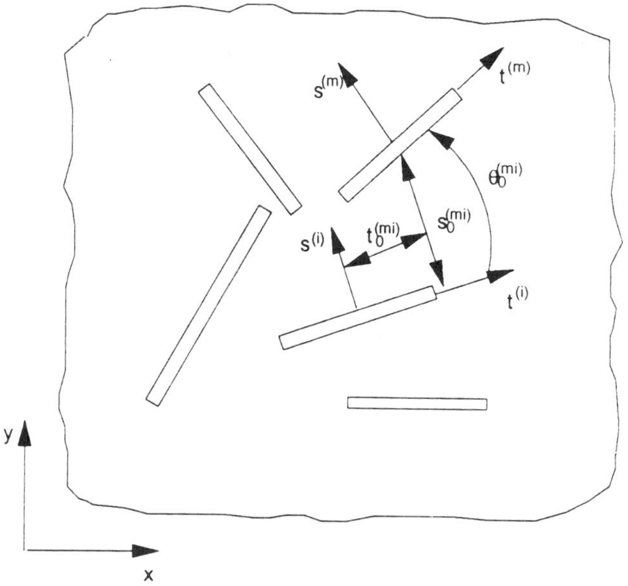

Figure 9.26. Schematic of multiple interacting cracks in an infinite body.

where G is the shear modulus. The constant $\kappa = (3 - 4\nu)$ for plane strain and $\kappa = (3 - \nu)/(1 + \nu)$ for plane stress. The details of the kernels H_{ij}, I_{ijk}, and J_{ijk} may be expressed as

$$H_{11}(x, y, \xi) = (1 - \nu)\tan^{-1}\frac{y}{x - \xi} + \frac{(x - \xi)y}{2[(x - \xi)^2 + y^2]} \tag{9.56a}$$

$$H_{12}(x, y; \xi) = \frac{1 - 2\nu}{4}\ln[y^2 + (x - \xi)^2] + \frac{y^2}{2[(x - \xi)^2 + y^2]} \tag{9.56b}$$

$$H_{21}(x, y; \xi) = -\frac{1 - 2\nu}{4}\ln[y^2 + (x - \xi)^2] - \frac{(x - \xi)^2}{2[(x - \xi)^2 + y^2]} \tag{9.56c}$$

$$H_{22}(x, y; \xi) = -(1 - \nu) + \tan^{-1}\frac{x - \xi}{y} - \frac{y(x - \xi)}{2[(x - \xi)^2 + y^2]} \tag{9.56d}$$

$$I_{111}(x, y; \xi) = (\kappa - 1)\frac{y}{r^2} + \frac{4(x - \xi)^2 y}{r^4} \tag{9.57a}$$

$$I_{112}(x, y; \xi) = (\kappa - 1)\frac{x - \xi}{r^2} + \frac{4(x - \xi)y^2}{r^4} \tag{9.57b}$$

$$I_{121}(x, y; \xi) = (\kappa + 3)\frac{x - \xi}{r^2} - \frac{4(x - \xi)y^2}{r^4} \tag{9.57c}$$

$$I_{122}(x, y; \xi) = (\kappa - 1)\frac{y}{r^2} + \frac{4(x - \xi)^2 y}{r^4} \tag{9.57d}$$

$$I_{211}(x, y; \xi) = -(\kappa - 1)\frac{x - \xi}{r^2} - \frac{4(x - \xi)y^2}{r^4} \tag{9.57e}$$

$$I_{212}(x, y; \xi) = -(\kappa + 3)\frac{y}{r^2} + \frac{4(x - \xi)^2 y}{r^4} \tag{9.57f}$$

$$I_{221}(x, y; \xi) = -(\kappa - 1)\frac{y}{r^2} - \frac{4(x - \xi)^2 y}{r^4} \tag{9.57g}$$

$$I_{222}(x, y; \xi) = (\kappa - 1)\frac{x - \xi}{r^2} + \frac{4(x - \xi)y^2}{r^4} \tag{9.57h}$$

and

$$J_{111}(x, y; \xi) = -\frac{[3(x - \xi)^2 + y^2]\ y}{r^4} \tag{9.58a}$$

$$J_{112}(x, y; \xi) = \frac{(x - \xi)[(x - \xi)^2 - y^2]}{r^4} \tag{9.58b}$$

$$J_{121}(x, y; \xi) = \frac{[(x - \xi)^2 - y^2](x - \xi)}{r^4} \tag{9.58c}$$

$$J_{122}(x, y; \xi) = \frac{y[(x - \xi)^2 - y^2]}{r^4} \tag{9.58d}$$

$$J_{221}(x, y; \xi) = \frac{[(x - \xi)^2 - y^2]y}{r^4} \tag{9.58e}$$

$$J_{222}(x, y; \xi) = \frac{(x - \xi)[(x - \xi)^2 + 3y^2]}{r^4} \tag{9.58f}$$

It is important to note here that the displacement gradient fields and the stress fields in equations (9.54)–(9.58) contain a singularity of the order of r as the distance r between a field point (x, y) and a source point $(\xi, 0)$ approaches zero. Such singular integrals, however, can easily be evaluated in a Cauchy principal value sense (e.g., Mukherjee 1982, Cruse 1988, Dundurs and Markenscoff 1989).

Following the approach outlined in sections 9.1–9.5 and neglecting any crack closure during the application of the external load, the governing equations for the mth crack in a system of M interacting cracks with arbitrary distributions may be expressed as

$$\sum_{i=1}^{M} \int_{-a^{(i)}}^{a^{(a)}} [K_{jkl}(t^{(m)}, t^{(i)}) b_l^{(i)}(t^{(i)}) + \sigma_{jk}^{(m0)}(t^{(m)})] \, dt^{(i)} = 0 \tag{9.59}$$

in the local tangential–normal coordinate system for the$m$th crack. The index j is identified as the normal direction for the mth crack. The index k refers to the local coordinates of the mth crack, while the index l refers to those of the *ith crack* $(k, l = 1, 2)$. Here, $b_l^{(i)}$ represents the unknown dislocations on the ith crack, and $\sigma_{jk}^{(m0)}$ denotes stress components at the location of the mth crack due to external loading, but in the absence of any crack. The governing equations (9.59) represent the traction-free conditions at crack surfaces. The equilibrium for the resulting stress fields is satisfied by virtue of K_{jkl}. Any tractions applied directly on the crack surfaces can also be handled by suitably modifying $\sigma_{jk}^{(m0)}$. Here, the kernels K_{jkl} in equation (9.59) may be obtained by transforming the kernels J_{jkl} to an appropriate local coordinate system for the mth crack and may be expressed as

$$\begin{aligned} K_{j11}(t^{(m)}, t^{(i)}) = & \frac{G}{\pi(\kappa+1)} \{\cos^2\theta^{(m)}[\sin\theta^{(i)} J_{111}(t^{(m)}, t^{(i)}) \\ & - \cos\theta^{(i)} J_{112}(t^{(m)}, t^{(i)})] \\ & + 2\sin\theta^{(m)}\cos\theta^{(m)}[\sin\theta^{(i)} J_{121}(t^{(m)}, t^{(i)}) \\ & - \cos\theta^{(i)} J_{122}(t^{(m)}, t^{(i)})] + \sin^2\theta^{(m)}[\sin\theta^{(i)} J_{221}(t^{(m)}, t^{(i)}) \\ & - \cos\theta^{(i)} J_{222}(t^{(m)}, t^{(i)})]\} \end{aligned} \tag{9.60a}$$

$$\begin{aligned} K_{j12}(t^{(m)}, t^{(i)}) = & \frac{G}{\pi(\kappa+1)} \{\cos^2\theta^{(m)}[\cos\theta^{(i)} J_{111}(t^{(m)}, t^{(i)}) \\ & + \sin\theta^{(i)} J_{112}(t^{(m)}, t^{(i)})] \\ & + 2\sin\theta^{(m)}\cos\theta^{(m)}[\cos\theta^{(i)} J_{121}(t^{(m)}, t^{(i)}) \\ & + \sin\theta^{(i)} J_{122}(t^{(m)}, t^{(i)})] + \sin^2\theta^{(m)}[\cos\theta^{(i)} J_{221}(t^{(m)}, t^{(i)}) \\ & + \sin\theta^{(i)} J_{222}(t^{(m)}, t^{(i)})]\} \end{aligned} \tag{9.60b}$$

$$K_{j21}(t^{(m)}, t^{(i)}) = \frac{G}{\pi(\kappa+1)} \{\sin\theta^{(m)}\cos\theta^{(m)}[\sin\theta^{(i)} J_{111}(t^{(m)}, t^{(i)})$$
$$- \cos\theta^{(i)} J_{112}(t^{(m)}, t^{(i)})] + [\sin^2\theta^{(m)} - \cos^2\theta^{(m)}]$$
$$\times [\sin\theta^{(i)} J_{121}(t^{(m)}, t^{(i)}) - \cos\theta^{(i)} J_{122}(t^{(m)}, t^{(i)})]$$
$$+ \sin\theta^{(m)}\cos\theta^{(m)}[\sin\theta^{(i)} J_{221}(t^{(m)}, t^{(i)})$$
$$- \cos\theta^{(i)} J_{222}(t^{(m)}, t^{(i)})]\} \tag{9.60c}$$

$$K_{j22}(t^{(m)}, t^{(i)}) = \frac{G}{\pi(\kappa+1)} \{\sin\theta^{(m)}\cos\theta^{(m)}[\cos\theta^{(i)} J_{111}(t^{(m)}, t^{(i)})$$
$$+ \sin\theta^{(i)} J_{112}(t^{(m)}, t^{(i)})] + [\sin^2\theta^{(m)} - \cos^2\theta^{(m)}]$$
$$\times [\cos\theta^{(i)} J_{121}(t^{(m)}, t^{(i)}) + \sin\theta^{(i)} J_{122}(t^{(m)}, t^{(i)})]$$
$$+ \sin\theta^{(m)}\cos\theta^{(m)}[\cos\theta^{(i)} J_{221}(t^{(m)}, t^{(i)})$$
$$+ \sin\theta^{(i)} J_{222}(t^{(m)}, t^{(i)})]\} \tag{9.60d}$$

where

$$I_{jkl}(t^{(m)}, t^{(i)}) \equiv I_{jkl}(x, y; \xi), \qquad J_{jkl}(t^{(m)}, t^{(i)}) \equiv J_{jkl}(x, y; \xi) \tag{9.61a}$$

with x, y, and ξ being substituted as

$$x = x_c^{(m)} + t^{(m)}\cos\theta^{(m)} \tag{9.61b}$$

$$y = t^{(i)}\sin\theta^{(i)} - t^{(m)}\sin\theta^{(m)} + y_c^{(m)} - y_c^{(i)} \tag{9.61c}$$

$$\xi = x_c^{(i)} + t^{(i)}\cos\theta^{(i)} \tag{9.61d}$$

Here, the center of the ith crack is denoted as $(x_c^{(i)}, y_c^{(i)})$ and the angle between x and $t^{(i)}$ is denoted as $\theta^{(i)}$.

It can be observed that equation (9.59) gives rise to $2M$ integral equations. In order to solve these equations completely for $b_l^{(i)}(l = 1, 2)$, however, we need to evaluate $2M$ additional constants of integration. The $2M$ additional equations may be obtained by considering the continuity of the opening shapes of the cracks. This yields

$$\int_{-a^{(i)}}^{a^{(i)}} b_l^{(i)}(t^{(i)})\, dt^{(i)} = 0, \qquad l = 1, 2 \tag{9.62}$$

As described before, equations (9.59)–(9.62) now provide an integral equation representation of the micro-scale effects containing M interacting microcracks in an infinite body. These governing integral equations can be solved very accurately and effectively (Hu and Chandra 1993a) using the Gauss–Chebychev quadrature scheme proposed by Erdogan et al. (1973) for singular integral equations.

The discretized system contains $2M(L-1)$ algebraic equations (where L is the number of Gauss points on a crack and M is the number of cracks) but $2ML$ unknowns. Equation (9.62), expressing the continuity of crack opening shapes, provides the additional $2M$ equations. Issues regarding the performance of such a polynomial scheme for close spacings of interacting cracks have been discussed in previous sections.

In order to evaluate the augmented fundamental solution for later incorporation into the macro-scale problem, the finite body with microcracks is first embedded in an infinite domain and is subjected to a point load at any desired location inside or on the boundary of the embedded finite body. As shown in figure 9.27, this problem may then be decomposed as the

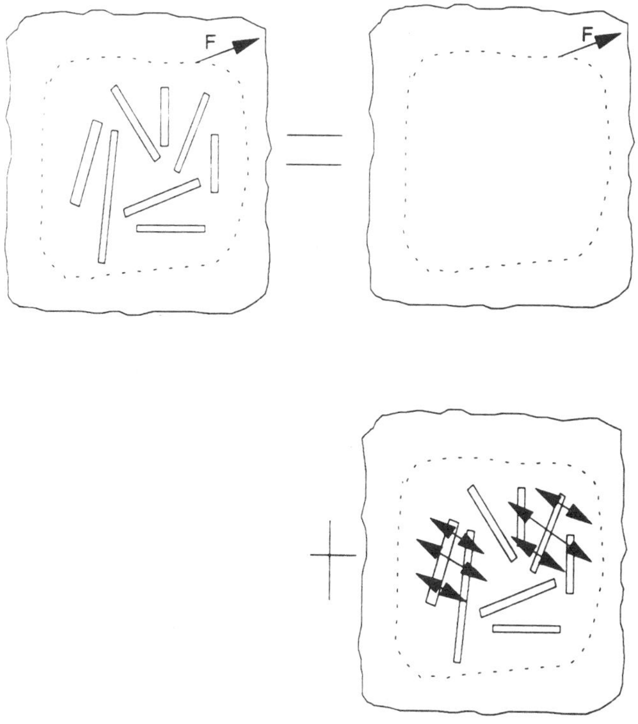

Figure 9.27. Decomposition for evaluating the augmented fundamental solution.

summation of two subproblems, one containing the point load but no cracks and the other containing the cracks along with the balancing tractions needed to maintain traction-free conditions along the crack surfaces for the overall problem.

The solution to the first subproblem, that of a point load in an infinite homogeneous elastic body, is well known in two as well as three dimensions (e.g., Banerjee and Butterfield 1981, Mukherjee 1982, Cruse 1988). As a first attempt, only two-dimensional applications will be considered here and the two-dimensional Kelvin solution for the first subproblem may be written as

$$u_i^{(R_1)}(p) = U_{ij}(p, q)\, e_j(q) \tag{9.63a}$$

and

$$\tau_i^{(R_1)}(p) = T_{ij}(p, q)\, e_j(q) \tag{9.63b}$$

in the absence of any body forces, where

$$U_{ij}(p, q) = \frac{-1}{8\pi(1-\nu)G}\left[(3-4\nu)\ln r\, \delta_{ij} - r_{,i} r_{,j}\right] \tag{9.64a}$$

and

$$T_{ij}(p, q) = \frac{-1}{4\pi(1-\nu)r}\left([(1-2\nu)\,\delta_{ij} + 2r_{,i}r_{,j}]\frac{\partial r}{\partial n} - (1-2\nu)(r_{,i}n_j - r_{,j}n_i)\right) \tag{9.64b}$$

In equations (9.63) and (9.64), p and q denote, respectively, a source and a field point at an internal point, while P and Q denote a source and a field point on the boundary. The distance between the source and the field points is denoted by r, and e_j represents a unit vector. Equation (9.64) is written for plane strain. For plane stress, ν should be replaced by $\overline{\nu} = \nu/(1+\nu)$. It should also be noted that U_{ij} contains a singularity of the order $\ln r$ and T_{ij} contains a singularity of the order $(1/r)$ as r approaches zero.

The integral equations (9.59)–(9.62) may now be used to evaluate the displacement and stress (hence the traction) fields in the second subproblem shown in figure 9.27. This solution is carried out using a Gauss–Chebychev polynomial interpolation scheme described previously. Knowing $u_i^{(R_2))}$ and $\tau_i^{(R_2)}$, as well as the unit vectors, the appropriate kernels $\overline{U}_{ij}$ and $\overline{T}_{ij}$ can easily be obtained from

$$u_i^{(R_2)}(p) = \overline{U}_{ij}(p, q)\, e_j(q) \tag{9.65a}$$

and

$$\tau_i^{(R_2)}(p) = \overline{T}_{ij}(p, q)\, e_j(q) \tag{9.65b}$$

for the desired sets of source and field points. The kernels $\overline{U}_{ij}$ and $\overline{T}_{ij}$ capture the effects of the interacting microcracks.

The augmented kernels may now be constructed as

$$\hat{U}_{ij}(p, q) = U_{ij}(p, q) + \overline{U}_{ij}(p, q) \tag{9.66a}$$

and

$$\hat{T}_{ij}(p, q) = T_{ij}(p, q) + \overline{T}_{ij}(p, q) \tag{9.66b}$$

where U_{ij} and T_{ij} are the conventional kernels obtained from the Kelvin solution discussed in chapter 2; $\overline{U}_{\mathrm{ij}}$ and $\overline{T}_{ij}$ are the kernels capturing the effects of the microcracks. The augmented kernels $\hat{U}_{ij}$ and $\hat{T}_{ij}$ can now be used directly in a macro-scale BEM analysis for elastic structures.

9.6.2 Micro–Macro BEM Formulation

Using Betti's reciprocal theorem (e.g., Banerjee and Butterfield 1981, Mukherjee 1982) or a weighted residual approach (Okada et al. 1990), a hybrid BEM formulation requiring only macro-scale discretization may be developed (Cruse 1988) as

$$u_j(p) = \int_{\partial T} [\hat{U}_{ij}(p,\ Q)\, \tau_i(Q) - \hat{T}_{ij}(p, Q)\, u_i(Q)]\, ds_Q \tag{9.67}$$

Equation (9.67) captures the effects of the interacting microcracks through the augmented kernels. A boundary integral equation for the unknown components of displacements and tractions in terms of the prescribed ones can be obtained by taking the limit as the internal source point p approaches a boundary point P. This leads to the boundary equation

$$C_{ij}(P)\, u_i(P) = \int_{\partial T} [\hat{U}_{ij}(P, Q)\, \tau_i(Q) - \hat{T}_{ij}(P, Q)\, u_i(Q)]\, ds_Q \tag{9.68}$$

It is interesting to note here that the kernels $\overline{U}_{ij}(P, Q)$ and $\overline{T}_{ij}(P, Q)$ will never be singular for microcrack clusters that are completely internal to the body. Hence, the singularity of $\hat{U}_{ij}(P, Q)$ and $\hat{T}_{ij}(P, Q)$ will essentially be contained in $U_{ij}\ (P, Q)$ and $T_{ij}(P, Q)$, respectively. The coefficients C_{ij} multiplying u_i

in the free term arise from the integration of $\hat{T}_{ij}(P,Q)u_i(Q)$. Hence, for two-dimensional problems, $C_{ij} = \frac{1}{2}\delta_{ij}$ if the boundary ∂B is locally smooth at P. Otherwise, C_{ij} can be evaluated in closed form for two-dimensional problems (Mukherjee 1982). Alternatively, proper combinations of $C_{ij}(P)$ and $\int_{\partial B}\hat{T}_{ij}(P,P)\,ds_Q$ can be obtained indirectly for general two-dimensional and three-dimensional bodies with sharp corners using rigid body modes (Cruse 1974, 1988, Lachat 1975). In the present work, the indirect approach is used to obtain desired combinations of $C_{ij}(P)$ and $\int_{\partial B}\hat{T}_{ij}(P,P)\,ds_Q$ in terms of the off-diagonal terms. It should be noted here that equations (9.67) and (9.68) involve boundary integrals only and require discretization only on the macro-scale boundary ∂B of the body. Thus, the traditional advantages of a linear elastic BEM approach for homogeneous problems are completely preserved.

The next step in the BEM formulation is to obtain the internal stresses. To achieve this goal, equation (9.67) is analytically differentiated at an internal source point p:

$$u_{j,\bar{l}}(p) = \int_{\partial T} [\hat{U}_{ij,\bar{l}}(p, Q)\tau_i(Q) - \hat{T}_{ij,\bar{l}}(p,Q)u_i(Q)]\,ds_Q \tag{9.69}$$

Here, $\bar{l}$ following a comma denotes differentiation with respect to a source point. The resulting displacement gradients at the source point can then be used in a Hooke's law to determine the internal stress

$$\sigma_{ij} = \lambda u_{k,k}\delta_{ij} + G(u_{i,j} + u_{j,i}) \tag{9.70}$$

The differentiated kernels $\hat{U}_{ij,\bar{l}}$ and $\hat{T}_{ij,\bar{l}}$ must now be evaluated. These may be written as

$$\hat{U}_{ij,\bar{l}} = U_{ij,\bar{k}} + \bar{U}_{ij,\bar{k}} \tag{9.71a}$$

and

$$\hat{T}_{ij,\bar{l}} = T_{ij,\bar{k}} + \bar{T}_{ij,\bar{k}} \tag{9.71b}$$

Here, $U_{ij,\bar{k}}$ and $T_{ij,\bar{k}}$ are the source point derivatives of U_{ij} and T_{ij}. Since U_{ij} and T_{ij} are two point kernels and

$$U_{ij,\bar{k}} = -U_{ij,k}, \qquad T_{ij,\bar{k}} = -T_{ij,k} \tag{9.72}$$

these may be evaluated easily using analytical techniques.

The kernels $\bar{U}_{ij,\bar{k}}$ and $\bar{T}_{ij,\bar{k}}$ depend on the sizes and distributions of the microcracks. Hence, they cannot easily be represented in terms of the corresponding field point derivatives. The source point derivatives of displacements may, however, be obtained by directly differentiating the displacement components $u_i(i = 1, 2)$ in equation (9.55a) at a source point. It is interesting to note that the kernels K_{j11} through K_{j22} in equation (9.59) do not depend on the macro-scale source points. Accordingly, the source point displacement gradients may be written as

$$u_{j,\bar{l}}(p) = H_{jk} b_{k,\bar{l}}(p) \tag{9.73}$$

Here, $b_{k,l}$ may be evaluated from direct differentiation of equation (9.59) as

$$\sum_{i=1}^{M} \int_{-a^{(i)}}^{a^{(i)}} [K_{jkl}(t^{(m)}, t^{(i)}) b^{(i)}_{l,\bar{m}}(t^{(i)}) \sigma^{(m0)}_{jk,\bar{m}}(t^{(m)})] \, dt^{(i)} = 0 \tag{9.74}$$

Once again, $K_{jkl}(t^{(m)}, t^{(i)})$ are independent of the macro-scale source points and $\sigma^{(m0)}_{jk,\bar{m}}(t^{(m)})$ represents the stress gradient at the location of the mth crack but without any cracks in the system. Accordingly, $b_{l,\bar{m}}$ can easily be evaluated by solving equation (9.74).

For efficient computation, equation (9.74) can actually be solved by retaining the same matrix decompositions used for solving equation (9.59) and modifying the right-hand side. Then, $u_{j,\bar{l}}(\,p)$ may be obtained from equation (973). The augmented kernels $\hat{U}_{ij,\bar{l}}$ and $\hat{T}_{ij,\bar{l}}$ may then be obtained from equation (9.71). The stresses at an internal point may now be evaluated through equations (9.69)–(9.70).

To evaluate the stress intensity factors (SIF) at the crack tip, the stresses are evaluated at several points near the crack tip and the SIF is interpolated. It has been observed by Owen and Fawkes (1983) that stress evaluations at $0.01a$, $0.04a$, and 0.16a (a being the half-length of the crack) are optimum for this purpose. It was also found during the course of the present work that the stress evaluation at $0.01a$ can yield the SIF to within 0.1 percent of analytical values in many cases, and such a scheme is used here to determine the crack tip SIFs. Alternatively, the crack opening displacements may be obtained from the internal equation (9.67) and the SIFs at crack tips may be evaluated once the crack opening profiles are known.

In order to complete a BEM formulation, an algorithm for determining the stresses on the boundary is required. The boundary stress algorithm first proposed by Rizzo and Shippy (1977) and used for nonlinear problems by Chandra and Mukherjee (1984, 1987) is adapted here.

For two-dimensional problems, the calculation involves four vectors and two tensors at the source point P on the boundary. The vectors are tractions τ_i, the tangential derivative of displacement $\partial u_i/\partial t$ on ∂B at P, as well as the unit normal and tangent vectors n and t at P. The tensors are the stresses σ_{ij} and displacement gradients $u_{i,\overline{J}}$ at P. The seven unknown tensor components, σ_{ij} and $u_{i,\overline{J}}$, can now be obtained from the following set of equations (seven scalar linear algebraic equations) at P:

$$\sigma_{ij} = \lambda u_{k,k}\delta_{ij} + G(u_{i,j} + u_{j,i}) \tag{9.75a}$$

$$\tau_i = \sigma_{ji} n_j \tag{9.75b}$$

$$\frac{\partial u_i}{\partial t} = u_{i,j} t_j \tag{9.75c}$$

In Equation (9.75*c*), the tangential gradients of displacement $(\partial u_i/\partial t)$ need to be evaluated numerically from the known boundary displacements. The tractions (τ_i) are also known, and equations (9.75) involving seven unknowns in σ_{ij} and $u_{i,j}$ can be solved from seven scalar linear algebraic equations.

9.6.3 Numerical Implementations for Hybrid Micro–Macro BEM

Numerical implementations of the BEM equations (9.67)–(9.69) for analyzing planar elastic problems containing multiple cracks are considered in this section. The boundary of the macro-scale problem is divided into N boundary segments (or elements). Geometric corners are accounted for through zero-length elements (Mukherjee 1982). Suitable shape functions must now be chosen for the variations of displacements and tractions on the boundary elements. In the present work, straight boundary elements are used with linear shape functions for both displacements and tractions. No other discretization is necessary for the macro-scale problem, and the effects of microcracks are introduced through the augmented fundamental solution.

For the micro-scale problem, six integration points are used on each crack for solving equation (9.59) and 30 integration points on each crack are used to solve equation (9.74). In each case, a Gauss–Chebychev polynomial scheme (Erdogan et al. 1973, Hu and Chandra 1993a,b,c) is utilized. Such a polynomial scheme requires only n quadrature points for accurate representations of polynomial functions of order $2n$. This makes it very effective for interacting cracks at close spacings. The issues regarding the effectiveness of such a scheme are discussed in section 9.6.1, and details on different applications are available in various references (e.g., Erdogan 1975, Theocaris and Ioakimidis 1977, Melin 1983, Li and Hills 1990, Rubinstein 1990, Hu and Chandra 1993b,c).

All integrations involving kernels U_{ij} and T_{ij} are carried out analytically. The kernels $\bar{U}_{ij}$ and $\bar{T}_{ij}$ are obtained numerically for the micro-scale problem. Accordingly, all integrations involving these kernels are also carried out numerically. It is interesting to note here that the kernels $\bar{U}_{ij}$ and $\bar{T}_{ij}$ are regular while the source point lies on the boundary. Hence, the accuracy of the proposed BEM scheme is not compromised by the numerical integration of these kernels.

Finally, the boundary equation (9.68) may be transformed into an algebraic system of the type

$$[A]\{u\} + [B]\{\tau\} = 0 \tag{9.76}$$

for the complete elastic micro–macro problem. In equation (9.76), only macro-scale discretizations are needed and the micro-scale effects are incorporated through the augmented fundamental solution (9.66). Equation (9.76) may now be solved easily for unknown displacements and tractions for a well-posed problem. It is important to note here that the proposed strategy requires only sequential micro- and macro-scale discretizations.

9.6.4 Numerical Results for Hybrid Micro–Macro BEM

The proposed hybrid micro–macro BEM scheme is first verified against known solutions involving multiple interacting cracks. For all cases except the indentation problem in figure 9.34, the stress intensity factors are normalized with respect to that for a single crack in an infinite plane $(\sigma_0\sqrt{\pi h})$, where $2a$ is the crack length and σ_0 is the remote stress. For the indentation problem in figure 9.34, the SIFs are normalized with respect to $P/\sqrt{\pi h}$, where h is the buried depth and P represents the indentation load.

The problem of a single crack of length $2a$ at the center of a plate of width $2b$ and height $2h$ under uniaxial tension is considered first. As shown in figure 9.28, the normalized mode I SIFs obtained from the hybrid micro–macro BEM scheme compare very well with the predictions reported by Tada et al. (1985). Three different plate geometries with $h/b = 0.5$, 1, and 5 are investigated. As h/b decreases, the top and bottom faces of the plate interact with the crack and increase the SIF. For increasing h/b values, the sides of the plate interact with the crack and increase the SIF, particularly at large values of a/b.

The problem of two collinear cracks in a square finite plate (as shown in figure 9.29) is considered next. The cracks are of length $2a$, with their centers placed at a distance of $b/2$ from the left and right edges, respectively, in a plate of width $2b$. The normalized SIFs obtained numerically from the

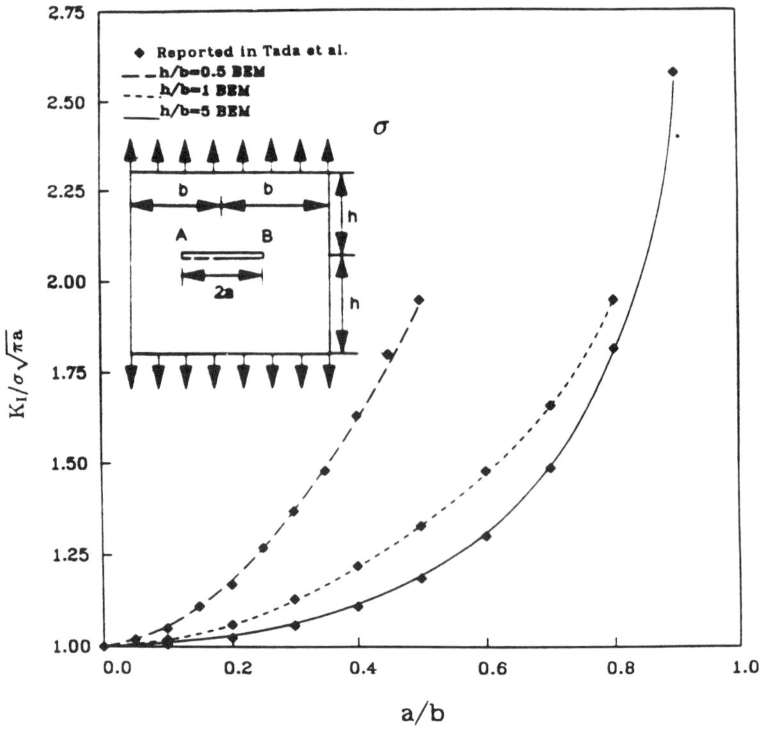

Figure 9.28. Comparison of normalized SIFs for a single crack under uniaxial tension.

proposed BEM formulation compare very well with those obtained by Liu and Altiero (1991). As the crack size increases, the interactions among the inner crack tips (B) become stronger. It is interesting to note, however, that the SIF at the outer tip (A) also increases, almost identically, with increasing a/b. This is due to the interactions between the outer tip and the traction-free edges of the plate.

The effects of variations in crack spacing (center distance $2c$) for two equal-length collinear cracks are considered next for different plate geometries with varying h/b ratios. Figures 9.30a and 9.30b show the variations of the normalized SIFs at the outer and inner crack tips, respectively, with increasing crack length for three different crack spacings of $c/b = 0.25$, 0.5, and 0.75. The aspect ratio of the plate is held fixed at $h/b = 0.5$. As expected, the normalized SIF increases with increasing a/b. It is interesting to note, however, that the rise in SIF (for both tips) with increasing a/b is the slowest for the case with $c/b = 0.5$. This is due to the presence of the finite geometry. At $c/b = 0.25$, the crack interactions dominate and cause the inner-tip SIF to rise dramatically. The crack-free edge

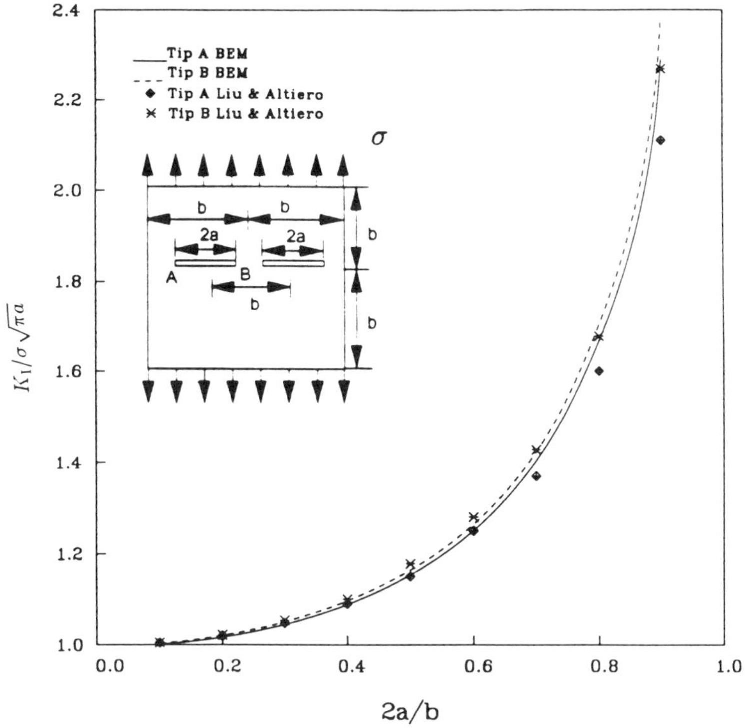

Figure 9.29. Normalized SIFs for two equal-length collinear cracks in a finite square plate.

interactions dominate at $c/b = 0.75$ and increase the outer-tip SIF. None of these interactions becomes severe at $c/b = 0.5$, causing the slowest growth in SIFs with increasing a/b.

Figures 9.31a and 9.31b depict the interactions among two equal-length collinear cracks and free edges at different crack spacings for a thin strip with an aspect ratio of $h/b = 5.0$. Once again, the slowest growth in SIFs at both outer and inner tips with increasing crack length is observed for $c/b = 0.5$. At very small crack sizes, the SIFs at both tips are the smallest for $c/b = 0.25$. With increasing crack length, the SIF at $c/b = 0.25$ rises dramatically for the inner tip and becomes the highest for $a/b > 0.14$. At the outer tip, this rise is less dramatic; however, the SIF for $c/b = 0.25$ still surpasses that for $c/b = 0.5$ for $a/b > 0.062$.

Liu and Altiero (1991) considered a problem involving offset cracks with arbitrary relative orientations. The inset in figure 9.32 shows the geometry of the problem. The crack length is $2a$, plate width $b/a = 9.0$, plate height $h/a = 30$, and the offset $c/a = 0.8$. It is observed, from figure 9.32, that the normalized mode I SIFs obtained from the BEM scheme at all four crack

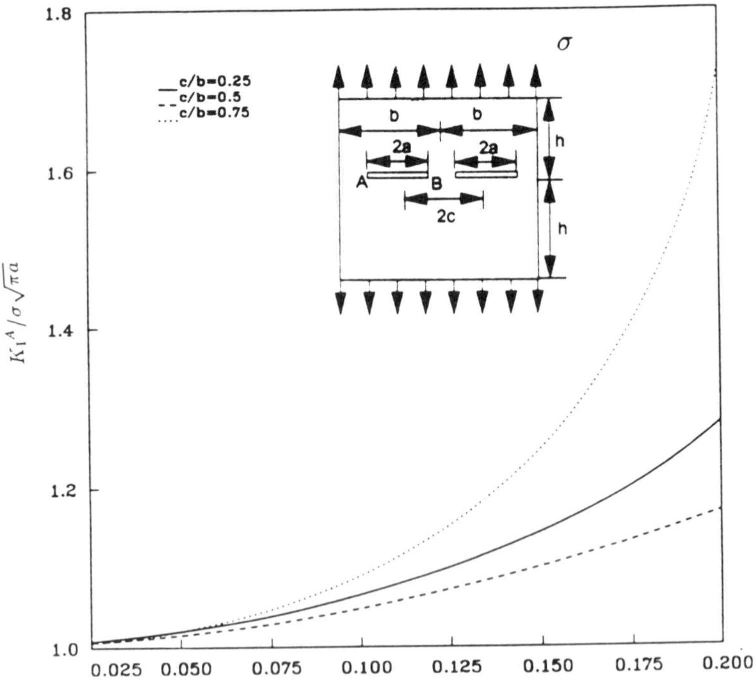

(a) a/b

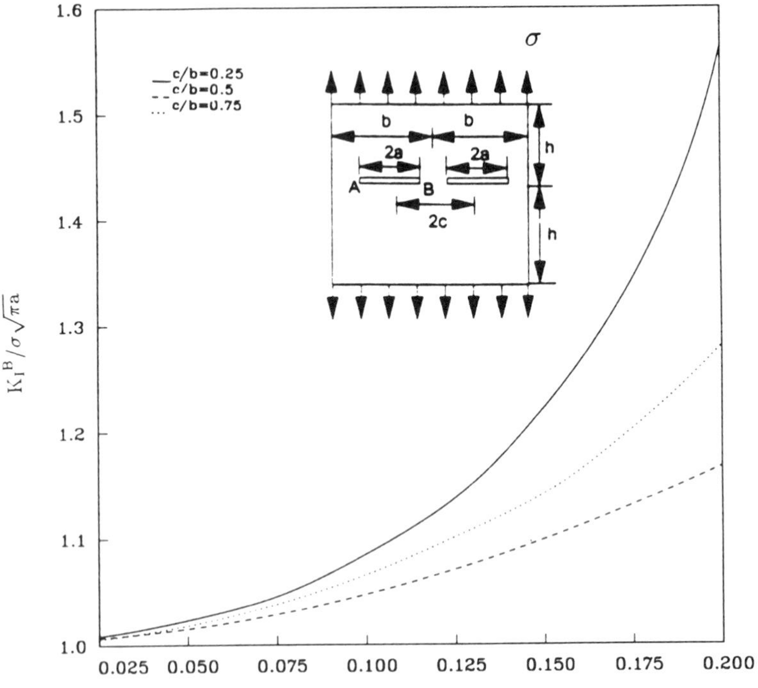

(b) a/b

Figure 9.30. Variations of normalized SIFs for two equal-length collinear cracks with different crack spacings h: (a) at the outer tip, A; (b) at the inner tip, B.

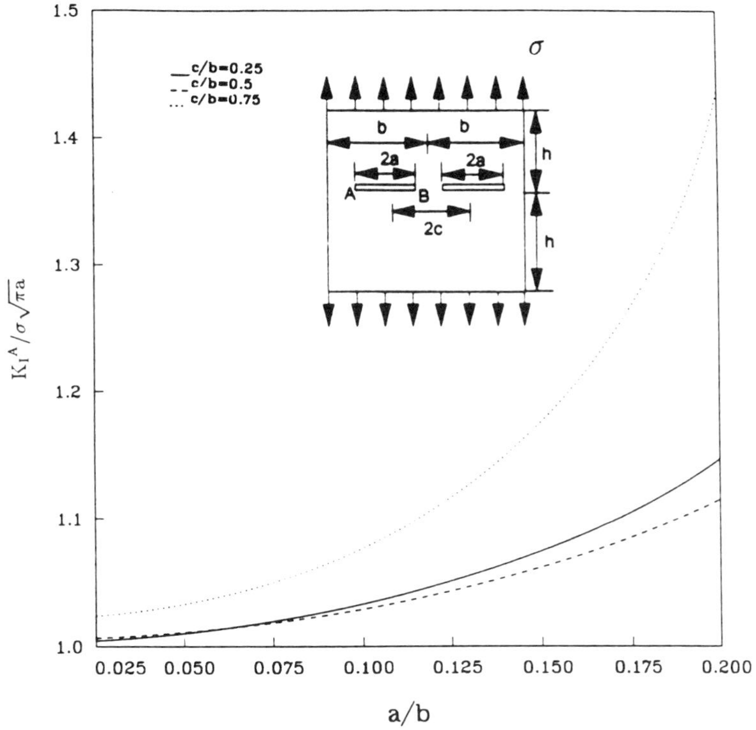

(a)

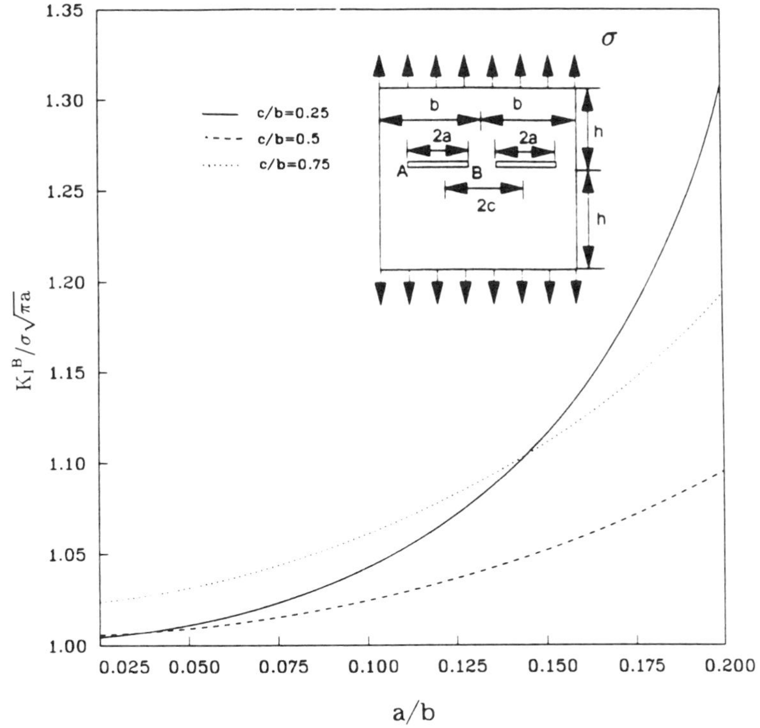

(b)

Figure 9.31. Variations of normalized SIFs for two equal-length collinear cracks with different crack spacings (h/b = 5.0): (a) at the outer tip, A; (b) at the inner tip, B.

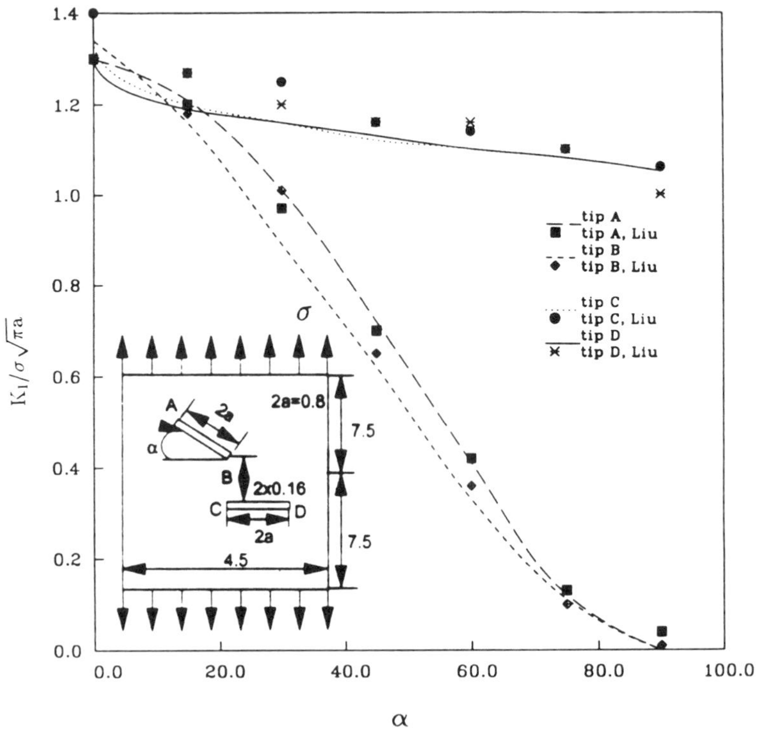

Figure 9.32. Comparisons of SIFs for two equal-length offset cracks with varying relative orientations ($b/a = 9.0$, $h/a = 30$, $c/a = 0.8$).

tips compare well with those obtained by Liu and Altiero over a large variation in relative orientation between the cracks.

A problem involving interactions among two cracks at an arbitrary angle to each other is considered next. The cracks are of equal length ($2a$) with centers at $b/2$ from the left and right edges, respectively, in a square plate of dimension $2b$. The plate is under uniaxial tension. One of the cracks is held fixed in its horizontal position while the other is rotated about is center. The normalized mode I SIFs with increasing crack size (increasing a/b) are shown in figure 9.33a for the outer tip of the horizontal crack, while figure 9.33b shows them at the inner. With increasing crack length, the crack–crack interactions dominate at the inner tips while the crack-free edge interactions dominate at the outer tips. Accordingly, the SIFs at both inner and outer tips rise as a/b is increased. However, the effects of the relative orientations between the cracks are much more predominant at the inner tips.

The issue of strength degradation in ceramic grinding operations (Hu and Chandra 1993a) is considered next. The ceramic grinding process is modeled

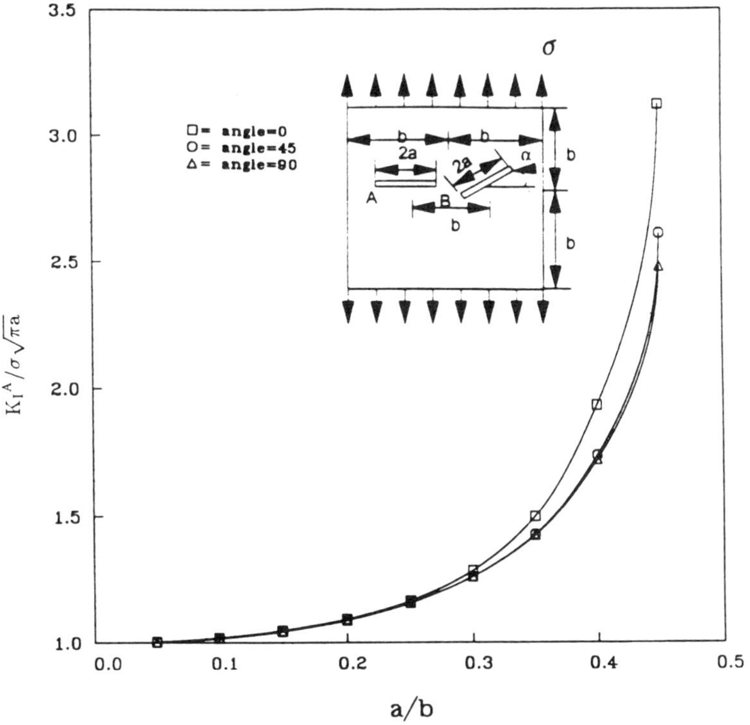

(a)

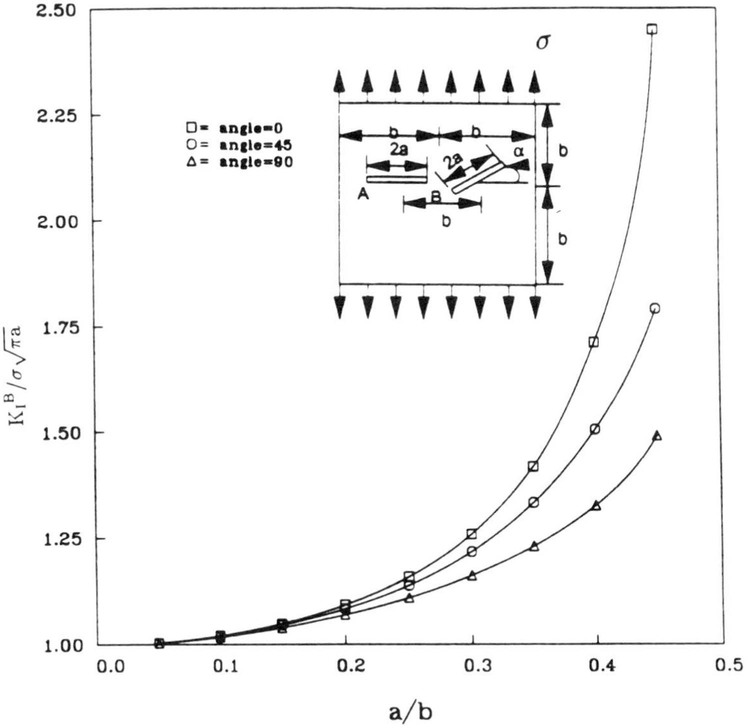

(b)

Figure 9.33. Variations of SIFs for interacting cracks at arbitrary relative orientations: (a) at the outer tip, *A*; (b) at the inner tip, *B*.

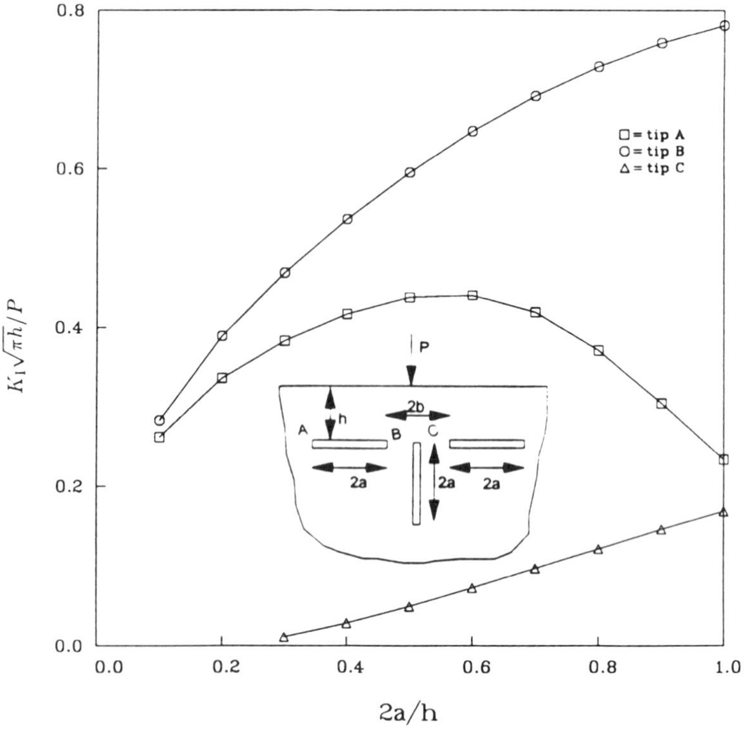

Figure 9.34. Normalized SIFs at radial and lateral crack tips in a ceramic indentation problem: (a) variations in mode I SIFs at tips A, B, and C with $2a/h$ ($b/h = 0.1, 0.1$); (b) variations in mode I SIFs at tips A, B, and C with $2a/h$ ($b/h = 0.1, 0.5$); (c) comparison of mode I and mode II SIFs at the outer tip A of the lateral crack ($b/h = 0.5$).

as a moving indentation problem. Under steady-state conditions, such a process may be represented by a system of radial (normal to the free surface) and lateral (parallel to the free surface) cracks. The proximity of the free surface to the interacting crack system makes the analysis of such problems difficult. Previous analyses of such systems (described in sections 9.4 and 9.5) required special kernels, and indentation loads had to be transferred to the crack face. The proposed BEM algorithm, however, allows the solution of complete boundary value problems for such indentation loadings. Such a problem with $b/h = 0.1$ and 0.5 is considered in figure 9.34. All radial and lateral cracks are assumed to be of equal size. In ceramic grinding, the propagation and eventual upward curvature of lateral cracks aids the material removal process while the radial cracks remain as surface defects in the finished product and can significantly compromise its strength. Accordingly, propagation of lateral cracks and retardation of the radial crack represent

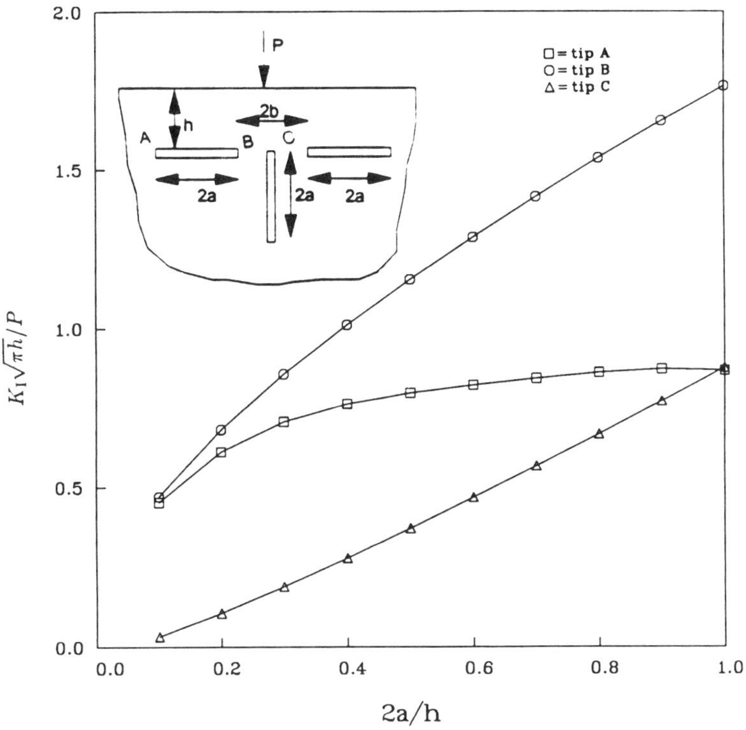

Figure 9.34—*continued*

the dual goal in optimizing the quality and efficiency of a ceramic grinding process. Figures 9.34a and 9.34b show the mode I SIFs for different clusterings ($b/h = 0.1$ and 0.5) at tips A and B of the lateral cracks and at tip C of the radial crack, with increasing a/h for a single-point indentation problem. Figure 9.34c shows the comparisons between mode I and mode II SIFs at the outer tip A of the lateral crack. It is observed for tips B and C that the mode I SIFs increase monotonically with increasing a/h and, as expected, the SIF values are higher for $b/h = 0.1$ than for $b/h = 0.5$. This is due to the fact that $b/h = 0.1$ represents closer clustering of the cracks, which results in stronger crack interactions. As the crack length is increased, the outer tip A of the lateral crack moves farther from the point of loading. This causes the mode I SIF at tip A to reach a maximum and decrease afterward with increasing a/h. As observed in figure 9.34c, however, the mode II SIF continues to increase monotonically. This will cause the lateral crack to grow

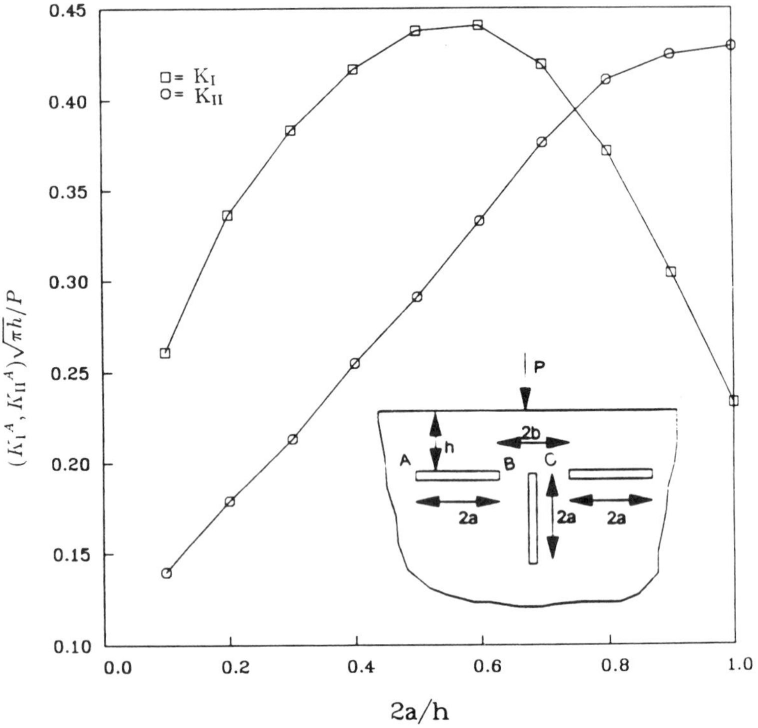

Figure 9.34—*continued*

in mode II and eventually curve upward to form a chip. The capability of the proposed hybrid micro–macro BEM formulation to decompose the overall problem into one involving only micro-scale features and another involving solely macro-scale variables provides a new avenue for introducing the effects of micro-scale features into macro-scale computations. The necessity for a fundamental solution is traditionally considered one of the inherent weaknesses of the BEM technique. The proposed scheme turns this feature of the BEM into one of its strengths and utilizes it as the conduit for introducing the analytical advancements into the BEM methodology for analyzing this class of problems.

References

Atkinson, C. (1972a). "The Interaction Between a Crack and an Inclusion," *Int. J. Eng. Sci.*, **10**, 127–136.

Atkinson, C. (1972b). "On Dislocation Densities and Stress Singularities Associated with Cracks and Pile Ups in Inhomogeneous Media," *Int. J. Engng. Sci.*, **10**, 45–71.

Atkinson, C. (1973). "Some Ribbon-Like Inclusion Problems," *Int. J. Eng. Sci.*, **11**, 243–266.1

Baeslack, W. A., Cieslak, M. J., and Headley, T. J. (1988). "Structure, Properties and Fracture of Pulsed Nd:YAG Laser Welded Ti-14.8 wt% Al-21.3 wt% Nb Titanium Aluminide," *Scripta Metall.*, **22**, 1155–1160.

Ballarini, R. (1987). "An Integral Equation Approach for Rigid Line Inhomogeneity Problems," *Int. J. Fracture*, **33**, R23-R26.

Banerjee, P. K. and Butterfield, R. (1981). *Boundary Element Method in Engineering Science*. McGraw-Hill, London.

Banerjee, P. K. and Henry, D. P. (1992). "Nonlinear Micro and Macromechanical Analyses of Composites by BEM," *Proc., HITEMP Conference*, vol. 2, pp. 42.1-42.11, NASA Lewis Res. Center, Cleveland, Ohio.

Barsoum, R. S. (1976). "On the use of Isoparametric Finite Elements in Linear Fracture Mechanics," *Int. J. Num. Meth. Eng.*, **10**, 25-27.

Bathe, K. J. (1982). *Finite Element Procedures in Engineering Analysis*. Prentice-Hall, Englewood Cliffs, N.J.

Becker, R. (1992). "An Analysis of Shear Localization During Bending of a Polycrystalline Sheet," *ASME J. Appl. Mech.*, **59**, 491–496.

Becker, P., Hsueh, C.-H., Angelini, P., and Tiegs, T. N. (1988). "Toughening Behavior in Whisker-Reinforced Ceramic Matrix Composites," *J. Am. Ceram. Soc.*, **71**, 1056–1061.

Broberg, K. B. (1987). "On Crack Paths," *Eng. Fracture Mech.*, **28**, 663–679.

Brussat, T. R. and Westmann, R. A. (1975). "A Westergaard-Type Stress Function for Line Inclusion Problems," *Int. J. Solids Structures*, **11**, 665–677.

Budiansky, B. (1965). "On the Elastic Moduli of Some Heterogeneous Material," *J. Mech. Phys. Solids*, **13**, 223–227.

Busch, D. M. and Prins, J. F. (1972). "A Basic Study of the Diamond Grinding of Alumina," *The Science of Ceramic Machining and Surface Finishing I*. NBS Special Publ. 348, pp. 73–87.

Chandra, A. and Mukherjee, S. (1984), "Boundary Element Formulations for Large Strain-Large Deformation Problems of Viscoplasticity," *Int. J. Solids Structures*, **20**, 41–53.

Chandra, A. and Mukherjee, S. (1987). "A Boundary Element Analysis of Metal Extrusion Processes," *ASME, J. Appl. Mech.*, **54**, 335–340.

Chandra, A. and Tvergaard, V. (1993). "Void Nucleation and Growth During Plane Strain Exclusion," *Int. J. Damage Mech.*, **2**, 330–348.

Chatterjee, S. N. (1975). "The Stress Field in the Neighborhood of a Branched Crack in an Infinite Elastic Sheet," *Int. J. Solids Structures*, **11**, 521–538.

Chen, Y. Z. (1984). "Solutions of Multiple Crack Problems of a Circular Plate or an Infinite Plate Containing a Circular Hole by Using Fredholm Integral Equation Approach," *Int. J. Fracture*, **25**, 155–168.

Cheung, Y. K. and Chen, Y. Z. (1987). "Solution of Branch Crack Problems in Plane Elasticity by Using a New Integral Equation Approach," *Eng. Fracture Mech.*, **28**, 31–41.

Chou, Y. T. and Wang, Z. Y. (1983). "Stress Singularity at the Tip of a Rigid Flat Inclusion," *Recent Developments in Applied Mechanics* (ed. F. F. Ling and I. G. Tadjbakhsh), pp. 21–30. Rensselaer Press, New York.

Christensen, R. M. and Lo, K. H. (1979). "Solutions for Effective Shear Properties in Three Phase Sphere and Cylinder Models," *J. Mech. Phys. Solids*, **27**, 315–330.

Chudnovsky, A., Dolgopolsky, A., and Kachanov, M. (1987). "Elastic Interaction of a Crack with a Microcrack Array—Part I and II," *Int. J. Solids Structures*, **23**, 1–10, 11–21.

Conway, J. C. and Kirchner, H. P. (1980). "The Mechanics of Crack Initiation and Propagation Beneath a Moving Sharp Indentor," *J. Mater. Sci.*, **15**, 2879–2883.

Cotterell, B. and Rice, J. K. (1980). "Slightly Curved or Kinked Cracks," *Int. J. Fracture*, **16**, 155–169.

Crouch, S. L. (1976). "Solution of Plane Elasticity Problems by the Displacement Discontinuity Method," *Int. J. Num. Meth. Eng.*, **10**, 301–343.

Cruse, T. A. (1974). "An Improved Boundary-Integral Equation Method for Three-Dimensional Elastic Stress Analysis," *Computers and Structures*, **4**, 741–754.

Cruse, T. A. (1988). *Boundary Element Analysis in Computational Fracture Mechanics*. Kluwer Academic Publishers, Dordrecht, The Netherlands.

Cruse, T. A. and Polch, E. Z. (1986). "Application of an Elastoplastic Boundary Element Method to Some Fracture Mechanics Problems," *Eng. Fracture Mech.*, **23**, 1085–1096.

Dundurs, J. and Hetenyi, M. (1961). "The Elastic Plane with a Circular Insert, Loaded by a Radial Force," ASME *J. Appl. Mech.*, **28**, 103–111.

Dundurs, J. and Markenscoff, X. (1989). "A Green's Function Formulation of Anticracks and Their Interaction with Load Induced Singularities," *J. Appl. Mech.*, **56**, 550–555.

Dundurs, J. and Mura, T. (1964). "Interaction Between an Edge Dislocation and a Circular Inclusion," *J. Mech. Phys. Solids*, **12**, 177–189.

Dundurs, J. and Sendeckyj, G. P. (1965). "Edge Dislocation Inside a Circular Inclusion," *J. Mech. Phys. Solids*, **13**, 141–147.

Erdogan, F. (1962). "On the Stress Distribution on Plates with Collinear Cuts under Arbitrary Loads," *Proc., 4th U.S. National Congress of Applied Mechanics*, pp. 547–553, ASME, New York.

Erdogan, F. (1975). "Mixed Boundary Value Problems in Mechanics." Report, Lehigh University, Bethlehem, Pa.

Erdogan, F. and Gupta, G. D. (1972). "Stress Near a Flat Inclusion in Bonded Dissimilar Materials," *Int. J. Solids Structures*, **8**, 533–547.

Erdogan, F. and Gupta, G. D. (1975). "The Crack Problem with a Crack Crossing the Boundary," *Int. J. Fracture*, **11**, 13–27.

Erdogan, F., Gupta, G. D., and Cook, T. S. (1973). "Numerical Solution of Singular Integral Equations," *Methods of Analysis and Solutions of Crack Problems* (ed. G. C. Sih), pp. 368–425. Noordhoff, Leyden.

Eshelby, J. D. (1957). "The Determination of the Elastic Field of an Ellipsoidal Inclusion, and Related Problems," *Proc. Royal Soc., London*, **A241**, pp. 376–396.

Eshelby, J. D. (1959). "The Elastic Field Outside an Ellipsoidal Inclusion," *Proc. Royal Soc., London*, **A252**, 561–569.

Evans, A. G., (1979). "Abrasive Wear in Ceramics: An Assessment," *The Science of Ceramic Machining and Surface Finishing II* (ed. B. J. Hockey and R. W. Rice). NBS Special Publ. 562, pp. 1–14.

Evans, A. G. (1984). *Fracture in Ceramic Materials*. Noyes, Park Ridge, N.J.

Evans, A. G. and Marshall, D. B. (1981). "Wear Mechanisms in Ceramics," *Fundamentals of Friction and Wear of Materials* (ed. D. A. Rigney), p. 439. ASME, New York.

Fielden, J. H. and Rubenstein, C. (1969). "The Grinding of Glass by a Fixed Abrasive," *Glass Technol.*, **10**, 73–81.

Fleck, N. A., Hutchinson, J. W., and Tvergaard, V. (1989). "Softening by Void Nucleation and Growth in Tension and Shear," *J. Mech. Phys. Solids*, **37**, 515–540.

Gallagher, R. H. (1975). *Finite Element Analysis: Fundamentals*. Prentice-Hall, Englewood Cliffs, N.J.

Goldstein, R. V. and Salganik, R. L. (1974). "Brittle Fracture of Solids with Arbitrary Cracks," *Int. J. Fracture*, **10**, 507–523.

Hagan, J. T. (1979). "Cone Cracks Around Vickers Indentation in Fused Silica Glass," *J. Mater. Sci.*, **14**, 462–466.

Haritos, G. K., Hager, J. W., Amos, A. K., Salkind, M. J., and Wang, A. S. D. (1988). "Mesomechanics: The Microstructure–Mechanics Connection," *Int. J. Solids Structures*, **24**, 1081–1096.

Hasebe, N., Keer, L. M., and Nemat-Nasser, S. (1984). "Stress Analysis of a Kinked Crack Initiating from a Rigid Line Inclusion; Part 1: Formulation," *Mech. Mater.*, **3**, 131–145.

Hashin, Z. (1983). "Analysis of Composite Materials—A Survey," *ASME J. Appl. Mech.*, **50**, 487–505.

Hashin, Z. (1988). "The Differential Scheme and Its Application to Cracked Materials," *J. Mech. Phys. Solids*, **36**, 719–734.

Hetenyi, M. and Dundurs, J. (1962). "The Elastic Plane with a Circular Insert, Loaded by a Tangentially Directed Force," *ASME J. Appl. Mech.*, **29**, 362–368.

Hill, R. (1965). "A Self-Consistent Mechanics of Composite Materials," *J. Mech. Phys. Solids*, **13**, 213–222.

Hills, D. A. and Ashelby, D. W. (1980). "On the Determination of Stress Intensification Factors for a Wearing Half-Space," *Eng. Fracture Mech.*, **13**, 69–78.

Hills, D. A. and Nowell, D. (1989). "Stress Intensity Calibrations for Closed Cracks," *J. Strain Anal.*, **24**, 37–43.

Hockey, B. J. (1972). "Observations on Mechanical Abraded Aluminum Oxide Crystals by Transmission Electron Microscopy," *The Science of Ceramic Machining and Surface Finishing I*, NBS Special Pub. 348, pp. 333–340.

Horii, H. and Nemat-Nasser, S. (1987). "Interacting Micro-Cracks Near the Tip in the Process Zone of a Macro-Crack," *J. Mech. Phys. Solids*, **35**, 601–629.

Horii, H. and Nemat-Nasser, S. (1986). "Brittle Failure in Compression: Splitting, Faulting and Brittle-Ductile Transition," *Phil. Trans. Royal Soc. Lond.*, **A319**, 337–374.

Hu, K. X. and Chandra, A. (1993a). "A Fracture Mechanics Approach to Modeling Strength Degradation in Ceramic Grinding Processes," *ASME J. Eng. Ind.*, **115**, 73–84.

Hu, K. X. and Chandra, A. (1993b). "Interactions Among General Systems of Cracks and Anticracks: An Integral Equation Approach," *ASME J. Appl. Mech.*, **60**, 920–928.

Hu, K. X. and Chandra, A. (1993c). "Interactions Among Cracks and Rigid Lines Near a Free Surface," *Int. J. Solids Structures*, **30**, 1919–1937.

Hu, K. X., Chandra, A., and Huang, Y. (1993a). "Fundamental Solutions for Dilute Distributions of Inclusions Embedded in Microcracked Solids," *Mech. Mater.*, **16**, 281–294.

Hu, K. X., Chandra, A., and Huang, Y. (1993b). "Multiple Void-Crack Interaction," *Int. J. Solids Structures*, **30**, 1473–1489.

Hu, K. X., Chandra, A., and Huang, Y. (1994). "On Interacting Bridged Crack Systems," *Int. J. Solids Structures*, **31**, 599–611.

Huang, Q. and Cruse, T. A. (1994). "On the Nonsingular Traction-BIE in Elasticity," *Int. J. Num. Meth. Eng.*, **37**, 2041–2062.

Huang, Y., Hu, K. X., Wei, X., and Chandra, A. (1994). "A Generalized Self-Consistent Mechanics Method for Composite Materials With Multiphase Inclusions," *J. Mech. Phys. Sol.*, **42**, 491–504.

Huerta, M. and Malkin, S. (1976a). "Grinding of Glass: the Mechanics of the Process," *ASME J. Eng. Ind.*, **98**, 459–467.

Huerta, M. and Malkin, S. (1976b). "Grinding of Glass: Surface Structure and Fracture Strength," *ASME J. Eng. Ind.*, **98**, 468–473.

Hughes, T. J. R. (1987). *The Finite Element Method*. Prentice-Hall, Englewood Cliffs, N.J.

Inasaki, I. (1986). "High Efficiency Grinding of Advanced Ceramics," *Ann. CIRP*, **35**, 211–218.

Ioakimidis, N. I. and Theocaris, P. S. (1979). "A System of Curvilinear Cracks in an Isotropic Half-Plane," *Int. J. Fracture*, **15**, 299–309.

Jha, S. C., Ray, R., and Clemm, P. J. (1988). "Microstructure of Ti3 Al + Nb Alloys Produced via Rapid Solidification," *Mater. Sci. Eng.*, **98**, 395–397.

Kachanov, M. (1985). "A Simple Technique of Stress Analysis in Elastic Solids with Many Cracks," *Int. J. Fracture*, **28**, R11–R19.

Kachanov, M. (1987). "Elastic Solids with Many Cracks: A Simple Method of Analysis," *Int. J. Solids Structures*, **23**, 23–43.

Kachanov, M. and Montagut, E. (1986). "Interaction of a Crack With Certain Microcrack Arrays," *Eng. Fracture Mech.*, **25**, 625–636.

Kamel, M. and Liaw, B. M. (1991a). "Boundary Element Analysis of Cracks at a Fastner Hole in an Anisotropic Sheet," *Int. J. Fracture*, **50**, 263–280.

Kamel, M. and Liaw, B. M. (1991b). "Boundary Element Formulation with Special Kernels for an Anisotropic Plate Containing an Elliptical Hole or a Crack," *Eng. Fracture Mech.*, **39**, 695–711.

Keer, L. M. and Bryant, M. D. (1983). "A Pitting Model for Rolling Contact Fatigue," *ASME J. Lub. Technol.*, **105**, 198–205.

Keer, L. M., Bryant, M. D., and Haritos, G. K. (1982). "Subsurface and Surface Cracking Due to Hertzian Contact," *ASME J. Lub. Tech.*, **104**, 347–351.

King, R. I. and Hahn, R. S. (eds.) (1986). *Handbook of Modern Grinding Technology*. Chapman and Hall, New York.

Kirchner, H. P. (1984). "Comparison of Single-Point and Multipoint Damage in Glass," *J. Am. Ceram. Soc.*, **67**, 347–353.

Kirchner, H. P., Gruver, R. M., and Richard, D. M. (1979). "Fragmentation and Damage Penetration During Abrasive Machining of Ceramics," *The Science of Ceramic Machining and Surface Finishing II* (ed. B. J. Hockey and R. W. Rice). NBS Special Publ. 562, pp. 23–42.

Koepke, B. G. and Stokes, R. J. (1979). "Effect of Workpiece Properties on Grinding Forces in Polycrystalline Ceramics," *The Science of Ceramic Machining and Surface Finishing II* (ed. B. J. Hockey and R. W. Rice). NBS Special Pub. 562, pp. 75–92.

Komanduri, R. (1971). "Some Aspects of Machining with Negative Rake Tools Simulating Grinding," *Int. J. Mach. Tool. Des. Res.*, **11**, 233–239.

Komanduri, R. and Maas, D. (eds.) (1985). *Proc. Milton C. Shaw Grinding Symposium*, PED-16, ASME, New York.

Krenk, S. (1975). "On the Use of the Interpolation Polynomial for Solutions of Singular Integral Equations," *Q. Appl. Math.*, **32**, 479–484.

Lachat, J. C. (1975). "A Further Development of the Boundary-Integral Technique for Elastostatics," Ph.D. thesis, University of Southampton, England.

Larchuk, T. J., Conway, J. C., Jr., and Kirchner, H. P. (1985). "Crushing as a Mechanism of Material Removal During Abrasive Machining," *J. Am. Ceram. Soc.*, **68**, 209–215.

Lawn, B. R. and Swain, M. V. (1975). "Microfracture Beneath Point Indentations in Brittle Solids," *J. Mater. Sci.*, **10**, 113–122.

Lawn, B. R., Evans, A. G., and Marshall, D. B. (1980). "Elastic-Plastic Indentation Damage in Ceramics: the Median/Radial Crack System," *J. Am. Ceram. Soc.*, **63**, 574–581.

Li, Y. and Hills, D. A. (1990). "Stress Intensity Factor Solutions for Kinked Surface Cracks," *Int. J. Strain Anal.*, **25**, 21–27.

Li, Q. and Ting, T. C. T. (1989). "Line Inclusions in Anisotropic Elastic Solids," ASME *J. Appl. Mech.*, **56**, 556–563.

Liu, N. and Altiero, N. J. (1991). "Multiple Cracks and Branch Cracks in Finite Plane Bodies," *Mech. Res. Com.*, **18**, 233–244.

Lo, K. K. (1978). "Analysis of Branched Cracks," *J. Appl. Mech.*, **45**, 797–802.

Lutz, E., Ingraffea, A. R., and Gray, L. J. (1992). "Use of Simple Solutions for Boundary Integral Methods in Elasticity and Fracture Analysis," *Int. J. Num. Meth. Eng.*, **35**, 1737–1752.

Madenci, E. (1991). "Slightly Open, Penny-Shaped Crack in an Infinite Solid Under Biaxial Compression," *J. Theor. Appl. Fracture Mech.*, **16**, 215–222.

Malkin, S. (1989). *Grinding Technology: Theory and Applications of Machining and Abrasives*. Ellis Horwood, Chichester, U.K.

Malkin, S. and Ritter, J. E. (1989). "Grinding Mechanisms and Strength Degradation for Ceramics," *ASME J. Eng. Ind.*, **111**, 167–174.

Marsh, D. M. (1963). "Plastic Flow in Glass," *Proc. Royal Soc. London, Series A.*, 420–435.

Marshall, D. B. (1984). "Geometrical Effects in Elastic-Plastic Indentation," *J. Am. Ceram. Soc.*, **67**, 57–60.

Marshall, D. B., Lawn, B. R., and Evans, A. G. (1982). "Elastic-Plastic Indentation Damage in Ceramics: the Lateral Crack System," *J. Am. Ceram. Soc.*, **65**, 561–566.

Marshall, D. B., Evans, A. G., Khuri-Yakub, B. T., Tien, J. W., and Kino, G. S. (1983). "The Nature of Machining Damage in Brittle Materials," *Proc. Royal Soc. London, Series A*, 385, pp. 461–475.

McClintock, F. A. and Argon, A. S. (1966). *Mechanical Behavior of Materials*. Addison-Wesley, Reading, Mass.

Melin, S. (1983). "Why Do Cracks Avoid Each Other?," *Int. J. Fracture*, **23**, 37–45.

Melin, S. (1986). "On Singular Integral Equations for Kinked Cracks," *Int. J. Fracture*, **30**, 57–65.

Montagut, E. and Kachanov, M. (1988). "On Modelling a Microcracked Zone by Weakened Elastic," *Int. J. Fracture*, **37**, R55–R62.

Mukherjee, S. (1982). *Boundary Element Methods in Creep and Fracture*. Elsevier Applied Science, London.

Müller, W. H. (1989). "The Exact Calculation of Stress Intensity Factor in Transformation Toughened Ceramics by Means of Integral Equations," *Int. J. Fracture*, **41**, 1–22.

Mura, T. (1987). *Micromechanics of Defects in Solids*, 2d ed. Martinus Nijhoff, Boston.

Mura, T. (1988). "Inclusion Problems," *ASME Appl. Mech. Rev.*, **41**, 15–20.

Muskhelishvili, N. I. (1953). *Some Basic Problems of the Mathematical Theory of Elasticity*. Noorhoff, Groningen, The Netherlands.

Nakamura, T. and Suresh, S. (1993). "Effects of Thermal Residual Stresses and Fiber Packing on Deformation of Metal-Matrix Composites," *Acta Metall. Mater.*, **41**, 1665–1681.

Needleman, A. (1987). "A Continuum Model for Void Nucleation by Inclusion Debonding," *ASME J. Appl. Mech.*, **54**, 525–531.

Needleman, A. and Tvergaard, V. (1991). "A Numerical Study of Void Distribution Effects on Dynamic Ductile Crack Growth," *Eng. Frac. Mech.*, **38**, 157–173.

Norris, A. N. (1985). "A Differential Scheme for the Effective Moduli of Composites," *Mech. Mater.*, **4**, 1–16.

Nowell, D. and Hills, D. A. (1987). "Open Cracks at or Near Free Edges," *J. Strain Anal.*, **22**, 177–185.

Oden, J. T. (1972). *Finite Elements of Nonlinear Continua*, McGraw-Hill, New York.

Okada, H., Rajiyah, H., and Atluri, S. N. (1990). "A Full Tangent Stiffness Field-Boundary Element Formulation for Geometric and Material Nonlinear Problems of Solid Mechanics," *Int. J. Num. Meth. Eng.*, **29**, 15–35.

Owen, D. R. J. and Fawkes, A. J. (1983). *Engineering Fracture Mechanics: Numerical Methods and Applications*. Pineridge Press, Swansea, U.K.

Petrovic, J. J., Dirks, R. A., Jacobson, L. A., and Mendiratta, M. G. (1976). "Effects of Residual Stresses on Fracture From Controlled Surface Flaws," *J. Am. Ceram. Soc.*, **59**, 177–178.

Raveendra, S. T. and Banerjee, P. K. (1992), "Boundary Element Analysis of Cracks in Thermally Stressed Planar Structures," *Int. J. Solids Structures*, **29**, 2301–2317.

Read, W. T. (1953). *Dislocations in Crystals*. McGraw-Hill, New York.

Ritter, J. E., Strezpa, P., and Jakus, K. (1984). "Erosion and Strength Degradation in Soda Lime Glass," *Phys. Chem. Glasses*, **25**, 159–162.

Ritter, J. E., Rosenfeld, and Jakus, K. (1985). "Erosion and Strength Degradation in Alumina," *Wear of Materials* (ed. K. C. Ludema), pp. 1–8. ASME, New York.

Rizzo, F. J. and Shippy, D. J. (1977). "An Advanced Boundary Integral Equation Method for Three-Dimensional Thermoelasticity," *Int. J. Num. Meth. Eng.*, **11**, 1753–1768.

Roscoe, R. (1952). "The Viscosity of Suspensions of Rigid Spheres," *Br. J. Appl. Phys.*, **3**, 267–269.

Rubinstein, A. A. (1990). "Crack-Path Effect on Material Toughness," *J. Appl. Mech.*, **57**, 97–103.

Samuel, R., Chandrasekhar, S., Farris, T. N., and Licht, R. H. (1988). "The Effect of Residual Stresses on the Fracture of Ground Ceramics," *Intersociety Symposium on Machining of Advanced Ceramic Materials and Components* (ed. S. Chandrasekhar, R. Komanduri, W. Daniels, and W. Rapp), pp. 81–98. ASME, New York.

Selvadurai, A. S. P. (1980). "The Displacements of a Flexible Inhomogeneity Embedded in a Transversely Isotropic Elastic Medium," *Fiber Sci. Technol.*, **14**, 251–259.

Sendeckyj, G. P. (1970). "Elastic Inclusion Problems in Plane Elastostatics," *Int. J. Solids Structures*, **6**, 1535–1543.

Sih, G. C. (1973). *Mechanics of Fracture*, **1**, Noordhoff, Leyden.

Stroh, A. N. (1958). "Dislocations and Cracks in Anisotropic Elasticity," *Phil. Mag.*, **3**, 625–646.

Stroh, A. N. (1962). "Steady-State Problems in Anisotropic Elasticity," *J. Math. Phys.*, **41**, 77–103.

Spur, G., Stark, C., and Tio, T. H. (1985). "Grinding of Non-Oxide Ceramics Using Diamond Grinding Wheels," *Machining of Ceramic Materials and Components* (ed. K. Subramanian and R. Komanduri), pp. 439–455. PED-17, ASME, New York.

Subramanian, K. and Keat, P. P. (1985). "Parametric Study on Grindability of Structural and Electronic Ceramics—Part 1," *Machining of Ceramic Materials and Components* (ed. K. Subramanian and R. Komanduri), pp. 25–41. PED-17, ASME, New York.

Tada, H., Paris, P. C., and Irwin, G. R. (1985). *The Stress Analysis of Cracks Handbook*. Paris Productions, St. Louis, MO.

Taya, M. and Chou, T.-W. (1981). "On Two Kinds of Ellipsoidal Inhomogeneities in an Infinite Elastic Body: An Application to a Hybrid Composite," *Int. J. Solids Structures*, **17**, 553–563.

Theocaris, P. S. and Ioakimidis, N. I. (1977). "Numerical Integration Methods for the Solution of Singular Integral Equations," *Q. Appl. Math.*, **35**, 173–182.

Timoshenko, S. and Goodier, J. N. (1973). *Theory of Elasticity*. McGraw-Hill, New York.

Tvergaard, V. (1982). "Material Failure by Void Coalescence in Localized Shear Bands," *Int. J. Solids Structures*, **18**, 659–672.

Tvergaard, V. (1989a). "Material Failure by Void Growth to Coalescence," *Advances in Applied Mechanics*, **27**, pp. 83–751. Academic Press, San Diego.

Tvergaard, V. (1989b). "Numerical Study of Localization in a Void Sheet," *Int. J. Solids Structures*, **25**, 1143–1156.

Vaidyanathan, S. and Finnie, I. (1972). "Grinding of Brittle Solids," *New Developments in Grinding, Proceedings of the International Grinding Conference*, pp. 813–826. Carnegie Press, New York.

Wang, Z. Y., Zhang, H. T., and Chou, Y. T. (1985). "Characteristics of the Elastic Field of a Rigid Line Inhomogeneity," *ASME J. Appl. Mech.*, **52**, 818–822.

Wang, Z. Y., Zhang, H. T., and Chou, Y. T. (1986). "Stress Singularity at the Tip of a Rigid Line Inhomogeneity Under Antiplane Shear Loading," *ASME J. Appl. Mech.*, **53**, 459–462.

Weng, G. J. (1984). "Some Elastic Properties of Reinforced Solids, with Special Reference to Isotropic Containing Spherical Inclusions," *Int. J. Eng. Sci.*, **22**, 845–856.

Wolfram, S. (1991). *Mathmatica, A System for Doing Mathematics by Computer*, 2d ed. Addison-Wesley, Redwood City, Calif.

Yahia, N. A. B. and Shephard, M. S. (1985). "On the Effect of Quarter-Point Element Size on Fracture Criteria," *Int. J. Num. Meth. Eng.*, **20**, 1629–1641.

Zang, W. L. and Gudmundson, P. (1989). "An Integral Equation Method for Piece-wise Smooth Cracks in an Elastic Half-Plane," *Eng. Fracture Mech.*, **32**, 889–897.

Zang, W. L. and Gudmundson, P. (1991). "Kinked Cracks in an Anisotropic Plane Modeled by an Integral Equation Method," *Int. J. Solids Structures*, **27**, 1855–1865.

Zienkiewicz, O. C. and Taylor, R. L. (1991). *The Finite Element Method, 2*, McGraw-Hill, Berkshire, U.K.

Index